Foundations of Mathematics and Physics One Century After Hilbert

Joseph Kouneiher
Editor

Foundations of Mathematics and Physics One Century After Hilbert

New Perspectives

Springer

Editor
Joseph Kouneiher
Nice and Sophia Antipolis University
Nice
France

and

Côte d'Azur University and Lab.
ARTEMIS UMR 7250
(OCA, UCA, CNRS)
Nice
France

ISBN 978-3-030-09728-8 ISBN 978-3-319-64813-2 (eBook)
https://doi.org/10.1007/978-3-319-64813-2

Cover photo: Archives of the Mathematisches Forschungsinstitut Oberwolfach

Printed on acid-free paper

This Springer imprint is published by the registered company Springer International Publishing AG part of Springer Nature
The registered company address is: Gewerbestrasse 11, 6330 Cham, Switzerland

In memory of Elie, the other me,
To the twins Ellie and Jezabel

Preface

From 1891 to 1933, David Hilbert gave a series of lectures on the foundations of mathematics and physics. Those unpublished lectures became available in a six-volume edition released by Springer Verlag. Hilbert's lectures and his personal interactions exercised a profound influence on the development of twentieth-century mathematics and physics.

In his address to the second International Congress of Mathematicians on Wednesday, the August 8, 1900, in Paris at the turn of the century, Hilbert began with the following words [1]:

> *Wer von uns wüurde nicht gern den Schleier luften, unter dem die Zukunft verborgen liegt, um einen Blick zu werfen auf die bevorstehenden Fortschritte unsrer Wissenschaft und in die Geheimnisse ihrer Entwickelung wäahrend der küunftigen Jahrhunderte! Welche besonderen Ziele werden es sein, denen die fuhrenden mathematischen Geister der kommenden Geschlechter nachstreben? welche neuen Methoden und neuen Thatsachen werden die neuen Jahrhunderte entdecken - auf dem weiten und reichen Felde mathematischen Denkens?*

> Who of us would not be glad to lift the veil behind which the future lies hidden; to cast a glance at the next advances of our science and at the secrets of its development during future centuries? What particular goals will there be toward which the leading mathematical spirits of coming generations will strive? What new methods and new facts in the wide and rich field of mathematical thought will the new centuries disclose?

Hilbert then went on to deliver a list of 23 problems for the twentieth century. The sixth problem is of particular concern for us. Indeed, it is in problem number 6 that Hilbert outlined his program for axiomatizing physics with the intent of putting it on the same level as axiomatized geometry:

> The investigations on the foundations of geometry suggest the problem: To treat in the same manner, by means of axioms, those physical sciences in which mathematics plays an important part . . .

> If geometry is to serve as a model for the treatment of physical axioms, we shall try first, by a small number of axioms, to include as large a class as possible of physical phenomena, and then, by adjoining new axioms, to arrive gradually at the more special theories. At the

> same time, Lie's principle of subdivision can perhaps be derived from the profound theory of infinite transformation groups. The mathematician will also have to take account not only of those theories that approach reality, but also, as in geometry, of all logically possible theories. He must be always alert so as to obtain a complete survey of all conclusions derivable from the system of axioms assumed.

Hilbert's work on the foundations of mathematics has its roots in his work on geometry of the 1890s, culminating in his influential textbook Foundations of Geometry (1899). Hilbert believed that the properly rigorous way to develop any scientific subject required an axiomatic approach. Through provision of an axiomatic treatment, the theory would be developed independent of any need for intuition, and it would facilitate an analysis of the logical relationships between the basic concepts and the axioms. Moreover, Hilbert's view of physics from a mathematician's perspective becomes quite explicit in remarks he made regarding the relationship between physics and geometry. Hilbert regarded geometry as a genuine branch of mathematics, so it had become mathematized, arithmetized, and eventually axiomatized, and was no longer subject to experimental examination. For Hilbert, this development was the proper advancement of science, and not simply an account of the factual historical development. An advancement should be furthered wherever possible.

Thus, as early as 1894, in a lecture on geometry that he gave while still in Königsberg, Hilbert said:

> Geometry is a science that essentially has developed to such a state that all its facts may be derived by logical deduction from previous ones [2, 3].

Later in this lecture, in the course of discussing the axiomatic foundations of geometry, he presented the axiom of parallels and discussed the alternatives of Euclidean, hyperbolic, and parabolic geometries. In this context, he remarked:

> Now, all other sciences are also to be treated following the model of geometry, first of all mechanics, but then optics and electricity theory as well [2, 3].

Many of the world's great scientific truths are based totally upon mathematical formulation. The extraordinarily results have left the originators obliged to admit to some mysterious and intimate connection between the physical world and its abstract mathematical counterpart. To quote Einstein himself:

> Here arises a puzzle that has disturbed scientists of all periods. How is it possible that mathematics, a product of human thought that is independent of experience, fits so excellently the objects of physical reality? [4].

These quotations demonstrate that, while the fields of mathematics and physics were considered separate, there was still a strong conjunction between them. The great upheavals in Physics in the first quarter of the twentieth century only deepened the relation between physics and mathematics. In his stunning 1931 paper (in which he predicted the existence of three new particles), Dirac was both eloquent and exuberant at the very outset:

> The steady progress of physics requires for its theoretical formulation a mathematics that gets continually more advanced ... What, however, was not expected by the scientific workers of the last century was the particular form that the line of advancement of the mathematics would take, namely, it was expected that the mathematics would get more and more complicated, but would rest on a permanent basis of axioms and definitions, while actually, the modern physical developments have required a mathematics that continually shifts its foundations and gets more abstract ... It seems likely that this process of increasing abstraction will continue in the future . . . [5].

Around the same time, Einstein expressed similar sentiments:

> Our experience up to date justifies us in feeling sure that in Nature is actualized the ideal of mathematical simplicity. It is my conviction that pure mathematical construction enables us to discover the concepts and the laws connecting them which give us the key to the understanding of the phenomena of Nature. Experience can, of course, guide us in our choice of serviceable mathematical concepts; it cannot possibly be the source from which they are derived; experience, of course, remains the sole criterion of the serviceability of a mathematical construction for physics, but the truly creative principle resides in mathematics. In a certain sense, therefore, I hold it to be true that pure thought is competent to comprehend the real, as the ancients dreamed [6].

Concerning the atomic physics and the fact that Quantum Mechanics, using radically new concepts, such as the linear superposition of states and the uncertainty principle, required an entirely new mathematical framework, Dirac wrote:

> Quantum mechanics requires the introduction into physical theory of a vast new domain of pure mathematics -the whole domain connected with non-commutative multiplication. This, coming on top of the introduction of the new geometries by the theory of relativity, indicates a trend which we may expect to continue. We may expect that in the future, further big domains of pure mathematics will have to be brought in to deal with the advances in fundamental physics [7].

Mathematics and Physics: A Common Matter?

Since Hilbert, conferences, physics, and mathematics have experienced great upheavals, with new ideas invading the two areas of study. Several ideas from physics have allowed for a better understanding of certain mathematical problems and their resolution. Indeed, over the past 50 years, a new type of interaction has taken place, as has happened frequently in the past, in which physicists, while exploring their new and still speculative theories, have stumbled across a whole range of mathematical discoveries.

The physicists' approach was derived by physical intuition and heuristic arguments, which are beyond the reach, as yet, of mathematical rigour, but which have withstood the tests of time and alternative methods. The impact of these discoveries on mathematics has been profound and widespread. Areas of mathematics such as topology and algebraic geometry, which lie at the heart of pure mathematics and appear very distant from the physics frontier, have been dramatically affected.

This development has led to many hybrid subjects, such as topological quantum field theory, quantum cohomology, and quantum groups, which are now central topics of research in both mathematics and physics. Remarkably, modern physical constructions such as quantum field theory and string theory, which are very far removed from everyday experience, have proven to be a similarly fertile setting for mathematical problems. Indeed, in many ways, quantum theory has turned out to be an even more effective framework for mathematics than classical physics. Particles and strings, fields and symmetries, they all have a natural role to play in mathematics. Understanding this is the great problem of our time.

Truth, Depth, and Beauty

Many mathematicians and physicists share the view that the beauty of mathematics is a guide toward a theory whose coherence and simplicity aids our comprehension of nature. Beauty is what guides the mathematician, while the physicist searches for truth, consistent with experiment. The mystery of the effectiveness of mathematics in fundamental physics is much deeper than just the miracle of its astonishing utility [9]. We aim to describe the microscopic laws in terms of simple mathematics, but, as we probe deeper, to microscopic scales, we require deeper mathematical structures. But beyond that, these mathematical structures are not just deep—they are also interesting, beautiful, and powerful. As Dirac put it:

> It seems to be one of the fundamental features of nature that fundamental physical laws are described in terms of great beauty and power and, as time goes on, it becomes increasingly evident that the rules that the mathematician finds interesting are the same as those that Nature has chosen [10].

On the relation between mathematics and Nature, Hermann Weyl wrote:

> There is inherent in nature a hidden harmony that reflects itself in our minds under the image of simple mathematical laws. That then is the reason why events in nature are predictable by a combination of observation and mathematical analysis. Again and again in the history of physics, this conviction, or should I say this dream, of harmony in nature has found fulfillments beyond our expectations [11, 12].

To appreciate mathematical beauty may require, as in music, extensive education and training, and it is always a subjective judgment. Nonetheless, there tends to be a large degree of consensus among mathematicians and physicists that the beautiful parts are those that explain the forces of nature as arising from principles of symmetry.

These are beautiful to physicists since, from a simple principle of symmetry, we deduce, in an almost unique fashion, via gauge theories, the nature of the fundamental forces and the existence of the carriers of these forces. The ugly parts are those that describe the strange spectrum of matter, which does not appear to follow from any symmetry principle. To agree with experiment, one requires far too many parameters to be put in by hand. Einstein's dream was that the ugly should be made beautiful, and that geometry should totally unify spacetime and matter. This is the task for us all, a task that may yet take the whole twenty-first century and beyond.

Supported by prominent scientists in mathematics and physics, this book celebrates the centenary of Hilbert's work on the foundations of physics and mathematics, and explores the rich new perspectives resulting from the deep interplay between mathematics and physics during the twentieth century. The result is a broad journey through the most recent developments in both mathematics and physics.

In mathematics, the journey takes us through differential and algebraic geometry, to topology, noncommutative geometry, and twistor theory.

In physics, the journey takes us through gauge and quantum field theories to string theory and quantum gravity.

Edinburgh, UK
2018

Michael Atiyah
Joseph Kouneiher

References

1. D. Hilbert, Mathematische Probleme. Vortrag, gehalten auf dem internationalen Mathematiker-Kongress zu Paris 1900. *Königliche Gesellschaft der Wissenschaften zu Göttingen. Mathematisch-physikalische* (Klasse, Nachrichten, 1900), pp. 253–297
2. Niedersächesische staats und Universitätsbibliothek Göttingen Handschriftenabteilung, Cod. Ms. Hilbert 541, p.7; Tilman Sauer, *The Relativity of Discovery: Hilbert's First Note on the Foundations of Physics*. arXiv:physics/9811050
3. D. Hilbert, *Uber die Grundlagen der Geometrie* (Gottinger Nachrichten, 1902), pp. 233–241
4. A. Einstein, *Sidelight on Relativity*, Forgotten Books ed., (1921), www.forgottenbooks.com
5. P.A.M. Dirac, *Quantised Singularities in the Electromagnetic Field*, Proc. Roy. Soc. A **133**, p. 60
6. A. Einstein, *On the Method of Theoretical Physics*, Philosophy of Science, vol. 1, no. 2, (The University of Chicago Press on behalf of the Philosophy of Science Association, 1934), pp. 163–169. http://www.jstor.org/stable/184387
7. P.A.M. Dirac, *The Relation between Mathematics and Physics*, Proceedings of the Royal Society (Edinburgh) vol. 59, 1938–39, Part II pp. 122–129, 1938–39.
8. M. Atiyah, in *Hermann Weyl Biographical Memoirs*, vol. 82, Nat. Acad. Sci. 2002.
9. D. Gross, *Mathematics and the Sciences*, Proc. Nati. Acad. Sci. vol **85**, (USA, 1988), pp. 8371–8375
10. P.A.M. Dirac (1939) Proc. R. Soc. Edinburgh Sect. A **59**, p. 122.
11. H. Weyl, *Philosophy of Mathematics and Natural Science* (Princeton Press, 1949)
12. J. Kouneiher, *Symmetry and Cohomological foundations of Physics* (Towards a new Philosophy of Nature, Hermann Sciences editions, 2010)

Acknowledgements

First of all, I would like to express my deep gratitude to all my friends contributors, who kindly accepted to be part and encouraged the project of this special book, which will remain a reference in the domain of mathematics and mathematical physics: Michael Atiyah, Jeremy Attard, Jeremy Butterfield, Ali Chamsddine, Alain Connes, Leo Corry, Jordan François, Misha Gromov, Sebastian de Haro, Serge Lazzarini, Colin MacLarty, Matilde Marcolli, Thierry Masson, Roger Penrose, Lee Smolin, John Stachel, and Edward Witten.

I wish to thank all my friends from the group geometry and physics for the many years of collaborations and discussions on mathematical physics issues and which found echo through the lines of the contribution "Where We stand today": Frédéric Hélein, Daniel Bennequin, Volodya Robtsov, Paul Baird, Franz Pedit, and Cécile Barbachoux.

I would like to express my very great appreciation for the fruitful encounters and discussions with my friends: Abhay Ashtekar, Robert Dijkgraaf, Carlo Rovelli, Alain Herreman, Jean Michel Alimi, Newton da Costa, Michel Paty, Jean Jacques Szczeciniarz, Dominique Lambert, Jean Luc Gautero, and Raffaele Pizano.

I want to extend my thanks to Salvatore Capozziello for his encouragement to undertake this project. I want to warmly thank also Nary-Catherine Man and Michel Boer from Artemis for their support.

I would like to thank Springer editions for their supports. I am particularly grateful to Kirsten Theunissen for her assistance and almost daily encouragement. Special thanks to Aldo Rampioni, for his help and collaboration from the beginning of the project of the book.

I wish to acknowledge the help and support provided by Cécile Barbachoux during the preparation of the manuscript. Special thanks should be given to my students for their active interactions during the elaboration of the book. I'm greatly indebted to my family for their indulgence.

Finally, I wish to express my sincere gratitude to Michael Atiyah, Alain Connes, Ali Chamsddine, Misha Gromov, Roger Penrose, and Edward Witten, for the long e-mails exchanges, encouragements, and support and especially the daily exchange with Michael about mathematics and physics and our heroes Hermann Weyl, James Clerk Maxwell, David Hilbert, Albert Einstein, and others.

Contents

Summary

The project of this book is the result of the desire of prominent scientists in mathematics and physics to, first, celebrate the centenary of Hilbert's work on the foundations of physics and mathematics, and second, to explore the rich new perspectives resulting from the deep interplay between mathematics and physics during the twentieth century. The papers published in this volume provide insight into their works, and analyze the impact of the breakthrough and the perspectives of their own contributions.

In his contribution, Leo Corry describes the motivations of Hilbert's Unified Foundations of Physics. He presents the two main pillars on which Hilbert's built his own theory as presented in Göttingen in November 1915. To do so, he discusses the contents of Mie's electromagnetic theory of matter and explains the context in which the theory needs to be understood as part of contemporary debates on gravitation in which also Einstein took part. He reviewed also the way in which Max Born mediated between Mie and Hilbert by presenting the formers work in a way that would be amenable to Hilbert's current scientific interests. Finally, he gives a brief account of Hilbert's talk of November 1915 and explains its contents against the background of the ideas of Gustav Mie's electromagnetic.

The common paper of Kouneiher and Stachel highlights the twenty-odd-year relationship between Einstein and Hilbert and traces the relationship between the two men during this period in the form of encounters, each of which characterizes a particular aspect of their relationship and their impact on the final form of Einstein equation and the way to derive it.

Colin Maclarty's contribution is a fine analysis of Grothendieck's *vast unifying vision* which provided new working and conceptual foundations for geometry, and even led him to logical foundations. Maclarty clarifies Grothendieck attitude of mind by favoring the words and commutative diagrams over pictures and refusing to think the geometry pictorially. This contrasts with the mainstream attitude where the majority prefer the pictures as illustrations of the geometry. We can see how, by using such approach, Grothendieck construct his mathematical and geometrical universe of topos and schemes.

In his contribution *A Dozen Problems, Questions and Conjectures about Positive Scalar Curvature* ($SC > 0$), Gromov invites us to explore and discover new paths and challenges us to open windows that give access to new facts in geometry of the domain ($SC > 0$). One of the beauties of the author's approach to geometry is his gritty hands-on method, dealing with basic concepts that one could explain to a nonexpert, rather than heading on a long trek of increasing abstraction. It is at the same time simple, but extraordinarily difficult to put into practice, and of course requires great insight to know where to look for answers. The author deals primarily with the difficult domain of positive scalar curvature. The paper reminds us that in spite of the remarkable advances in Riemannian geometry in recent years, there is still a wealth of fundamental unresolved problems.

In his contribution, Michael Atiyah presents a proof of a long-standing conjecture concerning the six-dimensional sphere and the possibility to have a complex structure. To do so, he uses two aspects: a hypothetical complex structure without any symmetry assumptions and he considers the case of conformal sphere $\mathbf{S}^6$ and not the round sphere S^6. More precisely, he uses the fact that $\mathbf{S}^6$ is a homogeneous space of the conformal group *Spiny*(7; 1), which preserves future and past. The proof is a master class move in mathematics. The author suggests also to shed new light on many problems of physics: *In the future I expect these ideas will provide a different perspective, with substantial benefits in all areas.*

The subject of Edward Witten's contribution[1] is one of the big breakthrough ideas in mathematical physics in the twentieth century, the strings theories, with a new paradigm based on the conception of elementary particle as one-dimensional string.[2] The aim of this contribution is to describe the minimum that any physicist should know about string theory, focusing on a few basic questions. How does string theory generalize standard quantum field theory? Why does string theory force us to unify general relativity with the other forces of nature, while standard quantum field theory makes it so difficult to incorporate general relativity? Why are there no ultraviolet divergences in string theory? And what happens to Einstein's conception of spacetime? For instance, as we know, in general, a string theory comes with no particular spacetime interpretation. The spacetime M emerges through the link between its metric tensor $G_{IJ}(X)$ and a particular 2D conformal field theory. That is the only way that spacetime entered the story.

Witten tries to answer all these questions with clarity and simplicity whenever it is possible.

In his contribution, Roger Penrose *Twistor Theory as an Approach to Fundamental Physics* describes the original motivations underlying the introduction of twistor theory which has been pioneered by him and others since the 1960s. The primary objective of twistor theory originally was—and still is—to find a deeper route to the working nature; so the theory should provide a mathematical

[1]An earlier and restraint version of this paper was published in physics today.

[2]Later on, people understood that some objects as Branes play also an essential and fundamental role in the theory conception.

framework with sufficient power and scope, to help us toward resolving some of the most obstinate problems of current physical theory. One of the original motivations was to unify general relativity and quantum mechanics in a nonlocal theory based on complex numbers. The application of twistor theory to differential equations and integrability has been an unexpected spin-off from the twistor program.

In capturing both relativity and quantum mechanics, twistor theory demands some modifications of both. For example, it allows for the introduction of nonlinear elements into quantum mechanics, which are in agreement with some current interpretations of the measurement process: The collapse of the wave function contradicts the principle of unitary time evolution, and it has been proposed that this failure of unitarity is due to some overtaking nonlinear gravitational effects. The main two ingredients of twistor theory are non-locality in spacetime and analyticity (holomorphy) in an auxiliary complex space, the twistor space.

Alain Connes's contribution has the ambition of answering the questions posed by the divers temptations to create a theory founded on the principle of quantum mechanics and general relativity and which goes beyond their limit to integrate the gauge theories and matter. *The point of view* adopted in this essay is to try to understand from a mathematical perspective, how the perplexing combination of the Einstein–Hilbert action coupled with matter, with all the subtleties such as the Brout–Englert–Higgs sector, the V-A and the see-saw mechanisms, etc., can emerge from a simple geometric model. The new tool is the spectral paradigm, and the new outcome is that geometry does emerge on the stage where quantum mechanics happens, i.e., Hilbert space and linear operators. In his contribution, Alain Connes introduce the noncommutative geometry and the spectral paradigm developed by the author since 1980s. It is based on the Hilbert space formalism of quantum mechanics and on mathematical ideas coming from K-theory and index theory. This new paradigm of geometry provided a new perspective on the geometric interpretation of the detailed structure of the standard model and of the Brout–Englert–Higgs mechanism.

With Ali Chamseddine, they understood that they could obtain the full package of the Einstein–Hilbert action of gravity coupled with matter by a fundamental spectral principle. In the language of NCG, this principle asserts that the action only depends upon the "line element", i.e., the inverse of the operator D. The presence of the other fields *forces, due to renormalization, the addition of higher derivative terms of the metric to the Lagrangian and this in turns introduces at the quantum level an inherent instability that would make the universe blow up*. The approach used in this contribution is based on the idea of "particle picture" for geometry, allowing to stay very close to the inner workings of the standard model coupled to gravity.

Ali Chamsddine's contribution forms a logical continuation to Alain Connes's one in this volume. Notice that all the material covered in Chamsddine review is a result of a long-time collaboration with Alain Connes. The author shows that starting with the axioms of noncommutative geometry supplemented by a minimal number of physical assumptions would result, unambiguously, in a unified theory of all fundamental interactions and matter content of spacetime. And so they will be able to establish a link between the quantization of volume of space at Planck energy and the constituents of matter and their symmetries. In addition, he uncovers the origin of the Higgs fields and symmetry breaking, and indicates possible solutions to long-standing problems such as resolving the singularities in GR, dark matter, and dark energy.

The author of *What are we missing in our search for quantum gravity?*, Lee Smolin, starts his contribution by considering the various approaches to quantum gravity and asks the question why, in spite of many efforts, we have not yet found the true theory of quantum gravity. He makes a succinct analysis of the causes of the failures of different approaches and suggests to consider them as different and complementary models of a single theory to be found by a common effort and an explicit choice of a scientific approach based on a more general physical principle. The idea is that, in the absence of a real experience at the Planck scale to guide us and inspired by various developed models, we can at least get back and make some reflections on what we may be missing in our search for quantum gravity.

In their contribution, Attard, François, Lazzarini, and Masson propose a review of gauge theory, one of the most profound breakthrough ideas of twentieth century in mathematical physics. More precisely, they describe another way to perform gauge symmetry reduction which they call the *dressing field method*. It is formalized in the framework of the differential geometry; it has a corresponding BRST differential algebraic formulation. The method boils down to the identification of a suitable field in the geometrical setting of a gauge theory that allows to construct partially of fully gauge invariant variables out of the standard gauge fields. This formalizes and unifies several works and approaches which found origins in Dirac's pioneering works.

Butterfield and De Haro's contributions propose a schema to understand duality between models in physics. Notice that the idea of the duality is one of the challenging ideas of the twentieth-century physics and mathematics. This paper is written for physicists as well as for philosophers of sciences. The approach uses a formalization of the notions of *theories*, *models* and what the mean of a *duality* in this framework. Discussions are given to illuminate some crucial points of this formal approach. The main proposed example concerns the duality known as "*bosonization*", which establishes an equivalence between two physical models: one based on bosons and the other on fermions. The authors insist, on one hand, on the fact that this boson duality, by contrast with other dualities in physics, is exact and on the other hand its role in both cases of isomorphic and non-isomorphic models.

In their contribution "*Syntactic Phylogenetic Trees*", M. Marcolli, K. Shu, S. Aziz, V. Y. Huynh, and D. Warrick try to apply some methods that came from mathematics and computational methods developed in the context of mathematical biology in the linguistic domain. They start by identifying several serious problems that arise in the use of syntactic data from the SSWL database for the purpose of computing phylogenetic trees of language families in the context of the field of historical linguistics.

They show that the most naive approach fails to produce reliable linguistic phylogenetic trees and identifies some of the sources of the observed problems. They describe also how the use of phylogenetic algebraic geometry can help in estimating to what extent the probability distribution at the leaves of the phylogenetic tree obtained from the SSWL data can be considered reliable, by testing it on phylogenetic trees established by other forms of linguistic analysis. They remark that after restricting to smaller language subfamilies and considering only those SSWL parameters that are fully mapped for the whole subfamily, the SSWL data match extremely well-reliable phylogenetic trees, according to the evaluation of phylogenetic invariants. This is a promising sign for the use of SSWL data for linguistic phylogenetics, which was their first motivation.

Where We Stand Today

Joseph Kouneiher

1 Hilbert and the Foundations of Mathematics and Physics

In his work of 1918, Hermann Weyl extended the general theory of relativity, which Albert Einstein had set forth in the years 1915–1916, to unify the two field phenomena known at this time, namely those described by electromagnetic and gravitational fields. But more was at stake. At the beginning of the paper in which Weyl worked out the mathematical foundations of the theory, he observed that:

> According to this theory, everything real, that is in the world, is a manifestation of the world metric; the physical concepts are no different from the geometrical ones. The only difference that exists between geometry and physics is, that geometry establishes, in general, what is contained in the nature of the metrical concepts, whereas it is the task of physics to determine the law and explore its consequences, according to which the real world is characterized among all the geometrically possible four-dimensional metric spaces. [124]

This work sounds like an echo of a work undertaken by Hilbert many years earlier.[1] Indeed, when Hilbert started studying the analysis of integral equations, he sought to achieve Poincaré's program unifying different aspects of mathematical analysis and physics. For him, the aim was to expose and simplify the known results, just like "formal" physicians such as Dirac, who sought to give physics a coherent mathematical basis. It was on this occasion that he developed the theory of quadratic forms to an infinite number of variables, work that would later lead to the birth of "Hilbert spaces" (and to spectral theory), and that consists in interpreting equations as terms of linear transformation of this space. He permitted, via his theory of spaces, a new

[1] See Corry's contribution in this volume.

J. Kouneiher (✉)
Côte d'Azur University and Lab. ARTEMIS UMR 7250 (OCA, UCA, CNRS), Nice, France
e-mail: Joseph.KOUNEIHER@unice.fr

J. Kouneiher
Nice and Sophia Antipolis University, Nice, France

J. Kouneiher (ed.), *Foundations of Mathematics and Physics One Century After Hilbert*,
https://doi.org/10.1007/978-3-319-64813-2_1

"geometrization" of physics, thanks to the invention of a new geometry that J. Von Neumann and F. Riesz axiomatized, and that became a powerful tool in mathematical physics.
In fact, Hilbert's approach concerning unification in physics can be seen as a problem of finding a consistent and satisfactory mathematical unification of the gravitational and electromagnetic fields, be it through modified field equations, a modification of the space-time geometry, or by increasing the number of space-time dimensions.
Hilbert's view of physics from a mathematician's perspective becomes quite explicit in remarks he made regarding the relationship between physics and geometry. Hilbert regarded geometry as a genuine branch of mathematics, so it had become mathematized, arithmetized and eventually axiomatized [59], and was no longer subject to experimental examination. However, originally, geometry was a natural science.
Hilbert's work on the foundations of mathematics and physics has its roots in his work on geometry from the 1890s, culminating in his influential textbook Foundations of Geometry (1899). Hilbert believed that the properly rigorous way to develop any scientific subject required an axiomatic approach. In providing an axiomatic treatment, the theory would be developed independent of any need for intuition, and it would facilitate an analysis of the logical relationships between the basic concepts and the axioms.

Thus, as early as 1894, in a lecture on geometry that he gave while still in Königsberg, Hilbert said:

> Geometry is a science that essentially has developed to such a state that all its facts may be derived by logical deduction from previous ones [58, 108].

Later in this lecture, in the course of discussing the axiomatic foundations of geometry, he presented the axiom of parallels and discussed the alternatives of Euclidean, hyperbolic and parabolic geometries. In this context, he remarked:

> Now, all other sciences are also to be treated following the model of geometry, first of all mechanics, but then optics and electricity theory as well [58, 108].

According to Hilbert, physics was but a four-dimensional pseudo-geometry, whose metric was connected, via his theory, to electromagnetic quantities, i.e., to matter. And with this knowledge, an old geometric problem could now be solved: whether and in what sense Euclidean geometry - about which we only know from mathematics that it is a logically consistent structure - is also valid in reality. After discussing Gauss's inability to verify empirically a non-Euclidean physics through angle measurement in a large triangle, Hilbert talked about how the physics of Einstein's general theory of relativity had a totally different relationship to geometry. The new physics started neither with Euclidean nor with any other fixed geometry in order to deduce the actual laws of physics. Instead, general relativity yielded, in one blow, the laws of geometry and physics through one and the same Hamiltonian principle, i.e., through the fundamental equations of his theory. Hilbert's conclusion was:

> Euclidean geometry is an action-at-a-distance law alien to modem physics: while the theory of relativity rejects Euclidean geometry as a general presupposition for physics, it teaches furthermore that geometry and physics are of a similar kind and rest, as one science (Wissenschaft), on a common foundation.

Weyl commented Hilbert's work on general relativity and unified field theories, noting that: "*Hopes in the Hilbert circle ran high at that time: the dream of a universal law accounting both for the structure of both the cosmos as a whole and of all the atomic nuclei seemed near fulfillment.*"
So, the idea of unification for Weyl and his contemporaries was understood not merely as a synthesis of the electromagnetic and gravitational fields, but also as a *unification of geometry and physics and as the quest for a universal world law accounting for the structure of both cosmos and matter*.
Later in the mid-1940s Weyl felt that it was insufficient to unite electromagnetism and gravitation, and that quantum and nuclear phenomena had to be taken into account as well. By focusing on the unification of the physical fields known at that time, Weyl continued the re-definition of the aims of the unification project that he had begun in the late 1920s with the advent of the new quantum mechanics. As the above passage from the Hilbert obituary shows, in the mid-1940s Weyl would not even discuss the union between geometry and physics that seemed so attractive in the 1910s.

Recall that the mathematical scene in Gottingen[2] in the first decade of the twentieth century was dominated by Felix Klein (1849–1925), David Hilbert (1862–1943), and Hermann Minkowski (1864–1909). Klein described the spirit that dominated at that time :

> Speaking, as I do, under the influence of our Göttingen traditions, and dominated somewhat, by the great name of Gauss, I may be pardoned if I characterize the tendency outlined in these remarks as a return to the general Gaussian program. A distinction between the present and the earlier period evidently lies in this: that what was formerly begun by a single mastermind, we now must seek to accomplish through united efforts and cooperation. [65]

As we know, many contemporary mathematicians envisaged a unified science at the time. Felix Klein's *History of the Development of Mathematics in the 19th century*[3], Kaluza, Einstein, Weyl and others are examples of this [109].

Note that Hilbert's perspective on the mathematical sciences as an integrated whole can be seen as an attempts to come to grips with the philosophical implications of an ever-increasing specialization in the natural sciences. So, by invoking *the axiomatic method (im Sinne der axiomatischen Methode*'), Hilbert was calling attention to a specifically epistemological method of investigation of mathematical

---

[2]In the early twentieth century, Göttingen was the location of an exceptionally vibrant community within which a belief in the mathematical comprehensibility of nature was widespread and facilitated very free exchanges between mathematicians.

[3]This can also be seen as a most interesting attempt to understand the inner organic unity of the corpus of mathematical knowledge [66].

theories (including those of physics) that he pioneered, and which he saw as being closely tied to the nature of thought itself [10]. Therefore, this term implicates more than a merely typical mathematical concern with the rigorous, explicit statement of a theory; it also connotes a specifically *logical and epistemological method* of investigation for *deepening the foundations* of the theory.
In Hilbert's mind, this is tributary to how cognition arises from the distinct sources of intuition, concepts and ideas. Therefore, the axiomatic method is conceived as a logical analysis that begins with certain facts' presented for our finite intuition or experience. Indeed, both pure mathematics and natural science alike begin with facts', i.e., singular judgments about something already given to us in representation (*in der Vorstellung*): "certain extra-logical discrete objects that are intuitively present as an immediate experience prior to all thinking".

In his 1930 paper entitled *Knowledge of Nature and Logic*', Hilbert commented on how modern science had led to the judgment that Kant had far overestimated the role and extent of a priori elements in cognition, and carried on to endorse the conception of such elements as *nothing more and nothing less than a basic point of view (Grundeinstellung) or expression of certain unavoidable preconditions of thinking and experience*'. He concluded that what remains of Kant's *synthetic a priori* is just this *a priori intuitive point of view*' that is presupposed in all theoretical concept construction in mathematics and physics. But Hilbert stressed that this was in full agreement with the basic tendency of Kantian epistemology:

> Thus, the most general and fundamental idea of Kantian epistemology retains its significance: namely, the philosophical problem of determining that a priori intuitive point of view (*jene anschauliche Einstellung a priori*) and thereby of investigating the conditions of the possibility of all conceptual knowledge and of all experience.

So, through this citation, we discover Hilbert's conviction of the existence of a third source of cognition (*Erkenntnisquelle*) outside of deduction and experience, what he called the "*a priori intuitive viewpoint*". Hilbert describes this intuitive viewpoint (*anschauliche Einstellung*) as *an a priori insight ... that the applicability of the mathematical way of reflection over the objects of perception is an essential condition for the possibility of an exact knowledge of nature*', an epistemological position, Hilbert goes on to state, that *seems to me to be certain*' [10, 37].
As an observation on Hilbert's program on the axiomatization of physics, Einstein[4] wrote:

> Our experience hitherto justifies us in believing that nature is the realization of the simplest conceivable mathematical ideas. I am convinced that we can discover by means of pure mathematical constructions the concepts and the laws connecting them with each other, which furnish the key to the understanding of natural phenomena. Experience may suggest the appropriate mathematical concepts, but they most certainly cannot be deduced from it. Experience remains, of course, the sole criterion of the physical utility of a mathematical construction. But the creative principle resides in mathematics. In a certain sense, therefore, I hold it true that pure thought can grasp reality, as the ancients dreamed [50].

---

[4]For the relation between Hilbert and Einstein, see Stachel and Kouneiher's contribution in this volume.

For Einstein, the problem of finding a mathematical representation that would provide a unification of the gravitational and electromagnetic fields was more than just a technical problem. This aspect of his work is expressed most convincingly in Einstein's own account of his lifelong research concerns, as given in his 1949 *Autobiographical Notes* [49]. Einstein, in his later work, followed a path that is not at all dissimilar to Hilbert's. Hilbert himself perceived Einstein as sharing his concern. Both Einstein and Hilbert belong to a tradition that attempts to integrate our human knowledge and to perceive an inner unity in science.

2 The Rise of Mathematical-Physics

A real interaction between mathematics and physics began to open up in the nineteenth century[5]. For example, in volume 2 of Nature, from 1870, we read of the following challenge from the pure mathematician Sylvester [86, 119]:

> What is wanting (like a fourth sphere resting on three others in contact) to build up the ideal pyramid is a discourse on the relation of the two branches (mathematics and physics) to, and their action and reaction upon, one another - a magnificent theme with which it is to be hoped that some future president of Section A will crown the edifice, and make the tetrology …complete.

James Clerk Maxwell, as president of the British Association, took up the challenge in a very interesting address in [83]. He modestly recommended his somewhat-neglected dynamical theory of the electromagnetic field to the mathematical community. According to [47], not many mathematicians paid attention, constituting one of the greatest Missed Opportunities of all time. Hertz commented on Maxwell's approach:

> Maxwell's theory consists of Maxwell's equations. One cannot escape the feeling that these equations have an existence and intelligence of their own, that they are wiser than we are, wiser even than their discoverers, that we get more out of them than was originally put into them.

In his address to the very first International Congress of Mathematicians in Zürich in 1897, Henri Poincaré chose as his topic *Sur les rapports de l'analyse pure et de la physique mathématique*, (On the relation of pure analysis to mathematical physics). He was particularly impressed by Maxwell's achievement:

> How was this triumph attained?
>
> Maxwell succeeded because he had become imbued with the idea of mathematical symmetry. Would he have triumphed so well had others before him not explored this symmetry for its own sake? […] Analysis was perhaps not among Maxwell's skills, but to him, it would have only been cumbersome and useless baggage. On the contrary, he was gifted with a profound sense of mathematical analogy. This is why he produced good mathematical physics. [96].

---

[5]For the history of the mathematization and geometrization of physics and the role of Euclid, Aristotle, Archimedes and the Greek philosophers, followed by Gallileo, Descartes, Newton, Leibniz, and, even later, Grassmann, Hamilton and Elie Cartan, see [73].

It is this realm of fundamental physics that is intimately intertwined with mathematical research at the frontiers of mathematical study. The relation between mathematics and physics is one with a long tradition going back thousands of years. This has been true from the beginning of modern physics, when Galileo first enunciated the proposition that the natural language of physics was mathematics. Newton, one of the greatest mathematicians of his day, invented the calculus of infinitesimals to calculate planetary orbits, as well as to solve pure mathematical problems. His universal law of gravitation explained everything from the fall of an apple to the orbits of the planets.

In the following centuries, there was little distinction between theoretical physics and mathematics, with many of the greatest contributors - Laplace, Legendre, Hamilton, Gauss, Fourier - being regarded as physicists and mathematicians at the same time.

The sophistication of Maxwell's equations in the nineteenth century in including the behaviour of electromagnetism induced an analogue process in the twentieth century through Einstein's theory of special relativity and then of general relativity. Einstein's choice to privilege symmetry over the laws of mechanics naturally implies the reformulation of gravitation and electromagnetism as field theories in four-dimensional space-time. This fusion of geometry and classical physics provided a strong stimulus to mathematicians in the field of differential geometry.

However, the twentieth century has witnessed two revolutions in physics and the completion of a theory of ordinary matter and its interactions. Once again, we have called on mathematics to supply the tools and framework for this task. When Einstein created general relativity, the dynamical theory of space and time, in 1915, the necessary tools of differential geometry were available. They had been created by Gauss and Riemann in the previous century. Riemannian geometry thus became a central topic of geometry.

By the 1920s, it had been realized that atomic physics in the form of quantum mechanics, and the use of radically new concepts, such as the linear superposition of states and the uncertainty principle, required an entirely new mathematical framework. Physics appeared to be diverging from classical mathematics and the hope of capturing the fundamental physical laws in terms of deep and elegant mathematics faded away. The development of quantum mechanics built on understanding of Hilbert spaces influenced the development of functional analysis. Early particle physics drew heavily on the theory of continuous groups, which itself was partly motivated by the desire to understand the spatial symmetry of crystalline structures. Nonetheless, during the middle part of that century, mathematics and fundamental physics developed in very different directions, with little significant interaction between them. This was due, in part, to an atmosphere of increased abstraction in the mathematics community, as well as an insistence on rigid formal rigor, as exemplified by the famous Bourbaki school.

However, much of the reason for this separation was due to developments in physics. First, the early development of quantum mechanics and the early applications of quantum mechanics to elucidating the structure of matter required little mathematical sophistication. During the first decades after World War II, the vistas of particle physics rapidly expanded. These times were dominated by experimental surprises, and theoretical model building required little more than traditional mathematical tools.

3 Gauges Theories, Dualities and Fiber bundles

The advent of the Yang-Mills equations in 1955 showed that particle physics could be treated with the same kind of geometry as Maxwell's theory [132], but with quantum mechanics playing a dominant role. Later, in the 1970s, it became clear that these non-Abelian gauge theories were indeed at the heart of the standard model of particle physics, which describes the known particles and their interactions within the context of quantum field theory. These non-Abelian gauge theories of strong, weak, and electromagnetic interactions are now universally accepted as yielding a complete description of all the interactions of matter at energies and distances that are experimentally accessible at present. This development was surely one of the most remarkable accomplishments of twentieth century science. Attention has more recently turned to even more ambitious attempts to construct unified theories of all the interactions of matter, together with gravity. In the development of these gauge theories, it has happened that many significant physical problems have led to equally significant concepts in modem mathematics.[6] Many of these concepts, in fact, were invented independently by physicists and mathematicians. It is a remarkable achievement that all the building blocks of this theory can be formulated in terms of geometrical concepts such as vector bundles,[7] connections, curvatures, covariant derivatives and spinors. This combination of geometrical field theory with quantum mechanics worked well for the structure of matter, but seemed to face a brick wall when confronted with general relativity and gravitation (for the next sections see [20, 93]).

3.1 Connections in a Fiber Bundle (Elie Cartan)

A notion that includes both Klein's homogeneous spaces and Riemann's local geometry is Cartan's generalized spaces (espaces generalisés). In modern terms, this

[6] For more details, see Masson et al's contribution in this volume.

[7] Paul Dirac in 1931 discussed the possible existence of elementary magnetic charges-magnetic monopoles [41]. He showed that in quantum mechanics, such magnetic monopoles made sense if, and only if, the product of their charge, g, with the electric charge of the electron, e, was an integer multiple of Planck's constant $\hbar$, precisely: $ge = n\hbar$. This was very exciting, since it meant that as long as there existed one magnetic monopole in the universe, all charges had to be quantized in units of $\hbar/g$. In mathematical terms, Dirac had discovered an integer that characterized the topological classification of vector bundles, mathematical constructs that were being invented at about the same time by mathematicians. These concepts have come to play a role of increasing importance in modern gauge theories.

is called "*a connection in a fiber bundle*." It is a straightforward generalization of the Levi-Civita parallelism, which is a connection in the tangent bundle of a Riemannian manifold. In general, we have a fiber bundle $\pi : E \longrightarrow M$, whose fibers $\pi^{-1}(x), x \in M$ are homogeneous spaces acted upon by a Lie group G. A connection is an infinitesimal transport of the fibers compatible with the group action by G.

In the case of a complex vector bundle, the fibers are complex vector spaces C_n of dimension n and $G = GL(n;\mathbb{C})$ [18]. The importance of complex numbers in geometry has a profound implication. It is well organized and complete. One manifestation is the simple behaviour of the group, $GL(n;\mathbb{C})$: its maximal compact subgroup $U(n)$ has no torsion and has, as a Weyl group the group of all permutations on n letters.

We shall call a frame an ordered set of linearly independent vectors $e_1, \ldots, e_n \in \pi^{-1}(x), x \in M$. In a neighborhood U, where a frame field $e_1(x), \ldots, e_n(x), x \in U$, is defined, a connection is given by the infinitesimal displacement

$$De_\alpha = \sum \omega_\alpha^\beta e_\beta, \quad 1 \le \alpha, \beta \le q, \tag{1}$$

where ω_α^β are linear differential forms in U. We call ω_α^β the connection forms and the matrix

$$\omega = (\omega_\alpha^\beta) \tag{2}$$

the connection matrix. Under a change of the frame field

$$e'_\alpha = \sum a_\alpha^\beta e_\beta\,, \quad A = (a_\alpha^\beta), \tag{3}$$

the connection matrix is changed as follows:

$$\omega' A = dA + A\omega\,. \tag{4}$$

We introduce the curvature matrix

$$\Omega = d\omega - \omega \wedge \omega\,, \tag{5}$$

which is a matrix of exterior two-forms. Through exterior differentiation of (4), we get

$$\Omega' = A\Omega A^{-1}, \tag{6}$$

It follows that the exterior polynomial

$$det\left(I + \frac{i}{2\pi}\Omega\right) = 1 + c_1(\Omega) + \cdots + c_n(\Omega), \tag{7}$$

in which $c_n(\Omega)$ is a 2α-form, is independent of the choice of the frame field, and is hence globally defined in M. Moreover, each c_α is closed, i.e.,

$$dc_\alpha = 0. \tag{8}$$

The form $c_\alpha(\Omega)$ has been called the αth Chern form of the connection, and its cohomology class $[c_\alpha(\Omega)]$, in the sense of de Rham cohomology, is an element of the cohomology group $H^{2\alpha}(M;\mathbb{Z})$ and is called the αth Chern class of the bundle E. These characteristic classes are the simplest and most fundamental global invariants of a complex vector bundle. They have the advantage of possessing a local representation, by curvature.

As in the Gauss-Bonnet formula, such a representation is of great importance, because the forms $c_\alpha(\Omega)$ themselves have a geometrical significance. Moreover, let $\pi' : P \longrightarrow M$ be the bundle of frames of the complex vector bundle. Then, the pull-back $\pi'^* c_\alpha$ becomes a derived form, i.e.,

$$\pi'^* c_\alpha = dTc_\alpha\,, \tag{9}$$

where Tc_α, a form of degree $2\alpha - 1$ in P, is uniquely determined by certain properties. This operation is called transgression, and Tc_α have been called the Chern-Simons forms [17]. These forms have played a role in three-dimensional topology and in the works of E. Witten on quantum field theory [128].

This theory can be developed for any fiber bundle (see [19]). The above provides the geometrical basis of gauge field theory in physics. Here M is a four-dimensional Lorentzian manifold, so that the Hodge $*$-operator is defined, and we define the codifferential

$$\delta = *d*, \tag{10}$$

There is a discrepancy of terminology and notation, as given by the following table:

Mathematics	Physics
Connection ω	Gauge potential A
Curvature Ω	Strength F

Maxwell's theory is based on a $U(1)$-bundle over M, and his field equations can be written as

$$dA = F, \quad \delta F = J, \tag{11}$$

where J is the current vector. Actually, Maxwell wrote the first equation as

$$dF = 0, \tag{12}$$

which is a consequence. For most applications, (12) is sufficient. But a critical study of an experiment proposed by Bohm and Aharanov and performed by Chambers shows that (11) are the correct equations [131]. A generalization of (11) to an $SU(2)$ bundle over M gives the Yang-Mills equations

$$DA = F, \quad \delta F = J. \tag{13}$$

It is indeed remarkable that developments in geometry have been consistently parallel to those in physics.

The following quote from Raoul Bott captures the spirit of the time[8]:

> Although we still often do not understand each other, the push and pull relationship of our two points of view has never been stronger and has invigorated both of us. Certainly in mathematics, the physically inspired aspects of the Yang-Mills theory has had a profound effect on our understanding of the structure of 4-manifolds, and I also think we mathematicians are only now learning to appreciate the rich mathematical structure of the Dirac sea - and indeed of the whole Fermion-inspired world of the physicists, as well as their mystical belief in supersymmetry. And on the other hand, the most modern achievements of mathematics - from cobordism to index theory and K theory - have by now made their way into some aspects of present day physics - I think to stay. [13].

We will illustrate this exchange between physics and mathematics described by Bott starting with the case of the three-dimensional topology [8].

3.2 *The Dawn of Mathematical Physics*

The prime example of a topological problem is that of knots in three-dimensional space,[9] In 1984, the world of knot theory underwent a remarkable new development when Vaughan Jones discovered a polynomial knot invariant (now named after him) that was different from the Alexander polynomial [64]. Crucially, it was chiral, that is, it could distinguish knots from their mirror images, which the Alexander polynomial could not. It soon emerged that this new invariant was part of a grand family of invariants based on Lie algebras and their representations. Shortly afterwards, Witten showed how to interpret the Jones polynomial in terms of a quantum field theory in three dimensions [126].
In fact, this relation between knot invariants and particles goes to the very beginning of relativistic quantum field theory as developed by Feynman and others in the 1940s. The basic idea is that, if we think of a classical particle moving in space-time, it will move in the direction of increasing time. However, within quantum theory, the rules are more flexible. Now, a particle is allowed to travel back in time. Such a particle going backwards in time can be interpreted as an anti-particle moving forwards in time. Once it is allowed to turn around, the trajectory of a particle can form, as it

[8]Supersymmetry, the link between bosons and fermions, is a closely related concept from physics that has also influenced differential geometry. As first noted by Edward Witten, supersymmetry applied within quantum mechanics is an elegant way to derive the basic principles of Morse theory [125]. Another application is in the development of hyper-Kähler geometry - the curved manifestation of Hamilton's quaternions. Although the definition has been in the differential-geometric literature since the 1950s, it was 30 years later, as a result of the infiltration of ideas from the supersymmetric sigma model, that a mechanism for constructing good examples was found.

[9]Henceforth in this section, we follow [8].

were, a complicated knot in space-time. The rules of quantum theory will associate, to each such trajectory a probability that describes the likelihood of this process actually taking place.

Relating quantum field theory to knot invariants along these lines had many advantages. First, it was fitted into a general framework familiar to physicists. Second, it was not restricted to knots in three-dimensional Euclidean space and could be defined on general three-dimensional manifolds. One could even dispense with the knot and get an invariant of a closed three-dimensional manifold. Such topological invariants are now called quantum invariants and have been extensively studied. The chirality of these invariants is then closely related to the chirality of Yang and Lee in weak-interaction physics. This example of quantum knot invariants makes clear that physics bridges the realms of geometry and algebra in a natural way. Indeed, quantization can be seen as taking a geometric object, say a curve in a Calabi -Yau space or a knot in a three-manifold, and associating an algebraic object with it, in these examples, the complex number that represents the probability amplitude of the corresponding quantum process.

The revitalization of the connections between mathematics and physics is especially true in the realm of elementary particle physics, leading to many hybrid subjects, such as topological quantum field theory, quantum cohomology and quantum groups, which are now central topics of research in both mathematics and physics.

For instance, modern physical constructions such as quantum field theory and string theory, which are very far removed from everyday experience, have proven to be a similarly fertile setting for mathematical problems. Indeed, in many ways, quantum theory has turned out to be an even more effective framework for mathematics than classical physics. Particles and strings, fields and symmetries, they all have a natural role to play in mathematics.

This influence manifests itself in two ways. In some sense, the easier one is for the mathematician to be presented with a clearly stated conjecture or problem. One then attempts a development of current conventional techniques to provide a rigorous solution. The construction of instantons in the 1970s was an example of this: a problem derived from quantum physics, but refined to one in conventional, but modern, differential geometry. The mathematics that developed from it, and its new viewpoints, was the work of Simon Donaldson [45].

Indeed, the Methods developed in quantum gauge theories, using so-called "instantons," were a sources of inspiration for Donaldson [44], Taubes, and Floer [51] in deducing some deep and astounding properties of the geometry of three - and four-dimensional spaces. Witten has reinterpreted Donaldson's theory in physical terms, using it to speculate on a new phase of quantum gravity [129].

In 1983, Simon Donaldson applied ideas from physics to make a spectacular breakthrough in our understanding of four-dimensional geometry. He showed that there were some very subtle invariants in four dimensions, not present in any other dimension, which were preserved under smooth deformation, but not under general continuous deformation [43, 44]. Indeed, dimension four is special not only for physics, in which it represents space-time, but also for geometry, in which there are unique phenomena associated with this particular dimension.

Donaldson's work[10] originated in physics in the same way that the Hodge theory of harmonic forms had been inspired by Maxwell's equations, but this connection appeared to be rather formal. More importantly, within Yang-Mills theory, the equations become nonlinear. Instead of studying the solutions themselves, Donaldson put in the so-called moduli space that parametrizes the family of solutions in a central position. In fact, one could regard the gauge-theoretic approach to four-dimensional differential topology as replacing the points of the four-manifold with nonlinear instanton solutions of the Yang-Mills equations.

Again, it was Witten who showed that Donaldson's invariants could be interpreted in terms of a quantum field theory and that this would have profound consequences. Moreover, this field theory was a close cousin of the standard theories used by particle physicists, except that it had a twist' that produced topological invariants, not dependent on the intricate details of the underlying geometry of space-time.

The interpretation by Witten[11] allowed Seiberg and Witten[12] to show that the corresponding physical theory could be solved in terms of a much simpler structure. Yang-Mills theory is fundamentally based on a choice of a non-Abelian Lie group, usually taken to be the group $SU(2)$. The non-commutativity of this symmetry group leads to the nonlinearities of the associated partial differential equations. However, in physics, it was known that in quantum theories, these non-Abelian symmetries often manifest themselves only at very short distance scales. At large distances, the symmetry can be broken into a much simpler Abelian group. For example, in the case

[10]Donaldson's work involved a melting pot of ideas, not so much from physics, but from the mathematical areas of nonlinear analysis of partial differential equations, differential and algebraic geometry, and topology. Nevertheless, the whole idea of studying a moduli space in this context- a space of connections up to gauge equivalence - had its essential origin in physics. In fact, Donaldson found his invariants by studying special solutions of the Yang-Mills equations, which had already been introduced by physicists under the name of instantons'. These solutions have the property that they are essentially localized to a small region in space-time, thereby describing an approximately instantaneous process. Instantons had a family resemblance to the solitons' or solitary waves first observed by John Scott Russell in the early nineteenth century.

[11]Remarkably, Donaldson theory is viewed as perturbative. The distinct, non-perturbative picture of the same theory yields Seiberg-Witten theory. The last one describes, in a different way, the same invariants as Donaldson did, they are two limiting forms of the same quantum field theory.
The point here is that the perturbative/non-perturbative physics view allowed for he resolution of some old problems. In another setting, that of Chern-Simons theory applied to the theory of knots and links, the perturbative view gives the Vassiliev invariants and the non-perturbative the Jones-Witten polynomial invariants.

[12]Seiberg and Witten were able to make this physical intuition precise for the class of twisted supersymmetric quantum field theories relevant for the Donaldson invariants [67, 112]. The resulting Seiberg-Witten invariants were based on a $U(1)$ gauge field interacting nonlinearly with a spinor field [127].
These invariants again involved characters familiar to the mathematician, Dirac operators and Spin structures, and this area was the focus of intense research activity in the 1990s. It seemed as if results that were difficult to prove using Donaldson's theory were easier here, and vice versa. The pay-off in mathematics from the appeal to the physicists''s intuition was clear: one had a new tool for studying four-dimensional manifolds.
On the other hand, to establish that really there was a link between the two theories in conventional terms proved to be an enormous task, one that was only recently accomplished.

of $SU(2)$, only the circle group $U(1)$ of electromagnetism would appear together, possibly with some charged matter particles.

The second mode of influence challenges mathematicians much more seriously, since it involves developing a sense of intuitive understanding parallel to that of every physicist, for whom this has been second nature since graduate school. It has spurred new ways of thinking in pure mathematics. Clearly, this approach is a long-term program with many ramifications, somewhat reminiscent of Hilbert's sixth problem: the axiomatization of all branches of science, in which mathematics plays an important part.

One example comes from enumerative geometry, a particularly active field in the nineteenth century.[13] In classical geometry, many problems concern counting the number of solutions. A slightly more complicated problem is to count the number of curves in the plane that satisfy certain constraints. One such problem for curves of degree d, which are rational, is to ask how many such curves go through n general points. The general formula was not known to classical geometers, and only emerged from physics! Moreover, the methods from physics are powerful enough to deal with more general curves, and also with more general problems of the same type [69, 107].

3.3 Algebraic Geometry, Cohomology and Strings theories

Algebraic geometry is a subject that somehow connects and unifies several aspects of mathematics, including, obviously, algebra and geometry, but also number theory, topology, and, recently, string theory in physics, etc. The power of the field arises from a point of view that was developed in the 1960s in Paris, by the group led by Alexandre Grothendieck.[14] The power comes from rather heavy formal and technical

[13]In the mid-nineteenth century Riemann introduced analytical methods into the algebraic geometry of curves. These were sometimes proved by appeal to physical principles such as the Dirichlet principle, a technique motivated by the physical tenet that nature works by minimizing actions and energy.

Yet the whole apparatus of differentials and theta functions enabled remarkable results to be proved or rendered obvious; special facts like the existence of precisely 28 bitangents to a quartic curve or 120 tritangent planes to a genus four curve are not so far removed in spirit from the remarkable count of rational curves on the quintic threefold by Candelas et al. [15], which is the most startling application of the string theorists's mirror symmetry in algebraic geometry. If one looks at the journals of the time, one will also see a very rapid succession of applications of these methods before a settling down at the end of the century to a mixture of techniques.

[14]Just prior to Grothendieck's entry into the subject, Weil had gotten important results in number theory through algebro-geometric arguments, and pointed the way to far more, but some of his methods went beyond existing rigorous foundations. He aimed to supply new foundations adequate to his ideas. Around the same time, Zariski and van der Waerden were also generalizing the foundations of algebraic geometry, and many others were introducing various innovations. In particular, the "Weil conjectures" suggested that topological methods of cohomology applied to algebraic geometry might have huge consequences for number theory – but neither Weil nor anyone else in,

machinery, in which it is easy to lose sight of the intuitive nature of the objects under consideration [122].
Grothendieck gave two simplifications.[15] The first concerned the Cohomology, which, in 1954, had numerous alternative forms in topology, geometry and analysis. It could be defined by covering spaces, or differential forms, or Cech covers, or simplicial decomposition, and much more. Grothendieck introduced the now-current ideas of the Abelian category and derived functor so that essentially all these cohomology theories used one simple kind of resolution: the injective. The second is the introduction of *schemes*. By the mid-twentieth century, leading algebraic geometers all saw that algebraic geometry needed a notion of algebraic space more general than spaces defined by polynomial equations over the complex numbers. Many people gave examples, led by Weil and, notably, Serre. In these, algebraic spaces could have coordinate functions much more general than polynomials with complex coefficients. They all stayed fairly close to taking coordinates in fields, generally, in algebraically closed fields. This was a serious obstacle to Weil's hopes for algebraic geometry being used as a tool in number theory, since the integers do not form a field at all. The rational numbers form a field, but it is as far from algebraically closed as it could be– if an integer polynomial with lead coefficient 1 has no integer roots, then it has no rational roots either, but, of course it has as many complex roots as its degree.
Grothendieck defined *schemes* so that every ring is the coordinate ring of a scheme. Rings that seem to be purely algebraic abstract entities still have geometric meaning in this theory. Some schemes by themselves have little apparent geometric sense, but they have eminently geometric relations to one another [39]. You can define schemes without knowing what polynomials are, let alone knowing what "algebraically closed" means. In lightly over-simplified terms, you only need to know the commutative, associative and distributive laws to learn what a commutative ring is, and thus define a scheme.

Another essential motivation for algebraic geometry is to study the manifolds. Real manifolds are things that locally look like bits of real n-space, and they are glued together to make interesting shapes. There is already some subtlety here - when you glue things together, you have to specify what kind of gluing is allowed. For example, if the transition functions are required to be differentiable, then you get the notion of a differentiable manifold. A great example of a manifold is a submanifold of $\mathbb{R}^n$ (consider a picture of a torus). In fact, any compact manifold can be described in such a way. You could even make this your definition, and without worrying about gluing.
This is a good way to think about manifolds, but not the best way. There is something arbitrary and inessential about defining manifolds in this way. Much cleaner is the notion of an abstract manifold, which is the current definition used by the mathematical community. There is an even more sophisticated way of thinking about manifolds.

---

say, 1954 could see how to actually do that. Serre believed more strongly than Weil himself did that such a cohomology could actually be created.

[15] For more details, see Colin MacLarty's contribution and [79].

A differentiable manifold is obviously a topological space, but it is a little bit more. There is a very clever way of summarizing what additional information there is, basically by declaring what functions on this topological space are differentiable. The right notion is that of a sheaf. It is true, but not obvious, that this ring of functions that we are declaring to be differentiable determines the differentiable manifold structure.

Very roughly, algebraic geometry, at least in its geometric guise, is the kind of geometry you can describe with polynomials. So, you are allowed to talk about things like $y^2 = x^3 + x$, but not $y = sinx$. So, some of the fundamental geometric objects under consideration are things in n-space cut out by polynomials. Depending on how you define them, they are called *affine varieties* or *affine schemes*. They are the analogues of the patches on a manifold. Then, you can glue these things together, using things that you can describe with polynomials, to obtain more general varieties and schemes. Thus, we'll have these algebraic objects that we call varieties or schemes, and we can talk about maps between them, and things like that.
In comparison with manifold theory, we've really restricted ourselves by restricting ourselves to the use of polynomials. But on the other hand, we have gained a huge amount too.
First of all, we can now talk about things that aren't smooth (that are singular), and we can work with these things. Algebraic geometry provides particularly powerful tools for dealing with singular objects. (One thing we'll have to do is to define what we mean by smooth and singular!). Also, we need not work over the real or complex numbers, so we can talk about arithmetic questions, such as: what are the rational points on $y^2 = x^3 + x^2$ (Here, we work with the field $\mathbb{Q}$.) More generally, the recipe by which we make geometric objects out of things to do with polynomials can be drastically generalize, and we can make a geometric object out of rings. This ends up being surprisingly useful. All sorts of old facts in algebra can be interpreted geometrically, and indeed, progress in the field of commutative algebra these days usually requires a strong geometric background.

An richer example is the counting of rational curves not on the plane, but on a quintic hypersurface X given by the equation image

$$x_1^5 + x_2^5 + x_3^5 + x_4^5 + x_5^5 = 0$$

in projective four-space. This equation can be seen to describe a manifold of three complex dimensions, or six real dimensions. The quintic X plays an important role in string theory. It is an example of a Calabi-Yau space, a special class of complex manifolds that allow for a solution of the Einstein equations of gravity in empty space. In string theory X can be used to compactify the 10-dimensional space-time down to the four dimensions of physics.

The idea is that there is a formulation of string theory that is able to capture the topology of string configurations [40]. Roughly, the idea is as follows: we introduce fermions fields θ^μ considered as spinors on the two-dimensional world-sheet. They have two components θ_L^μ, θ_R^μ, with the local action

$$\int d^2 z g_{\mu\nu}(x) \left(\theta_L^{\mu} \frac{D\theta_L^{\nu}}{\partial \bar{z}} + \theta_R^{\mu} \frac{D\theta_R^{\nu}}{\partial z} \right), \tag{14}$$

We assume the target space X is (almost) complex, so we can use holomorphic local coordinate x^i, $\bar{x}^{\bar{i}}$with a similar decomposition for the fermions. We consider the appropriate higher order terms to obtain a sigma model with $N = (2, 2)$ supersymmetry.

To produce the topological string, we change the spins of the fermionic fields. Or there are two inequivalent manners but which to do so, and we obtain two models A and B. However, the nature of the topological twisting determines wether if the path-integral of the sigma model localizes to a finite-dimensional space.

The A-model corresponds to the holomorphic maps

$$\frac{\partial x^i}{\partial \bar{z}} = 0.$$

The path-integral is reduced over all maps from Σ into M to a finite-dimensional integral over the moduli space $\mathcal{M}$ of holomorphic maps $f : \Sigma \longrightarrow M$ or, more precisely, the moduli space of paris (Σ, f), where Σ is a Riemann surface. The A-model depends only on the Kähler class $\omega \in H^2(M)$ of the manifold M.

Now, Quantum cohomology is a deformation of the De Rham cohomology ring $H^*(M)$ of a manifold. Classically, this ring captures the intersection properties of submanifolds. More precisely, if we have three cohomology classes, $\alpha, \beta, \gamma \in H^*(M)$, which are Poincaré dual to three subvarieties $A, B, C \subset M$, the quantity

$$I(\alpha, \beta, \gamma) = \int_M \alpha \wedge \beta \wedge \gamma \tag{15}$$

computes the intersection of the three classes, A, B and C, that is, it counts (with signs) the number of points in $A \cap B \cap C$.

In the case of the A-model where M is kähler, or at least a symplectic manifold with symplectic form ω, the stringy intersection product is related to the three-string vertex

$$I_q(\alpha, \beta, \gamma) = \sum_d q^d \int_{Maps_d} \alpha \wedge \beta \wedge \gamma \ , \tag{16}$$

where

$$q^d = exp\left[\frac{-1}{\alpha'} \int_{S^2} \omega\right] = e^{-(d.[\omega])/\alpha'} \ , \tag{17}$$

Remark that in Eq. (16), we integrate the differential forms over the moduli space of a pseudo-holomorphic map of degree d of a sphere into the manifold M. In the limit $\alpha' \to 0$, only the constant maps contribute, and we recover the classical definition of the intersection product by means of an integral over the space M. Geometrically, the quantum intersection counts the pseudo-holomorphic spheres inside M that intersect each of the three cycles A, B, C, and therefore there is no need for these cycles to intersect. It is enough to have a pseudo-holomorphic sphere with points a, b, c such that $a \in A, b \in B, c \in C$. That is, if there is a string world-sheet that connects the three. The B-model, which reduced to (almost) constant maps, depends only on the complex structure of M. The A-model and B-model can be interchanged by mirror symmetry. A powerful example of mirror symmetry is the calculation of Candelas et al. in $\mathbb{P}^4$ of the quintic Calabi-Yau manifold given by the equation

$$X = x_1^5 + x_2^5 + x_3^5 + x_4^5 + x_5^5 = 0\,. \tag{18}$$

In the case of the A-model, we have the expression

$$F(q) = \Sigma_d n_d q^d \ , \tag{19}$$

n_d computes the number of rational curves in M of degree d. It is very difficult to calculate these numbers. $n_1 = 2875$ was calculated in the nineteenth century, and $n_2 = 609250$, which counts the different conics, was calculated in 1980. But thanks to string theory and mirror symmetry we know all these numbers now. Indeed, mirror symmetry, relates the stringy invariants coming from the A-model on the manifold M to the classical invariants of the B-model on the mirror manifold $\hat{M}$. In particular, this leads to the Fuchsian differential equation for the function $F(d)$; the resolution of this equation give us the integers n_d.

An essential ingredient that helped in finding the physical solution is the existence of an equivalent formulation of this physical process using a so-called mirror Calabi-Yau manifold Y. As far as classical geometry is concerned, the spaces X and Y are very different; they do not even have the same topology. But in the realm of quantum theory, they share many properties. In particular, with a suitable identification, the string propagation in spaces X and Y are identical. The interchange of X with Y is called mirror symmetry [36, 62]. It is a typical example of a quantum symmetry.

Mirror symmetry is an example of a much broader area of mathematics influenced by physics: symplectic geometry. Gromov's theory of pseudoholomorphic curves in symplectic manifolds had already shown that, at the internal low-dimensional level, algebraic and symplectic geometry had common features, but the string theorists's notion of mirror symmetry brought the parallels between symplectic and algebraic geometry into much sharper focus. In Kontsevich's formulation of this phenomenon - homological mirror symmetry [68] - the natural subspaces on either side of the algebraic/symplectic divide (algebraic or Lagrangian subvarieties and bundles over them) are supposed to generate equivalent mathematical objects.

Supersymmetry, the link between bosons and fermions, is a closely related concept from physics that has also influenced differential geometry. As first noted by Edward Witten, supersymmetry applied within quantum mechanics is an elegant way to derive the basic principles of Morse theory [125]. Another application is in the development of hyper-Kähler geometry - the curved manifestation of Hamilton's quaternions. Although the definition has been in the differential-geometric literature since the 1950s, it was 30 years later, as a result of the infiltration of ideas from the supersymmetric sigma model, that a mechanism for constructing good examples was found.

3.4 Cohomology and Invariants

Homology[16] and cohomology have emerged as the main instruments of algebraic topology. Their influence goes far beyond the geometric topology. They provide the natural expressions for algebraic geometry, in differential geometry, and the algebraic theory of numbers. The cohomology depends on the local structure of the variety. It gives an account of forms, and defines them. It connects the continuous to the discontinuous. But the most remarkable aspect is its universality. There is a pleiad of cohomology leading to the same results. However, as its name implies, it is the dual of homology. The latter is based on the global properties of a variety. Some homological entities are known to all: the orientation of varieties, the connected component of a point in a topological space (an object in one piece). Another example: $H_1(\mathbb{T}^2)$, the set of homology classes of degree 1 of a torus of dimension 2.
In general, there are more algebraic structures with cohomology. For example, the cohomology group of degree 1 of the torus $\mathbb{T}^2$, $H^1(\mathbb{T}^2)$ is constructed from the representations: at each loop, assigning a real or complex number; with '*composition constraint*': when a path is obtained by putting two paths end to end, its number must be the sum of the numbers of the two components; and *deformation constraint*, when two close paths have the same number. The set of 'numerical assignments' and the "constraints" form $H^1(T^2, \mathbb{R})$ or $H^1(T^2, \mathbb{C})$, depending on the quality of the numbers employed. It is now an affine plan. The intersection of the paths on the torus in pairs provides this plane with an area unit. A secondary geometry appeared, and the group $SL_2(\mathbb{Z})$ is a distinguished part of its symmetries.

Homology can be considered as a general technique in mathematics used to measure the difficulty that certain sequences of morphisms have being exact. The idea is precisely to note that if a morphism α on a module M has $\alpha^2 = 0$, then $Im\ \alpha \in Ker\ \alpha$. Everything is in this remark! For we can then characterize the elements of M that are in the kernel of α (they are called cycles), but which are not in the image of α (these are called boundaries). So, we form the quotient of modules

$$H(M, \alpha) = \frac{Ker\ \alpha}{Im\ \alpha}$$

[16] It is in an article of 1895 that Poincaré [97–99] defines, for the first time, differential manifolds and chains (or sub-varieties), which he qualifies as homologous (see [97]). Its definition was somewhat imprecise, but the notion he used matches up exactly with the current acceptance: two closed chains are homologous if their difference is a boundary. However, Poincaré's text did not reveal the idea of Cohomology. The reason for this is that on a manifold, we can obtain completely cohomology from homology through Poincaré's duality. Roughly, Poincaré's duality connects the local statements of cohomology to the global statements of homology.

called the homology of M for α. This construction allows us to characterize the cycles that are not boundaries. It also allows to associate a sequence of Abelian groups or modules with a mathematical object like a topological space or a group.
The late 1930s and early 1940s witnessed the rise of homological algebra. This contributed largely to the emergence of notions of category and functor, ubiquitous notions in algebra and logic afterwards. Indeed, the tensor products of modules, the exact sequences and the functors Hom and Ext facilitated a remarkable amount progress both in calculating the homology group and conceptualizing what would eventually become the homological algebra in Henri Cartan and Eilenberg's works in the 1950s. Algebra topology, as its name indicates so correctly, proposes to study the topology of space by using algebraic concepts, such as homology groups, but also homotopy groups.
Another contribution that later led other paths is that of the axiomatization of simplicial homology by Eilenberg and Norman Steenrod in 1945. This work allowed us, on the one hand, to show that some of the other homologies defined in this context are isomorphic to simplicial homologies, and on the other hand, it has generated more generalizations, *the generalized homologies*, of which the K-theory is only an example.

In parallel with these developments in the algebraic topology domain, the works in algebra have conceptualized different essential notions, such as the extension of abelian groups: an extension of the Abelian group F by the Abelian group H is an Abelian group G containing F such that H is identified with the quotient G/F. In other words, we have an exact short sequence of Abelian groups

$$0 \longrightarrow F \longrightarrow G \longrightarrow H \longrightarrow 0$$

An essential aspect of modern mathematics and physics is the study of the invariants. Indeed, classifying the invariants became a central issue in physics and mathematics. In the introduction to his book The Principles of Quantum Mechanics, Clarendon Press, Oxford, 1930, the young Dirac (1902–1984) wrote:

> The important things in the world appear as invariants ...The things we are immediately aware of are the relations of these invariants to a certain frame of reference ...The growth of the use of transformation theory, as applied first to relativity and later to the quantum theory, is the essence of the new method in theoretical physics.

Two fundamental tools that will play an essential role finding and calculating those invariants are cohomology and homology. Indeed, cohomology[17] plays a fundamen-

[17] Usually, the non-vanishing of a cohomology class in algebra, geometry, and topology, express some sort "*failure*". Indeed, often in mathematics, you wish something were true, but in general, it is not. However, the quantification of how badly it fails help us in determining a more precise statement that holds generally. The size (or dimension) of the corresponding cohomology group is a measurement of how many ways things can go wrong. If it is nice or if you can understand it completely, then you may be able to analyze all the possible failure modes exhaustively, and use that to prove something interesting. This idea can be applied in an amazingly broad set of contexts. This somewhat explains the use of cohomology to describe quantization.

tal role in modern mathematics and physics. As we saw, cohomology is an example of a local - global structural connection that permeates mathematics.
Cohomology is used in physics[18] to compute the topological structure of gauge fields. This helps to explain why the Maxwell's equations in electrodynamics are closely related to cohomology, namely, de Rham cohomology based on Cartan's calculus for differential forms and the corresponding Hodge duality on the Minkowski space. Since the Standard Model in particle physics is obtained from the Maxwell's equations by replacing the commutative gauge group U(1) with the noncommutative gauge group $U(1) \times SU(2) \times SU(3)$, it should come as no great surprise that de Rham cohomology also plays a key role in the Standard Model in particle physics via the theory of characteristic classes (e.g., Chern classes, which were invented by Shing-Shen Chern in 1945 in order to generalize the Gauss–Bonnet theorem for two-dimensional manifolds to higher dimensions).
It is very clear now that the gauge-theoretical formulation of modern physics is closely related to important long-term developments in mathematics pioneered by Gauss, Riemann, Poincaré and Hilbert, as well as Grassmann, Lie, Klein, Cayley, Elie Cartan and Weyl. The prototype of a gauge theory in physics is Maxwell's theory of electromagnetism. The Standard Model in particle physics is based on the principle of local symmetry. In contrast to Maxwell's theory of electromagnetism, the gauge group of the Standard Model in particle physics is a noncommutative Lie group. This generates additional interaction forces, which are mathematically described by Lie brackets.
The presence of a cohomological nature in quantum field theory is confirmed by the modern treatment of quantum symmetries, gauge invariances, renormalization, anomalies, the BRST formalism and the numbers associated with the figures (diagrams) via the Feynman integrals. For the elliptic type, Witten sees it as a generalization of the characteristic classes, like that of Euler. He deduces the premises of a (rather infinite) geometric definition of the elliptic cohomology, which enters a hierarchy:

bordisme et cobordisme $\longrightarrow$ cohomologie elliptique $\longrightarrow$ K-theorie $\longrightarrow$ cohomologie ordinaire

In general, the topological quantum field theories in dimension 3 are cohomologies of a new type,[19] so that the spaces of states Φ_M of a quantum field theory seem to be close to cohomology as well. The spaces Φ_M would thus be a cohomology groups;

---

[18] In deciding to extend the concepts of homology and cohomology outside the ideal world of mathematics, we are led to accept the use of certain analogies. The homology appeared as a redoubling of abstraction; the homological forms have doubled the algebra of the geometric forms that they enveloped. We can quite clearly distinguish two movements: a birth of geometry or algebra followed by homological stabilization. From a logical point of view, a geometric object and homological object have the same nature.

[19] The cycles carried by a surface Σ are formal combinations of manifolds of dimension 3 bordered by Σ; The partition function Z defines a form of intersection on cycles, and homology occurs when we quotient by the kernel of this form.

the bundle $\mathcal{A}$ of the dynamic states would instead be something like an *algebra of operations*.

3.5 String Theory

String theory[20], which was originally discovered accidentally in an attempt to understand nuclear forces, has emerged in recent years as a promisingly realistic theory of all the interactions and, for the first time, a consistent theory of quantum gravity.
To some extent, string theory,[21] is a simple generalization of the ordinary framework of quantum field theory, in which the basic constituents of nature are not point-like but rather are extended one-dimensional objects-strings. Remarkably, this seemingly minor extension from point-like particles to extended strings, without modifying in any other way the fundamental principles of physics, leads to an incredible structure. This structure implies that the only forces that can exist are only those of the kind we can see: gravitational and gauge interactions. It can also produce the matter content of the world as we know it, as well as the specific pattern of forces that we observe. It also has bizarre implications, requiring that space-time be 10-dimensional. To agree with the crudest of observations, it must be the case that 6 of the spatial dimensions are curled up into a little closed space, so that we do not notice them. This can be achieved, since, as a generalization of Einstein's theory of general relativity, the theory incorporates the dynamics of space-time and possesses solutions with 6 compact, curled up, directions of space.
A string may be considered as a parametrized loop.[22] In this case, we study the manifold M through maps:

[20] See Witten's contribution in this volume.

[21] As we said, string theory makes use of deep structures in differential geometry and algebraic geometry, and connects to the theory of modular functions and finite groups. It even appears to have a place for branches of mathematics as number theory and knot theory.

[22] From the point of view of perturbative String Theory, we usually consider the classical motion of a fundamental string, so that the action is given by $S_{string} = -T_{string}V$, where $T = \frac{1}{2\pi\alpha}$ is the tension of the string, α is the Regge slope parameter and V is the area of the string world sheet. The action is called the Nambu-Goto action. Classically, the Nambu-Goto action is equivalent to the Polyakov action (the string sigma-model action):

$$S_\sigma = -\frac{1}{4\pi\alpha}\int_\Sigma d^2\sigma\sqrt{-h}h^{\alpha\beta}\eta_{\mu\nu}\partial_\alpha X^\mu\partial_\beta X^\nu \tag{20}$$

where σ and τ are coordinates on the world sheet, and $h_{\alpha\beta}(\sigma,\tau)$ is a world sheet metric, $h = deth_{\alpha\beta}$, $h^{\alpha\beta}$ is the inverse of $h_{\alpha\beta}$. Σ denotes the world sheet, and $d^2\sigma = d\sigma d\tau$. The functions $X^\mu(\sigma,\tau)$ describe the space-time embedding of the string world sheet. Quantum mechanically, we use the path integral to deal with the local symmetries and gauge fixing. Unfortunately, in this case, we have to handle the problem of anomalies, more specifically a conformal anomaly, unless the space-time dimension is $D = 26$. In superstrings (i.e. strings for which supersymmetry is added - either on the world sheet, as in the so-called RNS sector, or to the background space-time as in the GS sector), an analogous analysis gives a critical dimension $D = 10$.

$$x : S^1 \longrightarrow M,$$

that is, through the free loop space $\mathcal{L}M$. Quantization will associate a Hilbert space with this loop space. When a string moves in time, it sweeps a surface Σ. For a free string, Σ has the topology of $S^1 \times I$, but we can also consider at no extra cost interacting strings that join and split. In that case, Σ will be an oriented surface of arbitrary topology. Therefore, in Lagrangian formalism, we consider the maps:

$$x : \Sigma \longrightarrow M,$$

There is a natural action if we pick the Hodge star or conformal structure on Σ (together with the Riemannian metric g on M)

$$S(x) = \int_\Sigma g_{\mu\nu} dx^\mu \wedge *dx^\nu \ , \tag{21}$$

The critical points of $S(x)$ are the harmonic maps. In Lagrangian quantization formalism, one considers the formal path-integral over all the maps $x : \Sigma \longrightarrow M$:

$$\Phi_\Sigma = \int_{x:\Sigma \longrightarrow M} e^{-S/\alpha'} \ . \tag{22}$$

Here, the constant α' plays the role of Planck's constant on the string worldsheet Σ. It can be absorbed in the volume of the target M by rescaling the metric as $g \rightarrow \alpha'.g$. The semi-classical limit $\alpha' \rightarrow 0$ is therefore equivalent to the limit $vol(M) \rightarrow \infty$.

Two remarks are in order: in perturbative string theory, we study the loops in a space-time manifold. These loops can be thought to have an intrinsic length l_s, the string length. Because of the finite extent of a string, the geometry is necessarily "fuzzy". Within the limite $l_s \rightarrow 0$, the string degenerates to a point, and we recover the classical geometry. So, the parameter l_s controls the "stringiness" of the model. $l_s^2 = \alpha'$ plays the role of the Planck constant on the world sheet of the string. A second deformation of classical geometry has to do with the fact that strings can be split and joined, sweeping out a surface Σ of general topology in space-time. According to the general rules of quantum mechanics, we have to include a sum over all topologies. Such a sum over topologies can be regulated if we introduce a formal parameters g_s and the string coupling such that a surface of genus h gets weighted by a factor g_s^{2h-2}. Higher genus topologies can be interpreted as virtual processes wherein strings split and join - a typical quantum phenomenon. Therefore, the parameter g_s controls the quantum corrections. In fact, we can equate g_s^2 with Planck's constant in space-time. Only for small values of g_s can string theory be described in terms of loop spaces and sums over surfaces.

1.0.1. T-Duality The presence and remarkable power of dualities is one of the hallmarks of string theory.[23] Let us consider the case of particle or string on space-time

[23]For more details, see [40].

that is given by the n-dimensional torus $\mathbb{T}$, where

$$\mathbb{T} = \mathbb{R}^n/\Lambda,$$

with Λ beaing a rank n lattice. The quantum state of the particle on $\mathbb{T}$ is given by its momentum $p \in \Lambda^*$. The wavefunction $\Psi(x) = e^{ipx}$ forms a basis of $\mathcal{H} = L^2(\mathbb{T})$ that diagonalizes the Hamiltonian $H = -\Delta = p^2$. So, we can decompose the Hilbert space as

$$\mathcal{H} = \bigoplus_{p \in \Lambda^*} \mathcal{H},$$

where the graded pieces $\mathcal{H}_p$ are all one-dimensional. There is a natural action of the symmetry group

$$G = SL(n, \mathbb{Z}) = Aut\Lambda$$

on the lattice $\Gamma = \Lambda$ and the Hilbert space $\mathcal{H}$.
In the case of a string moving on the torus $\mathbb{T}$, states are labeled with a second quantum number, the winding number $\omega \in \Lambda$, which is simply the class in $\pi_1\mathbb{T}$ of the corresponding classical configurations. The winding number simply distinguishes the various connected components of the loop space $\mathcal{L}\mathbb{T}$, since

$$\pi_0\mathcal{L}\mathbb{T} = \pi_1\mathbb{T} \cong \Lambda \ .$$

We therefore have

$$\Gamma^{n,n} = \Lambda \oplus \Lambda^* \ ,$$

where $p \in \Lambda^*$ and $\omega \in \Lambda$.
This is an even self-dual lattice of signature (n, n) with the inner product

$$p = (\omega, k), \quad q^2 = 2\omega.k \ .$$

It turns out that all the symmetries of the lattice $\Gamma^{n,n}$ lift to symmetries of the full conformal field theory built up by quantizing the loop space. The elements of the symmetry group of the Narain Lattice

$$SO(n, n, \mathbb{Z}) = Aut\Gamma^{n,n}$$

are examples of T-dualities. A particular example is the interchange of the torus with its dual $T \leftrightarrow T^*$.
From the string theory point of view, T-duality on a circle maps modes of the string with momentum (which are heavy when R is small, where R is the radius of the circle) to modes of the string with winding (which are heavy when R is large).

T-dualities that interchange a torus with its dual can be also applied fiberwise. If the manifold M allows for a fibration $M \longrightarrow B$ whose fibers are tori, then we can

produce a dual fibration in which we dualize all the fibers. This gives a new manifold $\hat{M} \longrightarrow B$. Under suitable circumstances, this produces an equivalent supersymmetry sigma model. The symmetry that interchanges these two manifolds $M \longleftrightarrow \hat{M}$ is called mirror symmetry.[24]
Remark that most of the discovered dualities, thanks to the string theory, are quantum mechanics by nature. Therefore, quantum mechanics would make possible fundamental new symmetries, just as gravity, does in the light of General Relativity. Indeed, from gravity we learned about general covariance or the principle of equivalence - which forever changed our understanding of the role of gravity in nature. So, maybe the twenty-first century will be the arena of a similar process, but for the quantum mechanics side this time.

The latest developments in superstring theory, an ambitious theory that attempts to construct a unified quantum theory of matter and gravity, have begun to meet real mathematical frontiers. These theories have attracted much attention from mathematicians, since they give strong hints of connections between hitherto separate parts of mathematics.
Many physicists[25] believe that the final understanding of the structure of string theory will involve fundamental generalizations of geometry. Perhaps we are entering a golden era in the long history of cooperation between fundamental mathematics and physics.
The original, highly optimistic expectation that this theory would lead rapidly to new predictions and tests, has undergone sober reevaluation. It is not that there are any experimental contradictions, nor are there any indications of internal inconsistency, rather, it is clear that we do not yet know enough about the structure of the theory to control its dynamics sufficiently to make contact with experiment. Part of the problem is that we have stumbled onto this theory by accident, without knowing what the basic logical setting for the theory is or will be.
A more immediate problem is that in trying to discover the principles of this theory and applying it to the real world to test its validity, we are faced with the fact that the basic distance scale of the theory is very, very small. Unfortunately, this length scale is smaller by 17 orders of magnitude than the smallest distances that we can see with our most powerful microscopes, our most energetic particle accelerators. The fact that this number is so small bears responsibility for some of the most striking features of our universe. For example, the reason stars are so big is that, at the scale of the radius of ordinary atoms and nuclei, gravity is very weak (because this scale is 17 orders of magnitude below the Planck scale). In any case, it implies that string

[24]See Butterfield and de Haro's contribution in this volume.

[25]String theorists would freely admit that they don't know what the theory is, but they are fairly sure that what they have is a genuine theory. What they observe is its implications at different limits of coupling constants, where it makes contact with other areas of mathematics. The fundamental concepts in the terra incognito at its centre are unknown, yet its deep consistency unearths structures across a wide range of mathematics. They also admit that is harder than they initially thought when the possibilities opened up in the mid-1980s, but by being harder, it has drawn them closer to mathematics, and they are quite happy to use the predictive power within that domain, given that the physical experiments are currently impractical.

theory and loop quantum gravity are an attempt to extrapolate far, far beyond present day experiment. Even if we have an idea of the physics at these incredibly small Planckian distances, it is very hard to make our way up to the distances at which measurements are done at present.

3.6 Loop Quantum Gravity

Loop quantum gravity represents a research program that is underway, one that also entails a profound transformation of our traditional notions of space as the continuous background in which events occur, one that rivals that of Einstein's general theory of relativity, in which space-time was given a dynamical significance.[26]

Loop quantum gravity approach assumes Einstein's theory can be quantized non-perturbatively.

Loop quantum gravity is a canonical quantization approach for Hamiltonian formulation of Einstein's general theory of relativity. Abhay Ashtekar, in 1986 reformulated Einstein's general relativity in a way that facilitated overcoming the previous stumbling blocks of the canonical quantum gravity approaches [1]. in 1988, Carlo Rovelli and Lee Smolin built on Ashtekar's work to introduce the loop representation of quantum general relativity [116]. Since then, lots of progress has been made, and so far, no fatal flaws have been discovered. However, LQG suffers from a number of problems; perhaps the most frustrating is that we don't know if LQG becomes General Relativity as we move from the (quantized) Planck scale to the (continuum) scale at which our experiments and observations are done.

In general relativity, the space-time metric itself is the fundamental dynamical variable. There is no background. On the one hand, it is analogous to the Minkowski metric in Maxwell's theory; it determines space-time geometry, provides light cones, defines causality, and dictates the propagation of all physical fields (including itself). On the other hand, it is the analog of the Newtonian gravitational potential, and therefore the basic dynamical entity of the theory. In fact, the equivalence principle precisely codes this dual role of the metric. It is this feature that is largely responsible for the powerful conceptual elegance of general relativity.

The absence of background geometry makes it difficult to analyze singularities of the theory and to define the energy and momentum carried by gravitational waves. Since there is no a priori space-time, to introduce notions as basic as causality, time, and evolution, one must first solve the dynamical equations and construct a space-time. As an extreme example, consider black holes, whose definition requires the knowledge of the causal structure of the entire space-time. To find whether the given initial conditions lead to the formation of a black hole, one must first obtain their maximal evolution and, using the causal structure determined by that solution, ask if its future infinity has a past boundary [2].

[26] See Smolin's contribution in this volume.

On the quantum theory side, the problems become significantly more serious. Particles do not even have well-defined trajectories; time-evolution only produces a probability amplitude, $\Psi(x, t)$, rather than a specific trajectory, $x(t)$. This is due to the uncertainty principle. Similarly, in quantum gravity, one would not be left with a specific space-time (after evolving an initial state): how can we introduce notions such as causality, time, scattering states, and black holes in the absence of a space-time geometry?
The canonical approach essentially counts on the fact that the Hamiltonian formulation of general relativity is well-defined and attempts to use it as a stepping stone to quantization. The fundamental canonical commutation relations code the basic uncertainty principle. The motion generated by the Hamiltonian is to be thought of as time evolution. The causality is captured by the commuting of certain operators on the fixed (spatial) three-manifold. The emphasis is on preserving the geometrical character of general relativity, and on retaining the compelling fusion of gravity and geometry that Einstein created.

Ashtekar's canonical variables (a set of three vector fields[27] $E_i^a, i = 1, 2, 3$) provide what is called the connection representation of canonical general relativity. Ashtekar introduced this set of variables to represent an unusual way of rewriting the metric canonical variables on the three-dimensional spatial slices in terms of an $SU(2)$ gauge field [1]. This choice of variables led to the loop representation of quantum general relativity and, in turn, loop quantum gravity and quantum holonomy theory. This set of three vector fields $E_i^a, i = 1, 2, 3$ is orthogonal, that is,

$$\delta_{ij} = q_{ab} E_i^a E_j^b.$$

The E_i^a are called a triad or *dreibein*, and they can be thought of as the *square-root* of the metric. We usually consider

$$(\det(q))q^{ab} = \sum_{i=1}^{3} \tilde{E}_i^a \tilde{E}_i^b,$$

which involves the densitized dreibein $\tilde{E}_i^a$. In fact, $\tilde{E}_i^a$ and E_i^a contain the same information. However, the choice for $\tilde{E}_i^a$ is not unique, and in fact, one can perform a space local rotation with respect to the internal indices i without changing the (inverse) metric. This is the origin of the $SU(2)$ gauge invariance.

Let A_a^i be the configuration variable, where:

$$A_a^i = \Gamma_a^i + \beta K_a^i \,,$$

[27] There are now two different types of indices, "space" indices a, b, c that behave like regular indices in a curved space, and 'internal" indices i, j, k that behave like indices of flat-space (the corresponding 'metric' that raises and lowers internal indices is simply δ_{ij}).

where $\Gamma_a^i = \Gamma_{ajk}\epsilon^{jki}$ and $K_a^i = K_{ab}\tilde{E}^{bi}/\sqrt{det(q)}$. The densitized dreibein is the conjugate momentum variable of this three-dimensional $SU(2)$ gauge field (or connection) A_b^j, in that it satisfies the Poisson bracket relation

$$\{\tilde{E}_i^a(x), A_b^j(y)\} = 8\pi G_{\text{Newton}}\beta\delta_b^a\delta_i^j\delta^3(x-y).$$

The constant β is the Immirzi parameter, a factor that renormalizes Newton's constant G_{Newton}. The densitized dreibein can be used to reconstruct the metric as discussed above and the connection can be used to reconstruct the extrinsic curvature.

The radical aspect of the approach lies in the transformation of conventional notions of space and time as a continuous background in which events occur, as classically understood, to a notion of space-time as formed by loop-like states, which are essentially holonomies or structures formed by parallel transport along closed paths. Holonomies have long featured in gauge field theories as gauge invariant quantities, representing curvature in the gauge space. Within the approach of loop quantum gravity, such holonomies become quantum operators. The role SU(2) plays in loop quantity gravity is as the gauge group of the field of the holonomies. The choice by Rovelli and Smolin in 1995 of the "spin networks that Penrose had developed in the 1970s as a model of discrete quantum geometry (see [94, 116]) was key in establishing these ideas. While a spin basis network can be given for all compact gauge groups, the one that is relevant to quantum gravity is SU(2), in particular, its spinor structure.

3.7 Connections Versus Holonomies

The *loop representation* is an attempt to overcome the difficulties with the connection representation [89]. The transition between the connection and the loop representations was originally obtained via the *loop transform*, which can be thought of as a kind of functional Fourier transform [105]. Spin networks and spin foams - the modern formulation of LQG - are formulated in terms of holonomies.
Notice that those *holonomies* (variables) are gauge-covariant functionals[28] supported on one-dimensional links, or 'edges', usually designated by *e* (following established LQG notation). For a given edge, i.e., some (open) curve embedded in Σ, we set

$$h_e[A] = \mathcal{P}\exp\int_e A_m \mathrm{d}x^m\,, \quad \text{with } A \equiv A^a\tau_a\,. \tag{23}$$

Hence, $h_e[A]$ is a matrix-valued functional. The holonomy transforms under the action of SU(2) at each end of the edge e:

---

[28] Whereas in the connection representation, one works with functionals $\Psi[A]$, which are supported 'on all of Σ'.

$$h_e[A] \to h_e^g[A] = g(e(0))\, h_e[A]\, g^{-1}(e(1)) , \qquad \text{with}\, g(e(0)) ,\; g(e(1)) \in \mathrm{SU}(2) . \tag{24}$$

As the conjugate variable to $h_e[A]$, one takes the 'flux' vector

$$F_S^a[\tilde{E}] := \int_S \mathrm{d}F^a \tag{25}$$

through any two-dimensional surface S embedded in Σ.

3.8 *Quantisation*

We now need to find the appropriate commutation relations between the associated quantum operators [3, 4]. The essential assumption of LQG is that this quantisation should take place at the level of the bounded hermitean operators $h_e[A]$, rather than the connection A itself. This is analogous to ordinary quantum mechanics, when one replaces the Heisenberg operators x and p with Weyl operators e^{ix} and e^{ip}; the spin network representation actually uses the analog of a hybrid formulation with x and e^{ip}. The idea is that it makes no difference whether one quantises the Heisenberg or the Weyl algebra, i.e., that these quantisations are equivalent.
Recall that in the case of ordinary quantum mechanics, the matrix elements of the operators corresponding to $e^{i\alpha x}$ and $e^{i\beta p}$ are smooth functions of the parameters α and β (see Stone-von Neumann theorem [87, 106, 118]). In LQG the representations of operators do not satisfy this requirement.[29]

The discrete nature of space arises in this approach from spin networks forming space directly. Representations of $SU(2)$ label the edges of the spin network in three dimensions, and in this way, a mathematical description of the kinematics of a quantum gravitational field can be obtained in three spatial dimensions. On its significance, Rovelli remarks of the nature of a spin network; "*a spin network state is not in space: it is space. It is not localized with respect to something else: something else (matter) might be localized with respect to it* [104]. The evolution of a spin network in time, needed for a Feynman path integral formulation of the theory, generates a spinfoam structure. One of the consequences of this program (which still awaits completion), and the manner in which it has forged a place in competition with string theories, has been to raise deep conceptual questions about the nature of space and time.

[29]This is also the reason why the kinematical Hilbert space employed in loop quantum cosmology is already different from the standard one for a *finite* number of degrees of freedom [2]. When the number of degrees of freedom is infinite (as in quantum field theory), the Stone-von Neumann theorem does not apply anyhow.

4 Non-commutative Geometry

The Standard Model (SM) of elementary particles has been tremendously successful in explaining the world at the smallest scales of length that can currently be probed [117]. Yet, it leaves many feeling a bit uneasy, for some of its properties appear to be rather ad hoc; few people believe that we have fully *understood* the Standard Model. The application of noncommutative geometry [25] (NCG) to the subatomic realm might over time increase our understanding of the Standard Model. A line of thought that started with the Connes-Lott model [26] culminated in a geometric description [16] of the full Standard Model.

4.1 Preliminaries

This approach is rooted in the idea that any compact space X and the commutative algebra of continous functions on that space,

$$C(X; \mathbb{C}) = \{f : M \to \mathbb{C};\ \ f \textit{ is continuous}\}$$

contain the same information and they are consequently dual.[30] The essential point of this correspondence (or duality) is that various geometric properties of the space M can be translated into properties of the corresponding algebra $C(M)$, thereby establishing a link between two completely different fields of mathematics. NCG is a program with ambition to generalize this correspondence to noncommmutative algebras and to provide mathematical techniques in order to handle these noncommutative algebras. In physics, we are normally interested in the space M provided with extra structures. For instance, when M is a Riemannian manifold, i.e., a space that locally looks like the Euclidean space R^n (for some n) on which we define a Riemannian metric g. Notice that form the point of view of physics we are interested in the case in which M is a standard Minkowski space. Unfortunately, the minus sign of the metric is hard to handle.

At the very heart of NCG lies the notion of a *spectral triple*, describing a *noncommutative manifold*. It is a triple $(\mathcal{A}, \mathcal{H}, D)$, where $\mathcal{A}$ is a unital $*$-algebra that is represented as bounded operators on a Hilbert space $\mathcal{H}$ on which a *Dirac operator* D acts. The latter is an (unbounded) self-adjoint operator that has a compact resolvent and, in addition, satisfies $[D, a] \in B(\mathcal{H})\ \forall\, a \in \mathcal{A}$.

- We call a spectral triple *even* if there exists a grading $\gamma : \mathcal{H} \to \mathcal{H}$, with $[\gamma, a] = 0\ \forall\, a \in \mathcal{A}$ and $\gamma D = -D\gamma$.
- We call a spectral triple *real* if there exists an anti-unitary *real structure* $J : \mathcal{H} \to \mathcal{H}$ satisfying

$$J^2 = \pm 1, \qquad JD = \pm DJ.$$

[30] For more details, see Connes and Chamsddine's contributions in this volume.

The Dirac operator and real structure are required to be compatible via the *first-order condition* $[[D, a], Jb^* J^*] = 0 \; \forall \, a, b \in \mathcal{A}$.

- If a spectral triple is both real and even, there is the additional compatibility relation

$$J\gamma = \pm \gamma J.$$

The eight different combinations for the tree signs above determine the *KO-dimension* of the spectral triple. For more details, we refer to [22].

This is a rather abstract notion, which we will try to make more concrete by providing an example that plays a key role in the application of NCG to particle physics.

Example 1 (Canonical spectral triple) The triple $(\mathcal{A}, \mathcal{H}, D) = (C^\infty(M), L^2(M, S), \not{\partial} := i\nabla^S)$ serves as the motivating example of a spectral triple. Here, M is a compact Riemannian manifold that has a spin structure and $L^2(M, S)$ denotes the square-integrable sections of the corresponding spinor bundle. The Dirac operator $\not{\partial}$ is associated with the unique spin connection which, in turn, is derived from the Levi-Civita connection on M. This spectral triple can be extended by a real structure J_M ('charge conjugation') and —when dim M is even— a grading $\gamma_M \equiv \gamma^{\dim M+1}$ ('chirality'). The KO-dimension of a canonical spectral triple equals the dimension of M.

An essential ingredient that we will need here is a natural functional that can serve as the equivalent of the action we know from high energy physics. For that, we want something that only depends on the data that are present in the spectral triple. The most simplest that meets these requirements would be to count how many of the eigenvalues of DA are smaller than some mass scale. Now, for technical reasons, it turns out that taking this is not allowed, and we will have to settle for something similar:

$$Tr = [f(D_A^2/\Lambda^2] \,, \tag{26}$$

where the mass-scale Λ again appears, as does some (a priori arbitrary) function f. This is called the spectral action[31] postulate.

5 Twistor Theory

The motivation behind twistor theory[32] is the opinion that the space-time continuum picture of reality would prove inadequate on some small scale and that even at the much larger levels of elementary particles, or perhaps atoms, where quantum behaviour holds sway, the standard space-time descriptions have ceased to be the

[31] Using spectral action (26) we can get the Einstein-Hilbert action, which, in turn, gives us the field equations of General Relativity, including a cosmologial constant!

[32] For more details, see Penrose's contribution in this volume.

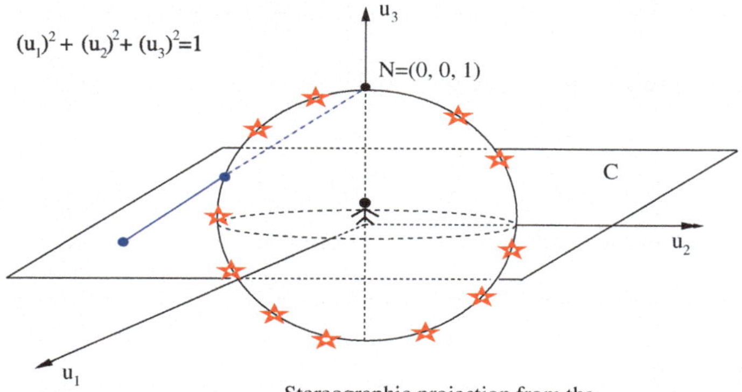

Fig. 1 The two-dimensional sphere is the simplest example of a non-trivial complex manifold

most physically appropriate ones, with some other picture of reality, though at that level equivalent to the space-time one, proving to be the more fruitful. The very fact that quantum behaviour is so hard to picture in the normal way had seemed to argue strongly that the normal space-time picture of things, even at that level, is inappropriate physically. Space-time descriptions of the normal kind can, of course, be used at the atomic or particle level, provided that the quantum rules are correctly applied, and they have implications that are extraordinarily accurate. Thus, this new geometrical picture must, at that level, be mathematically equivalent to the normal space-time picture - in the sense that some kind of mathematical transformation must exist between the two pictures.

The main two ingredients of twistor theory are non-locality[33] in space-time and analyticity (holomorphy) in an auxiliary complex space, the *twistor space*.

This auxiliary space can be thought of as the space of light rays at each point in space-time. Given an observer in a four-dimensional space-time at a point O, his *celestial sphere*, i.e., the image of planets, suns and galaxies he sees around him, is the backward light cone at O given by the 2-sphere

$$t = -1 \; and \; x^2 + y^2 + z^2 = 1 . \tag{27}$$

Two–dimensional sphere $S^2 \subset \mathbb{R}^3$ is a one-dimensional complex manifold with local coordinates defined by stereographic projection. Let $(u_1, u_2, u_3) \in S^2$. Define two open subsets covering S^2

[33] Non-locality of the fields in a physical theory is achieved by encoding the field information at a point in space-time into holomorphic functions on the twistor space. By choosing an appropriate description, one can cause the field equations to vanish on twistor space, i.e., holomorphy of a function on the twistor space automatically guarantees that the corresponding field satisfies its field equations.

$$U_0 = S^2 - \{(0, 0, 1)\}, \qquad U_1 = S^2 - \{(0, 0, -1)\},$$

Stereographic projection from the north pole (0, 0, 1) gives a complex coordinate

$$\mathcal{U} = \frac{u_1 + iu_2}{1 - u_3}.$$

Projecting from the south pole (0, 0, −1) gives another coordinate

$$\mathcal{U}' = \frac{u_1 - iu_2}{1 + u_3}.$$

The domain of $\mathcal{U}$ is the whole sphere less the North pole; the domain of $\mathcal{U}'$ is the whole sphere less the South pole.
On the overlap $U_0 \cap U_1$, we have : $\mathcal{U}' = 1/\mathcal{U}$, which is a holomorphic function, this makes S^2 into a complex manifold $\mathbb{CP}^1$ (Riemann sphere).
The double covering $SL(2, \mathbb{C}) \xrightarrow{2:1} SO(3, 1)$ can be understood in this context. If worldlines of two observers travelling with relative constant velocity intersect at a point in space–time, the celestial spheres these observers see are related by a Möbius transformation

$$\mathcal{U}' \to \frac{\alpha\mathcal{U} + \beta}{\gamma\mathcal{U} + \delta},$$

where the unit–determinant matrix

$$\begin{pmatrix} \alpha & \beta \\ \gamma & \delta \end{pmatrix} \in SL(2, \mathbb{C})$$

corresponds to the Lorentz transformation relating the two observers.
The fact that the null directions at a point have the holomorphic structure of a Riemann sphere (see [95]), and are closely related to the complex nature of Lorentzian spinors, were indications that spinors, and particularly Lorentzian 2-spinors, are more fundamental than Minkowskian world-vectors, and that the latter should be regarded as being derived from the former (2). In addition to this, complex numbers often have significant roles to play in the solutions of Einstein's vacuum equations. Penrose had been particularly impressed by the nature of the plane-fronted waves (these having originally been credited to Brinkmann 1923, but later rediscovered by Ivor Robinson [102] and also of their later generalizations to spherically fronted waves - credited to Robinson and Trautman [103]. For these waves, the behaviour along the null geodesics of propagation is fixed and there is completely arbitrary variation from wave-front to wave-front. But within each wave-front, the strength and polarization of the wave is governed by a single arbitrary holomorphic function.

> I was struck by the direct appearance of a free holomorphic function in the solution, the modulus and argument of this function both playing a direct role (as strength and polarization, respectively).

Twistor theory in the context of space-time is based on the association of a complex twistor space $\mathbb{CP}^3$ with the space of light rays in space-time.

The twistor space has four complex dimensions ($\mathbb{C}^4$). Consequently, it should contain more information than the "conventional space-time" with four real dimensions. A twistor Z is a point in this twistor space.

5.1 Renormalization and Hilbert's Twenty-First Problem

The most basic form of the *Riemann-Hilbert correspondence*[34] is a generalization of Hilbert's twenty-first problem to higher dimensions, which states: given a series of points in a complex plane and prescribed monodromies around these points, is there a Fuchsian ODE with these singularities and monodromies?
So, it codes the correspondence between certain systems of partial differential equations (linear and having very special properties for their solutions) and possible monodromies of their solutions. Nowdays, generalizations and refinements of this problem are called the Riemann-Hilbert problem [24].[35] The solutions and techniques used to find the corresponding ODE when it is possible are closely related to Riemann-Birkhoff factorization (realization of a holomorphic matrix function of a circle as a product of a matrix holomorphic on a neighborhood of closed disk, and a function of a matrix holomorphic on a neighborhood of an exterior of the disk including infinity and the circle itself). The correspondence between differential equations and monodromies can, in fact, be established and is true, in general, in the framework of sheaf theory.
The relation between the perturbative renormalization and the Riemann-Hilbert correspondence is codified in the Birkhoff factorization and, moreover, the existence of this unique decomposition is strongly related to an algebraic condition (namely, the Rota-Baxter property) of the couple regularization scheme and renormalization map. It shows that Rota-Baxter algebras play the role of a bridge between the study of ill-defined divergent Feynman integrals in QFT (by renormalization) and the extraction of finite values based on the Riemann-Hilbert problem in the study of a special class of differential system [27, 75]. Physical information of a given renormalizable QFT Φ are stored in Feynman diagrams equipped with related Feynman rules.

---

[34]The original setting of the Riemann-Hilbert correspondence concerned the Riemann sphere and the existence of regular differential equations with prescribed monodromy groups. However, a Riemann sphere can be replaced by an arbitrary Riemann surface, so in the case of higher dimensions, Riemann surfaces can be replaced by complex manifolds of dimension > 1. In this case, we have a correspondence between certain systems of (linear) partial differential equations and possible monodromies of their solutions.

[35] This is an equivalence from the category of flat connections on algebraic vector bundles on X with regular singularities to the category of local systems of finite-dimensional complex vector spaces on X in the case of regular singular connections. For X connected, the category of local systems is also equivalent to the category of complex representations of the fundamental group of X.

5.2 Renormalisation

The discovery of ultraviolet divergences in the 1930s caused many physicists to believe that the fundamental principles of physics had to be changed. Bethe, Feynman, Schwinger, Tomonaga, Dyson [110], and others proposed, in the late 1940s, a program of renormalization' that gave finite and physically sensible results by absorbing the divergences into redefinitions of physical quantities.[36] This led to calculations that agreed with experiment by up to 8 significant digits in QED, the most accurate calculations in all of science.

Let us restate the main features of the renormalization's approach:

(i) *Ultraviolet divergences*: When we perform a naïve calculation using local interactions (delta functions and their derivatives), we find that the results are generally inconsistent, due to short-distance divergences. The origin of these divergences is in the fact that quantum mechanics involves sums over a complete set of states, so quantum corrections are sensitive to the properties of high-momentum intermediate states.

(ii) *Regularisation*: To parameterize the sensitivity to a short distance, we modify the theory at a distance scale of order a (the cutoff), so that it is well-defined. We say that the theory has been regularized. In the theory with the cutoff, the ultraviolet divergences are replaced by sensitivity to a, in the sense that the physical quantities diverge in the limit $a \to 0$ with the couplings held fixed.

(iii) *Renormalization*: The regulated theory apparently has one more parameter than the naïve continuum theory, namely the cutoff. However, when we compute physical quantities, we find that they depend only on a combination of the cutoff and the other parameters. In other words, a change in the cutoff can be compensated for by a change in the couplings, so that all physical quantities are left invariant. We therefore finally obtain well-defined finite results that depend on the same number of parameters as the original local formulation.

There is usually considerable freedom in the choice of a regularisation procedure. Let us mention, among many others, the cut-off regularisation, which amounts to consideration of integrals over a ball of radius z (with $z_0 = +\infty$) and dimensional regularisation,[37] consisting, roughly speaking, in integrating over a space of

---

[36] In physics, it is very crucial to characterize the systems in interaction and to distinguish between bare parameters (such as mass, electric charge, acceleration, etc.), which are the values they would take if the interaction were switched off, and the actually observed parameters. Renormalisation is a "procedure" that is able to transform the bare parameters into the actually observed ones, which are called renormalized (i.e., with interaction taken into account).

[37] Dimensional regularization is a popular one, because it preserves many of the symmetries of the QFT in question. It involves rewriting physically interesting integrals over space-time as formal, but conceptually meaningless, integrals, in which the dimension of space-time becomes a complex number D. The integrals can then be written in terms of a Laurent series in a complex parameter, $z = D - d$, with a pole at $z = 0$. The point $z = 0$ corresponds to the original dimension of the problem, d. Finite values for divergent integrals are then extracted as residues taken around paths avoiding the singular points, by an application of Cauchy's theorem. This process of extraction is called minimal subtraction.

complex dimension z, with $z_0 = d$, the actual space dimension of the physical situation (for example, d = 4 for the Minkowski space-time). In this case, the function that appears is meromorphic in z with a pole at z_0 [38, 134]. However, naive dimensional regularization does not account for the fact that a complicated interaction may have additional sub-interactions that are also divergent. In the 1950s and 1960s, Bogoliubov, Parasiuk, Hepp [57], and Zimmermann [133] developed, corrected and proved the BPHZ algorithm for iteratively subtracting divergent sub-interactions [11, 12, 38, 134]. This algorithm applies to dimensional regularization and other regularization schemes.

Following Wilson's work [130], we know that the renormalized coupling contains all available information about the physics of the problem, i.e., the p dependence of the phase shift [61, 130]. In general, physical quantities depend on more than one dimensionful quantity, and the relation between renormalized couplings and physical amplitudes is not so simple, but we will see that they are still more closely related to physical quantities.

There is a beautiful physical picture that underlies these results, due to K.Wilson. The renormalized coupling is defined by decreasing the momentum cutoff Λ while changing the couplings to keep the low-energy physics the same. Because the theory with a lower cutoff has fewer degrees of freedom, this can be thought of as coarsegraining, or integrating out high-momentum fluctuations. In this way, we obtain a family of effective field theories that describe the same long-distance physics. The reason we can define such effective field theories is that physics at low momentum is sensitive to short-distance physics only through the value of the effective coupling. We can continue lowering the cutoff until it becomes on the order of the physical momentum p. At this point, almost all of the fluctuations have been integrated out, and the renormalized coupling contains essentially all the dynamical information in the theory.

The least intuitive part of this picture is that we can lower the cutoff Λ all the way to the physical scale p. In fact, if the cutoff and the physical scale are the same, there is no longer a small parameter that can make the effective field theory description valid. The reason we can take $\mu \sim p$ is that evolving the couplings from the scale to the scale μ using the renormalization group equation and boundary condition equation (28) does not affect the $O(1/2)$ corrections:

$$\Lambda \frac{d}{d\Lambda}(\frac{1}{c}) = -\frac{1}{\pi}, \tag{28}$$

where $\Lambda = \frac{1}{a}$ is a momentum cutoff, and c a dimensionless coupling constant. This is our first example of a renormalization group equation. The couplings $c(a)$ defined in this way are called running couplings. They define a family of effective theories with different cutoffs such that the low-energy physics is the same for all momenta p.

With its great success in dealing, for several decades, with the problem of divergencies in quantum field theory, the renormalization process is regarded as one of the greatest achievements in modern physics. Nevertheless, mathematicians have been skeptical about the soundness of the mathematical foundation of the renormalization process.

In general, QFT describes fluctuating systems with continuous degrees of freedom and is best understood through path integrals of the type

$$\int [D\phi]\, e^{-S[\phi]}, \tag{29}$$

where the integration is over a space of functions [74]. When considered as a perturbation of a Gaußian integral, this path integral is expanded over Feynman diagrams. Because of the continuous nature of the path integral, some of these diagrams yield divergent quantities, as dictated by dimensional analysis, and can only acquire a meaning through renormalization.
However, it is fair to say that this new calculus is not yet fully understood. In particular, it hides an algebraic structure analogous to diffeomorphisms, as uncovered by Connes and Kreimer a decade ago (see [28–30, 32, 33]) as well as ([24, 31, 35] for recent reviews). Connes and Kreimer reformulated earlier work by Kreimer and others on combinatorially-defined Hopf algebras of Feynman graphs in the language of loop groups [32, 33]. They then applied this new language to dimensional regularization to extract finite values from divergent integrals. Finally, they expressed the BPHZ renormalization process as the process of Birkhoff decomposition of loops into a Lie group defined by the Hopf algebra. Connes and Marcolli formulated dimensional regularization and BPHZ renormalization in terms of a connection on a principal bundle over a complex two-manifold B of complex renormalization parameters (corresponding to mass and space-time dimension) [35]. This bundle, along with the corresponding connection, seems to be a new object in both mathematics and physics.

5.3 Feynman Diagrams and Perturbative Renomalization

As far as renormalization is concerned, QFT is best studied in the framework of the Euclidian path integral.[38] In the simplest case, the fields are defined as functions ϕ from space-time $\mathbb{R}^D$ to $\mathbb{R}$ and the dynamics is governed by the action functional,

$$S[\phi] = \int d^D x \left(\frac{1}{2} \partial_\mu \phi\, \partial^\mu \phi + \frac{1}{2} m^2 \phi^2 + \frac{g}{N!} \phi^N \right), \tag{30}$$

where m is a mass and g a coupling constant for an N-particle interaction. At the quantum level, all the information is encoded in the Green's functions,

[38] For more details, see [74].

$$G(x_1, \ldots, x_n, m, g) = \mathcal{N} \int [D\phi]\, \mathrm{e}^{-\frac{S[\phi]}{\hbar}} \phi(x_1) \cdots \phi(x_n), \tag{31}$$

where the integration is over the space of all fields and $\mathcal{N}$ is a normalization constant to be defined later. $\hbar = 6.02\ 10^{-34}$ J·s is Planck's constant and measures the deviation of the quantum theory from the classical one. In this contribution, we choose a unit such that $\hbar = 1$. Note that this is only the simplest model of a scalar field theory.[39] To give a precise meaning to (31), it is convenient to expand $G(x_1, \ldots, x_n, m, g)$ as a power series in g using Feynman diagrams. The latter are best introduced on a simpler analogue, with the space of fields replaced by a finite dimensional vector space. Thus, the equivalent of (31) is

$$G_{i_1,\ldots,i_n}(A, V) = \mathcal{N} \int d\phi\, \mathrm{e}^{-S(\phi)} \phi_{i_1} \cdots \phi_{i_n}, \tag{32}$$

with the action

$$S(\phi) = \frac{1}{2}\phi \cdot A^{-1} \cdot \phi + V(\phi). \tag{33}$$

The quadratic term is defined by a positive definite symmetric matrix A,

$$\phi \cdot A^{-1} \cdot \phi = \sum_{i,j} (A^{-1})_{ij} \phi_j \phi_i, \tag{34}$$

and the interaction potential V is a polynomial in all the fields

$$V(\phi) = \sum_N \sum_{i_1,\ldots,i_N} \frac{g_{i_1,\ldots,i_N}}{N!} \phi_{i_1} \cdots \phi_{i_N}. \tag{35}$$

In this case, we choose $\mathcal{N}^{-1} = \det A/2\pi$ and expand $\mathrm{e}^{-V(\phi)}$ as a power series in the couplings $g_{i_1,\ldots,i_N}$. Thus, (32) amounts to the computation of the average of a monomial using a Gaußian weight and is given by Wick's theorem: it is a sum over all possible pairings of the variables in the monomial, each pairing of ϕ_i with ϕ_j being weighted by A_{ij}. For example,

$$\phi_i \phi_j \phi_k \phi_l \quad \to \quad A_{ij} A_{kl} + A_{ik} A_{jl} + A_{il} A_{jk}. \tag{36}$$

Then, each term of the expansion of (32) is associated with a diagram with n external legs and vertices of valence N,

[39] More complicated action functionals are required to account for the real world physics: Spinors ψ for fermionic particles and gauge connections A for their interactions, as is the case for QED and the Standard Model of elementary particles. Nevertheless, we restrict our attention in the sequel to the simplest example of a scalar field theory.

$$G_{i_1,\ldots,i_n}(A, V) = \sum_{\Gamma} \frac{G^{\Gamma}_{i_1,\ldots,i_n}(A, V)}{\mathrm{S}_\Gamma}. \tag{37}$$

The contribution of each diagram is computed using the Feynman rules:

- associate the indices $i_1, \ldots, i_n$ with the external legs and indices j_k to the internal half edges;
- associate a matrix element $A_{j_k j_l}$ with any edge connecting the indices j_k and j_l;
- associate a coupling $-g_{j_1,\ldots,j_N}$ to a N-valent vertex whose half edges have indices $j_1, \ldots, j_N$;
- sum over all the indices j_k.

Besides, one has to divide by the symmetry factor S_Γ, which is the cardinal of the automorphism group of the diagram, leaving the external legs fixed. For example,

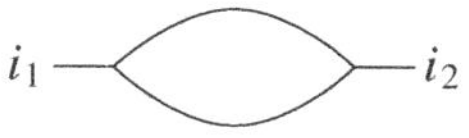

$$\rightarrow \quad \frac{1}{2} \sum_{\substack{j_1,j_2,j_3\\ j_4,j_5,j_6}} A_{i_1,j_1} g_{j_1,j_2,j_3} A_{j_2,j_4} A_{j_3,j_5} g_{j_4,j_5,j_6} A_{j_6,i_2}. \tag{38}$$

This simple finite dimensional model already captures some important algebraic aspects of perturbation theory, as will be discussed in the final section devoted to the Hopf algebras based on Feynman diagrams.

At a formal level, the Green's functions (31) can be computed as a power series in g by replacing ϕ_i with a function $x \mapsto \phi(x)$, the matrix element A_{ij} with the propagator

$$K(x, y) = \int_{\mathbb{R}^D} \frac{d^D p}{(2\pi)^D} \frac{\mathrm{e}^{\mathrm{i}p\cdot(x-y)}}{p^2 + m^2}. \tag{39}$$

and $V(\phi)$ with the interaction term

$$\begin{aligned} &\frac{g}{N!} \int_{\mathbb{R}^D} d^D x\, \phi^N(x) \\ &\quad = \frac{g}{N!} \int_{\mathbb{R}^D} \frac{d^D p_1}{(2\pi)^D} \cdots \frac{d^D p_N}{(2\pi)^D} (2\pi)^D \delta(p_1 + \cdots + p_N) \tilde{\phi}(p_1) \ldots \tilde{\phi}(p_N), \end{aligned} \tag{40}$$

with

$$\tilde{\phi}(p) = \int_{\mathbb{R}^D} d^D x\, \mathrm{e}^{-\mathrm{i}p\cdot x}\, \phi(x) \tag{41}$$

being the Fourier transform of $\phi(x)$. Heuristically, the Feynman diagrams can be thought of as quantum mechanical processes with particles on their external legs and virtual particles of momenta p propagating on the internal lines. It is important to notice that although momentum is conserved at each vertex and along each line, the particles that propagate along the loops may have arbitrary momenta.

To prevent the propagation of Fourier modes of momenta $\geq \Lambda$, let us alter the propagator by introducing a cut-off Λ,

$$K(x, y) \rightarrow K_\Lambda(x, y) = \int_{\frac{1}{\Lambda^2}}^{\infty} d\alpha \int_{\mathbb{R}^D} \frac{d^D p}{(2\pi)^D} e^{ip\cdot(x-y)} e^{-\alpha(q^2+m^2)}. \tag{42}$$

This procedure is known as regularization and can be performed in various ways. Besides the method used here, one could also discretize the theory on a lattice or evaluate the diagrams in complex dimension z and recover the divergences as poles when $z \rightarrow D$. In principle, all these methods are equivalent, but we restrict ourselves to the momentum space cut-off presented here, since it is suited to the Wilsonian point of view we adopt in this paper.

For a renormalizable theory like the ϕ^4 theory, one can trade the parameters g and m for some cut-off-dependent ones $g_0(\Lambda)$ and $m_0(\Lambda)$ and further introduce an additional wave function renormalization $Z(\Lambda)$ in such a way that $Z^{\frac{n}{2}}(\Lambda)G_\Lambda(x_1, \ldots, x_n, m_0(\Lambda), g_0(\Lambda))$ admits a finite limit when $\Lambda \rightarrow \infty$. To obtain definite physical predictions, the bare parameters $g_0(\Lambda)$ and $m_0(\Lambda)$ and the wave function renormalization $Z(\Lambda)$ must be determined in terms of normalization conditions involving renormalized parameters m_r and g_r measured at a low energy scale μ. Thus, we define the renormalized Green's functions as

$$\begin{aligned} &G_r(x_1, \ldots, x_n, m_r, \mu, g_r) \\ &\quad = \lim_{\Lambda\rightarrow\infty} Z^{\frac{n}{2}}(\Lambda, m_r, g_r, \mu)G_\Lambda(x_1, \ldots, x_n, m_0(\Lambda, m_r, g_r, \mu), g_0(\Lambda, m_r, g_r, \mu)) \end{aligned} \tag{43}$$

Note that we are dealing here with perturbative renormalization only, so that the previous equality must be understood as an equality between formal power series in g_r. In fact, $g_0(\Lambda)$, $m_0(\Lambda)$ and $Z(\Lambda)$ are themselves formal power series in g_r that can be computed in terms of Feynman diagrams using the Bogoliubov-Parasiuk-Hepp-Zimermann (BPHZ) formula [12, 57]. Roughly speaking, the contribution of a divergent diagram with 2 or 4 external legs to the renormalisation of the parameters is encoded in its counterterm $C(\Gamma)$, which is determined recursively by the relation

$$C(\Gamma) = T\Big(\sum_{\substack{\{\Gamma_i,\ldots\} \\ \Gamma_i\cap\Gamma_j=\emptyset}} \prod_i C(\Gamma_i) \frac{\Gamma}{\prod_i \Gamma_i}\Big), \tag{44}$$

with T taking the divergent part of the diagram. This sum runs over all sets , including the empty one, of disjoint, divergent, one-particle irreducible subdiagrams of Γ (i.e., diagrams that cannot be disconnected by cutting an arbitrary internal line). The reduced diagram on the RHS is obtained by shrinking each Γ_i to a single vertex and, finally, taking the divergent part of the whole sum, with a finite part determined by the normalization conditions at the low energy scale μ. In the framework of dimensional regularization, this operation is elegantly written as a Birkhoff decomposition for a loop in the space complex dimension, with values in a group associated with a commutative Hopf algebra (see the work of Connes and Kreimer [29, 30]).

5.4 Connes-Kreimer's Approach

Feynman graphs are Graphs[40] built from a fixed set of types of vertex and a fixed set of types of edges. A physicist is usually interested in numbers, and the Feynman rules associate a complex number with a Feynman graph Γ

$$\Gamma \to U(\Gamma) \in \mathbb{C},$$

However, these numbers are typically infinite and need to be renormalized. D. Kreimer described the combinatorics of Feynman graphs in terms of a Hopf algebra structure and formulated the hierarchical structure of subgraphs in terms of rooted trees. He showed that one can find these rules in very specific characters of the Hopf algebra $H_F(\Phi)$. So, the components of the Birkhoff factorization of a Feynman rules character are other characters such that determine renormalized values, counterterms, a renormalization group and a β-function [23, 33, 76]. This fact shows that these Birkhoff components have the ability to save physical meanings, and it can be interested to find situations in which these characters can play the role of integrals of motion for the given Feynman rules character. Later, A. Connes and Kreimer formulated the Hopf algebra directly in terms of Feynman graphs [33]. As an algebra, the Hopf algebra $\mathcal{H}$ is the free commutative algebra. It turns out that the collection of all Feynman rules constitutes a group. We start by considering the Feynman rules $\Gamma \to U(\Gamma) \in \mathbb{C}$ as characters on the free commutative algebra $\mathcal{H}$ generated by all Feynman graphs.

Theorem 1 (Connes-Kreimer). *There exists a co-unit, coproduct and antipode on the algebra $\mathcal{H}$ of Feynman graphs, turning $\mathcal{H}$ into a Hopf algebra (and G a group) [27]. The co-unit is*

$$\epsilon(\Gamma)\begin{cases} 1 & if\ \ \Gamma = 0 \\ 0 & otherwise \end{cases} \tag{45}$$

and the coproduct is defined by

$$\Delta(\gamma) = \Gamma \otimes 1 + 1 \times \Gamma + \sum_{\gamma \subseteq \Gamma} \gamma \otimes \Gamma/\gamma,$$

[40] A Feynman graph is a (non-planar) graph with a finite number of vertices and edges. An internal edge is an edge connected at both ends to a vertex (which can be the same in the case of a self-loop), an external edge is an edge with one open end, the other end being connected to a vertex. A Feynman graph is referred to by physicists as a vacuum graph, a tadpole graph, or a self-energy graph (respectively an interaction graph) if its number of external edges is 0, 1, 2, (*respectively* > 2). An edge can be of various types depending on which elementary particle it represents. A one-particle irreducible graph (in short, $1PI$ graph) is a connected graph that remains connected when we cut any internal edge. A disconnected graph is said to be locally $1PI$ if any of its connected components is $1PI$.

The above Hopf algebra $\mathcal{H}$ is the algebraic structure underlying the recursive procedure of renormalization.

– *Renormalization as a decomposition in G*

Bogoliubov, Parasiuk, Hepp and Zimmermann's renormalization procedure[41] is based on the following steps: Given a graph Γ, we replace the unrenormalized value $U(\Gamma)$ with a sum involving suitably defined subgraphs[42] $\gamma \subset \Gamma$ and the contracted graphs Γ/γ or cograph, obtained by collapsing each connected component of γ to a single vertex,

- *Regularization*: introduce a parameter $z \subset \mathbb{C}$ and define new Feynman rules U_z :

$$\Gamma \to U_z(\Gamma) \in \mathbb{C},$$

  The previous infinity becomes a pole at $z = 0$ of the Laurent series expansion in z.
- *Subtraction* : get rid of the whole pole part of the Laurent series expansion: this gives us the renormalized amplitude

$$\Gamma \to R_z(\Gamma) \in \mathbb{C},$$

This applies not only to the Feynman graph Γ, but also to its subgraphs: for a generic graph Γ : $R_z(\Gamma)$ defined by a recursive procedure.

In fact, for a character $U_z : \mathcal{H} \to \mathbb{C}$, there exists a character $C_z : \mathcal{H} \to \mathbb{C}$ (counterterm') defined for $z \neq 0$ as

$$C_z(\Gamma) = -T[U_z(\Gamma) + \sum_{\gamma \subseteq \Gamma} C_z(\gamma) U_z(\Gamma/\gamma)],$$

with T the projection onto the pole part in dimensional regularization, $C(\Gamma) = -T(\bar{R}_z)$, so that [23, 33]

$$\bar{R}_z = C_z * U_z = U_z(\Gamma) + \sum_{\gamma \subseteq \Gamma} C_z(\gamma) U_z(\Gamma/\gamma \ \ is\ \ finite\ \ at\ \ z = 0. \tag{46}$$

Consequently, the renormalized value of the graph Γ is given by

$$R(\Gamma) = \bar{R}(\Gamma) + C(\Gamma) = U(\Gamma) + C(\Gamma) + \sum_{\gamma \subseteq \Gamma} C_z(\gamma) U_z(\Gamma/\gamma, \tag{47}$$

---

[41]The mathematical description of the BPHZ method in renormalization is basically designed according to Atkinson's theorem. It provides inductive formulae (i.e., integral renormalization theorems) for components of the Birkhoff factorization of characters on rooted trees such that at this level, one can find the notion of a decomposition of determined Lie algebras with the Connes-Kreimer theory.

[42]Not necessarily connected.

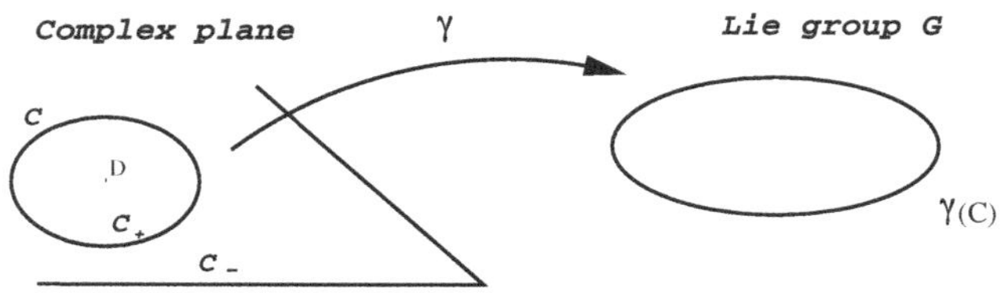

$\gamma(z) = \gamma_-^{-1}(z)\gamma_+(z), \quad z \in C$ where $\gamma_\pm : C_\pm \to G$ are holomorphic.

Fig. 2 Birkhoff decomposition

Birkhoff decomposition This gives a precise unexpected relation between renormalization and a basic geometric procedure called the Birkhoff decomposition, which originates in the problem of classifying holomorphic bundles on the sphere. A complex vector bundle E of dimension n on the Riemann sphere is obtained by clutching together two trivial bundles on the lower and upper hemispheres $C_\pm$, using a map $\gamma(z) \subset GL(n; \mathbb{C})$ defined on the common boundary C (Figure 2). When one replaces the group $GL(n; \mathbb{C})$ with a prounipotent simply connected complex Lie group G the Birkhoff decomposition of a map $\gamma(z) \subset G$ defined on the common boundary C, takes the simpler form [27]:

$$\gamma(z) = \gamma_-^{-1}(z)\gamma_+(z),$$

G is the group of characters of the Hopf algebra of Feynman diagrams, nice enough so it exists for any loop γ. The renormalisation condition $\gamma_-(\infty) = 1$ ensures the unicity of the decomposition $\gamma_\pm : C_\pm \to G$. γ defined on C_+ with a pole at D : $\gamma \to \gamma_+(D)$ is a natural principle for extracting finite value from the singular expression $\gamma(D)$.

Connes's et al.'s works show that the renormalized theory is just the evaluation at $z = D$ of the holomorphic part γ_+ of the Birkhoff decomposition of the loop γ, with values in G provided by the dimensional regularization.

Birkhoff decomposition and perturbative renormalization Let $\mathcal{A}$ be a complex function in $\mathbb{C}$ in D $(= 4)$, $\mathcal{A}_+$ a holomorphic function in $\mathbb{C}$, and $\mathcal{A}_-$ a polynômial in $\frac{1}{z-D}$ without a constant term. So, we have

$$\begin{cases} \textit{Feynman rules}: & \mathcal{H} \stackrel{U}{\Longrightarrow} \mathcal{A}, \\ \textit{Conterterms}: & \mathcal{H} \stackrel{C}{\Longrightarrow} \mathcal{A}_-, \\ \textit{Renormalized theory}: & \mathcal{H} \stackrel{R}{\Longrightarrow} \mathcal{A}_+, \end{cases} \tag{48}$$

where $C * U = R$.

Compose with character χ_z of $\mathcal{A}$,

$$\gamma(z) \doteq \chi_z \circ U, \quad \gamma_-(z) \doteq \chi_z \circ C, \quad \gamma_+(z) \doteq \chi_z \circ R, \tag{49}$$

where $\gamma(z)$, $z \subset C$ is a loop within the group G of characters of $\mathcal{H}$,

$$\gamma(z) = \gamma_-^{-1}(z)\gamma_+(z),$$

The renormalized theory is the evolution at D of the positive part of the Birkhoff decomposition of the bare theory.

Theorem 2 *(1) Let $\mathcal{H}$ be a graded connected Hopf algebra and $\phi : \mathcal{H} \to K$ be an algebra homomorphism. The Birkhoff decomposition of the corresponding loop is obtained recursively from the equalities*

$$\phi_-(X) = -T(\phi(X) + \sum \phi_-(X')\phi(X''))$$

and

$$\phi_+(X) = \phi(X) + \phi_-(X)\phi(X) + \sum \phi_-(X')\phi(X'')$$

(2) When $\mathcal{H}$ is the Hopf algebra of graphs and $\phi = U$ is the homomorphism associated wth the unrenormalized value of graphs in dimensional regularization then ϕ_- gives the counterterms C and ϕ_+ gives the renormalized value R.

So, we can conclude that the renormalization group, once properly formulated, appears as a perfect ambiguity group between solutions to a (physics) problem, and hence plays a very similar role to that of the galois group of an algebraic equation. And we thereby see a striking similarity between the ambiguity group occurring in physics and the ambiguities that occur in the resummation processes of divergent series (Stokes phenomenon, resurgence, etc.)
As stated by A. Connes:

> So, the recent years have witnessed the emergence of a much better mathematical understanding of perturbative renormalization, both in its Galois aspects related to the ambiguity inherent in the renormalization group and the role of the Birkhoff decomposition, as well as in the deep arithmetic nature of the numbers that appear as residues of Feynman graphs in the renormalization process.

He went on to add:

> The main lesson one learns from the above developments is that one should not consider the divergences of QFT as unwanted nuisances, but rather as the signature of subtle symmetries of Galois type that prevent one from making simple predictions unless they are carefully taken into account. It also shows that it is worthwhile to give a precise geometric support to the dimensional regularization and to understand in a more geometric manner the universal behavior of counterterms .... [27]

6 Dualities, QFT and Integrable Systems

6.1 Matter

The concept of a particle is a natural idealization of our everyday observation of matter. Dust particles and baseballs, under ordinary conditions, are stable objects that move as a whole and obey simple laws of motion. However, neither of these is actually a structureless object. That is, if sufficiently large forces are applied to them, they can readily be broken apart into smaller pieces.

The idea that there must be some set of smallest constituent parts, which are the building blocks of all matter, is a very old one. Democritus[43] is often credited for this idea, though his concept of the building block was quite different from ours today. He introduced the word that, in English, translates into 'atom' to describe the parts, whatever they might be.
History, however, plays tricks with language. The word 'atom' has acquired a meaning today that only partly matches Democritus' idea. Certainly, we know that matter is indeed composed of the objects we call atoms. Atoms were originally thought to be indivisible, that is, the smallest particle. However, we now understand that atoms are built up of smaller parts.

Therefore, the question concerning the nature of he particle is reminiscent of that relating the atom. So, we pull up the level and again ask: what are particles? How are they made? Are they just points? Or solid balls? or a kind of vortex in some invisible fluid, as imagined by Descartes for the stars? Or something else …

We are still very far from any answer. Indeed, the experiments of the nineteenth century have caused the classical definitions of particles and waves to be blurred.[44] Consequently, we were convinced for some time that we need to exclude the

[43]Born around 460 BC in Abdera, Thrace, Greece.

[44]From this perspective particles are discrete, their energy is concentrated into what appears to be a finite space, which has definite boundaries and contents that we consider to be homogenous (the same at any point within the particle). Particles exist at a specific location. If they are shown on a $3D$ graph, they have x, y, and z coordinates. They can never exist in more than one place at once, and to travel to a different place in space, a particle must move there under the laws of kinematics, acceleration, velocity, and so forth.

Interactions between particles have been studied for many centuries, and a few simple laws underpin how particles behave in collisions and interactions. The most primary of these are the conservation of energy and momentum, which allow us to simplify calculations between particle interactions on scales of magnitude that vary between planets and quarks.

Waves, unlike particles, cannot be considered a finite entity. Their energy cannot be considered to exist in a single place, since a wave, by definition, varies in both displacement and time. For example, a sound wave is a deformation in air pressure, and water waves a deformation of the water's surface.

point-like[45] and the solid ball conceptions for the particle. Moreover, when A. Einstein, after the pioneering work of M. Planck, realized one century ago that the light itself — understood as an oscillation of the electromagnetic field since J.C. Maxwell — must be quantized into photons, the problem became even more puzzling, and one was led, through the elaboration of the quantum field theory, to think of each particle as a quantized excitation of some field.

There is, however, a second mechanism by which a field analogous to the electromagnetic field could lead to structures that behave like a particle: the merging out of a vortex or a soliton as a solution of a non-linear partial differential equation. Maybe such an idea came into the mind of some scientist a long time ago, by observing the persistance of a vortex patch in water. When two centuries ago, we were able to modelize the motion of fluid with Euler equations or the Navier-Stokes equation, it was possible to check mathematically that, for an incompressible fluid (such as, in a good approximation the water) the vorticity was conserved. Eventually, this led Sir Thomson[46] to a sophisticated theory of rings and knots of vortices as the ultimate constituents of the matter.
The study of solitons is an important source of inspiration for theoretical physicists who seek to elucidate the structure of an elementary particle [56]. This question began to become relevant as early as the nineteenth century, when the hypothesis that matter is constituted of atoms was considered more and more seriously (and before the work of J. Perrin around 1900, when this idea was imposed). It soon became apparent that the naive idea that a particle could be concentrated at one point had to be abandoned, if only because the energy contained in the electric field in the vicinity of an electron should then be infinite. The simplest way to escape this infinity is to assimilate the electron into a sphere, a hypothesis used, in particular, by H. Poincaré. Obviously, such representation seems to be more of a working hypothesis, rather than a true model. Thus, other attempts at modeling the structure of the atom or electron were proposed at the time by imagining a sort of whirlwind or singularity of a fundamental physical field (see the theory of Lord Kelvin, using the nodes, or the theory of G. Mie). Within this approach, the KdV equation appears to be a particularly attractive toy, because, despite its relative simplicity, it hides a mechanism that permits the existence of solitons, these solutions concentrated in space and characterized by easily recognizable quantities, which one is assimilates willingly into a mass (energy) or an impulse.

In an area of space, unlike a particle, a wave can propagate until it exists in all locations and at all times; as a mathematical example we can use a pure sine wave, which has no beginning or end, but repeats every 2π. However, like particles, we can analyze a part or phase of the wave and obtain a value for its velocity within this area.

[45]Because then, charged electrons would have infinite energy.

[46]Kelvin suggested that molecules are knots in the aether. While we now know that there is no aether and that molecules are not the fundamental constituents of matter, the idea that matter has a topological origin remains beautiful and compelling.

Today, the question of the structure of the elementary constituents of matter can no longer be addressed without taking into account the upheavals brought about by quantum mechanics. Thus, the enigma has become even deeper, since Heisenberg's principle of uncertainty prohibits the simultaneous determination of the velocity and position of a particle.
Of course, the quantum theory revolution forced everybody to abandon such naive descriptions and the success of the atomic model temporarily satisfied people. But still, the question remained: what are electrons, neutrons, protons, or, as we should rephrase it now, quarks? They all are fermions, i.e., particles that possess strange properties, as imposed by Pauli's exclusion principle: two different fermions cannot be in the same quantum state, as there is only one place for one fermion in one quantum state, in contrast to bosons [71].

6.2 Solitons, Dualities and Integrable Systems

Tony Skyrme [55, 56, 113, 114], a British physicist, proposed [47] [80, 113, 115] going back to the idea of vortices merging out from a kind a fluid dynamics (or from a field dynamics) to modelize fermions.[48]
Such hypothesis would have more success at the end of the twentieth century than earlier, since, beside the examples of the vortices, other phenomena of "particles merging out" from a smooth field were known. These are called "solitons" by physicists, and we shall see that, as with vortices, solitons were observed more than 150 years ago in nature. Now, we are aware of many models of fields satisfying a partial differential equation that are soliton equations.

---

[47] The Skyrme model is based on a group-valued field from $\mathbb{R}^3$,

$$U : \mathbb{R}^3 \to G\,, \tag{50}$$

where the Lie group G is usually taken to be $SU(2)$, and $U(\boldsymbol{x}) \to 1$ as $|\boldsymbol{x}| \to \infty$. The degree of U as a map $S^3 \to SU(2)$ is identified with a baryon number. The minima of the Skyrme energy, for each baryon number, are called Skyrmions.
Skyrmions are free to rotate, both in physical space and through conjugation by elements of $SU(2)$. Quantising this motion gives the Skyrmions spin and electric charge. The proton and neutron, for example, are distinct quantum states of the essentially unique Skyrmion of degree 1. Therefore, the Skyrme model [9, 113] is a non-linear theory of pions whose topological soliton solutions are candidates for an effective description of nuclei, with an identification between soliton and baryon numbers. Indeed, in the Skyrme model, the basic idea is that a baryon number is identified with the degree of the map U in (2.1), or equivalently, with the instanton number (or second Chern class) of the $SU(2)$ bundle over $\mathbb{R}^4$.

[48] Recently, M. Atiyah et al., based on the skyrmion idea developed a geometrical model for matter [9].

It seems at first glance that the idea that particles, and in particular fermions, are solitons is in contradiction with the dogma of the quantum field theory[49] that particles are a quantization of the energetic excitation of a field (a dogma that is confirmed by every day physical experiment). Nevertheless, Hooft [61] showed that if we combine both ideas, it helps in solving the difficult problem of the renormalisation of the Yang-Mills-Higs model (and there is no other known way to do so).

The idea that fermions could be solitons was actually confirmed in theoretical models in 1975 by Coleman [21] in the case when the space-time is two-dimensional and with the sine–Gordon model.[50] More precisely, S. Coleman showed that two different classical models lead, when the quantum theory is constructed, to describing the same fermion particle. But in one model, the fermion is a quantum excitation of the field, and in the other model, the particle is a soliton.[51] Hence, both points of view can be reconciled!

So, in those field theories that have the striking feature that their classical dynamical equations have particle-like solutions known as solitons, there will be two different spectra. There is the spectrum of particle-like solitons and the spectrum of the actual particles. From this idea that, in certain field theories, the two spectra may be interrelated or inter-reliant emerges the particle-soliton duality.

In the quantized Sine-Gordon model, for example, there are two different particle spectra: a soliton spectrum and a spectrum arising out of quantization. What makes this equation so remarkable is the fact that there is a non-local transformation of the field that turns it into another one-dimensional equation known as the Thirring model. The transformation maps the soliton particles of the sine-Gordon equation onto the ordinary quantum excitations of the Thirring model [120], so the two types of particle are not so different after all. We say that there is a duality between the two models, the sine-Gordon [84] and the Thirring. They have different equations but they are really the same.

---

[49]Physicists use quantum field theories to describe fundamental particles. These quantum field theories are derived by quantizing classical field theories. Whereas classical field theories describe the dynamics of continuum fields, quantum field theories can be interpreted as describing the interactions of individual particles. Unlike the particles introduced by the quantization procedure, solitons are germane to the classical, continuum, theory. They owe their particle-like properties, not to quantization, but to the topology of the field theory itself.

[50]The sine-Gordon model was invented by Tony Skyrme, the name is a joke because it sounds like Klein-Gordon.

[51]The basic properties of solitons, like propagation and interaction without change in their velocity and shape, make it possible to treat them as robust localized objects. Solitons show their duality, having properties of both particles and waves. A soliton has the wave's nature and finite width, but it behaves like a particle when interacting with other solitons. That is why the solitons are often spoken of as quasiparticles.

This phenomenon is surprising and deep and occurs elsewhere in field theory. This idea has had important applications supersymmetric field theory and in superstring theory. It is similar to, and linked with, the T-duality.

It is a striking feature of some field theories that their classical dynamical equations have particle-like solutions known as solitons. Unlike the particles introduced by the quantization procedure, solitons are germane to the classical, continuum theory. They owe their particle-like properties not to quantization, but to the topology of the field theory itself.
The quantization of field theory with solitons exhibit two different spectra: the spectrum of particle-like solitons and the spectrum of the actual particles. Particle-soliton duality is the idea that, in certain field theories, the two spectra may be inter-related[52]. This phenomena is surprising and deep and occurs elsewhere in field theory and has had important applications in supersymmetric field theory and superstring theory. In this section, we investigate the conceptual foundation and the development of this duality and the role of the soliton in some new approaches in mathematical physics.

6.2.1 The Kinks

As a first example of a nonlinear field equation, let us consider

$$\frac{1}{c^2}\frac{\partial^2\phi}{\partial t^2} - \frac{\partial^2\phi}{\partial x^2} + 2\phi(\phi^2 - 1) = 0.$$

The solutions to this equation are critical points of the functional

$$\mathcal{A}[\phi] := \int_{\mathbb{R}\times\mathbb{R}} \left(\frac{1}{2c^2}\left|\frac{\partial\phi}{\partial t}\right|^2 - \frac{1}{2}\left|\frac{\partial\phi}{\partial x}\right|^2 - \frac{1}{2}(\phi^2 - 1)^2 \right) dtdx.$$

Let us try to find a solution of the type $\phi(t, x) = f(x - vt)$: it leads to the equation

$$f'' = 2\beta^2 f(f^2 - 1), \tag{51}$$

where $\beta := \left(1 - \frac{v^2}{c^2}\right)^{-\frac{1}{2}}$. We shall assume that $|v| < c$ in the following. One trick is to look first at solutions of the first order partial differential equation

$$f' + \beta(f^2 - 1) = 0, \tag{52}$$

since — as the readers can themselves check — any solution to (52) is automatically a solution to (51). Then, one easily sees that solutions to the first order equation (52) are of the form

[52]See Butterfield and de Haro's contribution in this volume.

$$f(s) = \tanh \beta s.$$

We can pause to stress the fact that this trick rests on finding Bogomol'nyi solutions to the second order equation (51). It is based on the fact that the functional $\mathcal{A}$ is part of a supersymmetric action functional and that the first order Bogomol'nyi condition is just asking the solution to be invariant by one supersymmetry generator (the Bogomol'nyi–Prasad–Sommerfeld states)[53] (see [55, 56] for more details).

Let us look now at the solution $\phi(t, x) = \tanh(\beta(x - vt))$, that we obtained. We can observe that:

- it is localised in space: the energy is concentrated in an area of size β^{-1} around the point $x - vt$ in space. This soliton (a kink) can be thought of as a kind of particle !
- the difference $Q := \frac{1}{2}(\phi(t, \infty) - \phi(t, -\infty))$ does not depend on t, it is just equal to 1. We could interpret this number as a charge.[54] One way to see that is to consider the 1-form $d\phi = \frac{\partial \phi}{\partial t} dt + \frac{\partial \phi}{\partial x} dx$: it is obviously closed and its integral over a constant time slice is equal to $2Q$ as soon as the field is assumed to be asymptotically constant in time at infinity.
- another solution is $-\tanh(\beta(x - vt))$. Its charge is -1. It is also a Bogomoln'nyi solution corresponding to the equation $f' - \beta(f^2 - 1) = 0$. One could think of this solution as an antiparticle. Moreover, we could imagine one kink and one antikink (in such a way that the total charge is 0) travelling towards one another. When they meet, they cancel each other out, like the disintegration of an electron–anti-electron pair.

All that suggests strongly that kinks behave like fermions. Note here that the total charge could only be equal to –1, 0 or 1, because of the conditions at infinity imposed by the finiteness of the energy

[53] A consequence of the supersymmetry involved is that the action has the form $\mathcal{A}[\phi] := \int_{\mathbb{R}\times\mathbb{R}} \left(\frac{1}{2c^2} |\phi_t|^2 - \frac{1}{2} |\phi_x|^2 - 2(W'(\phi))^2 \right) dt dx$, where here, $W(s) = \frac{1}{2}\left(\frac{1}{3}s^3 - s\right)$. It implies, in particular, that a function f of x is a solution to (51) if, and only if, it is a critical point of the functional $\mathcal{E}[f] := \int_{\mathbb{R}} \left((f')^2 + 4\beta^2 \left(W'(f)\right)^2 \right) dx$. Now, we can write this functional as $\mathcal{E}[f] = \int_{\mathbb{R}} \left((f' + 2\beta W'(f))^2 - 4\beta f' W'(f) \right) dx = \int_{\mathbb{R}} \left((f' + 2\beta W'(f))^2 - 4\frac{\beta d}{dx}(W(f)) \right) dx$. If we assume that $\lim_{x\to\pm\infty} f(x) = f_\pm$, then the last term on the right hand side is just $C := 4(W(f_-) - W(f_+))$. So, $\mathcal{E}[f] - C$ is the integral of the square of $f' + 2\beta W'(f)$ and a trivial solution is to set $f' + 2\beta W'(f) = 0$: this is exactly (52) (see [54]).

[54] In 1917, German mathematician Emilie Emmy Noether had shown that the mass, charge and other attributes of elementary particles are generally conserved because of symmetries. For instance, conservation of energy follows if one assumes that the laws of physics remain unchanged with time, or are symmetric as time passes. And conservation of electrical charge follows from a symmetry of a particle's wave function. Sometimes, however, as in our case, attributes may be maintained because of deformations in fields. Such conservation laws are called topological, because topology is that branch of mathematics that concerns itself with the shape of things.

$$\int_{\mathbb{R}} \left(\frac{1}{2c^2} \left| \frac{\partial \phi}{\partial t} \right|^2 + \frac{1}{2} \left| \frac{\partial \phi}{\partial x} \right|^2 + \frac{1}{2}(\phi^2 - 1)^2 \right) dx \,. \tag{53}$$

We recover Pauli's exclusion principle here: there is no place for two kinks on the same line!

6.3 The Sine–Gordon Equation

A more refined model is the following:

$$\frac{1}{c^2}\frac{\partial^2 \phi}{\partial t^2} - \frac{\partial^2 \phi}{\partial x^2} + \frac{\alpha}{\lambda}\sin \lambda\phi = 0.$$

The solutions $\phi : \mathbb{R} \times \mathbb{R} \longrightarrow \mathbb{R}$ of this equation are critical points of the functional

$$\mathcal{A}[\phi] := \int_{\mathbb{R}\times\mathbb{R}} \left(\frac{1}{2c^2} \left| \frac{\partial \phi}{\partial t} \right|^2 - \frac{1}{2} \left| \frac{\partial \phi}{\partial x} \right|^2 + \frac{\alpha}{\lambda^2}(\cos \lambda\phi - 1) \right) dt dx.$$

We are interested in solutions such that their energy $\int_{\mathbb{R}} \left(\frac{1}{2c^2} \left| \frac{\partial \phi}{\partial t} \right|^2 + \frac{1}{2} \left| \frac{\partial \phi}{\partial x} \right|^2 + \frac{\alpha}{\lambda^2}(1 - \cos \lambda\phi) \right) dx$ is finite. This requires that, that for any time, $\lim_{x\to\pm\infty} \phi(t,x) = \frac{2\pi}{\lambda} n_\pm$, for $n_\pm$ integer. Thus we could define a kind of topological charge $Q := n_+ - n_-$ as before, the only difference being that this charge could be any number in $\mathbb{Z}$ in principle (instead of being in $\{-1, 0, +1\}$). Besides the constant functions, the finite energy solutions of the type $\phi(t,x) = f(x - vt)$ have the form[55]

$$\phi(t,x) = \frac{2}{\lambda}\arccos\left(\pm \tanh \beta\sqrt{\alpha}(x - vt)\right) + \frac{2\pi}{\lambda}n, \quad \text{for } |v| < c,$$

where again $\beta := \left(1 - \frac{v^2}{c^2}\right)^{-\frac{1}{2}}$ and $n \in \mathbb{N}$. Again, it turns out that this solution behaves like a soliton and possesses the same properties as the kinks of the previous equation. So, we can think of these kinks as being fermions of charge ± 1. Note that there is no translating solution with a charge different from -1, 0 or 1. This reflects the exclusion principle for fermions. Superposition of several fermions would reveal the repulsion of fermions with the same charge and the attraction of fermions with opposite charges, with the possibility of annihilation.

[55] ϕ is also a Bogomol'nyi solution of the form $\phi(t,x) = f(x - vt)$, where $f' + 2\beta W'(f) = 0$ and $W(s) := -\frac{2\sqrt{\alpha}}{\lambda^2}\cos\frac{\lambda f}{2}$.

6.3.1 The Massive Thirring Model

The surprising result— already suggested by Skyrme[56]— was that these solitons really do behave like quantum fermions! This was proved in 1975 by Coleman [21]. For that purpose, we introduce the massive Thirring model: the fields are spinor fields ψ from the 2-dimensional space-time with two complex components ψ_1 and ψ_2. We consider the Lagrangian

$$L(\psi) := \overline{\psi}(i\gamma^{\mu}\frac{\partial}{\partial x^{\mu}} - m)\psi - \frac{1}{2}g\left(\overline{\psi}\gamma^{\mu}\psi\right)\left(\overline{\psi}\gamma_{\mu}\psi\right),$$

where

$$\gamma^0 = \sigma_1 = \begin{pmatrix} 0 & 1 \\ 1 & 0 \end{pmatrix}, \quad \gamma^1 = -i\sigma_2 = \begin{pmatrix} 0 & -1 \\ 1 & 0 \end{pmatrix},$$

and $\gamma_{\mu} = \eta_{\mu\nu}\gamma^{\nu}$ for $\eta_{00} = -\eta_{11} = 1, \eta_{01} = \eta_{10} = 0$. Using a perturbative approach, S. Coleman compared the quantum theories of the sine-Gordon equation with the quantum massive Thirring model. He found that both models are equivalent if we assume that

$$\frac{4\pi}{\lambda^2} = 1 + \frac{g}{\pi}. \tag{54}$$

And then, in the *sense of quantum operators*,

$$\overline{\psi}\gamma^1\psi\, dt + \overline{\psi}\gamma^0\psi\, dx = i\frac{\lambda}{2\pi}\left(\frac{\partial\phi}{\partial t}dt + \frac{\partial\phi}{\partial x}dx\right), \tag{55}$$

$$m\overline{\psi}\begin{pmatrix} 0 & 0 \\ 0 & 1 \end{pmatrix}\psi = -\frac{\alpha}{\lambda^2}e^{i\lambda\phi} \quad \text{and} \quad m\overline{\psi}\begin{pmatrix} 1 & 0 \\ 0 & 0 \end{pmatrix}\psi = -\frac{\alpha}{\lambda^2}e^{-i\lambda\phi},$$

where we have omitted a renormalisation constant that depends on the regularisation. Unfortunately, there is no rigorous mathematical proof of that, since both equations here are non-linear, and so the quantum theory of such equations has no solid mathematical base. So, the result of Coleman [21] is a verification that n-point Green functions (which are coefficients of the perturbative expansion) of both theories agree. This result was confirmed through other methods by Mandelstam [81] and by many other authors since.

---

[56] Skyrme's explanation was that, in the full quantum theory, it is possible to construct a new quantum field whose fluctuations are the solitons. The new field operator is obtained through an exponential expression in the original field ϕ

$$\psi_{\pm}(x) = e^{i\lambda(\phi \pm \int_{-\infty}^{x} dx' \frac{\partial\phi}{\partial t})} \tag{5}$$

with two spin components (and a normal ordering understood).
The construction (56) is an example of the vertex operator construction that would later become important in string theory and in the representation theory of infinite dimensional algebras (resembling quantum field theories).

Note that Eq. (55) relies on two forms which, for classical solutions, are closed, but each one for different reasons. The left hand side of this equation is the Noether current associated with a $U(1)$ invariance of the Thirring equation: it is divergence-free only if ψ is a solution to the equation; it describes the electric current density due to the motion of the fermions in space-time. The right hand side of (55) is always closed, even if ϕ is not a solution of the sine-Gordon equation; for that reason, it is called a topological current and integration of this current, on a constant time hypersurface produces the topological charge Q that we described earlier. It is also important to observe how relation (54) relates the coupling constants λ and g: if λ is close to 0, then the perturbative theory of the sine–Gordon is relatively accurate, but then g is very large and the perturbative theory of the Thirring model is almost meaningless. On the other hand, if $\lambda \leq \sqrt{4\pi}$ is close to $\sqrt{4\pi}$, then g is close to 0 and the situation is inversed: the perturbative theory for the sine-Gordon is not very good, whereas it is good for the Thirring equation

We are thus in a situation of duality between two models, particularly interesting because we could, for instance, use the perturbative quantum theory of the sine–Gordon model to obtain results on the Thirring model [120] in the case in which g is very large.

6.3.2 Reflexions

Many of the developments in field theory related to the development of current physics and mathematics are generally characterized by the presence of integrable systems. Quantum field theory is the application of quantum mechanics to fields and is one of the cornerstones of modern theoretical physics. These theories describe systems of several particles and usually have a large (often infinite!) number of degrees of freedom. For this reason, they can not be treated exactly, but rather by using perturbative methods, essentially limited developments in the powers of the coupling constant. One of the main problems encountered by field theory from the beginning was to find exact, non-perturbative approaches and methods for circumvent the limitations imposed by the perturbation theory. The concept of integrability proved very powerful in this regard. During the 1970s, these ideas, which were beginning to be well understood for the classical systems, were extended to quantum systems, for which the conserved quantities are generally associated with symmetries that, in some cases, are hidden. These developments were initially motivated by concrete physical problems and then led to powerful mathematical concepts, such as quantum groups in the context of massive integrable field theory and a deeper understanding of the algebras of Virasoro to the limit of zero mass, which are mostly conformal theories.

Among the many examples, we will mention three: in the late '80s, Maxim Kontsevich proposed a rigorous mathematical formulation of the conform fields theories in dimension two. He proved a remarkable conjecture of E. Witten linking the characteristic classes of spaces of moduli of stable curves with integrable systems. The demonstration included the first use in mathematics of the technique of Feynman diagrams. The second is the two-dimensional topological field theory, which was solved by Witten in the early 1990s and for which the null part of the partition function is a tau function of the Korteweg-de Vries hierarchy without dispersion. The third example is the complete solution of Seiberg and Witten of the Yang-Mills theory N = 2 super-symmetric with 4 dimensions. Here, the functions of effective low-energy partitions are tau functions of the Whitham hierarchy re-linked to integrable systems of the Hitchin type. In the same vein, analogous mechanisms exist in the dualities of quantum field theories and fully integrable systems. One of the simplest examples is the duality that appears in the sine-Gordon model in a two-dimensional space-time for bosonic fields.

We thus see that the search for models in which the duality between solitons and quanta seems to be realized, and the research of integrable systems leads to the same kind of equations. It is likely that this parallel between the two approaches goes deeper [55]. Thus, although we do not know how to prove the duality between the sine-Gordon equation and the massive Thirring model rigorously [120], we know how to establish an analogous duality for simpler models, called "Abelian", for which the kinks are replaced with constant functions per pieces, admitting some discontinuities. The proof of this result is based on the algebras of Vertex, a geometric description of which precisely uses the geometry of the loop groups (more precisely, the theory of the representations of their Central extensions, cf. [84, 100]). The duality between electricity and magnetism (though considered as conjecture), recognized by Dirac and later by Olive and Montonen, is another example of this convergence between duality and the existence of integrable systems. Notice that it is easier to verify such conjecture when we consider the supersymmetric version of the model, as noted by Witten and Olive [90]; the verification of this conjecture should be greatly facilitated if one considers the Supersymmetric Yang-Mills-Higgs fields version, with two ($N = 2$). This is based on a similar mechanism to the one we have seen. The first confirmation of these conjectures was obtained by Sen [111] with four systems of fermionic variables ($N = 4$). Then, another confirmation was obtained by Seiberg and Witten [112] for the theory $N = 2$. Again, these theories naturally involve fully integrable systems (Hyperkählerian manifolds, etc.) and supersymmetries.

Other dualities have been discovered, such as the T-duality, which is at the origin of the "mirror symmetry" that corresponds to two varieties of Calabi-Yau, by exchanging certain data characterizing the complex structures and symplectics (see [85] and the introduction in [70] for more examples and details). The 'M theory', currently under construction and supposedly able to unite all the known superstring theories (as well as supergravitation in dimension 11), is based on these various dualities.

A final parallel can be established between supersymmetric space-time models, and integrable systems. Indeed, several physical constraints of the relevant supersymmetric models (excluding spin particles greater than 2, which have as many bosons as fermions) are relatively small and the possible space-time dimensions can vary

only from 1 to 10 (or 11 for supergravitation). And among these dimensions, four are preferred: 1 + 2, 1 + 3, 1 + 5 and 1 + 9.
These are the dimensions for which the corresponding spin groups are Isomorphic to $SL(2, \mathbb{L})$, $SL(2, \mathbb{C})$, $SL(2, \mathbb{H})$ and $SL(2; \mathbb{Q})$ [55]. Geometrically, these dimensions correspond to the spaces-time in which the celestial sphere is the projective line on, (respectively) $\mathbb{R}$, $\mathbb{C}$, $\mathbb{H}$ and $\mathbb{Q}$. And then, the supergroups of Poincaré admit representations that are particularly simple and that restore most of the known examples of said supergroups (and their central extensions) by reduction to lower dimensions.
The resemblance with the integrable systems is that we can find an organization of the different theories by reductional dimension of the "fundamental" theories with 'maximal dimension' to derived theories, of smaller dimension (by comparison, in the present state of our knowledge, the most "integrable" systems live on a manifold of dimension 4 and are governed by $\mathbb{C}$ via the twistor space).
This analogy raises the question of the existence of "fundamental" integrable systems in dimensions 6 or 10 (in which quaternions or octonions play a role).
It is too early to know whether such speculations will succeed or not, and it is probable that the progress in understanding what lies behind the renormalization will someday completely modify these theories. The main point of these efforts is the search for theories based on the most perfect mathematical structures (supersymmetries, fully integrable systems, dualities).

7 Symmetry and the Foundations of Physics

The idea that the universe is governed by precise causal or dynamical laws,[57] is a very old one, and it largely due to Galileo, Descartes, Newton and others.
Newton, for example, had formulated a highly successful set of laws for material particles, known today as Newton's laws of motion and gravitation. So, it was natural for Newton to try to bring the behavior of light into this paradigm by posing the hypothesis that light consisted of material particles, called corpuscles. However, Newton's theory of light could not explain partial reflection, interference or diffraction, as is well known.

Physicists tried to solve these problems by abandoning Newton's ontology of corpuscles, while keeping his basic assumption that light obeyed deterministic laws, and replacing corpuscles with a wave. What made this appealing to them was that Huygens had formulated a law for the propagation of a wave, called Huygens' principle, according to which every point on a wave front acted as a source of secondary wavelets whose interference was sufficient to reconstruct the subsequent wavefronts. This was the first dynamical or causal law to govern the propagation of a wave, as opposed to Newton's laws that governed the propagation of material particles. And,

[57] Here, we should understand the word *laws* as a set of true principles that form a strong but simple and unified system that can be used to predict and explain. In other words, *it's a way to understand a great many complicated phenomena in a unified way, in terms of a few principles.*

using these laws and the new ontology that light is a wave, it was easy to explain all phenomena of light known at that time, including partial reflection, interference and diffraction.

The wave theory received a tremendous boost in the nineteenth century with the introduction of electric and magnetic fields by Faraday and Maxwell. These fields obeyed causal deterministic laws that were mathematically formulated by Maxwell. Moreover, light waves were recognized as special cases of this electromagnetic field, and Maxwell's laws justified Huygens' principle. The price to pay by keeping the paradigm of natural laws[58] was that the universe had to be regarded as a strange mixture of material particles and fields. Physicists lived with this dual ontology, even when an inconsistency was found between the two sets of laws that governed material particles and fields. This inconsistency, first clearly recognized by Einstein, was that the symmetries of the laws of mechanics that governed material particles were not the same as the symmetries of the laws of the electromagnetic field [72]. Einstein required that both symmetries should be the same, and asserted the primacy of fields over particles by requiring that the laws of mechanics should be modified so that they have the same Lorentz group of symmetries as the laws of the electromagnetic field. **This was the first time in the history of physics that symmetries took priority over laws** in the sense that the laws were modified to conform to the symmetries. Moreover, the existence of universal symmetries for all the laws of physics enabled the construction of a physical geometry having the same symmetries, namely the Minkowski space-time [14].

The idea of turning groups into basic building blocks for the geometric formulation of physics is simply the natural result of pushing ahead the old usage of imposing the compatibility of the observer in the same way Differential Geometry itself considers admissibility of a local chart. The requirement of a definite structure in the set of observers, or atlas, seriously delimits the nature of physical laws in that they must be formulated in terms of, say, $GL(n; R)$-tensors, although this requirement is not restrictive enough so as to actually predict dynamical laws. However, the condition of having defined an associative composition law in a set of active transformations of a physical system really predicts its dynamics in many cases, and can accordingly be considered as a basic postulate.

In physics, when a system is considered, we speak of a *symmetry* (and then about *invariance* with respect to this symmetry) by specifying *transformations* that leave some related quantities unchanged. Actually, we can define a symmetry as a change of coordinates or variables that leaves either the *action* invariant, or the equation of motion or field equations. Thereby, the first step, from the mathematical standpoint, arose with the theoretical definition of a *symmetry* as the invariance under a specific group of transformation, and was therefore followed by the group theory and the transformation group that appear in it. This movement culminates with the *Erlangen* program, rooted in Klein's insight [65]. F. Klein's vision indeed shows that geometry is perceived as the study of *structures* on spaces, considered with their *transformation*

[58]The laws of nature are supposed to be objective, independent of any interest and belief.

groups. Actually, the great insight of F. Klein has been to underline a unification principle to embrace different types of geometry. The fundamental point is then that the space of any geometry is defined with a transitive group action on it, revealing the invariance under group transformation.

8 The Modern Cosmology

Cosmology is the part of physics that has the whole universe as its area of research. As such, it covers a vast range of scales. Energy scales go from the present day temperature of 10^{-4}eV up to the Planck scale 10^{19}GeV. It aims to describe the evolution of the universe from its very beginning up to today, when it has an estimated age of the order of 10^{10} years. Due to its very nature of understanding the universe as a whole cosmology needs input from very different areas of physics. These naturally include astrophysics and theories of gravitation, but also plasma physics, particle physics and experimental physics.
Modern cosmology is based on two fundamental assumptions: first, the dominant interaction on cosmological scales is gravity, and second, the cosmological principle is a good approximation to the universe. The cosmological principle states that the universe, smoothed over large enough scales, is essentially homogeneous and isotropic.[59]
Observations such as the high degree of isotropy of the CMB indicate that, globally, the universe is well described by a spatially homogeneous and isotropic model. These are the Friedmann-Robertson-Walker solutions of general relativity.

How can the cosmological principle be justified? Obviously, the universe is not homogeneous and isotropic on scales as big as our Solar System, our Galaxy or even our Local Group of galaxies. Nevertheless, the cosmological principle has been invoked from the beginning of modern cosmology in the first half of the twentieth century, when almost nothing about the large-scale structure of the universe was known. The main reasons for its acceptance were simplicity and the Copernican principle. Applying the cosmological principle to general relativity yields rather strong constraints and leads to the simplest category of realistic cosmological models.

[59]These models follow from symmetry assumptions that dramatically simplify the task of solving Einstein's Fields Equations (EFE). They require that the space-time geometry is both homogeneous and isotropic. Roughly speaking, homogeneity requires that at a given moment of cosmic time, every spatial point looks the same, and isotropy holds if there are no geometrically preferred spatial directions. These requirements imply that the models are topologically $\Sigma \times \mathbb{R}$, visualizable as a stack of three-dimensional spatial surfaces $\Sigma(t)$ labeled by values of the cosmic time t. The worldlines of "fundamental observers", taken to be at rest with respect to matter, are orthogonal to these surfaces, and the cosmic time corresponds to the proper time measured by the fundamental observers. The spatial geometry of Σ is such that there is an isometry carrying any point $p \in \Sigma$ to any other point lying on the same surface (homogeneity), and at any point p, the three spatial directions are isometric (isotropy).

On the other hand, the Copernican principle, according to which we do not occupy any special place in the universe, fits the cosmological principle perfectly. If we perceive the universe around us isotropically, the Copernican principle asserts that other observers should also see the universe isotropically, since otherwise, we would occupy a special place in the universe. Since a universe that is isotropic everywhere is also homogeneous (in fact, isotropy around three distinct observers suffices), the cosmological principle is a relatively straightforward conclusion from an observed isotropy and the Copernican principle.

The discovery of cosmic background radiation (CBR) in 1964, together with the observed Hubble expansion of the universe, established hot big bang cosmology as a viable model of the universe. The success of the theory of nucleosynthesis in reproducing the observed abundance pattern of light elements, together with the proof of the black body character of the CBR, then established hot big bang as the standard cosmological model.
Actually, cosmology confronts a number of questions: the limits of scientific explanation, the nature of physical laws, and different types of underdetermination, for example. Due to the uniqueness of the universe and its inaccessibility, cosmology has often been characterized as more speculative than other areas of physics.

Cosmologists do, however, face a number of distinctive challenges. These challenges derive from different features of cosmology. One such feature is the finitude of the speed of light, a basic feature of relativistic cosmology that ensures that global properties of the universe cannot be established directly by observations. This is a straightforward limit on observational access to the universe. Another feature is the interplay between global aspects of the universe and local dynamical laws. Indeed, cosmology relies on extrapolating local physical laws so as to hold universally.
The Standard Model of cosmology is based on extrapolating local laws to the universe as a whole. Yet, there may be global-to-local constraints. The uniqueness of the universe implies that the normal ways of thinking about laws of physics and the contrast between laws and initial conditions do not straightforwardly apply. In other areas of physics, the initial or boundary conditions themselves are typically used to explain other things, rather than being the target of explanation. Many lines of research in contemporary cosmology aim to explain why the initial state of the Standard Model was obtained, but the nature of this explanatory project is not entirely clear. And due to the uniqueness of the universe, it is not clear what underwrites the assignment of probabilities.
Therefore, it is difficult to adjudicate this debate, due to the lack of independent access to the phenomena. The early universe is interesting because it is one of the few testing grounds for quantum gravity. Without a clear understanding of the initial state derived from such a theory, however, it is difficult to use observations to infer the dynamics governing the earliest stages of the universe's evolution.

9 Einstein's Fields Equations - EFE

In general relativity, physical space-time is modeled in terms of differential geometry as a Lorentzian manifold whose pseudo-Riemannian metric, or rather the Levi-Civita connection that corresponds to it encodes the field of gravity. The action functional describing the dynamics of this field is the Einstein-Hilbert action, in which the field of gravity enters in terms of the integral of the scalar curvature of the Levi-Civita connection over space-time [72].

However, from Einstein's point of view, space and time are different aspects of a single entity: space-time. Energy and momentum are united analogously. Einstein further realized that space-time is not a static stage on which physics unfolds, but a dynamic entity that can curve and bend. Gravitation is understood as a manifestation of space-time curvature, and space-time is built out of a gravitational field. One mathematical achievement from this idea is Einstein's equation for a (pseudo-)metric tensora $g_{\mu\nu}$ on a manifold M (the space-time):

$$R_{\mu\nu} - \frac{1}{2}g_{\mu\nu}R + \Lambda g_{\mu\nu} = 8\pi G T_{\mu\nu}, \tag{56}$$

where $R_{\mu\nu}$ is the Ricci curvature tensor of the metric and R is its scalar curvature. This equation codes the constraints imposed by the energy –momentum tensor $T_{\mu\nu}$, which encodes informations about the distribution of energy and momentum in space-time, on the metric tensor $g_{\mu\nu}$.

This new picture of space-time made it possible to conceive some ideas that were impossible to articulate in the Newtonian picture of the world. Consider the most important fact about cosmology: we live in an expanding universe. The distance between two galaxies grows with time. But the galaxies are not rushing apart from each other into some preexisting space, as though blown out of an explosion from some common center. Rather, more and more space is being generated between the galaxies all the time, so from the vantage point of any one galaxy, the others appear to be rushing away. This picture, impossible to imagine in Newton's universe, is an inevitable consequence of Einstein's theory.

Following Einstein's intuition the manifold M and the metric should be built simultaneously when solving Eq. (56). From this point of view, the only kinematic condition imposed is that, at each point of space-time, the tangent space is endowed with a Minkowski metric in the physical case of pseudo-Riemannian manifolds. Then, the field $(g_{\mu\nu})$ describes the way these metrics depend on the point in a smooth way and the Einstein equation (56) is the dynamical constraint on $g_{\mu\nu}$.

9.1 *Solutions of EFE and the Standard Model of Cosmology*

Einstein (1917) introduced a strikingly new conception of cosmology, as the study of exact solutions to general relativity that describe the space-time geometry of the

universe. One would expect gravity to be the dominant force in shaping the universe's structure at large scales, and it is natural to look for solutions of Einstein's field equations (EFE) compatible with astronomical observations.
Einstein's own motivation for taking the first step in relativistic cosmology was to vindicate Mach's principle and he also sought a solution that describes a static universe, that is, one whose spatial geometry is unchanging. He forced his theory to accommodate a static model by modifying his original field equations, with the addition of the famous cosmological constant. As a result, Einstein missed one of the most profound implications of his new theory: general relativity quite naturally implies that the universe evolves dynamically with time. Four of Einstein's contemporaries discovered a class of simple evolving models, the Friedman-Lemaître-Robertson-Walker (FLRW) models[60], that have proven remarkably useful in representing the space-time geometry of our universe. FLRW is usually reffered to as the Standard Model of Cosmology.

9.2 Friedmann Equation

In the standard model of cosmology, gravity is described by general relativity[61]. As in special relativity, space and time are united to describe a space-time. The invariant line element of special relativity is given by $ds^2 = \eta_{\mu\nu} dx^\mu dx^\nu$, where

$$\eta_{\mu\nu} = \begin{pmatrix} -1\,0\,0\,0 \\ 0\;\;1\,0\,0 \\ 0\;\;0\,1\,0 \\ 0\;\;0\,0\,1 \end{pmatrix} \tag{57}$$

is the Minkowski metric. In general relativity, this is determined by $ds^2 = g_{\mu\nu} dx^\mu dx^\nu$ for a general metric $g_{\mu\nu}$. Space-time is curved. The dynamics is determined by Einstein's equations[62],

---

[60]The main results of the FLRW model were first derived by the Soviet mathematician Alexander Friedmann in 1922 and 1924, but his work remained relatively unnoticed by his contemporaries [52, 53]. Albert Einstein, who, on behalf of Zeitschrift fr Physik, acted as the scientific referee for Friedmann's work, acknowledged the correctness of Friedmann's calculations, but failed to appreciate the physical significance of their predictions.
Friedmann died in 1925. In 1927, Georges Lemaître, arrived independently at results similar to those that Friedmann had and published them in the Annals of the Scientific Society of Brussels [78]. In the face of the observational evidence for the expansion of the universe obtained by Edwin Hubble in the late 1920s. Howard P. Robertson from the US and Arthur Geoffrey Walker from the UK explored the problem further during the 1930s. In 1935, Robertson and Walker rigorously proved that the FLRW metric is the only one on a space-time that is spatially homogeneous and isotropic [52, 53, 78, 101, 123].

[61]For more details, see [77].

[62]Repeated indices in pairs with one contravariant and one covariant are summed over. Greek indices take values between 0 and 3.

$$R_{\mu\nu} - \frac{1}{2} g_{\mu\nu} R + \Lambda g_{\mu\nu} = 8\pi G T_{\mu\nu}, \tag{58}$$

where the Ricci tensor is defined by $R_{\mu\nu} = \frac{\partial \Gamma^{\sigma}_{\mu\nu}}{\partial x^{\sigma}} - \frac{\partial \Gamma^{\sigma}_{\mu\sigma}}{\partial x^{\nu}} + \Gamma^{\sigma}_{\rho\sigma}\Gamma^{\rho}_{\mu\nu} - \Gamma^{\sigma}_{\rho\nu}\Gamma^{\rho}_{\mu\sigma}$, the Christoffel symbols are defined by $\Gamma^{\mu}_{\nu\rho} = \frac{g^{\mu\sigma}}{2}\left(\frac{\partial g_{\sigma\rho}}{\partial x^{\nu}} + \frac{\partial g_{\nu\sigma}}{\partial x^{\rho}} - \frac{\partial g_{\nu\rho}}{\partial x^{\sigma}}\right)$ and $R = g^{\mu\nu} R_{\mu\nu}$ is the Ricci scalar. $T_{\mu\nu}$ is the energy-momentum tensor that is conserved, satisfying $\nabla_{\nu} T^{\mu\nu} = 0$.

Einstein's equations encode the information that geometry determines matter distribution and evolution, and vice versa. Space-time is curved by the presence of matter. That is why, e.g., the trajectory of light from distant sources is deviated by the sun. This deflection angle is one of the classical tests of general relativity.
Einstein's equations are very complex. There are no general solutions known. It is always necessary to assume some degree of symmetry in order to find solutions. The Friedmann-Lemaître-Robertson-Walker solutions are isotropic and homogeneous and are described by the metric

$$ds^2 = -dt^2 + a^2(t)\left[\frac{dr^2}{1 - kr^2} + r^2\left(d\theta^2 + \sin^2\theta d\Psi^2\right)\right]. \tag{59}$$

In many cases, matter can be described by a perfect fluid with 4-velocity u^{μ} whose energy momentum tensor is given by

$$T_{\mu\nu} = [\rho(t) + P(t)]\, u_{\mu} u_{\nu} + p(t) g_{\mu\nu}, \tag{60}$$

where $\rho(t)$ and $p(t)$ are the energy density and pressure, respectively, which are only functions of time in a Friedmann-Lemaître-Robertson-Walker background [77].
Homogeneity and isotropy of the universe imply that the energy momentum tensor takes the diagonal form $(T_{\mu}^{\nu}) = \mathrm{diag}(-\rho, p, p, p)$, where ρ is the energy density of the universe and p the pressure. Energy momentum conservation $(T_{\mu\ ;\nu}^{\ \nu} = 0)$ then takes the form of the continuity equation

$$\frac{d\rho}{dt} = -3H(t)(\rho + p), \tag{61}$$

where the first term in the rhs describes the dilution of the energy due to the expansion of the universe and the second term corresponds to the work done by pressure. Eq.(61) can be given the following more transparent form:

$$d\left(\frac{4\pi}{3} a^3 \rho\right) = -p\, 4\pi a^2 da, \tag{62}$$

which indicates that the energy loss of a 'comoving' sphere of radius $\propto a(t)$ equals the work done by pressure on its boundary as it expands.

The cosmological principle tightly constrains the properties of the surfaces $\Sigma(t)$. These are three-dimensional spaces (Riemannian manifolds) of constant curvature, and all of the surfaces in a given solution have the same topology. If the surfaces are simply connected, there are only three possibilities for Σ: (1) spherical space, for the case of positive curvature; (2) Euclidean space, for zero curvature; and (3) hyperbolic space, for negative curvature.
Textbook treatments often neglect to mention, however, that replacing global isotropy and homogeneity with local analogs opens the door to a number of other possibilities. For example, there are models in which the surfaces Σ have finite volume, but are multiply connected, consisting of, roughly speaking, cells pasted together. Although isotropy and homogeneity hold locally at each point, above some length scale, there would be geometrically preferred directions reflecting the way in which the cells are connected. In these models, it is, in principle, possible to see around the universe and observe multiple images of a single object, but there is, at present, no strong observational evidence of such effects.

Note that imposing global isotropy and homogeneity reduces EFE - a set of 10 non-linear, coupled partial differential equations - to a pair of differential equations governing the scale factor $a(t)$ and $\rho(t)$, the energy density of matter. The scale factor measures the changing spatial distance between fundamental observers. The dynamics are then captured by the Friedmann-Lemaître-Robertson-Walker metric equation (59). Einstein's equations (58) then lead to the Friedmann equation

$$H^2 \equiv \left(\frac{\dot{a}(t)}{a(t)}\right)^2 = \frac{8\pi G}{3}\rho - \frac{k}{a^2} \; . \tag{63}$$

Averaging p, we write $\rho + p = \gamma\rho$. Eq. (61) then becomes $\dot{\rho} = -3H\gamma\rho$, which gives us $d\rho/\rho = -3\gamma da/a$ and $\rho \propto a^{-3\gamma}$. For a universe dominated by pressureless matter, $p = 0$, and thus $\gamma = 1$, which gives us $\rho \propto a^{-3}$. This is easily interpreted as mere dilution of a fixed number of particles in a 'comoving' volume due to the cosmological expansion. For a radiation-dominated universe, $p = \rho/3$, and thus $\gamma = 4/3$, which gives us $\rho \propto a^{-4}$. In this case, we get an extra factor of $a(t)$ due to the red-shifting of all wave-lengths by the expansion. Substituting $\rho \propto a^{-3\gamma}$ in the Friedmann equation with $k = 0$, we get $\dot{a}/a \propto a^{-3\gamma/2}$, and thus $a(t) \propto t^{2/3\gamma}$. Taking into account the normalization of $a(t)$ ($a(t_0) = 1$), this gives

$$a(t) = (t/t_0)^{2/3\gamma} \; . \tag{64}$$

For a matter-dominated universe, we get the expansion law $a(t) = (t/t_0)^{2/3}$. 'Radiation', however, expands as $a(t) = (t/t_0)^{1/2}$.
The universe, in its early stages of evolution, is radiation-dominated and its energy density is

$$\rho = \frac{\pi^2}{30}\left(N_b + \frac{7}{8}N_f\right)T^4 \equiv c\,T^4 \, , \tag{65}$$

where T is the cosmic temperature and N_b (N_f) is the number of massless bosonic (fermionic) degrees of freedom. The combination $g_* = N_b + (7/8)N_f$ is called the effective number of massless degrees of freedom. The entropy density is

$$s = \frac{2\pi^2}{45}\, g_*\, T^3 \,. \tag{66}$$

Assuming adiabatic universe evolution, i.e., constant entropy in a 'comoving' volume (sa^3 = constant), we obtain the relation aT = constant. The temperature-time relation during radiation dominance is then derived from the Friedmann equation (with $k = 0$):

$$T^2 = \frac{M_P}{2(8\pi c/3)^{1/2} t} \,. \tag{67}$$

We see that, classically, the expansion starts at $t = 0$ with $T = \infty$ and $a = 0$. This initial singularity is, however, not physical, since general relativity fails at cosmic times smaller than, roughly, the Planck time t_P. The only meaningful statement is that the universe, after a yet unknown initial stage, emerges at a cosmic time $\sim t_P$ with temperature $T \sim M_P$.

9.3 *Hubble Expansion*

One of the most remarkable discoveries in twentieth century astronomy was Hubble's (1929) observation that the red-shifts of spectral lines in galaxies increase linearly with their distance[63] [63]. Hubble took this to show that the universe is expanding uniformly, and this effect can be given a straightforward qualitative explanation in the FLRW models. The FLRW models predict a change in frequency of light from distant objects that depends directly on $R(t)$. There is an approximately linear relationship between red-shift and distance at small scales for all the FLRW models, and departures from linearity at larger scales can be used to measure spatial curvature. For cosmic times $t \gtrsim t_P \equiv M_P^{-1} \sim 10^{-44}$ s ($M_P = 1.22 \times 10^{19}$ GeV is the Planck scale) after the big bang, quantum fluctuations of gravity cease to exist. Gravitation can then be adequately described by classical relativity. Strong, weak and electromagnetic interactions, however, require relativistic quantum field theoretic treatment and are described by gauge theories.
As we know, the standard big bang (SBB) cosmological model [77] is based on the idea that the universe is homogeneous and isotropic (the *cosmological principle*). Under this assumption, the four dimensional space-time in the universe is described

[63] Hubble's distance estimates have since been modified, leading to a drastic decrease in the estimate of the current rate of expansion (the Hubble parameter, H_0). However, the linear redshift-distance relation has withstood scrutiny, as the sample size has increased from 24 bright galaxies (in Hubble 1929) to hundreds of galaxies at distances 100 times greater than Hubble's, and as astrophysicists have developed other observational methods for testing the relation (see [48, 91, 92]).

by the Friedmann-Lemaître-Robertson-Walker metric[64]

$$ds^2 = -dt^2 + a^2(t)\left[\frac{dr^2}{1-kr^2} + r^2\left(d\theta^2 + \sin^2\theta d\Psi^2\right)\right]. \tag{68}$$

The dimensionless parameter $a(t)$ is the *scale factor* of the universe and describes cosmological expansion. We normalize it by taking $a_0 \equiv a(t_0) = 0$, where t_0 is the present cosmic time.

The instantaneous radial physical distance is given by

$$R = a(t)\int_0^r \frac{dr}{(1-kr^2)^{1/2}} \cdot \tag{69}$$

For flat universe ($k = 0$), $\bar{R} = a(t)\bar{r}$ ($\bar{r}$ is a 'comoving' and $\bar{R}$ a physical vector in 3-space) and the velocity of an object is

$$\bar{V} = \frac{d\bar{R}}{dt} = \frac{\dot{a}}{a}\bar{R} + a\frac{d\bar{r}}{dt}, \tag{70}$$

where overdots denote derivation with respect to cosmic time. The second term on the right hand side (rhs) of this equation is the 'peculiar velocity', $\bar{v} = a(t)\dot{\bar{r}}$, of the object, i.e., its velocity with respect to the 'comoving' coordinate system. For $\bar{v} = 0$, Eq. (70) becomes

$$\bar{V} = \frac{\dot{a}}{a}\bar{R} \equiv H(t)\bar{R}, \tag{71}$$

where $H(t) \equiv \dot{a}(t)/a(t)$ is the Hubble parameter. This is the well-known Hubble law asserting that all objects run away from each other with velocities proportional to their distances and is considered as the first success of SBB cosmology.

10 Friedmann–Robertson–Walker (FRW) Cyclic Universe and Elliptic Curves

In this model, the space-time[65] can be represented as the direct product of a global time t-axis and a maximally symmetric three-dimensional space section with a metric of constant curvature k [82].

---

[64]Where r, Ψ and θ are 'comoving' polar coordinates, which remain fixed for objects that have no other motion than the general expansion of the universe. The parameter k is the 'scalar curvature' of the 3-space and $k = 0$, $k > 0$ and $k < 0$ correspond to flat, closed and open universe, respectively (for more understanding of the Manifolds with scalar curvature see Gromov's contribution in this volume).

[65]During one aeon: one cycle of universe.

We also choose a fixed time-like geodesic ("observer's history"), along which the metric is dt^2, and coordinatize each space section at the time t by the invariant distance r from the observer and two natural angle coordinates θ, Ψ on the sphere of radius r. By rescaling the radial coordinate, we may and will assume that the curvature constant k takes one of three values: $k = \pm 1$ or 0.
This rescaling produces the natural unit of length, when $k \neq 0$, and the respective unit of time is always chosen so that the speed of light is $c = 1$.
The FLRW metric of signature (1,3) is then given by the formula:

$$ds^2 = dt^2 - (R(t))^2(dr^2 + f_k^2(r)(d\theta^2 + \sin^2\theta d\phi^2)),$$

where, as usual, $f_k = \sin r, r, \sinh r$ according to $k = 1, 0, -1$.
The dynamics in this model is described [88, 121] by one scale factor (real function) $R(t)$: it increases from zero at the Big Bang of one aeon to infinity during this aeon, which becomes "almost zero time" of the next aeon. We scale $R(t)$ by putting $R = 1$ "now". If we introduce conformal time τ in terms of proper time t by $dt = Rd\tau$, the FLRW metric can be written as

$$ds^2 = (R(t))^2(d\tau^2 - dr^2 - f_k^2(r)(d\theta^2 + \sin^2\theta d\phi^2)) = (R(t))^2 d\tilde{s}^2, \tag{72}$$

and then the conformal metric $d\tilde{s}^2$ is cyclic: it extends smoothly through each Big Bang and future infinity. These are separated by infinite intervals of proper time, but finite intervals of conformal time.

As is conventional, we assume the matter content of the universe to be a mixture of dust and radiation, together with a positive cosmological constant Λ. The Einstein equations reduce to the Friedmann equation:

$$\dot{R}^2 = -k + \kappa(\rho_M + \rho_\gamma)R^2 + \frac{1}{3}\Lambda R^2, \tag{73}$$

where $\kappa = 8\pi G/3$, ρ_M and ρ_γ are the matter and radiation densities, and the overdot is d/dt with proper-time t, together with the perfect fluid conservation equation

$$3\dot{R}/R = -\dot{\rho}/(\rho + p).$$

The conservation equation for the two fluids can be solved to give

$$\rho_M = AR^{-3}, \quad \rho_\gamma = BR^{-4},$$

in terms of integration constants A, B. For simplicity, assume $k = 0$, though data given in, for example, [46] indicates that this term in the Friedmann equation is, in any case, very small.

Now, the Friedmann equation is just

$$\dot{R}^2 = \kappa(AR^{-1} + BR^{-2}) + \frac{1}{3}\Lambda R^2.$$

For the scale factor at the present time, write $R = R_0$ and introduce constant parameters a, b by

$$\Omega_M/\Omega_\Lambda = a, \quad \Omega_\gamma/\Omega_\Lambda = b,$$

at the present time, with the conventional meanings for Ω_M, Ω_Λ and Ω_γ. This translates to

$$\kappa\rho_M/(\frac{1}{3}\Lambda) = a, \quad \kappa\rho_\gamma/(\frac{1}{3}\Lambda) = b,$$

at the present time, and hence

$$\kappa A/(\frac{1}{3}\Lambda R_0^3) = a, \quad \kappa B/(\frac{1}{3}\Lambda R_0^4) = b.$$

Eliminate A, B, introduce $S = R/R_0$ and rationalise (73) as:

$$S^2\dot{S}^2 = \frac{1}{3}\Lambda(b + aS + S^4), \tag{74}$$

or otherwise write it as:

$$Y^2 = R^4 + aR + b, \tag{75}$$

This function is constrained by the Einstein–Friedmann equations (here with cosmological constant $\Lambda = 3$), which leads to the introduction of the elliptic curve given by the equation in the (Y, R)-plane. We shall be interested in conformal time τ rather than proper time t, so that $dt = Rd\tau$, and for a single aeon, we may choose the origins to coincide.
Besides the proper time t and the scale factor $R(t)$, global time may be measured by its conformal version τ given as the integral along a real curve on the elliptic curve (75):

$$\tau \cong \int_0^{R(t)} \frac{dR}{Y},$$

where $\tau = 0 = t$ at $R = 0$.
A physical interpretation of the coefficients a, b as characterising matter and radiation sources in (75) shows that, in principle, *a, b also depend on time*, although for asymptotic estimates, their values are usually fixed by current observations.
We close this subsection with the following qualitative summary:

In the FRW universe, the evolution of time is essentially described by a real curve on an algebraic surface (75) *that is a family of elliptic curves.*

11 Conclusion

Recent ideas from quantum gravity and string theory challenge the fundamental concepts of geometry at an even deeper level. Physical intuition tells us that the traditional pseudo-Riemannian geometry of space-time cannot be a definite description of physical reality. Quantum corrections in the theory of gravity will change this picture at distances on the order of the Planck scale. Familiar fundamental properties such as locality only appear at much larger scales. In fact, there is now much evidence within string theory - usually referred to as holography' - that, at the end, geometry itself is an emergent quantity. The classical laws of gravity only appear within the limit where the number of degrees of freedom of the underlying quantum theory is taken to infinity, very similar to the emergence of the macroscopic laws of thermodynamics out of the microscopic description of statistical mechanics. The definite mathematical formulation of such a concept of quantum geometry' [1] is, however, still far away.

One could also question whether there exists a single overarching mathematical structure that captures all these aspects of quantum theory, or whether one is simply dealing with a combination of different complementary points of view, like the charts and maps of a manifold. As a whole, the study of quantum geometry takes on the form of a rich mathematical program, very much like the Langland's program, with many non-trivial examples, strange relations, dualities and automorphic forms, tying together diverse fields, with vast generalizations, all in an open-ended project that seems to encompass more and more mathematics.

The lesson here expresses one given by Dirac in 1939:

> It would probably also be a good thing to give a preference to those branches of mathematics that have an interesting group of transformations underlying them, since transformations play an important role in modern physical theory, both relativity and quantum theory seeming to show that transformations are of more fundamental importance than equations [42, p. 125].

Notice that while physicists have been exploring their new and still speculative theories, they have stumbled across a whole range of mathematical discoveries'. These are derived by physical intuition and heuristic arguments, which are beyond the reach, as yet, of mathematical rigour, but which have withstood the tests of time and alternative methods. The impact of these discoveries on mathematics has been profound and widespread.

The rigorous study of quantum field theories is a very hard problem and has been slow in development, even for theories much simpler than those that impact on geometry [60]. Even though quantum field theory is used in physics every day, the mathematical foundations underlying the standard model of particle physics still have to be constructed. Indeed, the problem for mathematicians is that the functional integrals that form the basis of many of the above approaches, and in which the many exotic types of symmetry are more obviously present, are as yet not rigorously defined. Nevertheless, in the current phase of interaction, mathematicians are now becoming familiar with the physicists's way of wrapping up mathematical information in a partition function.

Mathematics has its own internal dynamics: some fields develop and brush against neighbouring areas, some settle down to steady progress for a few decades and then explode. Some of the growth areas of the 1960s, for example, when resources were poured into science, became quiescent twenty years later, but then sprang back onto the scene. Or, to take a longer-term view, one highlight Bernhard Riemann's work in the mid-nineteenth century on differential geometry; its subsequent development in higher dimensions by Gregorio Ricci-Curbastro in 1904 prepared it for its phenomenal expansion when it was seen as the language in which to express Einstein's general relativity. More recently, and certainly at the International Congress in Madrid, one experienced the shift from deterministic to stochastic methods, which have their origins in the nineteenth century physicists's study of thermodynamics. These movements sometimes originate from developments within the subject, sometimes from external influences[66] [60].

So, instead of reverting to the origins of these new ideas coming from physics, mathematicians are now developing their own axiomatic versions, in particular, of topological quantum field theories, to suit their own ends. This is not unusual in mathematics – instead of asking what the real numbers really are, we are happier to characterize them by their properties, which we can use on an everyday basis.

This mathematical approach has uncovered a rich structure. For example, a topological quantum field theory requires a hierarchy of concepts, the lower levels of which are quite familiar, but progress requires some tough mental activity in getting a feeling for new objects.

Perhaps a good demonstration of this, is the index theorem, one of the most important results in the twentieth century for unifying different branches of mathematics (see [60]).

"In 1962, Michael Atiyah and Isadore Singer began work on this theorem [5]. It began as a quest to explain why certain rational numbers in algebraic topology are integers - were they related to dimensions of vector spaces? In pursuing this aim, they rediscovered one of the fundamental differential operators of physics - the Dirac operator. Of course, the setting for this was somewhat different - they were working in Riemannian geometry rather than Einstein's space-time - but it was essentially the same operator. There began several proofs: the first two used ideas from two of the most active areas of mathematics at the time [5–7]. The first was a part of algebraic topology - Rene Thom's cobordism theory. Then came the second proof (with a wider range of applicability) using the far-reaching abstract ideas of Alexandre Grothendieck in algebraic geometry. Much later, in the mid-1970s, a third proof involving the heat kernel and differential geometry emerged.

Yet at the same time, physicists were in the process of rediscovering the theorem. For the physicists, who were studying what they called anomalies, the heat kernel expansions were commonplace. The new ideas for them were the links with algebraic topology. So, the evolution of their theorem was proceeding in the opposite direction, and only in the late 1970s, as both mathematicians and physicists began

[66] See Marcolli's contribution about the interaction between mathematics and computational linguistics.

to become interested in the Yang-Mills equations, did they really put their heads together. This was a crucial moment, when the mathematicians realized that physicists had uncovered a completely new way of looking at what they called connections, and physicists realized that the problems that had been bothering them for some time could be resolved through the use of some quite sophisticated mathematics, which was only then being developed. It was no longer true that the only mathematics a physicist needed to know was how to integrate by parts!"

Those mathematical results could be viewed as part of a much bigger quantum field theory, and the full force of the physicists intuition could be brought into play. There is, then, a difference between the current interactions and those of previous periods. It involves the scale of interactions, the range of mathematics being utilized and the changing dynamics of the subject. And still, the underlying irony is that the mathematical results that are being correctly predicted are often based on a non-rigorously posed quantum field theory.

Whatever the successes and failures of recent physical theories in the experimental domain, it is clear that the impact on mathematics, and on geometry in particular, is permanent, and the recent history shows how these distinct viewpoints work together for our mutual benefit, producing some of the most exciting and surprising results in mathematics. Given the rich grounds still to be explored, it seems likely that they will continue to do so for some time yet.

Acknowledgements First of all, I apologize to those whose works I forgot to mention and which helped to improve the quality of the paper. Secondly, I would like to thank all my friends who have contributed to this volume, and all the colleagues who, through exchanges, suggestions or their own writings, have enriched the content of this paper. I am particularly grateful to Michael Atiyah, Alain Connes, Edward Witten, Roger Penrose, Misha Gromov, Ali Chamsddine, Lee Somlin, Jeremy Butterfield and John Stachel for exchanges and suggestions.

References

1. A. Ashtekar, New variables for classical and quantum gravity. Phys. Rev. Lett. **57**, 2244 (1986)
2. A. Ashtekar, *Gravity, Geometry and the Quantum*, in Vers une nouvelle Philosophie de la nature, Joseph Kouneiher ed. Hermann, 2010
3. A. Ashtekar, J. Lewandowski, Quantum theory of geometry. I: area operators. Class. Quant. Grav.**14** (1997) A55–A82. http://xxx.lanl.gov/abs/gr-qc/9602046
4. A. Ashtekar, J. Lewandowski, Quantum theory of geometry. II: volume operators. Adv. Theor. Math. Phys.**1** (1998) 388. http://xxx.lanl.gov/abs/gr-qc/9711031
5. M.F. Atiyah, I.M. Singer, The index of elliptic operators on compact manifolds. Bull. Amer. Math. Soc. **69**(3), 422–433 (1963)
6. M.F. Atiyah, I.M. Singer, The index of elliptic operators I. Ann. Math. **87**(3), 484–530 (1968)
7. M.F. Atiyah, I.M. Singer, The index of elliptic operators V. Ann. Math. Second Ser. **93**(1), 139–149 (1971)
8. M. Atiyah, R. Dijkgraaf, N.l Hitchin, Geometry and physics. Phil. Trans. R. Soc. A **368**, 913–926 (2010)
9. M. Atiyah, N.S. Manton, B.J. Schroers, *Geometric Models of Matter*, arXiv:1108.5151

10. K.A. Brading, T.A. Ryckman, Hilbert's foundations of physics': gravitation and electromagnetism within the axiomatic method. Stud. Hist. Philos. Sci. B: Stud. Hist. Philos. Mod. Phys. **39**(1), 102–153 (2008)
11. N.N. Bogoliubov, D.V. Shirkov, *The Theory of Quantized Fields* (Interscience, New York, 1959)
12. N.N. Bogoliubov, O. Parasiuk, On the multiplication of the causal function in the quantum theory of fields. Acta Math. **97**, 227–266 (1957)
13. R. Bott, *On Mathematics and Physics*, Collected Works, vol. 4, p. 382
14. K. Brading, E. Castellani, *Symmetries in Physics: Philosophical Reflections*, 2003
15. P. Candelas, P. Green, L. Parke, X. de la Ossa, A pair of Calabi-Yau manifoldsas an exactly soluble superconformal eld theory. Nucl. Phys. B **359**, 21–74 (1991) https://doi.org/10.1016/0550-3213(91)90292-6
16. A.H. Chamseddine, A. Connes, M. Marcolli, Gravity and the standard model with neutrino mixing. Adv. Theor. Math. Phys. **11**, 991–1089 (2007)
17. S. Chern, J. Simons, Some cohomology classes in principal fiber bundles and their application to Riemannian geometry, Proc. Nat. Acad. Sci. USA **68**, 791–794, Or, characteristic forms and geometrical invariants. Ann. Math. **99**(48–69), 1974 (1971)
18. S. Chern, Vector bundles with a connection, Studies in Global Differential Geometry. Math. Asso. Amer. Studies No. **27**, 1–26 (1989)
19. S. Chern, *Complex Manifolds without Potential Theory*, 2nd edn. (Springer, Berlin, 1979)
20. S.-S. Chern, *What Is Geometry*? The American Mathematical Monthly, vol. 97(8), Special Geometry Issue, pp. 679-686 (1990)
21. S. Coleman, Quantum sine-Gordon equation as the massive Thirring model. Phys. Rev. D **11**, 2088 (1975)
22. A. Connes, M. Marcolli, *Noncommutative Geom.* (American Mathematical Society, Quantum Fields and Motives, 2007)
23. Alain Connes, Dirk Kreimer, Renormalization in quantum field theory and the Riemann-Hilbert problem II: the β-function, diffeomorphisms and renormalization group. Commun. Math. Phys. **216**, 215–241 (2001). arXiv:hep-th/0003188v1
24. A. Connes, M. Marcolli, *Renormalization, the Riemann-Hilbert correspondence and motivic Galois theory, Frontiers in number theory, physics and geometry*, vol. II (Springer, Berlin, 2007). pp. 617–713
25. A. Connes, *Noncommutative Geometry* (Academic Press, Cambridge, 1994)
26. A. Connes, J. Lott, Particle models and noncommutative geometry. Nucl. Phys. Proc. Suppl. **B18**, 29 (1989)
27. A. Connes *Geometry and Physics*
28. A. Connes, D. Kreimer, Hopf algebras, renormalization and noncommutative geometry. Commun. Math. Phys. **199**, 203 (1998). arXiv:hep-th/9808042
29. A. Connes, D. Kreimer, Renormalization in quantum field theory and the Riemann-Hilbert problem. I: the Hopf algebra structure of graphs and the main theorem. Commun. Math. Phys. **210**, 249 (2000). arXiv:hep-th/9912092
30. A. Connes, D. Kreimer, Renormalization in quantum field theory and the Riemann-Hilbert problem. II: the beta-function, diffeomorphisms and the renormalization group. Commun. Math. Phys. **216**, 215 (2001). arXiv:hep-th/0003188
31. A. Connes, M. Marcolli, *Noncommutative Geometry, Quantum Fields and Motives*. preliminary version available at http://www.alainconnes.org/en/downloads.php
32. A. Connes, D. Kreimer, Hopf algebras, renormalization and noncommutative geometry. Comm. Math. Phys. **199**, 203–242 (1998)
33. A. Connes, D. Kreimer, Renormalization in quantum field theory and the Riemann-Hilbert problem. I. The Hopf algebra structure of graphs and the main theorem. Comm. Math. Phys. **210**(1), 249–273 (2000)
34. A. Connes, D. Kreimer, Renormalization in quantum field theory and the Riemann-Hilbert problem. II. The β-function, diffeomorphisms and the renormalization group. Comm. Math. Phys. **216**(1), 215–241 (2001)

35. A. Connes, M. Marcolli, *From Physics to Number theory via Noncommutative Geometry, II: Renormalization, the Riemann-Hilbert correspondence, and motivic Galois theory, to appear in Frontiers in Number Theory, Physics, and Geometry*, vol. II. Preprint hep-th/0411114
36. D. A. Cox, S. Katz, *Mirror symmetry and algebraic geometry*. Mathematical Surveys and Monographs no. 68. Providence, RI: American Mathematical Society. 1999
37. L. Corry, *David Hilbert and the Axiomatization of Physics (1898-1918): From Grundlagen der Geometrie to Grundlagen der Physik* (Kluwer Academic Publishers, Dordrecht, 2004). p. 429
38. B. Delamotte, A hint of renormalization. Am. J. Phys., **72**(2) (2004)
39. P. Deligne, Quelques idées maîtresses de l'œuvre de A. Grothendieck, Matériaux pour l'histoire des mathématiques au XXe sicle, in *Proceedings of the workshop on the honour of Jean Dieudonné* (Nice 1996), France Mathematical society, pp. 11–19 (1998)
40. R. Dijkgraaf, *The mathematics of strings theory*, Séminaire Poincaré, 2004
41. P. A. M. Dirac, Quantised singularities in the electromagnetic field. Proc. Roy. Soc. A **133**, 60
42. P.A.M. Dirac, The Relation between Mathematics and Physics. Proc. R. Soc. (Edinburgh) 59(Part II), 122–129 (1939)
43. S. Donaldson, P. Kronheimer, *The Geometry of Four-Manifolds* (Oxford, 1990)
44. S. Donaldson, J. Diff. Geom. **18**, 269 (1983)
45. S. Donaldson, R. Friedman, Connected sums of self-dual manifolds and deformations of singular spaces. Nonlinearity **2**, 197–239 (1989)
46. R. Durrer, R. Maartens, Dark energy and dark gravity, Gen. Rel. Grav. **40**, 301–328 (2008) arXiv:0711.0077 (2007)
47. F.J. Dyson, The S-matrix in quantum electrodynamics. Phys. Rev. **75**, 1736. https://doi.org/10.1103/PhysRev.75.1736 (1949)
48. G. Efstathiou, in *The Physics of the Early Universe*, ed. by J.A. Peacock, A.F. Heavens, A. Davies (Adam-Higler, Bristol, 1990)
49. A. Einstein, Philosopher-Scientist, in *The Library of Living Philosophers*, ed. by P.A. Schilpp (Evanston, 1949), pp. 2–95
50. A. Einstein, Ideas and opinions, quoted from Schweber, *Einstein and Oppenheimer: the meaning of genius* (1954)
51. A. Floer, Bull. Am. Math. Soc. **16**, 279 (1987)
52. A. Friedman, Über die Krümmung des Raumes. Z. Phys. **10**(1), 377–386 (1922)
53. A. Friedman, Über die Möglichkeit einer Welt mit konstanter negativer Krümmung des Raumes. Z. Phys. **21**(1), 326–332 (1924)
54. D.S. Freed, G.W. Moore, Twisted equivariant matter. Ann. Henri Poincare **14**, 1927 (2013). arXiv:1208.5055 [hep-th]
55. F. Helein,*Dualités, supersymétries et systèmes complètement intégrables*, in Vers une nouvelle Philosophie de la nature, ed. by J. Kouneiher, Hermann edn. (2010)
56. F. Helein, J. Kouneiher, On the soliton-particle dualities, in *Geometries of Nature, Living Systems and Human Cognition*, ed. by L. Boi (World Scientific, 2005). pp. 93–128
57. K. Hepp, Proof of the Bogoliubov-Parasiuk theorem on renormalization. Commun. Math. Phys. **2**, 301–326 (1966)
58. D. Hilbert, Uber die Grundlagen der Geometrie. Gottinger Nachrichten **233–241** (1902)
59. D. Hilbert, *Lecture delivered before the International Congress of Mathematicians*, Paris France (1900). English translation appeared in Bull. Am. Math. Soc. **8** (1902), 437–479. A reprint of appears in *Mathematical Developments Arising from Hilbert Problems*, ed. by F. Brouder, Am. Math. Soc. 1976. The original address "*Mathematische Probleme*" appeared in Göttinger Nachrichten, 1900, pp. 253–297, and in Archiv der Mathematik und Physik, **3**(1), 44–63 and 213–237 (1901)
60. N. Hitchin, *Interaction between mathematics and physics*, ARBOR Ciencia, Pensamiento y Cultura, CLXXXIII 725, mayo-junio, pp. 427–432 (2007)
61. G. 't Hooft, M. Veltman, Regularization and renormalization of gauge fields. Nucl. Phys. B **44**, 189 (1972)

62. K. Hori, S. Katz, A. Klemm, R. Pandharipande, R. Thomas, C. Vafa, R. Vakil, E. Zaslow, Mirror symmetry. Clay Mathematics Monographs. Providence, RI: American Mathematical Society (2003)
63. E. Hubble, A relation between distance and Radial Velocity among Extra-Galactic Nebulae. Proc. Natl. Acad. Sci. **15**, 168–173 (1929)
64. V.F.R. Jones, A polynomial invariant for knots via von Neumann algebras. Bull. Am. Math. Soc. **12**, 103–111 (1985). https://doi.org/10.1090/S0273-0979-1985-15304-2
65. F. Klein, *Vergleichende Betrachtungen über neuere geometrische Forschungen*, in *Gesammelte mathematische Abhandlungen*, vol. i (1872) pp. 460–497
66. F. Klein, *Vorlesungen über die Entwicklung der Mathematik im 19* (Wissenschaftliche Buchgesellschaft, Jahrhundert. Darmstadt, 1979)
67. M. Kontsevich, Y. Soibelman, Stability structures, motivic Donaldson-Thomas invariants and cluster transformations, arXiv:0811.2435 [math.AG]
68. Kontsevich, M. Homological algebra of mirror symmetry. In Proc. Int. Congress of Mathematicians, Zrich, vols. 1-2 (Birkhuser, Basel, Switzerland, 1995). pp. 120–139
69. M. Kontsevich, M. Yu, Gromov-Witten classes, quantum cohomology, and enumerativegeometry. Commun. Math. Phys. **164**, 525–562 (1994). https://doi.org/10.1007/BF02101490
70. J. Kouneiher, in *Leibniz and the Dialogue between Sciences, Philosophy and Engineering, 1646-2016. New Historical and Epistemological Insights*, ed. by R. Pisano, M. Fichant, P. Bussotti, A.R.E. Oliveira (The College's Publications, London, 2017)
71. J.Kouneiher, *Conceptual Foundations of Soliton Versus Particle Dualities Toward a Topological Model for Matter. International Journal of Theoretical Physics*, vol. 55(6), pp. 2949–2968. 20p (2016)
72. J. Kouneiher, C. Barbachoux, Cartan's soldered spaces and conservation laws in physics. Int. J. Geom. Methods Mod. Phys. **12**(9) (2015)
73. J. Kouneiher, *Geometric Continuum and the Birth of the Mathematical-Physics*, to appear in IJGMMP, 2018
74. T. Krajewski, P. Martinetti, *Wilsonian renormalization, differential equations and Hopf algebras*, Talk given by T. Krajewski at the conference "Combinatorics and Physics" Max Planck Institut Für Mathematik Bonn, March 2007
75. D. Kreimer, On the Hopf algebra structure of perturbative quantum field theories. Adv. Theor. Math. Phys. **2**, 303 (1998). arXiv:q-alg/9707029
76. D. Kreimer, On the Hopf algebra structure of perturbative quantum field theories. Adv. Theor. Math. Phys. **2**, 303–334 (1998)
77. G. Lazardies, *Introduction to Cosmology*, arXiv:hep-ph/9904502
78. G. Lemaitre, Ann. Soc. Sci. Brux. **A53**, 81 (1933)
79. Colin MacLarty, How Grothendieck simplified algebraic geometry. Not. AMS **63**(3), 250 (2016)
80. N.S. Manton, *Skyrme fields and instantons, in "Geometry of Lowdimensional Manifolds : 1"*, ed. by S.K. Donaldson, C.B. Thomas, Lond. Math. Soc. Lec. Notes Ser. **150** (Cambridge University Press, Cambridge, 1990)
81. S. Mandelstam, Soliton operators for the quantized sine-Gordon equation. Phys. Rev. D **11**(10), 3026–30 (1975)
82. Y.I. Manin, M. Marcolli, Big Bang, blowup, and modular curves: algebraic geometry in cosmology. Symmetry Integrability Geom.: Methods Appl. SIGMA **10**, 73 (2014)
83. J.C. Maxwell, Mathematical and Physical Science, Section A (Mathematical and Physical Sciences) of the British Association, Liverpool, 1870. Nature **2**, 419–422 (1870)
84. T. Miwa, M. Jimbo, E. Date, *Solitons*, Cambridge Tracts in Math. 135 (Cambridge University Press, Cambridge, 2000)
85. C. Montonen, D. Olive, Magnetic monopoles as gauge particles, Phys. Lett. 72B(1) (1977)
86. G.W. Moore, *Physical Mathematics and the Future*, Preprint
87. J. von Neumann, Die Eindeutigkeit der Schrödingerschen Operatoren. Math. Ann. **104**, 570 (1931)

88. E.T. Newman, A fundamental solution to the CCC equations, Gen. Relativ. Gravit. **46**(5), 1717, 13p (2014). arXiv:1309.7271
89. H. Nicolai, K. Peeters, M. Zamaklar, *Loop Quantum Gravity: An Outside View*, 2014. arXiv:hep-th/0501114
90. D. Olive, E. Witten, *Supersymmetry algebra that include topological charges*, Phys. Lett. **78B**(1) (1978)
91. P.J.E., Peebles, The Large-scale Structure of the Universe (Princeton University Press, Princeton, 1980)
92. P.J.E. Peebles, D.N. Schramm, E.L. Turner, R.G. Kron, Nature **352**, 769 (1991)
93. Roger Penrose, *The Road to Reality: A Complete Guide to the Laws of the Universe* (A.A. Knopf, New York, 2005)
94. R. Penrose, Angular momentum: an approach to combinatorial space-time, in *Quantum Theory And Beyond*, ed. by T. Bastin (Cambridge University Press, Cambridge, 1971)
95. R. Penrose, W. Rindler, Spinors and space-time, in Two-Spinor Calculus and Relativitic Fields, vol. 1 (Cambridge University Press, Cambridge, 1984)
96. H. Poincaré, *Sur les rapport de l'analyse pure et de la physique mathématique*, Address to the 1897 ICM, Zurich
97. H. Poincaré, Analysis situs. J. Sec. Polyt. **1**, 1–121 (1895)
98. H. Poincaré, Sur la connexion des surfaces algébriques. C. R. Acad. Sc. **133**, 969–973 (1901)
99. H. Poincaré, Sur les cycles des surfaces algébriques; quatrième complement a l'Analysis situs. J. Math. Pures Appl. **8**, 169–214 (1902)
100. A. Pressley, G.B. Segal, *Loop Groups* (Oxford University Press, Oxford, 1986)
101. H.P. Robertson, On the foundations of relativistic cosmology. Proc. Natl. Acad. Sci. **15**(11), 822–829 (1929)
102. I. Robinson, Report to the Eddington Group, Cambridge, (1956)
103. I. Robinson, J. Math. Phys. **2**, 290 (1961)
104. C. Rovelli, *Quantum Gravity* (Cambridge University Press, Cambridge, 2010)
105. C. Rovelli, L. Smolin, Loop space representation of quantum general relativity. Nucl. Phys. B **331**, 80 (1990)
106. J. Rosenberg, A selective history of the Stone-von Neumann theorem', in Operator Algebras, Quantization, and Noncommutative Geometry: A Centennial Celebration Honoring John von Neumann and Marshall H. Stone, ed. by R.S. Doran, R.V. Kadison, *Contemporary Mathematics*, vol. 365 (American Mathematical Society, 2004)
107. Y. Ruan, G. Tian, A mathematical theory of quantum cohomology. J. Differ. Geom. **42**, 259–367 (1995). http://projecteuclid.org/euclid.jdg/1214457234
108. T. Sauer, *The Relativity of Discovery: Hilbert's First Note on the Foundations of Physics*. arXiv:physics/9811050
109. U. Majer, T. Sauer, *Hilbert's World Equations and His Vision of a Unified Science*. http://arxiv.org/abs/physics/0405110v1
110. S.S. Schweber, *Qed and the Men Who Made It: Dyson* (Schwinger, and Tomonaga, Princeton University Press, Feynman, 1994)
111. A. Sen, Dyon-monopole bound states, self-dual harmonic forms on the multi-monopole moduli space, and SL(2, Z) invariance in string theory. Phys. Lett. B **329**, 217–221 (1994)
112. N. Seiberg, E. Witten, Monopole condensation, and connement in N = 2 supersymmetric Yang-Mills theory. Nucl. Phys. B **426**, 19–52 (1994)
113. T.R.H. Skyrme, A Unified Theory for Mesons and Baryons. Nucl. Phys. **31**, 556 (1962); Proc. Roy. Soc. A **247**, 260 (1958)
114. T.R.H. Skyrme, Kinks and the Dirac equation. J. Math. Phys. **12**, 1735–42 (1971)
115. T.H.R. Skyrme, A non-linear field theory. Proc. Roy. Soc. **A260**, 127–138 (1961)
116. C. Rovelli, L. Smolin, Spin networks and quantum gravity. Phys. Rev. D **52**, 5743–5759 (1995)
117. H. Spiesberger, M. Spira, P.M. Zerwas. *The Standard Model: Physical Basis and Scattering Experiments*. Appears in: Scattering, ed. by R. Pike et al., vol. 2, (Academic Press, Cambridge, 2002). 1505–1533

118. M.H. Stone, Linear transformations in Hilbert space, III: operational methods and group theory. Proc. Nat. Acad. Sci. **16**, 172–175 (1930)
119. J.J. Sylvester, A plea for the mathematician, II. Nature **1**(10), 261–263 (1870)
120. W. Thirring, Ann. Phys. (N.Y.) **3**, 91 (1958)
121. P. Tod, Penrose's circles in the CMB and a test of inflation. Gen. Relativ. Gravitat. **44**, 2933–2938 (2012). arXiv:1107.1421
122. R. Vakil, *Foundations of Algebraic Geometry*. http://math.stanford.edu/~vakil/216blog/index.html
123. A.G. Walker, On Milne's theory of world-structure. Proc. Lond. Math. Soc., Ser. 2, **42**(1), 90–127 (1937)
124. H. Weyl, Reine Infinitesimalgeometrie, in *Weyl, Gesammelte Abhandlungen*, 4 vols., vol. II (Springer, Berlin, 1968). pp. 1-28, on p. 2
125. E. Witten, Supersymmetry and Morse theory. J. Differ. Geom. **17**, 661–692 (1982)
126. E. Witten, Quantum eld theory and the Jones polynomial. Commun. Math. Phys. **121**, 351–399 (1989)
127. E. Witten, Monopoles and four manifolds. Math. Res. Lett. **1**, 769–796 (1994)
128. E. Witten, *Quantum field theory and the Jones polynomial, Braid Group, Knot Group, and Statistical Mechanics*, ed. by C. N. Yang, M. L. Ke (World Scientific, 1989). pp. 239–329
129. E. Witten, Topological quantum field theory. Comm. Math. Phys. **117**(3), 353–386 (1988)
130. K.G. Wilson, The renormalization group: critical phenomena and the Kondo problem. Rev. Mod. Phys. **47**(4), 773 (1975)
131. C.N. Yang, Magnetic monopoles, fiber bundles, and gauge fields. Ann. NY Acad. Sci. **294**, 86–97 (1977)
132. C.N. Yang, R.L. Mills, Conservation of isotopic spin and isotopic gauge invariance. Phys. Rev. **96**, 191–195 (1954)
133. W. Zimmermann, Convergence of Bogoliubov's method of renormalization in momentum space'. Commun. Math. Phys. **15**, 208 (1968)
134. J. Zinn-Justin, *Quantum Field Theory and Critical Phenomena* (Oxford University Press, Oxford, 1999)

Mie's Electromagnetic Theory of Matter and the Background to Hilbert's Unified Foundations of Physics

Leo Corry

1 Introduction

On November 20, 1915, David Hilbert delivered a talk in Göttingen, presenting his new axiomatic derivation of the "basic equations of physics". This talk is often remembered because, allegedly, Hilbert presented in them, five days prior to Einstein, the correct, generally-covariant equations of gravitation that lie at the heart of the general theory of relativity (GTR).

The published version of Hilbert's talk opens with the following words [15, 395]:

> The tremendous problems formulated by Einstein, as well as the penetrating methods he devised for solving them, and the far-reaching and original conceptions by means of which Mie produced his electrodynamics, have opened new ways to the research of the foundations of physics. Hilbert [15]

Hilbert's 1915 talk in Göttingen had traditionally attracted the attention of historians interested in the work of Einstein and in the development of general relativity. A main issue of interest was the possible influences of Hilbert on Einstein's work and the question of priority concerning the formulation of the complete and explicit, fully-covariant equations of gravitation in the framework of the GTR. At the same time, much less attention was traditionally paid to the question of the place of Hilbert's talk, and his interest in GTR within the context of his overall scientific world and his own research programs.

In my book *David Hilbert and the Axiomatization of Physics (1898–1918): From Grundlagen der Geometrie to Grundlagen der Physik* [3], I summarized several years of my own historical research into the question of the place of physics in Hilbert's work and of the specific role of what he defined as "the axiomatic method" in mathematics and in the empirical sciences. As part of the broad and detailed historical picture that I aimed at presenting in the book, Hilbert's involvement with

L. Corry (✉)
Tel Aviv University, Tel Aviv, Israel
e-mail: corryleo@gmail.com; corry@post.tau.ac.il

J. Kouneiher (ed.), *Foundations of Mathematics and Physics One Century After Hilbert*,
https://doi.org/10.1007/978-3-319-64813-2_2

GTR is presented within the context of a long-standing involvement with current physical research, rather than as a sporadic incursion of a mathematician into the work of a physicist who was having trouble with a specific, technical issue in his theory.

One specific aspect of the contribution of my research was the uncovering of an important document that shed important light on the question of priority in the formulation of the field equations of GTR. This was the proof galleys of Hilbert's talk in the form that were sent to him on December 6, 1915, that is, about two weeks after the talk and several months before the talk was finally published in March of 1916. The galleys provide us with a closer look at the precise way in which Hilbert presented the equations in the talk. They make it clear that some crucial details were still missing when he originally delivered the talk. Later on, he had the opportunity to correct and complete these shortcomings in the published version. Most probably, he did so as anyone would correct deficiencies appearing in the proofs of article, after becoming aware of all the details of Einstein's published version of his own correct version [4]. The publication of the galleys gave raise to heated debates among historians interested in these questions, but for reasons of space I will not go into that debate in this article (see e.g., [29, 31]).

The purpose of this article is to discuss the two main pillars on which Hilbert built his own theory as presented in Göttingen on November of 1915. In doing so, my article repeats much of what appears in the relevant chapters of my book. Of course, a much clearer understanding of the general context of the story told here arises when these ideas are presented as part of the broader story told in my book, and the interested reader is referred to the book for further details.

The main sections of the article comprise the following: first, I discuss the contents of Mie's electromagnetic theory of matter; secondly I explain the context in which the theory needs to be understood as part of contemporary debates on gravitation in which also Einstein took part; thirdly, I explain the way in which Max Born mediated between Mie and Hilbert by presenting the former's work in a way that would be amenable to Hilbert's current scientific interests. Finally, I give a brief account of Hilbert's talk of November 1915 and explain its contents against the background of the ideas explained in the previous sections.

2 Gustav Mie's Electromagnetic Theory of Matter

Beginning in 1912, Gustav Mie (1868–1957) developed an original theory of matter which attempted to elaborate the main tenets of an "electromagnetic world-view electromagnetic view of nature" [19–21]. More specifically, it was aimed at developing the idea that the electron cannot be ascribed physical existence independently from the ether ether. Since the turn of the twentieth century, several physicist had promoted an electromagnetic world-view and had tried to work out detailed physical theories that would comply with that view. Prominent among them were Max Abraham (1875–1922), Walther Kaufmann (1871–1947) and Wilhelm Wien (1864–1928).

By 1910, however, the program had completely lost momentum, both for intrinsic and extrinsic reasons. The intrinsic reasons pertained to some technical problems that the program, and more specifically the electron theory electron theory, had not been able to fully solve. The extrinsic reasons pertained to the marginal position into which the program was pushed as physicists became increasingly interested in the new horizons opened up by relativity and by the early stages of quantum theory quantum theory.

Somewhat belatedly, Mie sought to relaunch the program and to formulate a theory of matter that would help achieve the ambitious goals of unification predicated by his predecessors. Unlike them, however, he did not reject relativity but rather the opposite: he took it as a main assumption for his theory. Indeed, Mie's theory became the most mathematically elaborate attempt put forward so far in order to achieve the desired electromagnetic unification. Mie had hoped that in the framework of his theory the existence of the electron with finite self-energy energy:self-energy could be derived from the field in purely mathematical terms. What is usually perceived as material particles, he thought, should appear as no more than singularities in the ether. Likewise, compact matter should be conceived as the accumulation of "clusters of world-lines ." Mechanics and electrodynamics electrodynamics would thus become the theory of the interaction of the field-lines inside and outside the cluster.

According to Coulomb's law, the field of a charged particle becomes infinite when its radius reduces to zero. Mie's equations generalized those of Maxwell's theory in such a way that the repulsive forces predicted inside the electron would be compensated by other forces of purely electrical nature. At the same time, the deviation of Mie's equations from Maxwell's becomes undetectable outside the electron. According to Mie, the recent development of quantum theory and the discoveries associated with it suggested the need to formulate some new equations to account for the phenomena that take place inside the atom. His theory was intended as a preliminary contribution in this direction. Together with an explanation of the existence of indivisible electrons in purely electromagnetic terms, Mie also sought to present the phenomenon of gravitation as a necessary consequence of his theory of matter. He intended to show that both the electric and the gravitational actions were direct manifestations of the forces that account for the very existence of matter.

Three explicitly formulated, basic assumptions were at the basis of this theory. The first one is that the electric and the magnetic fields are present inside the electron as well. This expresses the assumption that the electrons are an organic part of the ether ether, rather than foreign elements added to it, as was the common belief among certain physicists at the time. The electron is thus conceived as a non-sharply delimited, highly dense, nucleus in the ether that extends continually and infinitely into an atmosphere of electrical charge. An atom is a concentration of electrons, and the high intensity of the electric field around it is what should ultimately explain the phenomenon of gravitation. The second assumption is the universal validity of the principle of relativity (i.e., Lorentz covariance). The third one is that all phenomena affecting the material world can be fully characterized using the physical magnitudes commonly associated with the ether: the electric field $\boldsymbol{d}$, the magnetic field $\boldsymbol{h}$, the electric charge density ρ, and the charge current $\boldsymbol{j}$. While for Mie the validity of the principle of relativity was beyond any doubt, he considered his third assumption to be in need of further validation. Without stating it explicitly, Mie also assumed as obvious the validity of the energy conservation principle.

An additional constitutive element of Mie's theory is the separation of physical magnitudes into "quantity magnitudes" and "intensity magnitudes". This separation, which essentially can be traced back at least to Maxwell [34], appears as a central theme in Mie's conception of physics throughout his career, beginning with the first edition of his textbook on electricity and magnetism [18]. "Quantity magnitudes" may be measured by the successive addition of certain given units of the same kind: length, time duration, etc. Measuring "intensity magnitudes", on the contrary, is not accomplished by establishing a unit of measurement. Rather, one needs to establish a specific procedure according to which any given measurement of that magnitude can be attained. The foremost example of an intensity magnitude comes from the basic concept of mechanics: force. In the theory of elasticity the tension is an intensity quantity and the deformation is a quantity magnitude; in kinetic theory the corresponding pair would be pressure and volume [17].

This separation gives a certain coherence and symmetry to Mie's treatment of the electromagnetic theory of matter, but it does not really alter its actual physical content. The magnitudes mentioned in the third basic assumption of the theory, $\boldsymbol{h}$, $\boldsymbol{d}$, ρ, $\boldsymbol{j}$, are four quantity magnitudes. Against them Mie introduced four intensity magnitudes: the magnetic induction, $\boldsymbol{b}$, and the intensity of the electric field, $\boldsymbol{e}$, and two additional ones, φ, and $\boldsymbol{f}$. Mie did not assign any direct physical meaning to the latter two, and he simply stated that the four-vector $(\boldsymbol{f}, i\varphi)$ is in the same relation to $(\boldsymbol{j}, i\rho)$ as the six-vector $(\boldsymbol{b}, -i\boldsymbol{e})$ is to $(\boldsymbol{h}, -i\boldsymbol{d})$. The introduction of these four intensity magnitudes allowed Mie to present an alternative formulation of the third assumption, namely, that all physical phenomena can be described in terms of the ten values involved in the four intensity magnitudes $\boldsymbol{b}$, $\boldsymbol{e}$, φ, and $\boldsymbol{f}$. Mie thus formulated the Maxwell equations as follows:

$$\text{rot}\, h = \frac{\partial d}{\partial t} + j$$

$$div\boldsymbol{d} = \rho$$

$$\text{rot}\, e = -\frac{\partial b}{\partial t}$$

$$div\boldsymbol{b} = 0$$

The language of four- and six-vectors in which Mie couched his theory had originally been introduced by Hermann Minkowski (1864–1909) in his work on electrodynamics [25], and it had later been elaborated into the standard one for relativity theory by Arnold Sommerfeld (1868–1951) [32, 33]. In Mie's formulation, a possible connection with a tensorial theory of gravitation such as Einstein's was not particularly perspicuous; it only became so, after Born reformulated Mie's theory in 1913 with a more suggestive notation, as explained below.

From the issues discussed in Mie's theory of matter, two are especially relevant for discussing Hilbert's later work: the energy principle and gravitation. In Mie's theory, the concept of energy is formulated in terms of a scalar function W, the energy density, which, as a consequence of the Maxwell equations must satisfy the field equation

$$\frac{\partial W}{\partial t} = -\text{div}S,$$

$\boldsymbol{S}$ being the energy current vector. The energy conservation principle demands that dW be an exact differential, and Mie showed that this demand is fulfilled whenever W can be expressed in terms of the four parameters $\boldsymbol{d}, \boldsymbol{h}, \rho$ *and* $\boldsymbol{j}$. Moreover, this function can be determined in terms of a second scalar function H, of the same parameters, which must satisfy the equation

$$W = H + \boldsymbol{h}.\boldsymbol{b} + \boldsymbol{j}.\boldsymbol{f}.$$

Mie investigated several aspects of his theory of gravitation, such as the relations between the equations and the energy principle, the invariants that appear in the theory, the principle of action and reaction, and the relation between gravitational and inertial mass. A central point in this discussion was the status of the gravitational potential ω. Since the latter appears in the theory among the basic dynamic variables, it follows that the absolute value of the potential —rather than only potential differences— directly influences physical phenomena. Still, for regions of constant potential, the form of the equations guarantee that its effects can be fully taken into account by suitable rescaling all other dynamic variables. Thus, the effect of a constant gravitational potential could be made to become imperceptible for any given observer. The possibility of doing this is what Mie called "the principle of the relativity of the gravitational potential," which he explicitly formulated as follows [21, 63]:

> If two empty spaces differ from each other only in the fact that in the first one the average value ω_0 of the gravitational potential is very large while in the second one it is zero, then this difference has no influence whatsoever on the size and form of the electrons and of the other material particles, on their charge, on their laws of oscillation, and on other motion laws, on the speed of light, and in general on any physical relations and processes.[1]

The validity of this principle summarized for Mie the differences between his and other, contemporary theories of gravitation, especially those of Max Abraham and of Einstein.

Mie expressed the belief that his brief discussion was enough to prove that the basic assumptions of the theory led to no contradiction with experience, even in the case of gravitational phenomena. Preparing the way for a possible empirical confirmation of the law of gravitation, said Mie, was one of the main aims of his

[1]Unless otherwise stated, all translations from German are mine.

article, but he admitted that, at this stage, the results of his research did not really help at that. Two results derived from his theory, which in principle might be thought of as offering that possibility could not as a practical matter do so. The first was the relation obtained in the theory between inertial and gravitational mass. The two are identical, according to Mie, only if there are no motions inside the particle, and in general they are in a relation that depends on the temperature and on the atomic weight. The observable differences between the gravitational acceleration of two bodies of different masses would be, according to this account, of the order between 10^{-11} and 10^{-12} and therefore they would be of no help in constructing an actual experiment. The second result concerned the existence of longitudinal waves in the ether, also too small to be detectable by experiment [21, p. 64]).

Mie's theory contained several difficulties that he was never able to work out successfully, yet he never really abandoned his belief in its validity. The most serious shortcoming of the theory is connected with the fact that it depends on an absolute gravitational potential, and therefore the equations do not remain invariant when we replace the potential w by a second potential $w+const$. Under these conditions, a material particle will not be able to exist in a constant external potential field. Moreover, in retrospective it is also clear that Mie's theory did not account for either red shift or light bending, but these issues did not really become crucial until much later.[2]

3 Contemporary Debates on Gravitation

Mie published his electromagnetic theory of matter at a critical time from the point of view of the development of a relativistic theory of gravitation by Einstein and by others. Soon after the early formulation of Einstein's 1905 relativity article, several physicists were involved in attempts to develop relativistic theories for the various domains of physics: mechanics of discrete systems, thermodynamics, statistical mechanics, hydrodynamics, elasticity, and others. In most cases such attempts quickly led to satisfactory results. Considerable difficulties appeared, however, when they were directed towards the relativization of gravitational theories.[3] On the other hand, gravitation was perhaps the domain in which a relativistic treatment seemed more urgent, if only for the fact that Newtonian theory was based on a concept of force that is dependent on the distance between two bodies at a given point in time, and that acts instantly between them. By undermining the classical conception of simultaneity, Einstein's new theory of relativity posed a serious challenge to this fundamental aspect of Newtonian theory.

---

[2] For an historical account of light bending as a test for relativity, see [5]. For a parallel account of red shift, see [6].

[3] Readers interested in an updated, thorough account of the research on the history of general relativity, including the topics discussed here, should consult the collection [27].

Minkowski in 1907–1909 had been involved in separate efforts to formulate a Lorentz-covariant theory of gravitation. So was in 1908 Henri Poincare (1854–1912). Einstein soon came to deal with this question as well, and as early as 1907 he convinced himself that a relativistic treatment of gravitation would necessitate a broader kind of invariance. A major turning point in this context was marked by Einstein's arrival in Zurich in August 1912, to take a post at the ETH, A former fellow of his student days, Marcel Grossmann (1878–1936), was now professor of mathematics there. Together they started working on the mathematical and physical aspects of the problem and published a joint paper, the so-called *Entwurf* paper, containing the first serious, articulate effort to formulate a "generalized theory of relativity and a theory of gravitation" [12]. Following its publication in the summer of 1913, Einstein had ambivalent feelings concerning the value of this theory, but more often than not he was enthusiastic about the achievement. Over the next two years he worked almost exclusively, and with tremendous effort, on the details of the theory and its consequences, and gradually came to realize the essential difficulties involved in it. At the center of this whole effort lay the successful formulation of generally-covariant field equations gravitational field equations of the theory of gravitation. A satisfactory solution was not reached until November 1915, when Einstein presented to the Berlin Academy a famous series of four papers containing his definite formulation of these equations.

In the background to Einstein's efforts for generalizing relativity, one finds several important ideas interacting with each other. Of paramount importance was the "equivalence principle" between a gravitational field and an accelerating reference frame, an equivalence from which also the well-known, observed equality of inertial and gravitational mass could be derived. Einstein later described the formulation of this principle as "the most fortunate thought of my life" and he adopted it as a primary guideline for his research into this question as early as 1907. Unlike in Hilbert's axiomatic approach, Einstein's view of the role of such principles in physical theories was not as basic elements of a closed deductive system, but rather as open, heuristic guiding ideas for developing the theories in question. The equivalence principle arose within a very specific setting, namely, the attempt to modify Newton's theory of gravitation so as to make it fit the recently formulated theory of relativity. Within this specific setting, the heuristic value of the equivalence principle lay in allowing the replacement of a homogeneous, static gravitational field by a uniformly and linearly accelerated reference system. In the latter kind of system, Einstein believed, it would be much easier to develop the necessary theoretical treatment in terms of a generalization of the principle of relativity [7]. Still, it was only following his reencounter with Grossmann that Einstein came to learn the mathematical approach and tools needed to accomplish this task.

Einstein's collaboration with Grossman continued over the next two years. Over this time he was involved in debates and collaboration with other colleagues as well. One of them was the Finnish Gunnar Nordström. This physicist had suggested a theory of gravitation which, like Mie's, was also a scalar one. It was much simpler than Einstein's *Entwurf* theory, and Einstein acknowledged some its advantages. Nevertheless, he was not willing to accept its a-priori admittance of Euclidean geometry.

Another theory discussed at the time was that of Max Abraham. Abraham in 1912 had been involved in a caustic debate where he harshly attacked Einstein's early attempts to formulate a relativistic theory of gravitation, and the latter had a hard time finding the right arguments to defend himself. Still, Abraham was one of the few physicists whose opinion concerning his own theory, Einstein really valued [2].

In a famous talk of 1913 in Vienna, Einstein discussed the current state of his own research on gravitation, as well as that of some of his colleagues [8]. Einstein formulated four principles that, in his view, any general relativistic theory of gravitation gravitation should fulfil:

1. The principles of conservation of energy and of momentum are valid.
2. In closed systems, the inertial mass equals the gravitational mass.
3. The theory of relativity is valid in a restricted sense, i.e., the system of equations is invariant under generalized Lorentz transformations.
4. Laws that describe observable, natural phenomena do not depend on the absolute value of the gravitational potentials.

Einstein declared that, among the existing attempts to deal with gravitation theory, he favored Nordströ's theory most, because it complied with the above-mentioned physical principles. On the other hand, he did not even mention Mie's theory. In the discussion that followed the talk, and answering a question of Mie, Einstein explained that unlike the other theories, Mie's theory did not satisfy the principle of equivalence and therefore he had not really studied it in detail. More privately, in a letter written to his astronomer friend Erwin Finlay Freundlich (1885–1964) that same year, Einstein confided that Mie's theory was "fanciful and possesses, in my opinion, a vanishingly small intrinsic probability" of being right [13, Doc. 468].

Another remark that is worth mentioning here came from the Göttingen experimental physicist Eduard Riecke (1845–1915). Like most of his physicist colleagues in Göttingen, and contrary to the mathematicians of the same university, Riecke had never been really interested in the new horizons opened by Einstein's theory. In this occasion, he asked Einstein if the connection between the electromagnetic and the gravitational field was somehow explained by his theory. Einstein replied that, according to the theory, a mutual effect between both fields does exist, "but it seems futile to try to prove it experimentally." Only the bending of light rays by the gravitational field, Einstein said, may fall within the range of observable phenomena [14, Doc. 225]. Since the connection between the two fields will be central to Hilbert's unified theory one wonders whether Riecke and Hilbert had the opportunity to discuss this matter at Göttingen at this time or later on. There is no direct evidence indicating that they did so.

In December 1913, Mie wrote a detailed criticism of Einstein's theory. Among other things, he claimed that the relevant perspective from which to consider the invariance of Einstein's theory was that afforded by the principle of relativity of the gravitational potentials, rather than that of a generalized principle of the relativity of motion. Moreover, Mie stressed the difficulties implied by a tensorial theory over a scalar one, difficulties he considered not to be justified by any evident advantages of the former approach [22, 169–72]. Above all, Mie criticized the limited covariance of the *Entwurf* theory. Einstein replied that Mie had not understood him, but did not provide an actual rebuttal of his arguments [9].

In his early articles Mie was very explicit in stating that an explanation of gravitation would be an important by-product of his theory, but such an explanation was never his main aim. Still, Mie continued to lecture and publish on gravitation over the years [23] and, in fact, to relate to certain aspects of Einstein's work with a somewhat critical attitude. In a letter written to Hilbert on February 13, 1916, shortly after the latter's presentation of his unified theory in Göttingen, Mie still referred back to the discussions held in the 1913 Vienna meeting. He manifested his general skepticism towards the idea of a "general relativity", but at the same time he confessed that Hilbert's own ideas helped him realize that, after all, Einstein had perhaps been very close to the truth from the beginning. Still, Mie did not believe that Einstein would attain what he had announced as the aim of his research.[4] In 1921 Mie published a short monograph on Einstein's theory, where he admitted that the current development of the theory was satisfactory from his point of view. Concerning the validity of the postulate of invariance under arbitrary coordinate transformations coordinate systems, he wrote [23, 61]:

> I think that many of my [non-mathematical] readers will be astonished that it might be possible at all to satisfy that postulate. In fact, I believe that many professionals will have to concede that at the time when Einstein was still looking for the correct way to apply it, they doubted that he would possibly succeed. The author of this essay must confess that he himself belonged to these skeptics. It took Einstein many years until the problem had attained the clarity that led to its solution. Finally, however, he found the way to rely on the geometrical research of several mathematicians, and especially of the genial Riemann, that had worked out the most general geometries of many-dimensional continua. Einstein filled up the formerly pure mathematical thoughts of these researchers with physical contents and thus finally obtained his theory.

Mie's theory of matter, then, and his attempt to explain gravitation in electromagnetic terms, had a rather convoluted and unfortunate development. Still, it succeeded in attracting the attention of Hilbert from very early on. In fact, the sequence of events that led to Hilbert's foundational, unified physical theory started with his interest in Mie's theory as a viable theory of matter. It was only later when he sought to combine this theory with Einstein's quest for general covariance, that he was led to develop the theory that he considered to provide "The Foundations of Physics" in general. But Hilbert's encounter with Mie's theory came in a somewhat roundabout way, meditated by Max Born, who reformulated it in terms that would fit Hilbert's involvement with physics, and particularly with questions related with the structure of matter, at that time.

---

[4]Letter from Mie to Hilbert, February 13, 1916. The letter is preserved in the Hilbert Nachlass, NSUB Göttingen, Cod Ms David Hilbert 254/2.

4 Born's Formulation of Mie's Theory

Max Born (1882–1970) was the first among the Göttingen scientists to become interested in Mie's theory and to dedicate actual efforts to study and develop it. In fact, it was only through Born's reformulation of the theory, and perhaps through his personal mediation, that Hilbert got to adopt it as one of the central pillars of the unified foundation of physics that he was about to develop over the following years. Mie's theory connected naturally with Born's scientific concerns. After being a student in Göttingen between 1904 and 1907, Born had returned there in 1908 intent on working with Minkowski on relativity and on electron theory. Ever since his return, and particularly after Minkowski untimely death in 1909, Born attained scientific prominence. By that time he also became a close collaborator in Hilbert's inner circle electron theory. On December 17, 1912, Born presented for the first time Mie's theory at Göttingen at the meeting of the local Mathematical Society (GMG).[5] At the time, Hilbert was deeply immersed in research on kinetic theory and on radiation theory. The lecture notes of the courses Hilbert taught in the winter semester of 1912–13 ("Molecular Theory of Matter") and in the following semester ("Electron Theory") in spite of their obvious, direct connection with the issues addressed by Mie, show no evidence of a sudden interest in his theory or in the point of view developed in it. This may have been connected to the fact that Mie's strong electromagnetic reductionism was contrary to Hilbert's views at the time, which also favored reductionism, but still from a mechanistic perspective. Born, on the contrary, seems to have been immediately attracted to Mie's theory, since he continued to work on it by himself. Nearly one year later, on November 25, 1913, Born lectured again on Mie's theory at the meeting of the GMG.[6] On December 16, he presented to the same forum his own contribution to the theory, dealing with the form of the energy laws in it [1]. This time, he does seem to have caught Hilbert's attention.

Born was strongly influenced by Hilbert's views on physics, at least in what concerns the way physical theories have to be treated. Born stressed above all the role of the variational argument underlying Mie's theory, as well as the similarity of the latter with the classical, analytical approach to mechanics. Born's formulation was more general than Mie's, and his presentation was tensorial in spirit, although he did not explicitly use this word. Rather than speaking of the electromagnetic ether and its properties, as Mie had done, Born referred to a general four-dimensional continuum of the coordinates x_1, x_2, x_3, x_4, and to the deformations affecting it. The latter are expressed in terms of the projections u_1, u_2, u_3, u_4 (on a system of four orthogonal axes) of the displacements (*Verrückungen*) of the points of the continuum. The four basic electromagnetic magnitudes referred to by Mie, $\boldsymbol{h}$, d, $\boldsymbol{j}$ and ρ, appear in Born's article as no more than specific functions of the four coordinates. Born discussed the energy conservation principle in these terms and in doing so, he prepared the way

[5]See the announcement in *Jahresbericht der DMV* 22 (1913), 27. We have no direct evidence of the contents of Born's lecture at this time.

[6]See the announcement in *Jahresbericht der DMV* 22 (1913), 207.

for allowing the connection that Hilbert would eventually create between this theory and Einstein's GTR.

One point raised by Born right in his opening sentence could not have failed to attract Hilbert's attention: whereas Lorentz's theory of the electron was based on certain hypotheses concerning the nature of matter (e.g., the rigidity of the electron)—Born asserted—Mie attempted to derive mathematically the existence of electrons, and hence of atoms and matter in general, from a modified version of the Maxwell equations. In other words Maxwell equations, the basic properties of matter could be derived without having to start from any particular conception about the nature of physical phenomena.

Born explained the central ideas of Mie's theory by analogy with Lagrangian mechanics. The equations of motion of a mass system, he said, can be derived using the Hamiltonian principle, by stipulating that the integral

$$\int_{t_2}^{t_1} (T - U)\, dt$$

has to attain a minimal value. Here T - U is the Lagrangian Lagrangian function function, which is a function of the position q and of the velocity $\dot{q}$ of the system:

$$T - U = \Phi(\dot{q}, q)$$

The equations obtained from the variational principle are thus well-known:

$$\frac{d}{dt}\frac{\partial \Phi}{\partial \dot{q}} - \frac{\partial \Phi}{\partial q} = 0. \qquad (++)$$

In mechanics, Born explained, one has the relatively simple case of a quasi-elastic system, in which the function Φ has the form $\Phi = \frac{a}{2}\dot{q}^2 + \frac{b}{2}q^2$. One can also have, however, a more general case in which Φ is taken to be any arbitrary function satisfying the basic differential equation (++). Born saw the relation of Mie's theory to classical electrodynamics electrodynamics as parallel to that between these two possibilities in mechanics. Mie had shown how to derive the equations of electrodynamics from a variational principle similar to the Hamiltonian one, using only four functions of four variables and taking as Φ a well-determined quadratic form of the field magnitudes, which satisfies a differential equation analogous to (eq. 6.3). Born thus concluded [1, 24–25]:

> Mie's equations play the same role for electrodynamics that Lagrange's equation second-order equations do for the mechanics of systems of points: they provide a formal scheme that, through a suitable choice of the function Φ, can be made to fit the special properties of the given system. Very much as in earlier times the aim of the mechanistic explanation of nature was pursued by assuming a Lagrangian function Φ that describes the interactions among atoms, and from which all physical and chemical properties of matter could be derived, so has Mie set forward the task of choosing a specific "world-function " Φ, in such a way that, starting from that function and from the basic differential equation it satisfies, not only the very existence of the electrons and of the atoms might be derived, but also the totality of

> their interactions will emerge. I would like to consider this requirement of Mie as embodying the mathematical contents of that program that has set down as the main task of physics the erection of an "electromagnetic world-view."

Born was alluding here to several issues that were highly appealing to Hilbert's sensibilities. Firstly, the analogous conception of mechanics and electrodynamics in terms of a variational derivation. At least since he attended in 1905 Hilbert's lectures on the axiomatization of physics, Born had repeatedly heard the master's quest for pursuing unification of physical theories along these lines: the crucial step in all cases would be the choice of the suitable Lagrangian function, and the axioms of the theory would provide the constraints for choosing the adequate Lagrangian. Like Minkowski and like Hilbert, but unlike many other physicists, Born called this Lagrangian "world-function". Secondly, Born knew that Hilbert's preference for mechanical reductionism was subsidiary to the idea of mathematical simplicity. If it turned out that electromagnetic reductionism would be simpler in mathematical terms, then Hilbert would be inclined to adopt it. Finally, and connected to the latter, the last sentence of the quotation seems to allude to the famous concluding passage of Minkowski's talk "Space and Time" [24]. Born suggested that a consistent pursuit of the line of thought adopted by Hilbert and Minkowski—in which the mathematical and logical structure of the theory matters above all and in which any commitment to specific physical underlying assumptions should be avoided as much as possible—should naturally lead to closer attention to Mie's theory.

In the body of his treatment, and according to the tensor-like spirit of the presentation, Born introduced the notation

$$\frac{\partial u_\alpha}{\partial x_\beta} = a_{\alpha\beta},$$

and demanded that all the properties of the continuum might be deduced alone from the projections of the displacements u_α and their derivatives $a_{\alpha\beta}$. In this way, the variational principle is applied to an integral of the form

$$\int \Phi(a_{11}, a_{12}, a_{13}, a_{14}; a_{22}, ..., a_{44}; u_1, ...u_4) dx_1 dx_2 dx_3 dx_4.$$

If, in addition, one introduces the notation

$$\frac{\partial \Phi}{\partial a_{\alpha\beta}} = X_{\alpha\beta}, \qquad \frac{\partial \Phi}{\partial u_\alpha} = X_\alpha$$

then the variational principle leads to equilibrium equations that can be expressed as

$$\sum_\gamma \frac{\partial X_{\beta\gamma}}{\partial x_\gamma} - X_\beta = 0.$$

Born characterized Mie's theory as a particular application of the general variational principle, in which Φ is taken to depend on the magnitudes $a_{\alpha\beta}$ exclusively through the differences

$$a_{\alpha\beta} - a_{\beta\alpha} = \frac{\partial u_\alpha}{\partial x_\beta} - \frac{\partial u_\beta}{\partial x_\alpha}. \qquad (*)$$

These differences can be interpreted as the components of the infinitesimal rotation of a volume element of the continuum in the four-dimensional world. Born showed that in Mie's theory, these components appear as the coordinates of the six-vector vector $(\boldsymbol{b}, -i\boldsymbol{e})$, where $\boldsymbol{b}$ represents the magnetic induction and $\boldsymbol{e}$ the intensity of the electric field. The values of the rotational components are obtained from the determinant

$$(a_{\alpha\beta} - a_{\beta\alpha}) = \begin{vmatrix} 0 & -M_z & M_y & ie_x \\ M_z & 0 & -M_x & ie_y \\ -M_y & M_x & 0 & ie_z \\ -ie_x & -ie_y & -ie_z & 0 \end{vmatrix}$$

If Φ does not depend explicitly on the four coordinates x_1, x_2, x_3, x_4 then the energy conservation principle is valid in the theory and it can be reformulated as follows[7]:

$$\frac{\partial \Phi}{\partial x_\alpha} = \sum_\gamma \frac{\partial}{\partial x_\gamma}\left(\sum_\beta X_{\beta\gamma} a_{\beta\alpha}\right).$$

If one defines a 4×4 matrix T

$$T_{\alpha\beta} = \Phi\delta_{\alpha\beta} - \sum_\gamma a_{\gamma\alpha} X_{\gamma\beta}$$

then the principle takes the form

$$DivT = 0.$$

This general result can be specialized to the case of Mie's theory, given that its Lagrangian is assumed to be independent of the four coordinates x_i. This assumption, Born stated, "is the true mathematical reason for the validity of the energy momentum conservation principle" in the theory [1, 32]. On the other hand, Born also relied on the dependence of the Lagrangian function on the $a_{\alpha\beta}$ via the expressions (*) above. He thus defined a new 4×4 matrix S, $S = T + \omega$, where

$$\omega_{\alpha\beta} = \sum_\gamma a_{\alpha\gamma} X_{\gamma\beta} - u_\alpha X_\beta.$$

[7] Sauer [30, 553] points out that "this assumption distinguishes Mie's theory from the usual Maxwell theory with charges and currents as external sources are given by the usual Lorentz electron theory. This theory can formally be included into the general framework by letting Φ depend on external sources, however, then Φ would explicitly depend on the space-time variables."

Born showed easily now that $Div\omega = 0$, from which he obtained, finally,

$$DivS = 0.$$

In Hilbert's November 1915 talk this matrix S is alluded to as "Mie's stress-energy tensor", and it plays a central role in the theory. In defining it, Born was introducing a magnitude which is not dependent only on the field's strength, yet satisfies the energy equation. Remarkably, in the body of Born's article gravitation is barely mentioned, thus suggesting his awareness of the problematic status of this phenomenon in the framework of Mie's theory. Born declared that the theory, in the variational formulation he was proposing here, was an extension of Lagrange's "magnificent program": the theory attempts to find the appropriate world-function from which all the electromagnetic properties of the electrons and the atoms might be derived. All properties, that is except gravitation, which, as Born explained in a significant footnote added at this point, was left outside the scope of the article.

It is likely that Born had discussed these ideas with Hilbert way before the actual lecture was delivered at the GMG. On October 22, 1913, Mie wrote a letter to Hilbert expressing his satisfaction for the interest that the latter had manifested (in an earlier letter, which has not been preserved,) in his recent work.[8] Thus, it was probably not necessary for Born, at this stage, to phrase his introduction with the specific task in mind of convincing Hilbert of the importance of Mie's theory and of the power of its concomitant electromagnetic world-view. But it seems clear that under the formulation embodied in Born's presentation and for the reasons alluded in his introduction, Hilbert himself could not have failed to recognize the direct allure of Mie's theory to his own current concerns. Still, some time was needed until Hilbert came to adopt fully the view of physical reality presupposed by Mie's theory. In his lectures on electromagnetic oscillations, during the winter semester of 1913-14, we find clear indications that Hilbert had begun to think seriously about this theory, but until his talk of November 1915 on the fundamental equations of physics he never mentioned Mie's theory explicitly either in his published work or in the manuscript of the lectures that have been preserved.

5 Hilbert's Communication and Mie's Theory

We do not know with certainty when Hilbert finally adopted Mie's theory as a possible basis for a unified foundation of physics in general, but we do know that Born was instrumental in the process leading to it. The second pillar of Hilbert's theory was provided by Einstein's work on general relativity, about which Hilbert had at least some idea by the end of 1913. Einstein was invited to discuss the current state of his theory in Göttingen and he visited there between June 29 and July 7, 1915.

[8] Mie's letter is in Hilbert's *Nachlass*, Staats- und Universitäts Bibliothek, Göttingen Göttingen - Cod Ms David Hilbert 254/1.

Unfortunately, the exact content of his lectures in Göttingen is unknown to us,[9] and yet it is clear that he considered his visit to have been a complete success. He felt that his theory had been understood to the details and he was deeply impressed by Hilbert's personality [11, Doc. 96]. Hilbert, in turn, was likewise impressed by the younger Einstein [26, p. 193].

Einstein's trip to Göttingen came after more than two years of intense struggle with the attempt to formulate a generalized theory of relativity. He had initially abandoned the demand of general covariance as part of his theory, after coming to the conclusion that generally covariant field equations would necessarily lack any physical interest, because they would contradict the principle of causality. The ground for this conclusion was the so-called "hole argument", which he introduced first in the *Entwurf* paper of 1913, and later articulated most clearly in a summary of the latter, presented in October 1914 to the Berlin Academy of Sciences [9]. Quite certainly, Einstein's lectures in Göttingen did not depart significantly from what he had presented in this summary.

Einstein's quest for a relativistic theory of gravitation was eventually crowned with success only after he abandoned completely the 'hole argument', and adopted general covariance again as a leading principle of that theory. Einstein's confidence on the validity of the argument, however, did not begin to erode until mid-October 1915. He thus embarked in the effort that led him to present four consecutive papers at the weekly meetings of the Berlin Academy, starting on November 4. The fourth paper, presented on November 25, contained his final version of the generally covariant field equations of gravitation. Over this crucial month of November, Einstein and Hilbert engaged in an intensive correspondence in which they reported to each other, in "real time", about their current progress in developing their respective results. They also continued to correspond with each other after presenting their respective works. A detailed analysis of the interchange of ideas between Einstein and Hilbert, and of their possible mutual influence is, of course, an enormously interesting and important topic but for reasons of space I will not address it here. In this section I focus on the connection between Mie's and Hilbert's theory.

As already mentioned, Hilbert's communication appeared in print on March 1916 in the proceedings of Göttingen Academy of Science under the title: "The Foundations of Physics" [15]. This printed version, however, differs substantially from what he actually presented in his talk, as we learn from the galley proofs that I uncovered in 1994 in Hilbert's *Nachlass* and which sheds much light upon this entire story. The proof galleys are dated December 6, 1915. The most significant differences between the two versions are not marked on the proofs themselves, and they were probably introduced somewhat later, that is, after December 6. It is worth noticing, moreover, that Hilbert's article was republished again in 1924 in the *Mathematische Annalen* with some additional, interesting changes, and yet once again with additional

[9] Actually, I made some efforts to gather documents related to this visit, unfortunately without much success. Nevertheless, I did find in Hilbert's *Nachlass* in Göttingen the handwritten notes taken by an unidentified person at the first of Einstein's lectures (Staats- und Universitäts Bibliothek, Göttingen, Cod Ms D Hilbert 724). These notes have now been published in [10, 586–90].

editorial comments in 1932, in the third volume of Hilbert's collected works. Typically, Hilbert did never mention any of the major changes he introduced between the various versions. In [15, 1], for instance, Hilbert explained that he was basically reprinting what had appeared in the past in two parts, with only minor editorial changes. Again, for lack of space the detailed analysis of these interesting changes is beyond the scope of this article.

Hilbert's theory took from Einstein the account of the structure of spacetime in terms of the metric tensor. Mie's theory served as a basis for explaining the structure of matter in terms of the electromagnetic fields. To these two elements Hilbert applied powerful mathematical tools taken from the calculus of variations and from Riemannian geometry.

The first axiom of Hilbert's theory of gravitation (which he called: "Axiom I: Mie's axiom of the world-function") is based on a variational argument. The axiom is formulated for a scalar Hamiltonian function[10] $H(g_{\mu\nu}, g_{\mu\nu l}, g_{\mu\nu lk}, q_s, q_{sl})$, whose parameters are the ten gravitational potentials $g_{\mu\nu}$, together with their first and second derivatives

$$g_{\mu\nu l} = \frac{\partial g_{\mu\nu}}{\partial \omega_l}, \, g_{\mu\nu lk} = \frac{\partial^2 g_{\mu\nu}}{\partial \omega_l \partial \omega_k} (l,k = 1, 2, 3, 4)$$

and the four electromagnetic potentials q_s, together with their first derivatives q_{sl}. The gravitational potentials $g_{\mu\nu}$ are the components of a symmetric tensor and, like in Einstein's theory, they constitute the metric tensor of a four-dimensional manifold. The electromagnetic potentials behave like vectors with respect to the four world-parameters ω_l $(l = 1, 2, 3, 4)$. The Hamiltonian is used to derive the basic equations of the theory, starting from the assumption that, under infinitesimal variations of its parameters, the variation of the integral

$$\int H\sqrt{g}d\omega$$

(where $g = |g_{\mu\nu}|$, and $d\omega = d\omega_1 d\omega_2 d\omega_3 d\omega_4$) vanishes for any of the potentials. In fact, instead of the covariant magnitudes $g_{\mu\nu}$ and their derivatives, Hilbert used consistently the contravariant tensor $g^{\mu\nu}$ and their derivatives throughout the argument. The second basic axiom of the theory (Axiom II: axiom of general invariance) postulates that H is invariant under arbitrary transformations of the coordinates ω_l.

Hilbert added two interesting footnotes that explain the relation of the two first axioms with Mie's and with Einstein's works respectively. First, he said, Mie himself had not included the electromagnetic potentials and their derivatives in the world-function, but rather this had been a contribution of Born. Thus, from this point on, it is clear that Hilbert will be referring to Born's version of the theory, rather than to

---

[10] In present-day terms, this function would be more properly called a Lagrangian function, while the term "Hamiltonian" usually refers to functions involving momenta and representing the total energy of the system considered. For the purposes of the present article and for the sake of historical precision, however, it seems more convenient to abide by the original terminology.

Mie's original one. Still, what characterizes Mie's theory, Hilbert explained, is the very introduction of the world-function as part of the Hamiltonian. Second, Mie had postulated the demand of orthogonal, rather than general covariance. But whereas in Einstein's work the Hamiltonian principle plays only a secondary role, Axiom II expresses in the simplest way his demand for general covariance.

Besides the two basic axioms, the core of Hilbert's derivation is based on a central mathematical result (Theorem I), which Hilbert initially described as the *Leitmotiv* of the theory. According to this theorem the number of equations that can be obtained from the variational integral is in fact smaller than the fourteen that one would expect to attain on the face of it. More specifically, in the first printed version of the theory Hilbert formulated the theorem as follows [15, 397]:

> **Theorem I.** Let J be a scalar expression of n magnitudes and their derivatives that is invariant under arbitrary transformations of the four world-parameters, and let the Lagrange variational equations corresponding to the n magnitudes be derived from the integral
>
> $$\delta \int J \sqrt{g} d\omega = 0.$$
>
> Then, in the system of n differential equations on n variables obtained in this way, four of these equations are always a consequence of the other n - 4, in the sense that four linearly independent combinations of the n differential equations and their total derivatives are always identically satisfied.

The variational principle introduced above yields ten equations for the gravitational potentials and four for the electromagnetic ones:

$$\frac{\partial \sqrt{g} H}{\partial g^{\mu\nu}} - \sum_k \frac{\partial}{\partial \omega_k} \frac{\partial \sqrt{g} H}{\partial g_k^{\mu\nu}} + \sum_{k,l} \frac{\partial^2}{\partial \omega_k \partial \omega_l} \frac{\partial \sqrt{g} H}{\partial g_{kl}^{\mu\nu}} = 0 \, (\mu, \nu = 1, 2, 3, 4)$$

$$\frac{\partial \sqrt{g} H}{\partial q_h} - \sum_k \frac{\partial}{\partial \omega_k} \frac{\partial \sqrt{g} H}{\partial q_{hk}} = 0 \, (h = 1, 2, 3, 4).$$

Hilbert denoted the left-hand sides of these equations as $[\sqrt{g}H]_{\mu\nu}$ and $[\sqrt{g}H]_h$, and called them the fundamental equations of gravitation and of electrodynamics respectively. Theorem I was obviously conceived with the intention of being applied to these equations, thus leading to the claim that four of them are in fact consequences of the other ten. In particular, Hilbert concluded, the four equations $[\sqrt{g}H]_h = 0$, are a consequence of the ten gravitational ones, $[\sqrt{g}H]_{\mu\nu} = 0$. This latter conclusion amounted, Hilbert suggested, to nothing less than a definitive explanation of the intimate interconnection between the two kind of physical phenomena involved:

> Based on the above theorem we can advance the following claim: *in the already indicated sense the electrodynamic phenomena are an effect of gravitation.* By recognizing this, I discern the simple and very surprising solution of the problem of Riemann, who was the first to search for a theoretical connection between gravitation and light. Hilbert [15, 397–398]footnoteHilbert was presumably referring here to a short paper on gravitation and light taken from Riemann's *Nachlass* [28, 532–38].

Hilbert did not prove this theorem as part of his exposition of the theory, but he claimed that the necessary proof would appear in a different place. As it happened, however, the mathematical conclusions Hilbert drew from the theorem were erroneous: in fact, the validity of the theorem would imply that four among the equations are dependent on the other ten, but this in no way warrants that precisely the four electromagnetic ones are dependent on the gravitational ones, as Hilbert asserted here. Theorem I was an early version of what later came to be known as Noether's theorem (Noether 1918), but Hilbert's conclusions went way beyond what the theorem actually allows. Over the coming years, Hilbert's theory gave rise to a vivid debate among the Göttingen mathematicians, and the problematic status of his Theorem I and its implications came to be at the focus of that debate [16].

The main point of connection between Mie's and Hilbert's theory comes to the fore in the treatment of the concept of energy. This is also a point where we find truly significant difference between the proofs and the printed version. In each case Hilbert defined a certain magnitude that is a sum of formal expressions involving the Hamiltonian H with some additional differential relations among the various potentials, plus an arbitrary contravariant vector p^l. The expressions defined in both cases were quite different from each other, but in both cases Hilbert performed very complex mathematical derivations that led to the conclusion that the magnitude in question has zero divergence, thus justifying their choice as representing energy in the theory.

A complete formulation of the theory required additional assumptions necessary for determining the specific form of the world-function H. Hilbert stipulated that the Hamiltonian be composed of two parts: $H = K + L$. The first term K accounts for the gravitational part of the world-function. Like Einstein, Hilbert made K to depend on the gravitational potentials and their first and second derivatives, in order to produce a theory as close as possible to Newton's. K is then, in fact, the Riemann curvature scalar $K = \sum_{\mu,\nu} g^{\mu\nu} K_{\mu\nu}$, where $K_{\mu\nu}$ is the Ricci tensor.

The second term, L, is also an invariant, and it accounts for the electromagnetic part. For simplicity, Hilbert assumed that it depends on q_s, q_{sl} and $g^{\mu\nu}$, but not on the derivatives of the latter. Using again a formal mathematical theorem (Theorem II) —a correct result which he did not prove here, yet claimed that proving it would be an easy task— Hilbert showed that, under the assumptions stated above, L must satisfy the following relation:

$$\frac{\partial L}{\partial q_{sk}} + \frac{\partial L}{\partial q_{ks}} = 0.$$

He thus concluded that the derivatives of the electromagnetic potentials appear in the equations only as part of the relation:

$$M_{ks} = q_{sk} - q_{ks}, \qquad (**)$$

from which he deduced that, as a consequence of the basic assumptions of the theory, L depends only on $g\mu\nu$, q_s, and curl q_s (but not simply on the derivatives of q_s

as originally assumed). Hilbert claimed that this conclusion was among the most significant results of his theory, since, as he said, it "is a necessary condition for establishing the Maxwell equations," and here it was obtained as a direct consequence of the assumption of general covariance alone. It is in passages like this, that Hilbert's reliance on Born's version, rather than on Mie's own presentation of the theory, becomes directly manifest. In fact, we saw above that Born had stressed as a main characteristic of the theory, that its Lagrangian depends only on differences which are equivalent to those appearing in (**).

Based on Theorem II Hilbert also deduced the form of the electromagnetic energy in the theory, which in the proofs was

$$-2\sum_{\mu}\frac{\partial\sqrt{g}L}{\partial q^{\mu\nu}}g^{\mu m}=\sqrt{g}\left\{L\delta_{\nu}^{m}-\frac{\partial L}{\partial q_m}q_{\nu}-\sum_{s}\frac{\partial L}{\partial M_{ms}}M_{\nu s}\right\}. \qquad (***)$$

Hilbert now claimed that in the limiting case —$g_{\mu\nu}=0$ (for $\mu\neq\nu$), $g_{\mu\mu}=1$ (i.e., when no gravitational field is present)— his expression for the stress energy tensor equals that of Mie's theory. This fact led him to conclude, with evident satisfaction, that:

> Mie's electromagnetic energy tensor is none but the generally covariant tensor obtained by derivation of the invariant $L$with respect to the gravitational potentials $g^{\mu\nu}$ in the limit. This circumstance first indicated me the necessary, close connection between Einstein's general theory of relativity and Mie's electrodynamics, and also convinced me of the correctness of the theory developed here. (p. 404)

What Hilbert meant with these claims would be rather obscure, unless we recalled that he was actually referring to Born's rendering of Mie's theory, rather than to the latter's own. In Born's formulation, the stress energy tensor of Mie's theory was given, as we saw above, by Div $S=0$. When this is specialized to the flat case, its connection with (***) (or with the corresponding equation that Hilbert wrote in the printed version) becomes apparent, although it still needs to be spelled out in detail.

6 Concluding Remarks

By the end of 1912, the question of the structure of matter had come to occupy a central place among Hilbert's scientific concerns. Mie's theory of matter, however, does not seem to have attracted his attention until Born reformulated it in terms more akin to his scientific sensibilities. Eventually, Hilbert became convinced that the theory showed good prospects for helping erect, based on it, a foundation for a unified theory that would account for all physical phenomena. Hilbert's interest in Einstein's theory came later. What startled Hilbert from Einstein's ideas, and directly motivated the consolidation of his own theory, was the possibility of embedding Mie's theory into a space-time formalism, that rendered evident a new, significant relation between gravitation and two important elements of the theory (the stress-energy tensor and the electromagnetic Lagrangian). At the same time the metric tensor was ostensibly put to the service of the explanation of the structure of matter, which was Hilbert's main focus of interest over the preceding years. Thus, inspired by Einstein's introduction of the metric tensor as a basic idea in the discussion of gravitation, Hilbert was led to consider Born's version of Mie's theory from a new perspective, under which new insights came to light that were not perspicuous in the flat case.

It is noticeable that neither in Born's nor in Hilbert's articles we find any direct or implicit reference, to Mie's *gravitational* theory. As already mentioned the latter presented considerable difficulties that Mie himself never really came to terms with. Born and Hilbert simply seem to have ignored this part of the theory in the framework of their discussions. Mie's gravitational theory was a scalar one and Born did not attempt to find a way to embed it in his own tensor-like presentation of the electromagnetic theory. Moreover, Born was most certainly aware of the criticism directed towards the theory in the Vienna meeting of 1913 or in its sequel, and he had no intention to counter this criticism when elaborating Mie's electromagnetic theory of matter. Then, in Hilbert's article, Mie is only mentioned with reference to the electromagnetic part of the theory presented. Hilbert did not generalize Mie's scalar gravitational theory into a tensorial, generally covariant version of it, but *rather, he used Mie's electrodynamic account of matter as a basis for his own unified field theory*.

On the other hand, Hilbert's idiosyncratic, and perhaps somewhat narrow, way of approaching Einstein's ideas precluded him from seeing the whole *physical* situation involved here. Hilbert did not discuss in any detail the main physical questions that had perplexed Einstein over the preceding years, and had delayed for so long the formulation of his generally-covariant equations. Moreover, in those places where Hilbert did elaborate on the physical implication of his theory, some of his claims are quite problematic. For instance, after formulating the field equations and commenting on the relation between Einstein's and Mie's theories, Hilbert returned to the interconnection —already suggested at the beginning of his argument— between the electromagnetic and the gravitational basic equations, and in particular concerning the linear combinations between the four electromagnetic equations and their derivatives. These linear combinations Hilbert deduced to be of the following form:

$$\sum_m \left(M_{mr} \left[\sqrt{g}L\right]_m + q_r \frac{\partial}{\partial \omega_m} \left[\sqrt{g}L\right]_m \right) = 0.$$

This formula embodied, in Hilbert's view, "*the exact mathematical expression of the claim formulated above in general terms, concerning the character of electrodynamics as a phenomena derived from gravitation*" (p. 406. Italics in the original). But in fact this conclusion turned out to be quite problematic and in the future versions of the theory, Hilbert had to reconsider the significance of the relation between these two kinds of physical phenomena.

The opening passage of the printed version of Hilbert's communication explains the background to the theory by giving credit first to Einstein and only then to Mie. It is remarkable that the proofs show the reverse order, as follows:
The far reaching and original conceptions by means of which Mie produced his electrodynamics, and the tremendous problems formulated by Einstein, as well as the penetrating methods he devised for solving them, have opened new ways to the research of the foundations of physics.

In the events following the publication of his theory we find many reasons why Hilbert chose to publish in the order he actually did. But in light of the historical context described in the foregoing pages, it seems to me that the order chosen in the original version (first Mie and only then Einstein) reflects more faithfully the way in which he had actually arrived at his theory. The same can be said about the relative importance that both components must be attributed as the actual motivations behind his efforts.

References

1. M. Born, Der Impuls-Energie-Satz in der Elektrodynamik von Gustav Mie, in *Nachrichten von der Königlichen Gesellschaft Der Wissenschaften zu Göttingen, Mathematische-Physikalische Klasse*, 23–36 (1914)
2. C. Cattani, M. De Maria, Max Abraham and the reception of relativity in Italy: His 1912 and 1914 controversies with Einstein, in *Einstein and the History of General Relativity*, vol 1, ed. by D. Howard, J. Stachel (Birkhäuser, Basel, 1989), pp. 160–174
3. L. Corry, *David Hilbert and the Axiomatization of Physics (1898–1918): From Grundlagen der Geometrie to Grundlagen der Physik*, 1st edn (Kluwer, Dordrecht 2004)
4. L. Corry, J. Renn, J. Stachel, Belated decision in the Hilbert-Einstein priority dispute. Science **278**(5341), 1270–1273 (1997)
5. J. Earman, C. Glymour, Relativity and eclipses: The British eclipse expeditions of 1919 and their predecessors. Hist. Stud. Phys. Sci. **11**, 49–85 (1980)
6. J. Earman, C. Glymour, The gravitational red shift as a test of general relativity: History and analysis. Stud. Hist. Philos. Sci. **11**, 175–214 (1980)
7. A. Einstein, Über das Relativitätsprinzip und die aus demselben gezogenen Folgerungen. Jahrb. Radioaktivität Elektron. **4**, 411–62 (1907)
8. A. Einstein, Zum gegenwärtigen Stande des Gravitationsproblems. Phys. Z. **14**, 1249–1266 (1913)
9. A. Einstein, Prinzipielles zur verallgemeinerten Relativitätstheorie und Gravitationstheorie. Phys. Z. **15**, 176–180 (1914)
10. A. Einstein, in *The Collected Papers of Albert Einstein*, ed. by A.J. Kox, M.J. Klein, R. Schulmann. The Berlin Years: Writings, 1914–1917, vol 6 (Princeton University Press, Princeton, 1996)
11. A. Einstein, D. Howard, in *The Collected Papers of Albert Einstein*. The Swiss Years: Writings, 1912–1914, vol 4 (Translated by Anna Beck, Princeton University Press, Princeton, 1996)
12. A. Einstein, M. Grossman, Entwurf einer verallgemeinerten Relativitätstheorie und einer Theorie der Gravitation. Z. Math. Phys. **62**, 225–261 (1913)
13. A. Einstein, D. Howard, *The Collected Papers of Albert Einstein, Volume 5: The Swiss Years: Correspondence, 1902-1914* (Translated by Anna Beck) (Princeton University Press, Princeton, NJ, u.a., 1995)

14. A. Einstein, D. Howard, *The Collected Papers of Albert Einstein, Volume 4: The Swiss Years: Writings, 1912–1914* (Translated by Anna Beck) (Princeton University Press, Princeton, NJ u.a., 1996)
15. D. Hilbert, Die Grundlagen der Physik (Erste Mitteilung). in *Nachrichten. Königliche Gesellschaft der Wissenschaften zu Göttingen. Mathematische-Physikalische Klasse*, pp. 395–407 (1916)
16. D. E. Rowe, The Göttingen Response to General Relativity and Emmy Noether's Theorems, in *The Symbolic Universe. Geometry and Physics 1890-1930*, ed by J. Gray (Oxford: Oxford University Press, 1999), pp. 189–234
17. H. Hönl, Intensitäts- und Quantitätsgrößen: In Memoriam Gustav Mie zu seinem hundertsten Geburtstag. Phys. J. **24**(11), 498–502 (1968)
18. G. Mie, *Lehrbuch der Elektrizität und des Magnetismus* (Ferdinand Enke Verlag, Eine Experimentalphysik des Weltäthers für Physiker, Chemiker, Elektrotechniker, Stuttgart, 1910)
19. G. Mie, Grundlagen einer Theorie der Materie. Ann. Phys. **37**, 511–534 (1912)
20. G. Mie, Grundlagen einer Theorie der Materie. Zweite Mitteilung. Ann. Phys. **39**, 1–40 (1912)
21. G. Mie, Grundlagen einer Theorie der Materie. Dritte Mitteilung. Ann. Phys. **40**, 1–66 (1913)
22. G. Mie, Bemerkungen zu der Einsteinschen Gravitationstheorie. Phys. Z. **15**:115–176 (1914)
23. G. Mie, *Die Einsteinsche Gravitationstheorie. Versuch einer allgemein verständlichen Darstellung der Theori* (Hirzel, 1921)
24. H. Minkowski, Raum und Zeit. Phys. Z. **10**, 104–111 (1909)
25. H. Minkowski, Die Grundgleichungen für die elektromagnetischen Vorgänge in bewegten Körpern. Math. Ann. **68**(4), 472–525 (1910)
26. L. Pyenson, *The Young Einstein, The Advent of Relativity*. 1st edn. (CRC Press, Bristol, 1985)
27. J. Renn, T. Sauer, M. Janssen, J.D. Norton (eds.), *The Genesis of General Relativity—Sources and Interpretations*, 4 vols (Springer, Dordrecht, 2007)
28. B. Riemann, R. Narasimhan, R. Dedekind, H. Weber, *Bernhard Riemann: Gesammelte mathematische Werke, Wissenschaftlicher Nachlass und Nachträge: Collected Papers*. (Springer, Berlin, New York; BSB B.G. Teubner Verlagsgesellschaft, Leipzig, 1990)
29. D.E. Rowe, 'Zwei Wirkliche Kerle': Neues zur Entdeckung der Gravitationsgleichungen der allgemeinen Relativitätstheorie durch Albert Einstein und David Hilbert, Wuensch Daniela. Termessos, Göttingen (2005). Hist. Math., Special Issue on Geometry and its Uses in Physics, 1900–1930, **33**, 500–508 (2006)
30. T. Sauer, The relativity of discovery: Hilbert's first note on the foundations of physics. Arch. Hist. Exact Sci. **53**, 529–575 (1999)
31. T. Sauer, Einstein equations and Hilbert action: What is missing on page 8 of the proofs for Hilbert's first communication on the Foundations of physics? Arch. Hist. Exact Sci. **59**(6), 577–590 (2005)
32. A. Sommerfeld, Zur Relativitätstheorie. I. Vierdimensionale Vektoralgebra. Ann. Phys. **337**(9), 749–776 (1910)
33. A. Sommerfeld, Zur Relativitätstheorie. II. Vierdimensionale Vektoranalysis. Anna. Phys. **338**(14), 649–689 (1910)
34. M.N. Wise, The mutual embrace of electricity and magnetism. Science **203**(4387), 1310–1318 (1979)

Hilbert and Einstein

Joseph Kouneiher and John Stachel

Abstract Highlights of the twenty-odd-year relationship between Einstein and Hilbert are reviewed. We trace the relationship between the two men during this period in the form of encounters, each of which characterizes a particular aspect of their relationship. We begin with the encounter that never took place (1912) when Einstein declined Hilbert's invitation to Göttingen; the fateful encounter (1915–1916) leading to a dispute over the final formulation of general relativity; The tragic-comic encounter (1928–29) over editorship of the Annalen der Mathematik leading to what Einstein called "The battle of the Frogs and Mice"; L'envoi (1932) Einstein's final letter of congratulations to Hilbert on his 70th birthday.

1 Introduction

On 25 November 1915, Einstein's paper on Mercury's perihelion was published. The calculations within this paper are related to the anomaly in Mercury's motion that had remained an unsolved puzzle in the context of the Newtonian theory of gravity. The paper had been submitted only a week earlier in turn, but it did not yet contain the final gravitational field equations that would become the core of the general theory of relativity.

That Einstein correctly worked out Mercury's perihelion before arriving at the final gravitational field equations indicates that the latter sit at the center of an intricate theoretical web of mathematical tools, physical assumptions and techniques, which together form general relativity. In the Mercury paper, Einstein used an approximation of what would later be called the Schwarzschild solution to the Einstein field

J. Kouneiher (✉)
Nice S. A. University, Nice, France
e-mail: joseph.kouneiher@unice.fr

J. Stachel
Center for Einstein Studies, Boston University, Boston, USA

J. Kouneiher (ed.), *Foundations of Mathematics and Physics One Century After Hilbert*,
https://doi.org/10.1007/978-3-319-64813-2_3

equations to describe the gravitational field of the sun. The main idea he needed was that the sun's gravitational field could be modeled by the so-called metric tensor. He then assumed that the gravitational field of the sun would be static, i.e. not change over time; spherically symmetric; and fall off to zero infinitely far from the sun. With these assumptions, he could find the approximation to a metric tensor that adequately described the gravitational field of the sun. Einstein then assumed that Mercury would move on the geodesics of this metric, i.e. on the straightest possible lines allowed for by the spacetime geometry defined by the metric. Together, these two assumptions made possible one of the most significant empirical confirmations of the new theory of general relativity.

On November 19, 1915 Hilbert sent a polite letter in which he congratulated Einstein "*on overcoming the perihelion motion. If I could calculate as rapidly as you [...]*". In fact Einstein did not calculate the result that rapidly. He presented his work on the Perihelion Motion[1] of Mercury on the November 18; but the basic calculation was done two years earlier with Michele Besso in the Einstein-Besso manuscript [1]. Einstein transferred the basic framework of the calculation from the Einstein-Besso manuscript and corrected it according to his November 11 field equations.

On November 25, 1915 Einstein submitted one of the most remarkable scientific papers of the twentieth century to the Prussian Academy of Sciences in Berlin. The paper presented the final form of what are called the Einstein Equations, the field equations of gravity which underpin Einstein's General Theory of Relativity. Thus this year marks the centenary of that theory. Within a few years this paper had supplanted Newton's Universal Theory of Gravitation as our explanation of the phenomenon of gravitation, as well as overthrown Newton's understanding of such fundamental concepts as space, time and motion. As a result Einstein became, and has remained, the most famous and celebrated scientist since Newton himself.

Ten days prior to his submission of the final field equations, Einstein wrote to David Hilbert that he was suffering from exhaustion and abdominal pains. His intense work on general relativity and poor nutrition caused by the ongoing war had clearly taken their toll on Einstein's health. Still, when Einstein submitted the field equations on 25 November 1915, he was aware that he had reached his goal[2]: the discovery of a law of gravity more accurate than Newton's, consistent with the results of special relativity, and indeed a generalisation of the latter theory. From the very beginning of searching for this new law of gravity, Einstein took the lesson from special relativity that mass and energy are equivalent as one of his starting points; or rather the idea that both mass and energy have to produce gravitational fields. The main question was

[1]The success of Einstein's calculation was also based on his November 11 theory. The condition $\sqrt{g} = -1$, implied by the assumption of an electromagnetic origin of matter, was essential for this calculation, which Einstein considered a striking confirmation of his audacious hypothesis on the constitution of matter, definitely favoring this theory over that of November 4. Thus when writing the Perihelion paper Einstein was still influenced by Hilbert's electromagnetic theory of matter.

[2]The history of the long quest towards the final equations has been described by M. Janssen, J. Norton, J. Renn, T. Sauer and J. Stachel, all of whom have been part of the Einstein Papers Project editorial team. The winding story of the discovery of the Einstein Field Equations has recently been summarized by Janssen and Renn in an article for Physics Today.

what the resulting gravitational fields would be represented by, what the "left-hand side" of the Einstein field equations would be, given that their "right-hand side" was mass-energy.

Hilbert ended his November 19, 1915 letter by asking Einstein to continue and keep him up to date on his latest advances but, he did not tell Einstein about a particularly important talk he planned to give the day afterwards. Hilbert presented on November 20 a paper to the Göttingen Academy of Sciences, "*The Foundations of Physics*", including his version to the gravitational field equations of general relativity. Five days later on November 25, Einstein presented to the Prussian Academy his version to the gravitational field equations.

So, in 1915 Hilbert played a crucial role in the history of those equations. Indeed, it was during Einstein visit to Gottengingen that *Hilbert convinced him that the goal of a fully general relativistic theory was achievable, something Einstein had nearly convinced himself could not be done*. Einstein returned to work, and by November, he had found the field equations which give General Relativity its final form. However, Hilbert also worked on the ideas Einstein had discussed with him and published a paper discussing how Einstein's theory fitted in with his own ideas on the role of mathematics in physics.

In the same November 20 Hilbert submitted a paper written by him which included the Einstein equations, derived from fundamental principles. Hilbert even sent Einstein a copy which probably reached Einstein before he submitted his own paper. Unfortunately, there was a confusion and dispute concerning the priority of the elaboration of first those equations.

Einstein felt himself to be the injured party in this short-lived priority dispute. He complained to a friend that Hilbert was trying to "nostrify" his theory, to claim a share of the credit. Einstein complained to Hilbert himself indeed, and some of the changes made in proofs by Hilbert included the addition of remarks giving credit for the basic ideas behind the theory to Einstein. However, Einstein tried not to let proprietary feelings color his feelings of gratitude for Hilbert. He recalled well that Hilbert had played an important role in encouraging Einstein to return to his theory at a time when Einstein had, to some extent, given up on his original goals. On December 20, 1915, he wrote to Hilbert:

> There has been a certain resentment between us, the cause of which I do not want analyze any further. I have fought against the feeling of bitterness associated with it, and with complete success. I again think of you with undiminished kindness and I ask you to attempt the same with me. It is objectively a pity if two guys that have somewhat liberated themselves from this shabby world are not giving pleasure to each other. (translated and quoted in Corry et al. 1997).

In 1990s Leo Corry[3] made a remarkable discovery. He found a copy of the proofs of Hilbert's paper, with a printers stamp dating it to December 6, 1915. These proofs show that Hilbert made significant changes to the paper after this date. In addition, the proofs do not contain the Einstein equations. The proofs have been cut up here and there (probably by the printers themselves as they worked), so it is possible

[3]See his contribution to this volume.

that the equations would be there if we had the missing pieces. But it is also quite possible that amidst the changes Hilbert made to the paper, he took the opportunity to include the final form of the equations from Einstein's paper. Indeed some of the changes he made after December 6 were to update his argument from earlier versions of Einstein's theory to the later version.

2 The Encounter That Never Took Place (1912)

On 30 March 1912, David Hilbert, the eminent Göttingen mathematician, wrote Albert Einstein, then still living in Zurich:

> Highly esteemed colleague,
> I would be very happy if I had your theoretical works on gas theory and radiation theory in my possession [Stachel's translation – trying to reproduce the slightly pompous tone of Hilbert's German] [2, p. 439] [4]

What Hilbert had in mind is made clear by Einstein's letter of 4 October 1912 [2, p. 502], politely declining Hilbert's request that he deliver a lecture on the kinetic theory of matter at Göttingen. Einstein gave two reasons: He had nothing new to say on the subject, which was not quite true; and he was completely occupied with other matters, which was quite true. The nature of these "other matters" is made clear in a letter of 1 November 1912 from Arnold Sommerfeld to Hilbert:

> Einstein is apparently so deeply mired in gravitation that he is deaf to everything else (see [2, p. 506], note 6).

Well, not quite everything else. He made a trip to Berlin in April of that year, during which he was offered a position at the *Physikalisch-Technische Reichsanstalt*. From the people he met during this trip and the institution named, it clear that, as in the case of the Göttingen invitation, it was not primarily his work on relativity and gravitation, but his work on the quantum theory of solids that led to the Berlin offer. He declined this offer, but this Berlin trip was the beginning of a connection that a year later led to his appointment as a full member of the Prussian Academy of Sciences, and his subsequent move to the German capital.
The attractions of Berlin were not exclusively intellectual. During his 1912 visit, he had renewed acquaintances with his uncle Rudolf Einstein ("the rich" as Einstein called him), now retired, his wife Fanny; and their daughter Elsa Löwenthal, Einstein's cousin. Divorced in 1908, she and her two daughters had joined her parents in Berlin. Thus began an affair that ultimately led to Einstein's 1919 divorce from his first wife, Mileva Marić, and marriage to Elsa.

---

[4] All Einstein documents mentioned in this article can be found in J. Stachel et al., eds., The Collected Papers of Albert Einstein, Princeton U. Press (1987). Hereafter we use the notation CPAE X; Y, with X the volume number and Y the document number.

3 The Fateful Encounter (1915–1916):

Before getting to the first meeting between Hilbert and Einstein, we must mention something that did not happen, but served to create a bond of sympathy between the two: Neither agreed to sign the notorious "Manifesto of the 93 [German Intellectuals] To the World of Culture." This document tried to justify, in the name of German "Kultur," the German invasion of neutral Belgium in 1914, soon after the outbreak of WWI. Einstein joined in a futile effort to launch a pacifist counter-manifesto "To the Europeans." It garnered only three signatures, and was not published until much later. Hilbert was not among the three, but Einstein valued him highly for his independence of judgment, writing in mid-1915:

> One is doubly overjoyed in these times by the few men, who stand quite above the situation and do not let themselves be driven by the sad currents of the time. One such is Hilbert, the Göttingen mathematician. Hilbert now regrets doubly, as he told me, out of negligence not having cultivated more international connections (AE to Heinrich Zangger, 7 July 1915, [3], pp. 144–145).

The occasion for this letter was an account of Einstein's first meeting with Hilbert:

> I spent a week in Göttingen, where I learned to know and love him [Hilbert]. I gave six two-hour lectures there about my now very much clarified gravitational theory and experienced the joy of completely convincing the mathematicians (ibid.).

Einstein's joy at the sympathetic reception of his work by the Göttingen mathematicians, whom he had earlier scorned for what he regarded as mathematical pedantry, contrast sharply with his disappointment at the response of his colleagues in physics:

> Physicist humanity reacts rather passively to the gravitational work Laue is inaccessible to arguments of principle, Planck also is not, Sommerfeld is a bit. A free, unconstrained outlook is really alien to the (adult) German (Einstein to Michele Besso, after 1 January 1914, [2], pp. 588–589).

At this point, a word of caution is necessary. Although Einstein is already beginning to speak of "general relativity" around this time, the theory he discussed in Göttingen was not the one that we now denote by that name. Rather, it was a variant of the so-called Einstein-Grossmann Entwurf theory, first formulated in 1912–1913, the field equations of which are not generally covariant. As we shall see, it was only in four papers dating from the late fall and early winter of 1915 that Einstein propounded the generally-covariant theory that we all know and (at least some of us) love as general relativity.

Among the enthusiasts attending Einstein's Göttingen lectures was Hilbert. He owed much of his already-considerable fame to his rigorous axiomatization of Euclidean geometry, and for several years had been attempting to apply his axiomatic method to the foundations of physics (see Corry [4]). With the help of Max Born, then a *Privatdozent* at Göttingen, he was working on Gustav Mie's electromagnetic theory of matter. On the basis of a set of non-linear (and non-gauge-invariant) electromagnetic equations, this theory aimed to offer a field-theoretical explanation of the existence, structure and stability of the electron. Hilbert conceived the idea of enlarging Mie's program by combining it with Einstein's. He wanted to conjoin Mie's electromagnetic four-potential q_i with Einstein's ten gravitational potentials $g_{\mu\nu}$ in a set of non-generally covariant field equations to produce what we would now call a unified field theory of gravitation and electromagnetism (see Renn and Stachel [5]).

While Hilbert was embarking with great enthusiasm on this program, Einstein was becoming more and more unhappy with the Einstein-Grossmann theory; and in mid-1915 returned to his earlier search for a generally-covariant gravitational theory. But the common ground between Einstein and Hilbert's programs soon led to some friction between the two men, with Einstein accusing Hilbert of wanting to "nostrify" [i.e., make his own] Einstein's work. We shall not here rehearse the details of the dispute except to note that it was quickly and happily resolved (see Stachel [6]).

On November 26, 1915 a day after Einstein presented the final version of the field equations he wrote his close friend Zangger:

> The general relativity problem is now finally dealt with. The perihelion motion of Mercury is explained wonderfully by the theory. [] The theory is beautiful beyond comparison. However, only one colleague has really understood it, and he is seeking to clearly "nostrify" it (Abrahams expression[5]). In my personal experience I have hardly come to know the wretchedness of mankind sometimes better than this theory and everything connected to it. But it does not bother me [8].

On 5 December 1915 Hilbert and four colleagues proposed Einstein for membership in the Royal Society of Göttingen, to which he was duly elected on 18 December. Two days later, in a letter thanking Hilbert, Einstein wrote:

> I am taking advantage of this opportunity to tell you something else, which is more important to me. There has been a certain strained atmosphere [Verstimmung] between us, the cause of which I shall not analyze. I have fought against the feeling of bitterness associated with it, and indeed with complete success. I think of you again with unclouded friendliness and beg you to attempt the same with me. It is objectively too bad if two real guys, who have made something out of this shabby world, should not mutually enjoy each other (20 December 1915, [3], p. 222).

It is perhaps worth noting that Hilbert did not reply to this letter, and indeed had no further correspondence with Einstein until 27 May 1916, when he responded to a post card from Einstein with some questions about Hilbert's paper on his new theory, which had just been published. Hilbert invited Einstein to visit Göttingen again and stay with him; but in spite of several invitations over the next few years, this third visit to Göttingen never took place, perhaps because of Einstein's poor health during the last years of WWI (he had a stomach ulcer). However, they continued to correspond over issues connected with Hilbert's paper.

As Jürgen Renn and Stachel have indicated [5], Hilbert made several mathematical errors in the course of work on his new theory. Some have questioned whether a great

---

[5]In 1912, Max Abraham had blamed Einstein's theory of relativity and Einstein as well. Abraham thought that Einstein borrowed expressions from his new gravitation theory. On March 26, 1912, Einstein wrote to Michele Besso : "*Abraham's theory was created of the top of his head, i.e., from mere mathematical beauty considerations, torn off and completely untenable. In fact, "nostrification" was Einstein's expression and not Abraham's*" [7]. That was almost Abraham's opinion of Einstein's theory, except for the mathematical beauty.

mathematician like Hilbert could really have made the mistakes we suggested. We shall let another great mathematician, Gian-Carlo Rota, answer. In a section of Rota [9] entitled "Do not worry about your mistakes," he writes:

> When the Germans were planning to publish Hilbert's collected papers and present him with a set on the occasion of one of his later birthdays, they realized that they could not publish the papers in their original versions because they were full of errors, some of them quite serious. Thereupon they hired a young unemployed mathematician, Olga Taussky-Todd, to go over Hilbert's papers and correct all mistakes. Olga labored for three years; it turned out that all mistakes could be corrected without any major changes in the statement of the theorems. There was one exception ...At last on Hilbert's birthday a freshly printed set of Hilbert's collected papers was presented to the *Geheimrat*. Hilbert leafed through them carefully and did not notice anything (p. 201).

We might add: or at least he did not say anything!

4 The Tragic-Comic Encounte—(1928–29):

There is a special folder in the Einstein Papers that bears the following caption, typed by Helen Dukas, his assistant:

> Professor Einstein wanted this correspondence kept in a special folder under the title: *Der Frosch-Mäuser Krieg*

Our account is based in large part on the contents of this folder.

We remind you that "The Battle of the Frogs and Mice" is the name of a Greek mock-epic poem, *Batrachomyomachia*, dating from classic times, which was written to make fun of the style and content of real epic poems, such as the *Iliad* (see Chapman [10] for a classic English translation). Evidently, even in ancient times there were scoffers and cynics who used satire as their preferred weapon.
At what "epic" struggle was Einstein scoffing? It was the well-known (to those who know it well!) controversy over the foundations of mathematics raging in the nineteen-twenties between Hilbert, the fighting formalist and Luitzen Egbertus Jan Brouwer, the ingenuous intuitionist.

> The Brouwer-Hilbert debate grew increasingly bitter and turned into a personal feud. The last episode was the "Annalenstreit" [battle over the Mathematische Annalen –JS], or, to use Einstein's words, "the frog-and-mouse battle." It followed the unjustified and illegal dismissal of Brouwer from the editorial board of the Mathematische Annalen by Hilbert in 1928 and led to the disbanding of the old Annalen company and the emergence of a new Annalen under Hilbert's sole command but without the support of its former chief editors, Einstein and Carathéodory [11, p. 3].

We will not enter into mathematical or philosophical aspects of this battle, but shall only touch on its human side and in particular Einstein's role in it. This seems especially worthwhile since this episode is not even mentioned in any of the many Einstein biographies we consulted.

By the end of 1927 Brouwer's program of re-constructing mathematics had run aground. He lost almost all international support, especially the occasion to install a mathematics institute in Amsterdam. Hilbert's annextion of *meta-mathematics* "without mention of autorship", and his public attack on Brouwer considered as direct attack against Intuitionism. Brouwer emotional reaction concerning the paternity claim of the notion "meta-mathematics" and the improvements in the formalist program due to his work of the validity of the principle of the Excluded Middle was [12]:

> Formalism has received nothing but benefits from Intuitionism and can expect further benefits. The formalist school should therefore show due recognition instead of war-mongering against Intuitionism in sneering tones, never once making proper reference to authorship. Moreover, Formalism should remember that in the Formalist structure so-far nothing mathematical has been achieved (we are still waiting for a proof of the non-contradictority of its axion system), whereas Intuitionism pn the nasis of its constructive definition of sets and the Fundamental Property of Finite Sets has already erected new structures in real mathematics of unshakable certainty. [13, p. 4]

The controversy concerning the International Congress of Mathematics at Bologna in September 1928 was the last stages in the semi political battle between Hilbert ad Brouwer.
In 1928, Brouwer public call on all German mathematicians to boycott the Congress, was interpret by Hilbert as interference in German affair and an attempt to prevent him from attending the congress as head of the German delegation. Notice, that Hilbert's address to the Bologna Congress was his last public appearance before retirement.
However, Hilbert did not consider the matter was closed. For him those events show up Brouwer's ambition and personal influence. Hilbert fearing an eventual influence on the editorial Board of the *Mathematisch Annalen*, he dismissed Brouwer from the editorial Board without the agreement of the other chiefs editorial board, Einstein and Carathéodory.
The editorial board was of course surprised, but they were anxious to avoid any unpleasantness. Carathéodory, ask Laren to persuade Brouwer not to take immediate action, because "Hilbert is desperately ill and will regret his step in a few weeks". Brouwer felt that his co-editors were prepared to sacrifice him. However, to him the dismissal from the Editorial Board was an injustice and the end of his career.
How did Einstein become involved in the controversy? It started in 1919 with a gesture of friendship and confidence on Hilbert's part: He invited Einstein to join the editorial board of the *Mathematische Annalen*, then the premier German mathematics journal. In 1928, as a result of increasing tension between Hilbert and Brouwer, Hilbert wrote to the other editors of the *Annalen*, asking their sanction for the removal of Brouwer from the editorial board. At first Einstein tried to pass off the matter with a joke:

> Herr Brouwer is an involuntary champion of Lombroso's theory of the close connection between genius and madness (Einstein to Hilbert, 19 October 1928).

He refused to sign the letter of expulsion and, in a letter to Constantin Carathéodory, a fellow member of the board, attributed this move to a momentary fit of pique on the part of Hilbert.

> It would surely be best to pay no attention at all to this Brouwer-matter. I would never have thought that Hilbert could be capable of such outbursts of emotion (Einstein to Caratheodory, 19 October 1928).

Caratheodory replied in confidence, explaining to Einstein that the expulsion seems to have been a carefully calculated move on the part of Hilbert and several others board members. Caratheodory added that he was resigning from the board, but asked Einstein to keep his resignation secret for the present because he did not want to appear to be taking sides against Hilbert.
Carathéodory felt so strongly about this "dishonorable affair" that he left Europe and accept a chair at Stamford.
Blumental and Courant were more concerned about Hilbert's reputation and tried to persuade Carathéodory to accept and spread their agreed version :

> Hilbert feared that Brouwer's personality might be damaging and dangerous for the future of the Annalen. It is not "an interpretation constructed after the event" if one emphasizes this factual motive, even if Hilbert's action at first might give a different impression. For Hilbert's sake we cannot allow a version of his reasons to become public which does not do him full justice. If you already accept such a version what can we axpect from the public at large? (Courant to Carathéodory, 23.12.28)

In mid-1929 Max Born intervened in the ongoing dispute, hoping to induce Einstein to side with Hilbert, largely on the grounds that Hilbert was a dying man and his last days should not be clouded by this controversy (Max Born to Albert Einstein, 2 August 1929). Curiously, Born did not include Einstein's letter of refusal in his collection of their correspondence, published thirty-odd years later.
Hilbert was indeed seriously ill at the time, but recovered and did not die until 1943, at the age of 81, and then it was from complications of a broken arm.
The last item in the "Frog-Mouse War file" is Einstein's reply to a 1930 letter from Jacques Hadamard, the eminent French mathematician, with whom he was on very friendly terms. After his expulsion from the Annalen, Brouwer had approached Hadamard, among others, to join him in founding a new mathematics journal, and Hadamard turned to Einstein for advice. Einstein replied:

> It was a vile dispute between Brouwer and Hilbert, for which nevertheless in my opinion Hilbert bears the main guilt (15 November 1930).

Nevertheless, in spite of Brouwer's ill treatment by Hilbert and his good works and intentions, in view of Brouwer's well-known and well-deserved reputation for querulousness Einstein advised Hadamard to keep hands off.

5 L'envoi (1932):

Luckily, we can end our story on a more pleasant note. The last letter from Einstein to Hilbert is dated 26 February 1932, congratulating him on his seventieth birthday:

> If I cannot allow myself to follow you along all your daring avenues of thought, yet I am able to form for myself a picture of the strength and beauty of your thought, and am obliged to you for sessions of cloudlessly beautiful experience.

This is a good place to end our tale, before the horrors of the Third Reich finally engulfed Germany—along with almost all the rest of Europe—only ending in the /it Götterdämmerung of 1945, which Hilbert was blissfully spared.
Although known to be upset by the actions of the regime, and seeing many of his closest collaborators such as Max Born, Richard Courant and Emmy Noether forced into exile, Hilbert stayed on at Göttingen until his death in 1943. Einstein of course left Germany in 1933, never to return and, as far as is known, never again to be in direct contact with Hilbert.

References

1. A. Einstein, *Erklärung der Perihelbewegung des Merkur aus der allgemeinen Relativittstheorie*. Königlich Preüische Akademie der Wissenschaften (Berlin, 1915c). Sitzungsberichte, pp. 831–839. 831
2. CPAE 5, *The Collected Papers of Albert Einstein*, vol. 5, ed. by M.J. Klein, A.J. Kox, R. Schulmann. The Swiss Years: Correspondence 1902–1914 (Princeton University Press, Princeton, 1993)
3. CPAE 8a, ibid., vol. 8, ed. by R. Schulmann, A.J. Kox, M. Janssen, J. Illy. The Berlin Years: Correspondence, 1914–1918, Part A 1914–1917 (Princeton University Press, Princeton, 1998)
4. L. Corry, *David Hilbert and the Axiomatization of Physics, 1898–1918* (Kluwer, Dordrecht, 2004)
5. J. Renn, J. Stachel, Hilbert's foundation of physics: from a theory of everything to a constituent of general relativity, in *The Genesis of General Relativity*, vol. 4, ed. by J. Renn, M. Schemmel. Gravitation in the Twighlight of Clasical Physics: The Promise of Mathematics (2007), pp. 857–973
6. J. Stachel, New light on the Einstein Hilbert priority question. J. Astrophys. Astron. **20**, 91–101 (1999)
7. Einstein to Besso, March 26, 1912, CPAE 5, Doc. 377
8. Einstein to Zangger, November 26, 1915, CPAE 8, Doc. 152
9. G.-C. Rota, *Indiscrete Thoughts* (Birkhäuser, Boston/Basel/Berlin, 1997)
10. G. Chapman, *Homer's Batrachomyomachia, Hymns and Epigrams*, 2nd ed., London: John Russell Smith 1888. Available online at https://ia700300.us.archive.org/29/items/homersbatrachomy00chapuoft/homersbatrachomy00chapuoft.pdf
11. W.P. van Stigt, Brouwer's Intuitionist Program, in *From: Brouwer to Hilbert/The Debate on the Foundations of Mathematics In the 1920s* ed. by P. Mancosu (Oxford University Press 1988), pp. 1–22
12. W.P. van Stigt, The Brouwer-Hilbert controversy and the annalen affair, in *Brouwer's Intuitionism* vol. 2 (1990), pp. 100–103
13. J. van Heijenoort, (2nd printing with corrections), *From Frege to Gödel: A Source Book in Mathematical Logic*, 18791931, (Harvard University Press, Cambridge Massachusetts, 1976); Brouwer, Intuitionistic reflections on formalism, in *Philosophy of Mathematics: Selected Readings*, ed. by H. Putnam, P. Benacerraf (Englewood Cliffs, N.J.: Prentice-Hall, 1964)

Grothendieck's Unifying Vision of Geometry

Colin McLarty

If one thing has fascinated me in mathematics since childhood, it is this power to identify in words, and perfectly express, the essence of certain mathematical things which on first approach present themselves as elusive or mysterious beyond words.

— A. Grothendieck *Récoltes et Semailles* p. 14

These notes attempted to show something that was still very controversial at that time: that schemes really were the most natural language for algebraic geometry and that you did not need to sacrifice geometric intuition when you spoke "scheme."

— David Mumford *The Red Book* (1988, p. VIII)

Abstract Grothendieck's "vast unifying vision" provided new working and conceptual foundations for geometry, and even led him to logical foundations. While many pictures here illustrate the geometry, Grothendieck himself favored apt words and commutative diagrams over pictures and did not think of geometry pictorially.

GEOMETRY, FOUNDATIONS, AND THE SCOPE OF GROTHENDIECK'S VISION
Geometry has touched on number theory at least since the Babylonians linked trigonometric calculations to integer solutions of $a^2 + b^2 = c^2$ some 4000 years ago. But the links went little beyond that until Fermat hinted at much more. His followers were few, albeit stellar, including Carl Friedrich Gauss, until André Weil put geometrization of arithmetic at the top of the agenda for pure mathematics [51, 52]. Weil simultaneously traced his own vision back to Leopold Kronecker at mid-19th century, and tied it to cutting edge topology.

By 1950 the project was led by the leaders of Bourbaki: Weil, Claude Chevalley, and Jean Dieudonné. It was closely tied to Emil Artin, Bartel van der Waerden, Oscar Zariski, Henri Cartan, Samuel Eilenberg, Saunders Mac Lane, John Tate—

C. McLarty (✉)
Department of Philosophy, Case Western Reserve University, Cleveland, OH, USA
e-mail: colin.mclarty@case.edu

J. Kouneiher (ed.), *Foundations of Mathematics and Physics One Century After Hilbert*,
https://doi.org/10.1007/978-3-319-64813-2_4

and the cognoscenti all knew the rising star Jean-Pierre Serre. Serre won a 1954 Fields Medal, at age 27 the youngest Fields Medalist ever, and around that year he recruited Grothendieck to the project.

The geometric aspect has not always been evident. Mumford rightly says even insiders did not all see the geometry of Grothendieck's *schemes* at first. And even today many mathematicians are not too comfortable with derived functor cohomology as geometry. No such doubt plagued Grothendieck:

> This vast unifying vision can be described as a new geometry. It seems to be what Kronecker dreamed of in the last century. But reality (which a bold dream may let us guess or foresee, and encourage us to discover) is always richer and more resonant than even the most reckless or profound dream. (*ReS* [21, p. P28])

The vision is so vast and so unifying as to reach foundations in three senses: working, and conceptual, and even logical foundations.

Working foundations are the specific tools and theorems which practitioners all master and routinely use. So line integrals and the Cauchy integral theorem have long been the working foundation of complex analysis. In 1950 the working foundations of algebraic geometry were completely up in the air. The established theory of varieties over the complex numbers, or indeed over any algebraically closed field, was clearly too narrow for the number-theoretic goals. Many generalizations were on offer. None was established.

Conceptual foundations orient some work though they might not be widely known in detail and might not even be fully developed. Weil hinted at a conceptual foundation for his geometry when he posed his famous *Weil conjectures* as an astonishing analogy between number theory and *cohomology theory* in topology [51, pp. 498, 507]. So it would not be a geometry of distances and rigid figures. It would be a geometry of continuity, connectedness, dimension, and genus, all described in Sect. 2. But as Grothendieck says:

> No one had the least idea how to define such a cohomology and I am not sure anyone but Serre and I, not even Weil if that is possible, was deeply convinced such a thing must exist. (*ReS* [21, p. 840])

The conceptual foundation of Grothendieck's geometry began when he fundamentally re-conceived cohomology, in his Tôhoku paper (1957), by axioms for *Abelian categories* and *derived functors*. His unprecedented easy agile use of functors simplified and extended the links between topology and algebra. And that is a radical understatement. He used no topological or algebraic particulars of any existing cohomology theory. Of course the particulars can be fitted to this framework for use towards particular ends. This had an immediate, concrete pay-off as it led to the *Grothendieck-Riemann-Roch Theorem* [6, 7]. But the axioms are far simpler than any particulars. Grothendieck was sure this was the right way to think of all existing cohomology theories and all that would exist—including some future cohomology adapted to the Weil conjectures.

Today Grothendieck's derived functor cohomology is standard in the conceptual foundations of algebraic geometry. Its concepts and central theorems are widely taught and used. But some of its proofs use tools most geometers never need in their further work. So textbooks use those theorems without proof. The still-leading text on cohomological algebraic geometry does exactly this in "the technical heart of the book" [25, p. xiv], as do its more recent rivals. There is a complication here concerning logical foundations, to which we return at the end of this section.

Soon after Grothendieck clarified cohomology, his *schemes* became the standard working foundation of algebraic geometry. Some of the audience at the Stockholm International Congress of Mathematicians thought Serre was being a bit narrowly Parisian in his definition of algebraic geometry: "I take this term in the sense it has had for several years now: the theory of schemes" [47, p. 190]. Within a few years schemes were the world-wide standard spaces for all algebraic geometry beyond the classical theory of complex varieties—and for advanced theorems on complex varieties. Mumford's photocopied lecture notes played a big role in that as they circulated for years before being printed as [41].

Grothendieck got schemes along with a full conceptual foundation:

> The two crucial driving ideas (*idées forces*) in launching and developing the new geometry were that of *scheme*, and that of *topos*. Appearing almost at the same time and in close symbiosis with each other, they were as if a single sinew in the spectacular flight of the new geometry, and this from the very year they appeared. (*ReS* [21, p. P31])

Schemes were not only meant to *contain* arithmetic information but to *reveal* it by supporting a suitable cohomology theory. Grothendieck made that happen by treating a scheme as a *generalized topological space* and more precisely as a topos. Section 4 goes into this.

Grothendieck was bitterly aware "it has been good manners in 'high society' to look down on those who dare pronounce the word 'topos' " (*ReS* [21, p. 182]).[1] This repugnance is linked with logical foundations because Grothendieck's idea of topos poses problems for naive set theory. So do his ideas of Abelian category, derived functor, and *a fortiori* their characterizations by universal properties, all of which are standard textbook material today. But the issue is associated with topos theory since Grothendieck wrote about it in *Théorie des Topos et Cohomologie Etale des Schémas* [3]. Grothendieck cared to have a precise logical foundation. So he asked Pierre Samuel[2] to write 30 pages on set theoretic *universes*, signed N. Bourbaki, as an appendix to [23]. Many geometers regret this.

Grothendieck used a strong logical foundation precisely to assure rigor without needing to check details at each step. He did not worry about finding the weakest logic for his tools. Section 5 gives the current state of the art on that. Meanwhile Connes' contribution to this volume draws on the topos perspective in geometry without laboring the logical foundations [10].

---

[1]While this holds of many geometers, [14] puts toposes among Grothendieck's best ideas. Only Deligne emphasizes what Grothendieck also knew: you can think with topos intuitions while officially using only small sites in proofs. Cf. [40, p. 263].

[2]This information from Pierre Cartier, discussion February 2015.

1 Pictorial Visualization I: Schemes and Crystals

Mumford [41] includes famous drawings of arithmetic schemes. He gave graphic geometric intuition to things like split and ramified primes, and others now use his conventions e.g. [15]. It seems Grothendieck did not draw pictures of schemes.

Deligne draws pictures of what Grothendieck taught him. When I asked if Grothendieck also drew them, Deligne paused and looked up as if searching the past. Refocussing his eyes on the present, he said:

> In talks, no, I don't think so. In speaking with him he would easily recognize them and communicate by them but he did not draw them. (Conversation, IAS, March 10, 2005)

Deligne is so visual he looked 40 years back to see if his teacher was drawing pictures! On the other hand, while Grothendieck drew commutative diagrams, and our Sect. 3.1 describes two compass drawings by him, his algebraic geometry was little pictorial in his own mind. Grothendieck's geometry, in his own mind, captured geometric intuitions in words. The difference in mental picturing mattered so little to their communication that Deligne hardly noticed it until asked.

2 This Geometry Before Grothendieck

> It becomes substantially easier to conceive of a complex variable extended over a connected two dimensional domain when it is linked with spatial intuition. [42, p. 3]

2.1 Genus

Bernhard Riemann geometrized complex analysis by sometimes ignoring details of differentials and integrals, in favor of continuous maps between surfaces. For example he studied integrals such as

$$\int_0^t \frac{1}{\sqrt{(1-z^4)}}\,dz$$

by considering continuous maps to a sphere from a torus.

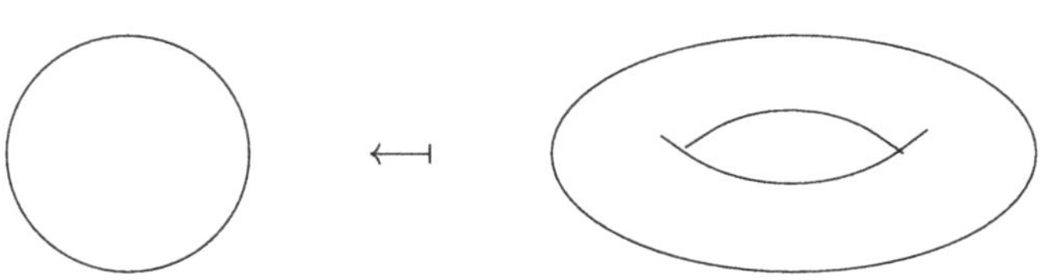

He trivially observed the sphere can represent the complex plane plus a point at infinity. He saw far from trivially how a torus wrapped twice around the sphere can represent the two values $\pm(1-z^4)^{1/2}$ at each point z. He saw the values are organized the same way for any equation $y^2 = P(z)$ with a degree 3 or 4 polynomial $P(z)$ with no multiple roots. The values $\pm P(z)^{1/2}$ at each point z make a torus wrapped twice around the sphere.

Riemann saw how much complex analysis follows from rudimentary algebra plus geometric ideas like 2-dimensionality and connectedness. Today we say the mere *topology* of a Riemann surface S determines much about analysis on S.

One key idea was the *genus* of a surface. A sphere has genus 0, a torus has genus 1. An n-torus, which is like n copies of a plain torus combined into one surface, has genus n. Here are a 2- and a 3-torus:

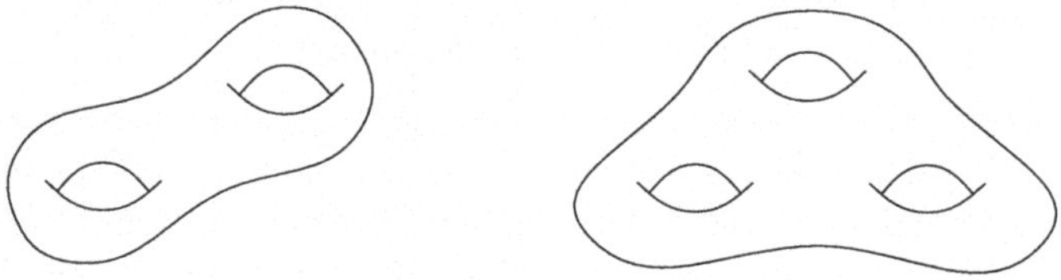

These correspond to higher degree polynomial equations $Q(y, z) = 0$ on y, z.

Geometric intuitions work here. You would not expect to make a 2-torus cover a 3-torus. The two handles of the first can only wrap around two handles of the other. You would not expect to map a 3-torus into a 2-torus, without collapsing one of the handles or twisting two of them together. These intuitions lead to the *Riemann-Hurwitz formula* putting sharp limits on maps between Riemann surfaces, and thus on complex analysis for algebraic functions in all degrees. See the vividly pictorial "scissors and paste" proof by McKean and Moll [35, p. 40].

2.2 *Counting Solutions Without Finding Them*

All calculus students know a continuous function $f\colon[0,1]\to\mathbb{R}$ with $f(0) < 0 < f(1)$ has $f(x) = 0$ for some $0 < x < 1$. There could be any number of solutions, even infinitely many. But if there are only finitely many then stronger conclusions follow.

When there are a finite number of solutions $f(x) = 0$ then each one has a *separating interval*. That is an interval $x \in U_x \subseteq [0, 1]$ such that x is the only solution in U_x. Figure 1 illustrates the three things that can happen:

(1) x is a *rising solution* if all $y \in U_x$ with $y < x$ have $f(y) < 0$; and all $y \in U_x$ with $x < y$ have $0 < f(y)$.
(2) x is a *falling solution* in the opposite case.
(3) x is a *tangent solution* if all $y \in U_x$ have $f(y) \leq 0$, or all have $0 \leq f(y)$.

A numerical conclusion follows:

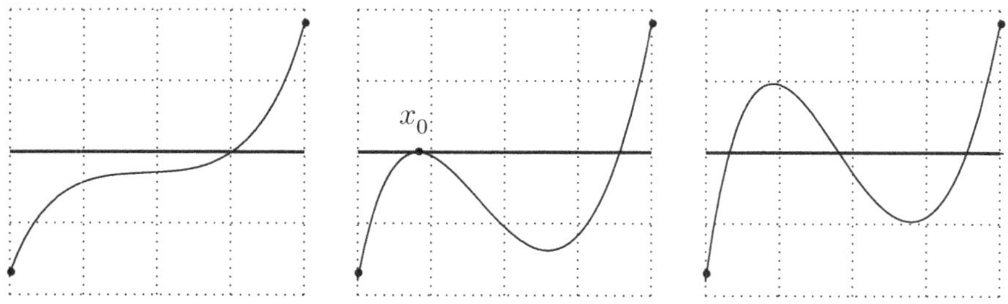

Fig. 1 Three graphs with $f(0) < 0 < f(1)$

(1) There is one more rising solution to $f(x) = 0$ than falling, and so
(2) if there are no tangent solutions, there is an odd number of solutions.

One more thing matters here. The middle example in Fig. 1 has one tangent solution $f(x_0) = 0$. Slightly raising that graph at x_0 will split the solution in two, and a slight lowering will eliminate it as a solution. Tangent solutions are not *stable*. Indeed, no graphic can assure the apparent tangency is real. Finer resolution might show the curve crosses slightly over the axis there, or slightly undershoots.

More sophisticated topology gives the *Lefschetz fixed point theorem*. See the deft popularization by Atiyah [4]. A fixed point $x \in M$ for a map $g : M \to M$ from a space M to itself is a solution to $g(x) = x$. Here M could be any of the n-tori discussed in Sect. 2.1 or some higher dimensional manifold.

A given map $g : M \to M$ may have infinitely many fixed points. But if it has finitely many then the fixed point theorem together with some purely topological information about g puts limits on the number without in any way determining where the fixed points are. Given some stability assumptions it may even give the exact number of fixed points.

2.3 *Local and Global*

Cohomology expresses the basic geometric ideas of connectedness, genus, and dimension in terms of the passage between *local* and *global*.[3]

2.3.1 Constant Functions and Contour Maps: Dimension and Genus

A real valued function $f : M \to \mathbb{R}$ on any space M is *locally constant* if each $x \in M$ is surrounded by some open ball $x \in U \subseteq M$ with f constant on U. Of course if

[3]The following is based on *de Rham cohomology*. Local contour maps represent closed 1-forms, while actual contour maps represent exact 1-forms.

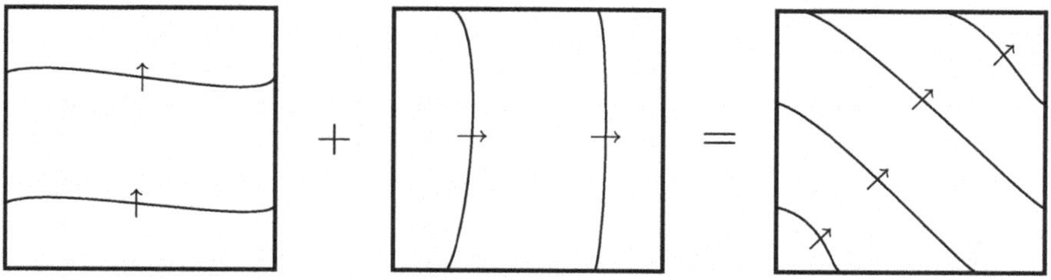

Fig. 2 Addition of local contour maps on a square

$x, y \in M$ are linked by some chain of overlapping balls, with f constant on each of those balls, then $f(x) = f(y)$.

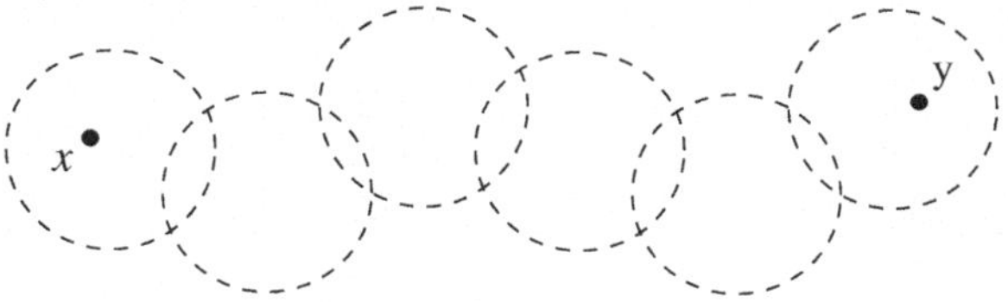

It is intuitively plausible, and it is a theorem in topology, that M is *connected* if and only if every locally constant function on M is actually constant. If M is a union of disjoint open parts $M_1 \cup M_2$, so M is not connected, then a locally constant f could have $f(x) = 0$ for all $x \in M_1$ and $f(x) = 1$ for all $x \in M_2$.

An example one step richer was in fact a key inspiration for the new ideas. Figure 2 shows three square charts. Think of them as contour maps of a square of ground, except that the lines do not represent any specific land elevation. They represent rise in elevation with small arrows to indicate the direction of rise.[4] The leftmost chart shows the land rising two feet from south to north. The middle shows the land rising two feet from west to east. The rightmost is a sum of those, with the land rising a total 4 ft from southwest to northeast.

Now look at an annulus, the plane region between two circles. Each chart in Fig. 3 makes sense locally. Each tells how the terrain rises or falls in any small region. Call them both *local contour maps*.

The left hand chart α in Fig. 3 could be contour lines on an annular region in a topographical map. Going once around α, in either direction, means rising a total of three feet and falling a total of three feet and so a net change of 0. But chart β on the right makes no sense globally. Going around that one once, counterclockwise, leaves a net rise of 6 ft. The terrain cannot meet itself! Call α but not β a *contour map*; or an actual, global contour map in contrast to a local contour map.

[4]The geometric point is that, like level lines on a topographical map, contour lines never cross or merge with each other; and never start or stop except at the edges of the chart.

Fig. 3 Two local contour maps on an annulus

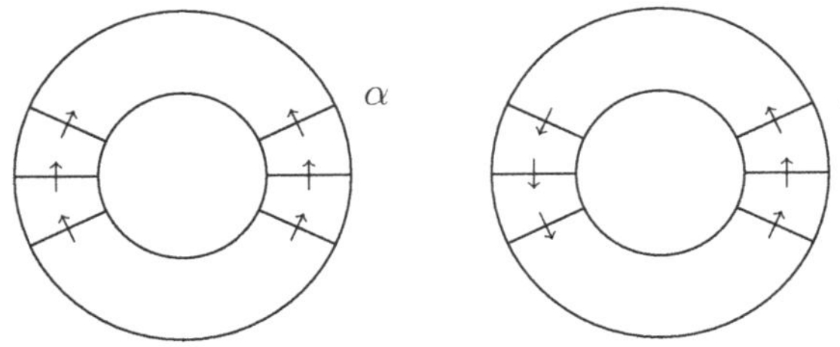

This is already enough to say something about dimension. A space M has *cohomological dimension* at least 1 if some local contour map on M is not global. So the annulus has cohomological dimension ≥ 1.

The 1-dimensional cohomology of any space M measures how many essentially different ways a local contour map on M can fail to be global. On the annulus there is only one way: the total change in elevation on any route around the annulus. This rests on a salient fact.

(1) Every local contour map on an annulus that gives total elevation change 0 on some round trip around the annulus is global. It could actually be the contour lines on a topographical map.

Combining this with the idea of adding contour maps illustrated in Fig. 2 gives the key to algebraizing cohomology:

(2) For every local contour map γ on an annulus there is a unique real number i_γ, called the *period* of γ, such that the difference $\gamma - (i_\gamma \cdot \beta)$ is global.[5]

The parameter i_γ in fact 2 has been called the *period* of γ at least since Riemann. The period i_γ is not unique to γ of course, but it does uniquely identify the *equivalence class* of γ in the *group of local contour maps modulo actual, global contour maps* on the annulus. This is the 1-dimensional cohomology group of the annulus, $\mathrm{H}^1(A)$. By definition it is the quotient group, and by fact 2 it is isomorphic to the additive group of real numbers $\mathbb{R}$.

2.3.2 Cohomology Groups and the Lefschetz Fixed Point Theorem

The existence of local but non-actual contour maps reveals the hole in the annulus. This works for other spaces as well. Consider the local contour maps in Fig. 4. α, β on a torus, and their sum $\alpha + \beta$. Dashes show the contour lines passing behind the torus.

Local contour map α reveals the hole in the center of the torus because, using α, a round trip counterclockwise around the outermost edge of the torus means a total rise in elevation of 1 foot. So α is not an actual contour map. A trip vertically around the tube shows β is not either—it surrounds the hole inside the hollow tube. Either

[5]Periods are often defined as line integrals, $i_\gamma = \oint \gamma$. This agrees with our definition when the line of the integral goes once around the annulus and α has $\oint \alpha = 1$.

one of these trips also works to show $\alpha + \beta$ is not global. And two salient facts for the torus correspond to the two given above for the annulus:

1′ Every local contour map on a torus that has total elevation change 0 on both the round trip around the periphery of the torus and the vertical trip around the tube is global.

2′ For every local contour map γ on a torus there are unique real numbers i_γ, j_γ such that the difference $\gamma - (i_\gamma \cdot \alpha + j_\gamma \cdot \beta)$ is global.

The 1-dimensional cohomology group of the torus, $\mathrm{H}^1(T)$, is defined as the quotient group of local contour maps modulo the actual, global ones. Fact 2′ shows it is isomorphic to the additive group $\mathbb{R}^2$ of real number pairs $\langle i, j \rangle$.

The local contour maps α, β on the usual torus have analogues on each handle of the n torus:

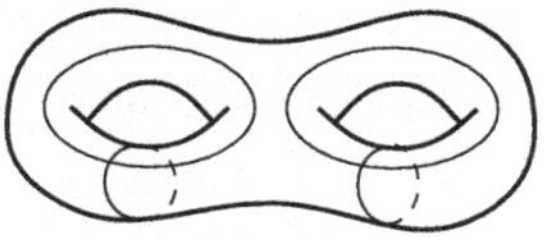

The 1-dimensional cohomology group of the n-torus, $\mathrm{H}^1(T^n)$, is defined by the quotient group of local contour maps modulo the actual, global ones. It is isomorphic to the additive group $\mathbb{R}^{2n}$ of $2n$-tuples of real numbers. The genus of a Riemann surface is characterized by its cohomology.

Higher dimensional cohomology groups $\mathrm{H}^k(M)$ for any space M are defined as quotient groups measuring how various other constructions can work locally on M while failing to work globally. They suffice to describe the dimension of M, higher dimensional analogues of genus, and everything needed to state and prove the Lefschetz fixed point theorem sketched in Sect. 2.2.

2.3.3 Sheaves, the Not at All Secret Key

The systematic way to define local structures on a space M is to define compatible global structures on open subspaces $U \subseteq M$ of M. The local but non-global contour map on the annulus in Sect. 2.3.1 is covered by global contour maps on these overlapping parts.

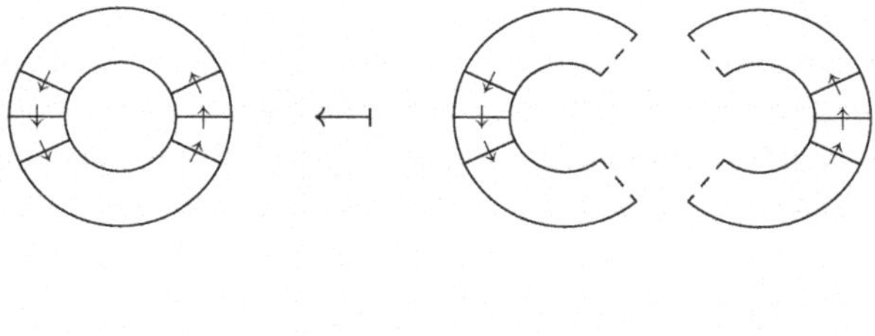

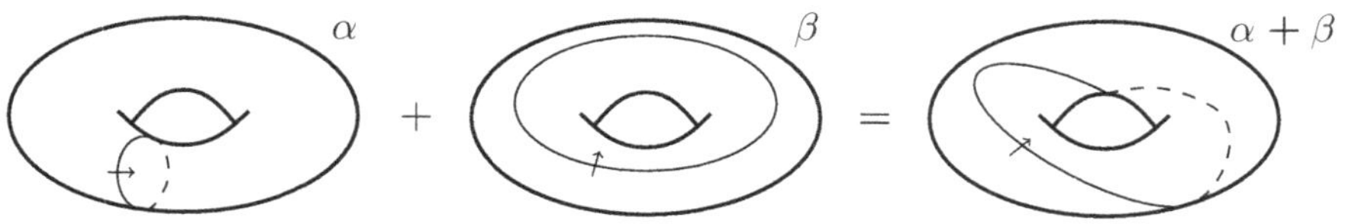

Fig. 4 Local contour maps on a torus

On each part the ground simply rises three feet from one end to the other. Where they overlap, both agree the ground does not rise.

Sheaves collect all local structures of some kind, on one space M, by organizing the actual structures of that kind on all open subsets $U \subseteq M$ of M.[6] There are now as in 1950 two materially different definitions which are equivalent in effect.

One says a sheaf $\mathcal{F}$ is a local homeomorphism $\mathcal{F}: E \to M$. That means $\mathcal{F}$ stacks some space E above M so that all small enough open parts $U \subseteq E$ map isomorphically onto their images. For example a spiral E maps to a circle M:

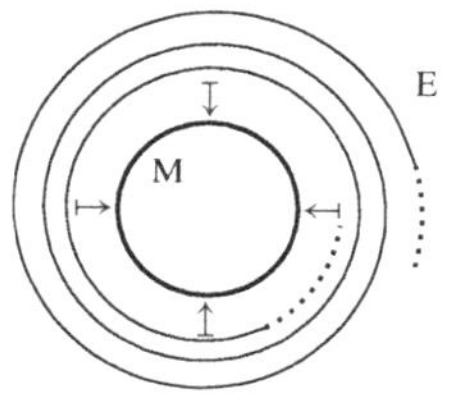

The other says a sheaf $\mathcal{F}$ on M assigns some structure $\mathcal{F}(U)$ to each open subset $U \subseteq M$. The relevant sheaves here are sheaves of Abelian groups. So the *sheaf of continuous functions* on M, or $\mathcal{O}^0_M$, assigns to each open subset $U \subseteq M$ the additive group $\mathcal{O}^0_M(U)$ of all continuous functions $f: U \to \mathbb{R}$. Here addition is defined pointwise: for each $U \subseteq M$ the sum of any $f, g: U \to \mathbb{R}$ is defined by

$$\text{For all } x \in U, \ (f + g)(x) = f(x) + g(x).$$

The *sheaf of contour maps* on M would assign to each open subset $U \subseteq M$ the group of all actual contour maps on U, with addition as suggested in Figs. 2 and 4. One precise version would be the sheaf of 1-forms on a manifold M.

Looking ahead to Sect. 3.2.1 note $\mathcal{O}^0_M(M)$ is the set of global real-valued functions on M. The sheaf of contour maps on M would assign to M itself the set of global contour maps on M.

Both formal definitions of sheaf are too long to give here. Grothendieck in *ReS* [21] merely analogizes sheaves to "meter sticks" measuring spaces and does not even name examples (p. P38). After all, his *Tôhoku* paper (1957) worked with sheaves in terms of simple relations in categories of sheaves rather than details of how sheaves

---

[6][32, pp. 252ff] has a gentle introduction, for more see Tennison [49].

are made. He had no reason to choose between sheaves as local homeomorphisms $\mathcal{F}\colon E \to M$ and sheaves as assignments of values $\mathcal{F}(U)$. The two give equivalent categories of sheaves over any space M. See Sect. 3.2.

3 Cohomology in Grothendieck's Words

3.1 Pictorial Visualization II: Compass Drawings

Grothendieck's experience with a compass relates more to philosophy than to geometry:

> I was interned in the concentration camp of Rieucros (near Mende). It is there that I learnt, from another prisoner, Maria, who gave me free private lessons, the definition of the circle. It impressed me by its simplicity and its evidence, whereas the property of "perfect rotundity" of the circle previously had appeared to me as a reality mysterious beyond words. It is at this moment, I believe, that (without being able to formulate it in these terms) I caught a glimpse of the creative power of a "good" mathematical definition. [22, p. 280].

Grothendieck explains he was drawing six-fold rosettes with a compass, as in the left hand figure above. This divides the circumference of the central circle in six parts with endpoints one radius apart. His schoolbook said the circumference is $2\pi R$, so he concluded $\pi = 3$ even though the book said it was around 3.14:

> As is typical, I discovered my error (which was to confuse the length of an arc with that of the chord connecting its endpoints) when I expressed my astonishment at the ignorance of my predecessors to someone else (a prisoner Maria who gave me free individual lessons in math and French) as I was about to show her why the circumference must be $6R$. (*ReS* [21, p. 263])

The contrast of arc and chord certainly has a visual component, but drawing the arc had not made him notice it. Maria's definition of the circle did. Telling the story years later Grothendieck concluded a child's ability to question authority is precious, as is the ability to recognize a mistake.

Around the time he recorded that story, he used the same compass method to draw the "12 petalled flower" or "sun" above on the right.[7] He labelled its vertices by cosmic concepts such as mystery, order, and freedom, to display their relations (*ReS* [21, p. PU46]). Like a commutative diagram, this drawing was made to organize conceptual information—not to visualize shapes. As De Toffoli and Goyvaerts put it

[7] He plays with the fact that the French *soleil* means both *sun* and *sunflower*.

these diagrams "do not rely on any kind of topological notions: what counts is their combinatorial structure" [11, p. 6].

3.2 The Sea Rises Around Cohomology

Grothendieck approached Weil cohomology by his typical *method of the rising sea*. He never says this is the one way to do math. To the contrary, he often says he could not have done what he did without the benefit of Serre's style which Grothendieck calls the *method of hammer and chisel*.[8] But Grothendieck passionately described this as his way see McLarty [37]. Without specifically seeking a Weil cohomology, he reflected on the problem. The first result was that cohomology itself became simple to him:

> The viewpoint of sheaves was the sure and silent guide, the effectual (and by no means secret) key. (*ReS* [21, p. P38])

But each sheaf contains unmanageably much information, most of it unrevealing. Rather than specific sheaves, Grothendieck looked at

> the set of all sheaves on a given topological space or, if you like, the prodigious arsenal of all "meter sticks" that measure it ... as equipped with its most evident structure, the way it appears so to speak "right in front of your nose"; that is what we call the structure of a "category." (*ReS* [21, p. P38])[9]

Members of Cartan's seminar [9] already defined the cohomology of a space M as a connected series of functors from the category $\mathbf{AbSh}_M$ of all sheaves of Abelian groups on M to the category **AbGrp** of ordinary Abelian groups.

$$\mathbf{AbSh}_M \xrightarrow{\mathrm{H}^i} \mathbf{AbGrp} \qquad\qquad i \in \mathbb{N}.$$

Each value $\mathrm{H}^i(\mathcal{F})$ is called the *i-th cohomology group of M with coefficients in $\mathcal{F}$*.

The seminar relied on "drawings (called 'diagrams') full of arrows covering the board" (*ReS* [21, p. 19]).[10] The vertices could represent sheaves, or Abelian groups,

---

[8] Marble sculptures are made by hammer and chisel.

[9] MacLane [33, Ch. VII] is a masterful account of Abelian categories. For homological algebra through derived functors and spectral sequences see Lang [27, Ch. XX].

[10] In e-mails of June and July 2004 Serre argues that Grothendieck mis-remembered the events. Certainly Grothendieck was wrong to say the 1948–49 seminar discussed spectral sequences (*ReS* [21, p. 19]) as Serre did not know of them then. Serre suggests Grothendieck did not often attend the seminar, whose contents would not have interested him at the time: "Grothendieck spent the year 48–49 in Paris (straight from his "licence" at Montpellier) and he stated in print several times that he attended the Cartan seminar of that year. I don't doubt this, but I have no memory of him then.... He probably got the written texts; I am not even sure he looked at them before 53 or even 54. They only influenced him in retrospect—just as a book one reads and finds interesting" (26 June 2004).

and the arrows represent morphisms. The diagram would show that if some morphisms have some properties then others exist with other properties. The most famous, because it is constantly used, is the *Snake Lemma*:

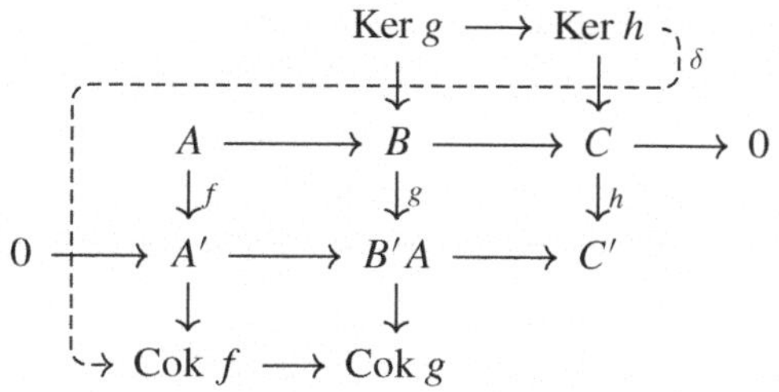

Proofs are given in numerous books, websites, and YouTube videos.

Grothendieck saw clearly what everyone in the seminar saw somehow: the diagrams for sheaves were the same as for plain Abelian groups. Everyone saw categories of sheaves are a lot like **AbGrp**. He asked precisely how that likeness makes cohomology work. While [8, 30, 31] asked a similar question, Grothendieck took it to unprecedented heights [37, pp. 305–11].

Grothendieck [18] gave a short list of axioms such that any category **A** satisfying those axioms will have a sequence of functors supporting all the general theorems of cohomology.

$$\mathbf{A} \xrightarrow{\mathrm{H}^i} \mathbf{AbGrp} \qquad\qquad i \in \mathbb{N}.$$

The standard tools of cohomology, notably including *spectral sequences*, all apply. This is *derived functor cohomology* and is a staple of current algebraic geometry, group cohomology, and much of algebraic topology.

Such a category is called an *Abelian category* and specifically an *AB5 category*. For every topological space M the sheaf category $\mathbf{AbSh}_M$ is AB5 and gives the now-standard cohomology of M. The module category over any ring is AB5 and for suitable rings it gives the now-standard cohomology of groups.

An expert in cohomology could feel, as many did in 1957, that AG's axioms are a mere abstraction from specific known theories, albeit a surprisingly apt and concise one. It made classical results radically easier to prove. But it also led to new results already in the Tôhoku paper. And the astonishing thing, even to Grothendieck, is how perfectly it later suited the then undreamt of theory of topos cohomology:

> Certainly, for more than one aspect of this new geometry (if not for all) no one, on the very eve of the day it appeared, could have dreamed of it—the worker himself no more than others. (*ReS* [21, p. P28])

3.2.1 Classical Topology and the Universal Property of Cohomology

Grothendieck in effect defines cohomology as the universal measure of the difference between local and global structures.[11]

First, it is intuitively clear that when a space $M = M' \cup M''$ is a union of overlapping open parts M', M'' then the passage between local and global structures on M can be divided into

(1) the passages on M' and M'' separately, and
(2) connections between those and the passage on their overlap $M' \cap M''$.

The division has been well understood since [50].

Fortunately the details are not important here. We only need say it is expressed by the *Mayer-Vietoris sequence* putting the cohomology groups of M and $M' \cap M''$ into an infinite exact sequence with the sum of groups $\mathrm{H}^i(M') \oplus \mathrm{H}^i(M'')$:

$$\begin{aligned} &\mathrm{H}^0(M) \to \mathrm{H}^0(M') \oplus \mathrm{H}^0(M'') \to \mathrm{H}^0(M' \cap M'') \to \mathrm{H}^1(M) \to \\ &\quad \mathrm{H}^1(M') \oplus \mathrm{H}^1(M'') \to \mathrm{H}^1(M' \cap M'') \to \mathrm{H}^2(M) \to \mathrm{H}^2(M') \oplus \mathrm{H}^2(M'') \to \cdots \end{aligned}$$

Every algebraic topology textbook shows this is a naturally intuitive approach to the cohomology of manifolds.

A *δ-functor* $F^i : \mathbf{A} \to \mathbf{AbGrp}$ on any AB5 category $\mathbf{A}$ is any infinite sequence of functors that gives Mayer-Vietoris-like exact sequences with sheaves $\mathcal{H}$, $\mathcal{K}$, $\mathcal{K}/\mathcal{H}$ in place of spaces whenever $\mathcal{H}$ is a subsheaf of $\mathcal{K}$ and $\mathcal{K}/\mathcal{H}$ is their quotient:

$$\begin{aligned} &F^0(\mathcal{H}) \to F^0(\mathcal{K}) \to F^0(\mathcal{K}/\mathcal{H}) \to F^1(\mathcal{H}) \to \\ &\qquad\qquad F^1(\mathcal{K}) \to F^1(\mathcal{K}/\mathcal{H}) \to F^2(\mathcal{H}) \to F^2(\mathcal{K}) \to \cdots \end{aligned}$$

Now let us say the formal candidates for the cohomology of a space M are the δ-functors F^i on $\mathbf{AbSh}_M$ with F^0 the global section functor. So F^0 assigns to any sheaf $\mathcal{H}$ the group $\mathcal{H}(M)$. That is, a formal candidate must apply to sheaves on M and must yield something formally like Mayer-Vietoris sequences. This very weak requirement does not by itself force F^i to be much like the cohomology of M.

The miracle is that derived functor cohomology on M is the unique (up to isomorphism) *initial object* in the category of formal candidates. It is the one formal candidate that maps uniquely into every formal candidate. In this precise sense, cohomology is the sequence of functors on $\mathbf{AbSh}_M$ that best "avoids positing extraneous elements" and "avoids imposing extraneous equations" among all sequences that could possibly express the passage between local and global.

[11] Grothendieck [18, p. 141], Tennison [49, p. 128], Hartshorne [25, p. 206], et alia.

4 The Spectacular Flight of the New Geometry

Grothendieck told the Edinburgh International Congress of Mathematicians:

> Serre's idea that a 'reasonable' algebraic principal fiber space with structure map G, defined on a variety V, if it is not locally trivial, should become locally trivial on some covering of V unramified over a given point of V....[12] has been the starting-point of a definition of the Weil cohomology ... which gives clear suggestions how Weil's conjectures may be attacked by the machinery of Homological Algebra. [19, p. 104]

For him the machinery of Homological Algebra was Tôhoku (1957).

With clear ties to classical topology and Galois theory, [46, p. 124] defined a new kind of 1-dimensional cohomology groups for algebraic varieties that gave the 0- and 1-dimensional part of Mayer-Vietoris-like sequences suited to the Weil conjectures. Serre saw great obstacles to higher dimensional cohomology of this kind. But Grothendieck was certain, as Deligne later put it, "given any category of sheaves a notion of cohomology groups results" [14, p. 16]. And a category of sheaves just meant any AB5 category. The problem for Grothendieck was to build an AB5 category out of Serre's *isotrivial covers* for any one scheme.

4.1 The Proper Object of Topology

Prima facie the method did not need to look like classical topology at all. It only had to produce an AB5 category. Yet the actual result was so close to classical that

> as the term "topos" itself is precisely intended to suggest, it seems reasonable to the authors of this seminar to take as the object of topology the study of toposes (and not only of topological spaces). [24, p. 3]

By the Spring of 1961 Grothendieck saw how to do topology, and specifically sheaf theory, using Serre's isotrivial covers of a scheme $f: E \to S$ instead of open subsets $U \subseteq S$ [20, p. 298 § 4.8].

> I come to the second couple I wanted to speak of, the notions of scheme and the closely related one of topos...[formalizing] the topological intuition of *localization*.... The flagrant needs of algebraic geometry led me to introduce one after the other schemes and toposes. This couple of notions had in them the power to produce a vast rebirth of algebraic geometry and also arithmetic, and topology, by a synthesis of these too-long separated "worlds" in one common geometric intuition. (*ReS* [21, p. 180])

The objects of any one topos act like sheaves on one space.[13] Yet Grothendieck also knew one topos is like one universe of sets in which much of mathematics can

---

[12]Grothendieck cites [45] which loosely presages unramified covers. The truly germane [46] was not yet in print.

[13]Sometimes Grothendieck distinguishes a *petit topos* of sheaves on a generalized space, from a *gros topos* which is a category of generalized spaces. Lawvere [28, 29] has developed this idea further. This distinction has nothing to do with set theoretic size. Gros and petit Grothendieck toposes are both proper classes in naive set theory.

be interpreted (see discussion in [36], pp. 358ff). The Abelian groups in this universe correspond to Abelian sheaves on a topological space or on a scheme. Grothendieck applies this viewpoint in several ways to cohomology (besides that William Lawvere and others have pursued much wider applications).

For example Grothendieck and Verdier [24, p. 207ff] show how various facts about base change of sheaves along a scheme morphism $f: S \to S'$ appear as simply linear algebra in the étale toposes $\mathbf{Sh}_S^{\text{ét}}$ and $\mathbf{Sh}_{S'}^{\text{ét}}$ on those schemes.

For two reasons internalization does not radically change the proofs:

(1) Working correctly inside a topos requires a bit of care.
(2) Internalization formalizes intuitions the experts already have about how sheaves of algebras, for example, are like plain algebras.

As to 1 though, internalization sometimes offers real simplifications. And as to 2 Grothendieck found real value in knowing these expert intuitions are not just metaphors. They are theorems of topos theory. And so, although toposes can be eliminated from proofs, the anonymous author(s) of the Introduction to SGA 6 "would advise the reader nonetheless to learn the topos language which furnishes an extremely convenient unifying principle" [5, p. VII].

4.2 *The Long Awaited Marriage of Geometry and Arithmetic*

However, the rebirth needed specifics. In Grothendieck's words the couple, schemes and topos, or arithmetic and topology, had to marry (*Res* [21, p. 24], and much passim). This is where Michael Artin came in.

In 1961 Michael Artin proved the first concrete theorem on higher dimensional étale cohomology [2, p. 359]. It was never published and no one I can contact is confident of recalling it. David Mumford suggests it was that the plane with origin deleted has non-trivial 3-dimensional cohomology H^3. This was a classical result for any classical cohomology theory using real number coefficients, when the "plane" is taken to be $\mathbb{C}^2$ for the complex numbers $\mathbb{C}$. Topologically, deleting the origin in $\mathbb{C}^2$ just gives 4-dimensional real space $\mathbb{R}^4$ with its origin deleted. That space contracts smoothly onto the 3-sphere so they have the same cohomology.

$$S^3 = \{\langle x, y, z, w\rangle \mid x^2 + y^2 + z^2 + w^2 = 1\} \subset \mathbb{R}^4.$$

Like every orientable 3-dimensional manifold, S^3 has non-trivial H^3. Specifically,, every orientable 3-dimensional manifold M has $\mathrm{H}^3_{\mathbb{R}}(M) = \mathbb{R}$.

Likely Artin proved his result also for $\mathbb{C}^2$ with deleted origin; but for finite étale coefficient groups. And he probably did not calculate the 3-dimensional cohomology for any coefficients. More likely he used a spectral sequence for a Künneth formula to show the H^3 are not always zero.

Whatever it was exactly, Artin's proof showed étale cohomology will not only work in principle, somehow. He made it prove a classical theorem of cohomology—

considerably changed no doubt, but little enough changed to be discoverable, and to work as desired. Grothendieck invited him to collaborate in the seminar creating *Théorie des topos et cohomologie étale* [3].

Then Deligne joined the seminar and extended the general theory of topos cohomology and the specific étale analogues of classical theorems. Volume 3 of SGA 4 [3] is devoted to Artin's and Deligne's progress on this, aiming for a Lefschetz fixed point theorem to count solutions to Weil's equations.

This was incredible. Many good experts did not believe it could work. The Lefschetz theorem, as sketched in Sect. 2.2, depends entirely on continuity of the manifold *M*. The arithmetic of whole numbers, let alone finite fields, should not be expected to produce continuous topologies. Indeed the relevant schemes have countable point sets. They cannot be continuous in anything like the sense that the classical Lefschetz theorem relies on. And yet it worked.

SGA 5 [26] proves a Lefschetz Theorem. Deligne [13] gives a summary of the proof plus improvements. The rest of the proof of the Weil conjectures did not go exactly as Grothendieck intended [12]. But it did hang on the Lefschetz Theorem in étale cohomology.

Grothendieck writes:

> The two consecutive seminars together, SGA 4 and SGA 5 (which are like one single "seminar" for me) develop simultaneously, from scratch, the powerful tool of synthesis and discovery which is the language of topos, and the perfectly adapted, perfectly efficacious tool which is étale cohomology—which is currently better understood in its essential formal properties than is the cohomology of ordinary topological spaces.... These two seminars are inseparably linked for me. In their unity they represent at once the vision, and the tool—toposes and a complete formalism of étale cohomology.... algebraic geometry in its most fascinating aspect for me—the "arithmetic" aspect, apprehended by intuition, concepts, and techniques all of "geometric" nature. (*Res* [21, p. 372f])

5 Logical Foundations and Mathematical Progress

Grothendieck's idea of a scheme is simpler than Weil's idea of an abstract variety in exactly the same way as the idea of a commutative ring is simpler than the idea of an algebraically closed field of infinite transcendence degree over its prime field. Those are the algebraic structures they start with. Of course detailed assumptions are sometimes needed. To prove a theorem that only holds over algebraically closed fields Grothendieck can look at that case. But for much basic algebraic geometry such assumptions are useless complications. For advanced work they can be damaging restrictions.

Grothendieck's typical method shows in his proof of a Riemann-Roch theorem for continuous families of varieties over any field. This goes beyond Hirzebruch's version for single varieties over the complex numbers:

> Grothendieck has generalized the theorem to the point where not only is it more generally applicable than Hirzebruch's version but it depends on a simpler and more natural proof. [7]

Grothendieck uses *base change* to turn any given continuous family to another simpler one. But, even starting with a single complex variety, that simpler family will not be a single variety. The simpler proof requires the more general idea of continuous families of varieties.

Grothendieck found simplicity came with unity, while unity and generality "are two aspects of one quest. Unity represents the profound aspect, and generality the superficial" (*Res* [21, p. PU 25]).

So when Grothendieck states theorems for all sheaves on a space, or all δ-functors on a category, generality is the superficial aspect. Unity and simplicity are the point. Avoiding irrelevant restrictions makes things simpler. The totality of all sheaves on one space forms a unifying context, a topos. The totality of all δ-functors on some AB5 category allows a unified conceptual definition of cohomology, by a universal property which in fact is constantly used in proofs.

As a byproduct, Grothendieck often invokes "sets" too large to exist on ordinary set theoretic foundations. Many mathematicians after him do the same even when they do not care to notice the fact. This is why Pierre Cartier says "Nowadays, one of the most interesting points in mathematics is that, although all categorical reasonings are formally contradictory, we use them and we never make a mistake" [16, p. 33]. He knows several fixes for the problem but also knows the most common approach today is to ignore it.

Grothendieck cared to get it right. He extended ordinary set theory by using *universes* [23]. McLarty [38] describes what difference this makes, how often it is used in the literature, and places where the issue arises and is ignored.

It is cryingly obvious that specific known proofs in number theory like those for the Weil conjectures, or Fermat's Last Theorem, do not truly need the logical strength of Grothendieck universes. But it is also obvious that, for example, antiques costing thousands of dollars at a top gallery can be bought for a tenth of that—if you have weeks to spend at estate sales. Using higher logical strength (like spending more money) saves time and effort.

In fact, though, the very things Grothendieck bases on universes, such as toposes, the derived categories of toposes and so on, can be had in the same way at far lower strength. The key point here was necessarily unavailable to Grothendieck at the time. It appears in hindsight knowing what theorems actually occur in the SGA and related works.

None of their proofs but one ever uses the *axiom scheme of replacement*. That one lies at the base of Tôhoku (1957, p. 135) and is constantly used in the SGA. It is the proof that AB5 categories have *enough injectives*. But it can be rewritten without replacement. With that change, none of Grothendieck's cohomology uses replacement or even *unbounded separation*.

So the entire SGA could be simply re-edited changing nothing but the few pages defining universes, so the same proofs still work verbatim, with all the conceptual unity Grothendieck created, at the logical strength of *higher order arithmetic* [39]. Naively put, there are natural numbers, sets of natural numbers, sets of those sets, and so on through any finite iteration of sets of sets—but only finite iterations. This is also the strength of Simple Type Theory, or Bounded Zermelo Set Theory, or the Elementary Theory of the Category of Sets. It is a very natural strength for ordinary abstract mathematics. It is overtly the weakest foundation that can preserve the unifying theory of toposes since it has the logical strength of saying there is one elementary topos with a natural number object.

On the other hand, this is still far stronger than most concrete arithmetic and analysis truly need. Aside from things like Gödel sentences, most existing concrete arithmetic and analysis is currently known to require no more logical strength than Peano Arithmetic (PA) see Friedman and Simpson [17, 48]. Strictly speaking PA has only numbers and not even sets of them. But it can interpret a very limited amount of set theory. Working at the strength of PA means using cohomology only in ways that can be eliminated in principle. That requires putting numerical bounds on specific steps of individual proofs. Number theorists often do explore such bounds quite apart from any concern with logic, because bounds can be interesting and difficult to prove. Macintyre [34, Appendix] gives extensive, wide ranging evidence that all the proofs in current cohomological number theory can be so bounded. For what my opinion is worth I expect that is true. But Macintyre cites numerous steps where it is not now actually known. It is just very plausible. He predicts it would take a great deal of effort, including substantial new theorems of number theory, to prove the relevant bounds exist.

Grothendieck did not care about that. Using large sets is easier. But he linked working, conceptual, and logical foundations all to progress. He was discouraged as a student in Montpellier when "Professor Soula assured me that the last problems to have been posed in mathematics had been solved 20 or 30 years ago by one Lebesgue" (*Res* [21, p. 33]). But he heard the Parisians would know if anything was new.[14] Preparing to go there he learned some set theoretic foundations and wrote about it to a friend. He concluded:

> As you have seen it has been a slow process for mathematicians to work their way to the fundamentals of their concepts; almost against their will, they have for a time turned away from the unresolved formalism of classical algebra and analysis, in order to break down their concepts, theories, and results into their truly elementary parts. One may say that they have succeeded and that some among them have at last developed the mind set which allows them to seek the fundamentals of every definition, and then to investigate the essential formal elements in every theory and theorem, which may permit them to restructure what has been observed of an already known theory, or to extend it to more general conclusions. (letter of July 29, 1948 quoted in [43], p. 163).

References

1. The abbreviation *ReS* in citations refers to Grothendieck (1985–1987)
2. M. Artin, Interview, in *Recountings: Conversations with MIT Mathematicians*, ed. by J. Segel (A K Peters/CRC Press, Wellesley, MA, 2009), pp. 351–74

[14] See the work in progress [44].

3. M. Artin, A. Grothendieck, J.-L. Verdier, *Théorie des Topos et Cohomologie Etale des Schémas*. Séminaire de géométrie algébrique du Bois-Marie, 4. (Springer, Berlin, 1972). Three volumes, cited as SGA 4
4. M. Atiyah, Bakerian lecture, 1975: global geometry. Proc. R. Soc. Lond. Ser. A **347**(1650), 291–99 (1976)
5. P. Berthelot, A. Grothendieck, L. Illusie, *Thórie des intersections et théorème de Riemann-Roch*. Number 225 in Séminaire de géométrie algébrique du Bois-Marie, 6. Springer, Berlin. Generally cited as SGA 6 (1971)
6. A. Borel, J.-P. Serre, Le théorème de Riemann-Roch. Bull. Soc. Math. Fr. **86**, 97–136 (1958)
7. R. Bott, Review of A. Borel and J-P. Serre, Le théorème de Riemann-Roch. Bull. Soc. Math. Fr. **86**, 1958, 97–136 (1961). *Mathematical Reviews*, (MR0116022 (22 #6817))
8. D.A. Buchsbaum, Exact categories and duality. Trans. Am. Math. Soc. **80**, 1–34 (1955)
9. H. Cartan, Séminaire Henri Cartan, vol. I. (Secrétariat Mathématique, École Normale Supérieure, Paris, 1949)
10. A. Connes, Geometry and the quantum. In this volume (2017)
11. S. De Toffoli, I. Goyvaerts, Aspects of diagrammatic reasoning in category theory. Draft in progress (2017)
12. P. Deligne, La conjecture de Weil I. In *Publications Mathématiques*, **43**, 273–307. Institut des Hautes Études Scientifiques (1974)
13. P. Deligne (eds), *Cohomologie Étale*. Séminaire de géométrie algébrique du Bois-Marie; SGA 4 1/2. Springer, Berlin. Generally cited as SGA 4 1/2, this is not strictly a report on Grothendieck's Seminar (1977)
14. P. Deligne, Quelques idées maîtresses de l'œuvre de A. Grothendieck, in *Matériaux pour l'Histoire des Mathématiques au XXe Siècle* (Nice, 1996), pp. 11–19. Soc. Math. France (1998)
15. D. Eisenbud, J. Harris, *The Geometry of Schemes* (Springer, Berlin, 2000)
16. J. Fresán, The Castle of groups interview with Pierre Cartier. Newsl. Eur. Math. Soc. **74**, 31–34 (2009)
17. H. Friedman, Mathematically natural concrete incompleteness. On-line at u.osu.edu/friedman.8/files/2014/01/Putnam062115pdf-15ku867.pdf (2015)
18. A. Grothendieck, Sur quelques points d'algèbre homologique. Tôhoku Math. J. **9**, 119–221 (1957)
19. A. Grothendieck, The cohomology theory of abstract algebraic varieties, in *Proceedings of the International Congress of Mathematicians* (Cambridge University Press, Cambridge, 1958), pp. 103–18
20. A. Grothendieck, *Revêtements Étales et Groupe Fondamental*. Séminaire de géométrie algébrique du Bois-Marie, 1. (Springer, Berlin, 1971). Generally cited as SGA 1
21. A. Grothendieck, *Récoltes et Semailles*. Université des Sciences et Techniques du Languedoc, Montpellier. Published in several successive volumes (1985–1987)
22. A. Grothendieck, Esquisse d'un programme, in ed. by L. Schneps, P. Lochak, *Geometric Galois Actions: 1. Around Grothendieck's Esquisse d'un Programme*, (Cambridge University Press, Cambridge, 1997), pp. 5–48. French original, 243–84 English translation
23. A. Grothendieck, J.-L. Verdier, Préfaisceaux, in ed. by M. Artin, A. Grothendieck, J.-L. Verdier *Théorie des Topos et Cohomologie Etale des Schémas*, vol. 1 of *Séminaire de géométrie algébrique du Bois-Marie, 4*, (Springer, Berlin, 1972a), pp. 1–218
24. A. Grothendieck, J.-L. Verdier, Topos, in M. Artin, A. Grothendieck, J.-L. Verdier *Théorie des Topos et Cohomologie Etale des Schémas*, vol. 1 of *Séminaire de géométrie algébrique du Bois-Marie, 4*, (Springer, Berlin, 1972b), pp. 299–519
25. R. Hartshorne, *Algebraic Geometry* (Springer, Berlin, 1977)
26. L. Illusie, A. Grothendieck, *Formule de Lefschetz, Cohomologie l-adique et Fonctions L. SGA5*. Number 589 in Séminaire de géométrie algébrique du Bois-Marie, 5. Springer, Berlin. Generally cited as SGA 5 (1977)
27. S. Lang, *Algebra*, 3rd edn. (Addison-Wesley, Reading, MA, 1993)

28. F.W. Lawvere, Categories of spaces may not be generalized spaces as exemplified by directed graphs. *Revista Colombiana de Matematicas*, **XX**, 179–186. Republished in: *Reprints in Theory and Applications of Categories*, No. 9 (2005) pp. 1–7, available on-line at http://www.tac.mta.ca/tac/reprints/articles/9/tr9abs.html (1986)
29. F.W. Lawvere, Categories of space and quantity, in *The Space of mathematics*, ed. by J. Echeverria, A. Ibarra, T. Mormann (de Gruyter, New York, 1992), pp. 14–30
30. S. MacLane, Groups, categories and duality. Proc. Nat. Acad. Sci. U.S.A. **34**, 263–267 (1948)
31. S. MacLane, Duality for groups. Bull. Am. Math. Soc. **56**, 485–516 (1950)
32. S. MacLane, *Mathematics: Form and Function* (Springer, Berlin, 1986)
33. S. MacLane, *Categories for the Working Mathematician*, 2nd edn. (Springer, New York, 1998)
34. A. Macintyre, The impact of Gödel's incompleteness theorems on mathematics, in *Kurt Gödel and the Foundations of Mathematics: Horizons of Truth*, pp. 3–25. Proceedings of Gödel Centenary, Vienna, 2006 (2011)
35. H. McKean, V. Moll, *Elliptic Curves: Function Theory, Geometry, Arithmetic* (Cambridge University Press, Cambridge, 1999)
36. C. McLarty, The uses and abuses of the history of topos theory. Brit. J. Philos. Sci. **41**, 351–75 (1990)
37. C. McLarty, The rising sea: Grothendieck on simplicity and generality I, in *Episodes in the History of Recent Algebra*, ed. by J. Gray, K. Parshall (American Mathematical Society, Providence, RI, 2007), pp. 301–326
38. C. McLarty, What does it take to prove Fermat's Last Theorem? Bull. Symbolic Logic **16**, 359–77 (2010)
39. C. McLarty, The large structures of grothendieck founded on finite order arithmetic. Preprint on the mathematics arXiv, arXiv:1102.1773v3 (2011)
40. C. McLarty, How Grothendieck simplified algebraic geometry. Not. Am. Math. Soc. **63**(3), 256–65 (2016)
41. D. Mumford, *The Red Book of Varieties and Schemes* (Springer, Berlin, 1988)
42. B. Riemann, *Grundlagen für eine allgemeine Theorie der Functionen einer veränderlichen complexen Grösse*. Inauguraldissertation, Universität Göttingen. Reprinted in R. Dedekind and H. Weber eds. *Bernhard Riemann's Gesammelte mathematische Werke*. Leipzig: B.G. Teubner, 1876, pp. 3–45 (1851)
43. W. Scharlau, *Anarchy*, vol. 1 of *Who is Alexander Grothendieck?: Anarchy, Mathematics, Spirituality, Solitude*. Books on Demand (2011)
44. L. Schneps, (to appear). *Mathematics*, vol. 2 of *Who is Alexander Grothendieck?: Anarchy, Mathematics, Spirituality, Solitude*. online at grothendieckcircle.org
45. J.-P. Serre, Sur la topologie des variétés algébriques en caractristique p, in *Symposium Internacional de Topologa Algebraica (1956)*, pp. 24–53. La Universidad Nacional Autónoma de Mexico y la UNESCO (1958a)
46. J.-P. Serre, Espaces fibrés algébriques. In *Séminaire Chevalley*, chapter exposé no. 1. Secrétariat Mathématique, Institut Henri Poincaré (1958b)
47. J.-P. Serre, Géométrie algébrique, in *Proceedings International Congress of Mathematicians* (Stockholm, 1962) pp. 190–196. Inst. Mittag-Leffler, Djursholm (1963)
48. S. Simpson, *Subsystems of Second Order Arithmetic* (Cambridge University Press, Cambridge, 2010)
49. B. Tennison, *Sheaf Theory* (Cambridge University Press, Cambridge, 1975)
50. L. Vietoris, Über die Homologiegruppen der Vereinigung zweier Komplexe. Monatshefte für Mathematik und Physik **37**(1), 159–62 (1930)
51. A. Weil, Number of solutions of equations in finite fields. Bull. Am. Math. Soc. **55**, 487–95 (1949)
52. A. Weil, Number theory and algebraic geometry, in *Proceedings of the International Congress of Mathematicians* (Cambridge, MA, 1950), pp. 90–100. American Mathematical Society (1952)

Understanding the 6-Dimensional Sphere

Michael Atiyah

Dedicated to the memory of Shing-Shen Chern

Abstract In [1] I gave a proof of the long-standing conjecture that the 6-dimensional sphere has no complex structure. In this paper I will present the proof in a more transparent manner. I use the example of the 6-sphere to shed new light on many problems of physics. In the future I expect these ideas will provide a different perspective, with substantial benefits in all areas.

1991 Mathematics Subject Classification. Primary 53A55 · Secondary 53B15

1 Introduction

In [1] I gave a short proof of the conjecture that the 6- dimensional sphere has no complex structure. The proof, though short, was intricate and it failed to convince the many experts in the field. Indeed there were many sound reasons to be sceptical, notably

(1.1) the integrability of the hypothetical complex structure was never used,

(1.2) another elderly distinguished mathematician, S. S.Chern, had claimed to have found a proof which was subsequently shown by Bryant [11] to prove something weaker and it was not clear how much symmetry the new purported proof assumed, which suggested a gap like that in Chern's proof.

M. Atiyah (✉)
Trinity College, Cambridge and the University of Edinburgh, Edinburgh, UK
e-mail: M.Atiyah@ed.ac.uk

M. Atiyah
School of Mathematics, University of Edinburgh, James Clerk Maxwell Building, Peter Guthrie Tait Road, Edinburgh EH9 3FD, UK

J. Kouneiher (ed.), *Foundations of Mathematics and Physics One Century After Hilbert*,
https://doi.org/10.1007/978-3-319-64813-2_5

Having listened carefully to all these objections I have found a new and even shorter proof which I will explain in this paper. In particular I will explicitly use a hypothetical complex structure without any symmetry assumptions, thus avoiding Chern's error. But in homage to a great man who, despite old age, was not afraid to tackle difficult problems in Geometry, I decided to dedicate this paper to his memory.

I will use one new idea which greatly simplifies the presentation, making it more transparent. This is to consider not the round sphere S^6, but the conformal sphere $\mathbf{S}^6$. I will begin in Sect. 2 by recalling the geometry behind relativistic physics.

2 Relativistic Physics

The conformal 6-sphere $\mathbf{S}^6$ is the base of the light-cone in 8-dimensional Minkowski space and is the boundary of hyperbolic 7-space. As such it has no natural Riemannian metric, but it has a natural conformal structure which is the limit of space-like Riemannian manifolds $S(c)$ with natural hyperbolic metrics. Here c is the velocity of light and the limit is that in which c tends to infinity. For small c, $S(c)$ represents slow motion. There is also the light-like region where $c < 0$. Natural always means compatible with the appropriate symmetry group, which here is the Lorentz group of automorphisms of Minkowski space $\mathbb{R}(7, 1)$. This is a non-compact Lie group as opposed to the compact symmetry group SO(8) of the round sphere. The failure to distinguish clearly between the roles of these two groups is ultimately the source of much confusion which I will now strive to dispel. There is another difference between Minkowski and Euclidean space: the automorphism group of Minkowski space has 4 components, coming from time reversal T and spatial orientation reversal(parity) P. There is also charge reversal C which, in a Kaluza-Klein frame-work, is a circle reversal. The famous CPT theorem asserts formally that $CPT = 1$. This holds in both 4 and 8 dimensions and is elementary. Outwith the Kaluza-Klein framework it just defines C as the inverse of PT. Consider Minkowski 8-space with coordinates $\mathbf{x}, \mathbf{y}, t$ where $\mathbf{x}$ is a 4-vector, $\mathbf{y}$ a 3-vector and t a scalar. The quadratic equation, for positive real t,

$$\|\mathbf{x}\|^2 + \|\mathbf{y}\|^2 = c^2 t^2 \tag{2.1}$$

defines $S(c)$, the hyperbolic approximation to the conformal $\mathbf{S}^6$, if c is real and positive (with t being time and c the velocity of light). If $c' < c$ then $S(c')$ is inside $S(c)$, and all $S(c)$ are diffeomorphic. It follows that any differential-topological invariant is the same for all positive c and hence for the limit manifold $\mathbf{S}^6$. Our aim is to understand, in a fundamental way, why the base of the standard Minkowski light cone S^2 has a complex structure while its 6-dimensional analogue does not.

3 Finite Symmetries

The proof in [1] was difficult to comprehend partly because it did not use the homogeneity of $\mathbf{S}^6$, and would have applied to other 6-manifolds satisfying weaker conditions. Now I will use the fact that $\mathbf{S}^6$ is a homogeneous space of the conformal group $\mathrm{Spin}^\dagger(7, 1)$, which preserves future & past. More explicitly, any two distinct points on $\mathbf{S}^6$ can be transformed by the connected group $\mathrm{Spin}^\dagger(7, 1)$ into the standard anti-podal pair on the 3-sphere given by (2.1) with $y = 0$. Thus $\mathbf{S}^6$ is naturally the homogeneous space

$$\mathrm{Spin}^\dagger(7, 1)/(\mathbb{Z}/2 \times \mathrm{Spin}(6))$$

with a compact isotropy group having two components. The $\mathbb{Z}/2$ factor interchanges (x, t) and $(-x, -t)$, switching the two poles. This group acts on $\mathbf{S}^6$ intransitively with the 1-parameter family of 5-dimensional orbits indexed by c described in Sect. 1. It is important, as in all physics since Dirac, to use the Spinor groups instead of the orthogonal groups. Let me now review briefly the relation between representation theory and K-theory. We can start with finite groups where it is classical algebra, but which I will formulate geometrically, so that a G-vector bundle V on G/H is the same as a representation of H. More precisely the action of H on V at the base point (restriction) is naturally isomorphic to the Poincaré dual, push-forward (induction). The same is true for complex reductive Lie groups where the push-forward is holomorphic induction. This is the Bott-Borel-Weil Theorem BBW [10] and can be understood via the Hirzebruch-Riemann-Roch Theorem (HRR). For the round sphere S^6, the compact group Spin(7) acts transitively so that index theory takes values in the representation ring $R(\mathrm{Spin}(7))$ but, for the conformal sphere $\mathbf{S}^6$, the compact group $\mathbb{Z}/2 \times \mathrm{Spin}(6)$, is not transitive so the index takes values in the smaller representation ring $R(\mathbb{Z}/2 \times \mathrm{Spin}(6))$. In fact, for our purposes, we can restrict even further to the maximal 2-group of $\mathrm{Spin}^\dagger(7, 1)$, which is the quaternion group Γ of order 8. HRR is a special case of the Atiyah-Singer index theorem AS [3] and it has an equivariant form for compact groups [4]. The main features of AS, as opposed to the classical HRR are

(i) It is metric-independent
(ii) only an almost complex structure is needed.

For these reasons index theory leads to complex cobordism invariants, for almost complex manifolds as shown by Milnor and Quillen. Using equivariant K theory, index theory will do the same for equivariant cobordism. I will only use equivariance for the maximal 2-group concerned, which for $\mathbf{S}^6$ is the quaternion group Γ of order 8. It is crucial to note that Γ acts on the conformal sphere $\mathbf{S}^6$ **without invoking any additional symmetry**. This is clear from the background physics where C, P, T are universally applicable. Geometrically, at each point, the 3 complex variables can be conjugated. To understand this action in more detail let me recall that Γ is a central extension by $\mathbb{Z}/2$ of the abelian quotient $\mathbb{Z}/2 \times \mathbb{Z}/2$, giving the exact sequence

$$0 \to \mathbb{Z}/2 \to \Gamma \to \mathbb{Z}/2 \times \mathbb{Z}/2 \to 0 \tag{3.1}$$

Geometrically the abelian quotient is just the conjugation action on $\mathbb{C}^3$, preserving orientation. The central $\mathbb{Z}/2$ is not local but global, switching the base-point and its opposite, compatibly with the local complex structures. Note that this global $\mathbb{Z}/2$ action is complex conjugation in the sense of [2]. An alternative and more sophisticated physical way of understanding Γ is to use both orientations of Minkowski space and work with $\mathbb{Z}/2$ graded objects (Fermions & Bosons). This incorporates Spin (or Pin) and the central extension. Representations can be either Fermionic (odd) or Bosonic (even). The index, denoted by Trace$(-1)^F$, is a topological invariant. For Minkowski space $\mathbb{R}(7, 1)$ this index is odd, while a complex structure would have even index. This way of explaining the Theorem may be more attractive to physicists, but this paper like [1] is intended for mathematicians, usually wary about physical arguments. With these preliminaries out of the way I will, in Sect. 4, explain how to prove the theorem.

4 Proof of the Theorem

Theorem The 6-dimensional sphere has no complex structure.

Proof I apply the Γ-equivariant Atiyah-Singer Index theorem to the hypothetical complex structure on $\mathbf{S}^6$ with its Γ action. The index is a representation of Γ. Since the central $\mathbb{Z}/2$ acts trivially on the complex structure the index is an abelian representation. This has used the hypothetical complex structure. On the other hand, forgetting about this complex structure and just using its manifold structure, we can choose any Γ invariant metric to calculate the Γ index. Naturally we choose the round metric and see at once that Γ acts freely, and its index is the regular representation. This is faithful and so non-abelian.
The contradiction is now evident and the Theorem is proved.

Notes:

4.1 Because Γ is intrinsic to $\mathbf{S}^6$, any complex structure on $\mathbf{S}^6$ has to have an action of Γ.
4.2 Because of property (ii) in Sect. 3, the complex integrability can be dispensed with as in [1].
4.3 In [1] I used $KR(p, q)$ theory. In this paper, standard K-theory is adequate, because the representations of Γ embody the same refinements.
4.4 The reason why $\mathbf{S}^2$ differs from $\mathbf{S}^6$ is that the analogue of Γ is now of order 2 and so is abelian.

5 Future Generalization

This volume was designed to look at the physics/mathematics interface pointing towards the future. The problem about $\mathbf{S}^6$ is a good example of a frontier problem, since it is the base of the light-cone in Minkowski 8-space. My approach to this problem is a good illustration of new techniques, which have old roots but whose full potential has not been thoroughly exploited. In fact the applications of these techniques are extensive and central to the physics/mathematics interface. Here is a brief list of some topics to be explored in the future.

5.1 Index theory and atoms [5]
5.2 Equivariant index theory for real forms of reductive Lie groups and its relation to the work of Harish-Chandra [6, 13]
5.3 Index theory, monopoles and the Yang-Baxter equations on curved backgrounds [9]
5.4 de Broglie-Bohm pilot waves and solitons
5.5 Duality between the discrete (Bohr) and the continuous (Einstein)
5.6 Quaternionic index theory and the work of Kazhdan-Lusztig [8]
5.7 Index theory, the magic square and the Hopf-Kervaire invariants [7, 12].

References

1. M. Atiyah, *The nonexistent 6-sphere*. arXiv:1610.09366
2. M. Atiyah, K-theory and reality. Quart. J. Math., Oxford (2), **17**, 367–386 (1966)
3. M. Atiyah, I.M. Singer, The index of elliptic operators on compact manifolds. Bull. Amer. Math. Soc. **69**, 422–433 (1963)
4. M. Atiyah, G.B. Segal, The index of elliptic operators II. Ann. of Math. **87**2, 531–545 (1968)
5. M. Atiyah, *Geometric models of Helium*. arXiv:1703.02532
6. M. Atiyah, W. Schmid, A geometric construction of the discrete series for semisimple Lie groups. Invent. Math. **42**, 1–62 (1977)
7. M. Atiyah, J.F. Adams, K-theory and the Hopf invariant. Quart. J. Math. Oxford (2) **17**, 31–38 (1966)
8. M. Atiyah, R. Bielawski, Nahm's equations, configuration spaces and flag manifolds. Bull. Bras. Math. Soc. (N.S.) **33**, 157–176 (2002)
9. M. Atiyah, Magnetic monopoles and the Yang-Baxter equations. Topological methods in quantum field theory (Trieste, 1990). Internat. J. Modern Phys. A **6**(16), 2761–2774 (1991)
10. R. Bott, Homogeneous vector bundles. Ann. of Math. **66**2, 203–248 (1957)
11. R. Bryant, *S.S. Chern's study of almost-complex structures on the six-sphere*. arXiv:1405.3405
12. M. Hill, M.J. Hopkins, D. Ravenel, On the nonexistence of elements of Kervaire invariant one. Ann. of Math. (2) **184**(1), 1–262 (2016)
13. R. Parthasarathy, Dirac operator and the discrete series. Ann. of Math. (2) **96**, 1–30 (1972)

A Dozen Problems, Questions and Conjectures About Positive Scalar Curvature

Misha Gromov

Abstract Unlike manifolds with positive sectional and with positive Ricci curvatures which aggregate to modest (roughly) convex islands in the vastness of all Riemannian spaces, the domain $\{\mathcal{SC} > 0\}$ of manifolds with *positive scalar curvatures* protrudes in all direction as a gigantic octopus or an enormous multi-branched tree. Yet, there are certain rules to the shape of $\{\mathcal{SC} > 0\}$ which limit the spread of this domain but most of these rules remain a guesswork. In the present paper we collect a few "guesses" extracted from a longer article, which is still in preparation: *100 Questions, Problems and Conjectures around Scalar Curvature*. Some of these "guesses" are presented as questions and some as conjectures. Our formulation of these conjectures is not supposed to be either most general or most plausible, but rather maximally thought provoking.

1 Definition of Scalar Curvature

The scalar curvature of a C^2-smooth Riemannian manifold $X = (X, g)$, denoted $Sc = Sc(X) = Sc(X, g) = Sc(g)$ is a continous function on X, written as $Sc(X)(x)$ and $Sc(g)(x)$, $x \in X$, which is uniquely characterised by the following four properties.

$\bullet_1$ *Additivity under Cartesian-Riemannian Products.*

$$Sc(X_1 \times X_2, g_1 \oplus g_2) = Sc(X_1, g_1) + Sc(X_2, g_2),$$

where this equality is understood point-wise,

$$Sc(X_1 \times X_2)(x_1, x_2) = Sc(X_1)(x_1) + Sc(X_2)(x_2).$$

$\bullet_2$ *Scale covariance.*

M. Gromov (✉)
Institut des Hautes Etudes Scientifiques, Les Ulis, France
e-mail: gromov@ihes.fr

J. Kouneiher (ed.), *Foundations of Mathematics and Physics One Century After Hilbert*,
https://doi.org/10.1007/978-3-319-64813-2_6

$$Sc(X, \lambda^2 \cdot g) = \lambda^2 \cdot Sc(X). \text{ for all real } \lambda > 0.$$

Thus, for instance, since $(\mathbb{R}^n, g_0)$ is isometric to $(\mathbb{R}^n, \lambda^2 \cdot g_0)$ for the Euclidean metric g_0,

$$Sc(\mathbb{R}^n) = 0 \text{ for all } n = 1, 2, 3, \ldots.$$

$\bullet_3$ *Volume Comparison.* If the scalar curvatures of n-dimensional manifolds X and X' at some points $x \in X$ and $x' \in X'$ are related by the strict inequality

$$Sc(X)(x) < Sc(X')(x'),$$

then the Riemannian volumes of small balls around these points satisfy

$$vol(B_x(X, \varepsilon)) > vol(B_{x'}(X', \varepsilon))$$

for all sufficiently small $\varepsilon > 0$.

This volume inequality, in agreement with $\bullet_1$, is *additive under Riemannian products* if

$$vol(B_{x_i}(X, \varepsilon)) > vol(B_{x'_i}(X_i', \varepsilon)), \text{ for } \varepsilon \leq \varepsilon_0,$$

and *for all points* $x_i \in X_i$ and $x'_l \in X'_i$, $i = 1, 2$, then

$$vol_n(B_{(x_1, x_2)}(X_1 \times X_2, \varepsilon_0)) > vol_n(B_{(x'_1, x_2')}(X'_1 \times X'_2, \varepsilon_0)$$

for all $(x_1, x_2) \in X_i \times X_2$ and $(x'_1, x'_2) \in X'_1 \times X'_2$.

This follows from the Pythagorean formula

$$dist_{X_1 \times X_2} = \sqrt{dist^2_{X_1} + dist^2_{X_2}}.$$

and the Fubini theorem applied to the "fibrations" of balls over balls:

$$B_{(x_1, x_2)}(X_1 \times X_2, \varepsilon_0)) \to B_{x_1}(X_1, \varepsilon_0) \text{ and } B_{(x'_1, x'_2)}(X'_1 \times X'_2, \varepsilon_0)) \to B_{x_1}(X'_1, \varepsilon_0),$$

where the fibers are balls of radii $\varepsilon \in [0, \varepsilon_0]$ in X_2 and X'_2.

$\bullet_4$ *Normalisation/Convention for 2-spheres.* The unit sphere $S^2 = S^2(1)$ has constant scalar curvature 2 (twice the sectional curvature).

It is an elementary exercise to prove the following.

$\bigstar_1$ *The function $Sc(X, g)(x)$ which satisfies $\bullet_1$-$\bullet_4$ exists and is unique;*

$\bigstar_2$ *The unit spheres and the hyperbolic spaces with $sect.curv = -1$ satisfy*

$$Sc(S^n(1)) = n(n-1) \text{ and } Sc(H^n(-1)) = -n(n-1).$$

Thus,

$$Sc(S^n(1) \times H^n(-1)) = 0 = Sc(\mathbb{R}^n),$$

which implies that the volumes of the small balls in $S^n(1) \times H^n(-1)$ are "very close" to the volumes of the Euclidean $2n$-balls.

$\bigstar_3$ *The scalar curvature of a Riemannian manifold X is equal to the sum of the values of the sectional curvatures at the bivectors of an orthonormal frame*[1] *in X,*

$$Sc(X)(x) = \sum_{i,j} c_{ij}, i, j = 1, \ldots, n.$$

For example, all *compact Riemannian symmetric spaces X*, except for the n-torus $\mathbb{T}^n$, have $Sc(X) > 0$, while $\mathbb{T}^n$, being covered by $\mathbb{R}^n$, has $Sc(\mathbb{T}^n) = 0$.

It may be tempting to take the above $\bullet_1 - \bullet_4$ for a definition of scalar curvature for singular metric spaces X. In fact, it may work for X with moderate singularities, e.g. for *Alexandrov's spaces with sectional curvatures bounded from below* (see [1]), where the properties of the so defined scalar curvature must be comparable to what is observed in the smooth case (see Sect. 7).

Yet, volumes of balls do not touch the heart of the scalar curvature; we suggest an alternative in Sect. 7.

2 Soft and Hard Facets of Scalar Curvature

We are not so much concerned with the scalar curvature $Sc(X)$ per se, but rather with the effect of *lower scalar curvature bounds* on the geometry and the topology of X, where, for instance, the inequality "$Sc(X) > 0$" can be **defined** by saying that

all sufficiently small balls $B_x(\varepsilon) \subset X$, $\varepsilon \leq \varepsilon_0(x) > 0$, have volumes smaller than the volumes of the equividimensional Euclidean ε-balls.

Then "$Sc(X) \geq 0$" is defined as

$$Sc(X) > -\varepsilon\text{" for all } \varepsilon > 0.$$

Similarly

"$Sc(X) \geq \sigma$", $\sigma > 0$, *is equivalent the volumes of $B_x(\varepsilon)$ in X being smaller than the volumes of the ε-balls in the Euclidean spheres $S^n(R)$ of radii $R > \sqrt{(n(n-1)/\sigma)}$,*

and $Sc(X) \geq -\sigma$ is expressed by

the bound on the volumes of $B_x(\varepsilon)$ by those of the ε-balls in the hyperbolic spaces with constant the sectional curvatures $< -\sigma/n(n-1)$.

---

[1] Remarkably, this sum is independent of the frame by the Pythagorean theorem.

Alternatively, "$Sc(X) \geq -\sigma$" can be defined with *no reference* to hyperbolic spaces by *the reduction to the case* $\sigma = 0$ and appealing to the relation

$$Sc(X \times S^m(R)) \geq 0 \text{ for } R = \sqrt{(m(m-1)/\sigma},$$

where one may use any $m \geq 2$ one likes.

Although the key role of the scalar curvature in general relativity was established by Hilbert's variational derivation of the Einstein equation more than a century ago (see [2]) the significance of $Sc(X)$ in the global geometry and in topology remained obscure until 1963, when André Lichnerowicz (see [3]) showed that the inequality $Sc(X) > 0$ imposes non-trivial constraints on the topology of X.

For instance, Lichnerowicz' theorem implies that

if m is even, then smooth complex projective hypersurfaces $X \subset \mathbb{C}P^{m+1}$ (these have real dimension $dim(X) = 2m$) *of degrees* $\geq m+2$, *e.g.* $X \subset \mathbb{C}P^3$ *given by the equation*

$$x_1^4 + x_2^4 + x_3^4 + x_4^4 = 0,$$

admit no metrics with $Sc > 0$.

This follows from the Atiyah-Singer formula for the (Atiyah-Singer)-Dirac operator D confronted with (what is now called) the *Schroedinger-Lichnerowicz-(Weitzenboeck-Bochner) identity.*

In fact, the index formula implies that the index of D on these manifolds *does not vanish*,[2] and, consequently, *there are non-zero harmonic spinors on these X* (i.e. solutions s of $D(s) = 0$), while the *Schroedinger-Lichnerowicz-(Weitzenboeck-Bochner) identity*

$$D^2 = \nabla\nabla^* + \frac{1}{4}Sc,$$

shows that closed manifolds with $Sc > 0$ admit *no harmonic spinors.*

Eleven years later, Nigel Hitchin (see [4]) used a more sophisticated 1971 version of the Atiyah-Singer index theorem which yields harmonic spinors on some *exotic spheres* Σ^n (which are homeomorphic but not diffeomorphic to the ordinary spheres S^n) of dimensions $n = 8k+1$ and $n = 8k+2$ and which, together with the Schroedinger-Lichnerowicz' identity, implies that

there is no metrics with $Sc > 0$ *on these* Σ^n.

Then Stefan Stolz, elaborating on the earlier work by several authors, showed that there are

no further obstructions to the existence of metrics with $Sc > 0$ *on* simply connected *manifolds of dimension* ≥ 5 *besides those delivered by the index theorem* [5].

For instance

[2]This formula says in the present case that $Ind(D) = \hat{A}(X)$ where $\hat{A}(X)$ is a particular Pontryagin number of X.

all simply connected manifolds of dimensions $n = 3, 5, 6, 7 \mod 8$ *admit metrics with positive scalar curvatures.*

The proof of this theorem, which relies on *surgery of manifolds with* $Sc > 0$ and on the *cobordism theory*, suggests that manifolds with positive scalar curvature are almost as *soft* as smooth manifolds with no geometric constraints imposed on them. But the grand picture of scalar curvature in all its beauty unravels when one looks beyond this "almost".

(The opposite inequality $Sc(X) < 0$ is truly and fully soft and, unlike $Sc > 0$, has no influence on the topology and global geometry of X what-so-ever (see [6])).

A manifestly *rigid* property of $Sc > 0$ can be already seen in the following corollary to Schoen-Yau solution of the Riemannian positive mass conjecture in relativity (see [7]).

Solution of the Geroch Conjecture.[3] *The Euclidean metric* g_0 *on* $\mathbb{R}^3$ *(which has* $Sc(g_0) = 0$*) admits no non-trivial compactly supported perturbations* g *with* $Sc(g) \geq 0$.

Namely, if a smooth Riemannian metric g *on the Euclidean space* $\mathbb{R}^3$ *has* $Sc(g) \geq 0$ *and if* g *is equal to* g_0 *outside a compact subset in* $\mathbb{R}^3$*, then* $Sc(g) = 0$*; moreover,* g *is Riemannian flat, that is* $(\mathbb{R}^3, g)$ *is isometric to* $(\mathbb{R}^3, g_0)$.

This result has been refined and generalised in a variety of directions (see below and also [13, 21] at the end of the next section and references therein) but the rigidity of $Sc > 0$ we are after, albeit related to the above, is of different nature. In fact what we look for is

a structurally organised set of (desirably sharp) *geometric inequalities satisfied by manifolds with* $Sc > 0$*, more generally, with* $Sc \geq \sigma$.

Also, we search for a general category (or categories) of spaces, or other kind of objects, which would satisfy (certain classes of) such inequalities.

Additional Remarks and References

Geroch conjecture has been validated in all dimensions:

The Euclidean metrics on $\mathbb{R}^n$ *for all* n *admit no non-trivial compactly supported perturbations with* $Sc \geq 0$.

This (trivially) follows, for instance, from *non-existence of metrics with* $Sc > 0$ *on the* n*-tori* where the latter can be most easily proved by applying the index theorem to suitably "twisted" Dirac operators.

Witten suggested a different way of using the Dirac operator in the context of the positive mass problem, where the index theorem is replaced by a direct proof

[3] Attribution of this simplified positive mass conjecture to Robert Geroch is made in the above cited paper by Schoen and Yau.

In fact, the full Riemannian positive mass conjecture which describes possible asymptotic behaviours of metrics with $Sc > 0$ on $\mathbb{R}^3$ (and on $\mathbb{R}^n$ for this matter) which are close (rather than equal) to the Euclidean metric at infinity follows from this Geroch conjecture according to *J. Lohkamp, Scalar curvature and hammocks, Math. Ann. 313 (1999), 385–407.*

of *harmonic stability of parallel spinors* on $\mathbb{R}^n$ under certain perturbations of the Euclidean metric.

By a similar method, Min-Oo (see [8]) proved that

the hyperbolic metric g_0 on the real hyperbolic space $H^n_{\mathbb{R}}$ admits non non-trivial compactly supported perturbations g with $Sc(g) \geq -n(n-1) = Sc(g_0)$.

Apparently, it is unknown

if other symmetric spaces of non-compact types admit compactly supported perturbations of their Riemannian metrics which would increase scalar curvature.

3 Bounds on the Uryson Width, Slicing Area and Filling Radius

A. Conjecture. *Let X be an n-dimensional Riemannian manifold with scalar curvature bounded from below by*

$$Sc(X) \geq n(n-1) = Sc(S^n).$$

Then the $(n-1)$-dimensional Uryson width *of X is bounded by a universal constant.*

This means that there exists a continuous map from X to an $(n-1)$-dimensional polyhedral space P,

$$f : X \to P = P^{n-1},$$

such that the pullbacks of all points have controllably bounded diameters, namely,

$$diam_X(f^{-1}(p)) \leq const \text{ for all } p \in P.$$

for some universal constant $const > 0$ possibly (and undesirably) depending on n.

This conjecture says, in effect, that that n-dimensional manifolds X with $Sc(X) \geq \sigma > 0$ "topologically spread" in at most $n-1$ directions.

In fact, one expects that these X spread only in $n-2$ direction which can be formulated as follows.

A_+. Conjecture. *The above X admits a continuous map f to an $(n-2)$-dimensional polyhedral space P, such that $diam_X(f^{-1}(p)) \leq const_+$ for all $p \in P$.*

But the most attractive (and least tenable) is the conjecture A_{++} below which claims that *closed* manifolds with $Sc \geq \sigma > 0$ can be *sliced by surfaces with small areas* according the following definition.

Slicings and Waists. An *m-sliced n-cycle*, $m \leq n$, is an n-dimensional pseudo-manifold $P = P^n$ partitioned into *m-slices* $P_q \subset P$, which are the pullbacks of the points of a simplicial map $\varphi : P \to Q$ where Q is an $(n-m)$-dimensional pseudomanifold and where all pullbacks $P_q = \varphi^{-1}(q) \subset P$ have $dim(P_q) \leq m$, $q \in Q$.

(Sometimes one insists that φ must be *proper*, hence, with compact pullbacks $\varphi^{-1}(q)$, even if P is non-compact.)

The m-waist (mod 2), denoted $waist_m(h)$, of a homology class $h \in H_n(X; \mathbb{Z}_2)$ is

the infimum of the numbers w,

such that X receives a Lipschitz map from a compact m-sliced cycle, $\phi : P^n \to X$, which represents h, i.e.

$$\phi_*[P] = h$$

and

the images of all slices in X have m-volumes $\leq w$,

where these "volumes of the images" are counted with multiplicities (which is unneeded for generically 1-1 maps.)

A$_{++}$. Conjecture. Let X be a *closed* n-dimensional Riemannian manifold the scalar curvature of which is bounded from below as earlier:

$$Sc(X) \geq n(n-1)(= Sc(S^n)).$$

Then the slicing area of the fundamental homology class $[X] \in H_n(X; \mathbb{Z}_2)$ is bounded by

$$waist_2[X] \leq const_{++}.$$

(Ideally, one expects

$$const_{++} = waist_2(S^n)$$

where $waist_2[S^n] = area(S^2) = 4\pi$ by an Almgren's theorem.)

The above conjectures can be interpreted as saying that X contains "many" small subsets of dimensions 1 and/or 2.

For instance, **A** implies that X contains a *topologically significant/representative* family of 1-*dimensional* subsets (graphs) with diameters $\lessapprox \frac{1}{\sqrt{\sigma}}$.

This suggests the following.

(a) Conjecture. *If $Sc(X) \geq \sigma > 0$ and if X is a* closed *(compact without boundary) manifold, then X contains a closed minimal geodesic of length $\leq \frac{const_n}{\sqrt{\sigma}}$, or, at least, a stationary one-dimensional $\mathbb{Z}_2$-current of diameter (better length) $\leq \frac{const_n}{\sqrt{\sigma}}$.*

And A$_{++}$ actually implies the following.

(a$_{++}$) Conjecture. *Closed manifolds X with $Sc(X) \geq \sigma > 0$ contain closed minimal surfaces (i.e. stationary two-dimensional $\mathbb{Z}_2$-currents) of areas $\leq \frac{const_n}{\sigma}$.*

Below is a weaker version of **A** which already imposes non-trivial topological constraints on X.

A_-. Conjecture. *If $Sc(X) \geq n(n-1)$ then the* filling radius *of X is bounded by*

$$fil.rad(X) \leq const_-.$$

Definition of fil.rad. If $X = (X, g)$ is *closed* Riemannian manifold then the *filling radius* is equal to the infimum of $R > 0$, such that the cylinder $X^\times = X \times [0, 1)$ admits a Riemannian metric $g^\times$ with the following three properties.

$\bullet_1$ The restriction of $\hat{g}$ to $X = X_0 \times \{0\} \subset X \times [0, 1) = X^\times$ is equal to g; moreover,

$$dist_{g^\times}|X = dist_g.$$

This means that the g-shortest curves in X between all pairs of points in X minimise the $g^\times$-lengths of such curves in $X^\times \supset X$.

$\bullet_2$ All points in $X^\times$ lie within distance at most R from X,

$$dist_{g^\times}(x^\times, X) \leq R \text{ for all } x^\times \in X^\times.$$

$\bullet_3$ The n-dimensional volumes of the submanifolds $X \times \{t\} \subset X \times [0, 1) = X^\times$, $t < 1$, with respect to $g^\times$ vanish in the limit for $t \to 0$,

$$vol(X \times \{t\}) \to 0 \text{ for } t \to 1.$$

(The equivalence of this definition to the usual one follows from the *filling volume inequality* see [9] and references therein).

Then the filling radius of a compact manifold X with boundary—our manifolds may, a priori, have boundaries and/or to be incomplete—is defined as $fil.rad$ of the double of X along the boundary and $fil.rad$ of an open X is defined via exhaustions of X by compact submanifolds.

It is obvious that $\mathsf{A}_+ \Rightarrow \mathsf{A} \Rightarrow \mathsf{A}_-$ and that A_+ is optimal in a way.

Indeed, the product $X_r = X_0 \times S^2(r))$, where X_0 is, a compact manifold and $S^2(r))$ is the 2-sphere of small radius $r \to 0$, (these spheres have $Sc(S^2(r)) = \frac{2}{r^2}$), has $Sc(X_r) \geq (\frac{2}{r^2} - const_{X_0}) \to +\infty$, while the $(n-2)$-dimensional size/spread of X_r is as large as that of X_0.

Also one knows (see [17] at the end of this section and references therein) that

$$\mathsf{A}_{++} \Rightarrow \mathsf{A}_-.$$

(It is plausible in view of [18] that $\mathsf{A}_{++} \Rightarrow \mathsf{A}$.)

On the other hand, it is not hard to show that *if the isometry group of a Riemannian manifold $\hat{X}$ acts* cocompactly *on $\hat{X}$, i.e $\hat{X}/isom(\hat{X})$ is compact, and if $\hat{X}$ is* contractible, *then*

$$fil.rad(\hat{X}) = \infty.$$

Therefore, **A**$_-$ yields the following topological $Sc > 0$-non-existence corollary.

B. Conjecture. *Closed manifolds X with contractible universal coverings $\tilde{X}$ admit no metrics with $Sc > 0$.*

(Granted **B**, the non-strict inequality $Sc(X) \geq 0$ implies that X *Ricci flat* by Kazdan-Warner's perturbation theorem (see [10][4]). And since $\tilde{X}$ is contractible, the universal covering $\tilde{X}$ is *isometric to the Euclidean space* $\mathbb{R}^n$, $n = dim(X)$, by the *Cheeger-Gromoll splitting theorem.*)

Remarks and References

However plausible, none of the **A**-conjectures (above dimension 2) has been confirmed except for **A**$_+$ for 3-manifolds X with (apparently non-sharp) constant $const_+ = 2\pi\sqrt{6}$ (see [14] below).

On the other hand, **B** is known to hold for many manifolds X, starting from the case of n-tori due to Schoen and Yau. Later B was proven by a use of *twisted Dirac operators*[5] for several classes of manifolds with "large" universal coverings including those X which admit metrics with non-positive sectional curvatures.

Below are a few relevant papers where one can find further references.

In [11], the authors introduced their method of *induction descent by minimal hypersurfaces* and proved non-existence of metrics with $Sc > 0$ on the n-tori[6] and, more generally, on n-dimensional manifolds X which admit smooth maps $X \to \mathbb{T}^{n-2}$, such that the homology classes in $H_2(X)$ represented by the pull backs of generic points are *non-spherical.*

Originally, this method was limited to $n \leq 7$, but the techniques developed in [12, 13] apparently remove this limitation.

In [14] besides above mentioned **A**$_+$ for 3-manifolds, we *rule* out complete metrics with $Sc > 0$ on certain classes of manifolds, including

closed orientable n-dimensional spin[7] *manifolds X which admit continuous maps to complete manifolds Y with non-positive sectional curvatures, such that the fundamental classes $[X] \in H_n(X)$ go to non-zero classes in $H_n(Y)$ under these maps.*

The paper [15] presents a geometric perspective on the Dirac operator and soap bubble methods in the study of scalar curvature and related problems.

---

[4]If a metric g_0 *with* $Sc \geq 0$ *can't be perturbed to* g *with* $Sc(g) > 0$, *then* $Ricci(g) = 0$

[5]This means: *Dirac operators with coefficients in some (possibly infinite dimensional) vector bundles.*

[6]This trivially implies non-existence of compactly supported perturbations with $Sc > 0$ of the Euclidean metric on $\mathbb{R}^n$.

[7]A manifold of dimension $n \geq 3$ is spin if the restrictions of the tangent bundle $T(X)$ to all immersed surfaces in X are trivial bundles.

Most (all?) known non-existence results for $Sc > 0$ obtained for *spin manifolds* more or less automatically generalize to manifolds whose *universal coverings are spin*, i.e where $T(X)$ trivializes on all immersed 2-spheres in X.

A chapter in this book [16][8] offers a friendly introduction to the Dirac operator methods in the $Sc > 0$ problems.

The two papers [17] and [18] references therein give a fair idea of results and ideas around the filling radius.

The authors of these two papers [19] and [20] are concerned with topological versions of $\mathbf{A}_+$ for certain classes of manifolds X.

Rosenberg [21] is survey of topological obstructions to metrics with $Sc > 0$ on spin manifolds X expressed in terms of indices of *Dirac operators twisted with C^*-algebras* of $\pi_1(X)$.

Also obstructions for 4-*dimensional* manifolds X with non-vanishing *Seiberg-Witten invariants* due to Taubes and Le Brun are described in this paper.

www.ihes.fr/~gromov/PDF/Morse-Spectra-April16-2015-.pdf

Gromov [22] is an overview of waists and related invariants which may bear some relevance to $Sc \geq \sigma$.

4 Extremality and Rigidity with Positive Scalar Curvature

The proof(s) of the above A-conjectures (let them be only approximately true) would require *constructions* of certain maps or spaces which makes these conjectures difficult.

What is easier is getting upper bounds on the "size" of an X with $Sc(X) \geq \sigma > 0$ by proving *lower bounds* on dilations of *topologically significant* maps from X to (more or less) standard manifolds Y.

The first *sharp* bound of this kind was proved in [23] followed by [24] and [25] where further references can be found.

What is proven in these papers can be expressed in the the following terms.

Extremality/Rigidity. A Riemannian metric $\underline{g}$ on a manifold Y is called *length extremal* if it *can't be enlarged without making the scalar curvature smaller somewhere.* Namely, the inequalities

$$Sc(g) \geq Sc(g_0) \text{ and } g \geq g_0$$

for a Riemannian metric g on Y imply

$$Sc(g) = Sc(\underline{g}).$$

Then the stronger implication

$$[Sc(g) \geq Sc(g_0)]\&[g \geq g_0] \Rightarrow [g = \underline{g}]$$

[8]Also see Min-Oo, *K-Area, mass and asymptotic geometry*, http://ms.mcmaster.ca/minoo/mypapers/crm.

is qualified as *length rigidity* of $\underline{g}$.[9]

CY-Example. If a closed manifold Y admits no metric with $Sc > 0$, then all g_0 with $Sc(g_0) = 0$[10] are extremal according to this definition.

Instances of such *scalar flat* manifolds are flat Riemannian manifolds (with universal coverings $\mathbb{R}^n$) and also (simply connected) hypersurfaces $Z \subset \mathbb{C}P^{n+1}$ of degree $n+2$ and even n, with *Ricci flat Calabi-Yau metrics,* where non-existence of metrics with $Sc > 0$ on these Z follows from the Lichnerowicz's theorem.

Next, define *area extremality* and *area rigidity* by relaxing the inequality $g \geq g_0$, which says in effect that $lenght_g(C) \geq lenght_{\underline{g}} f(C)$ for all smooth curves $C \subset Y$), to

$$area_g(\Sigma) \geq area_{\underline{g}}(f(\Sigma))$$

for all smooth surfaces $\Sigma \subset Y$, where the extremality and rigidity requirements remains the same: $Sc(g) = Sc(\underline{g})$ and $g = \underline{g}$.

Stronger versions of these extremalities and rigidities allow modifications of the topology as well as geometry of Y, where the role of "topologically modified" Y are played by a Riemannian manifold $X = (X, g)$ and a map $f : X \to Y$, where the above inequalities are understood as

$$Sc(g)(x) \geq Sc(\underline{g})(f(x)), lenght_g(C) \geq lenght_{\underline{g}} g(f(C))$$

and

$$area_g(\Sigma) \geq area_{\underline{g}}(f(\Sigma))$$

correspondingly.

Accordingly, the required conclusion for *extremality* is

$$Sc(g)(x) = Sc(\underline{g})(f(x)),$$

while both, the *length* and the *area rigidities*, signify that

$$lenght_g(C) = lenght_{\underline{g}}(f(C)).$$

for all smooth curves $C \subset X$.

Of course, these definitions makes sense only for particular topological classes of manifolds X and maps f, such for instance as the class $\{\mathcal{DEG} \neq 0\}$ of orientable manifolds of dimension $n = dim(Y)$ and C^2-*smooth* maps with *non-zero degrees.*

C. Problem. Find verifiable criteria for extremality and rigidity, decide which manifolds admit extremal/rigid metrics and describe particular classes of extremal/rigid manifolds.

---

[9]Extremal manifolds define, in a way, the boundary of the domain $\{\mathcal{SC} \geq 0\}$ of manifolds with $Sc \geq 0$.

[10]The condition $Sc(g_0) = 0$ implies g_0 $Ricci(g_0) = 0$ on these Y by the Kazdan-Warner perturbation theorem, see [10] in Sect. 3.

For instance,

do all closed manifolds which admits metrics with $Sc \geq 0$ also admit (length) extremal metrics?

More specifically, prove (disprove?) the following.

C$_1$. Conjecture. *All compact Riemannian symmetric spaces are area extremal in the class $\{\mathcal{DEG} \neq 0\}$ and those which have $Ricci > 0$ (this is equivalent to absence of local $\mathbb{R}$ factors, and to is finiteness of fundamental group) are area rigid in this class.*

This conjecture was proved by Llarull (see [23] above) in the case $Y = S^n$, under the additional assumption of X being *spin*.[11]

Then Min-Oo [24] proved *area extremality* for *Hermitian symmetric spaces* in the class
$\{\mathcal{SPIN}, \mathcal{DEG} \neq 0\}$, where the maps $f : X \to Y$, besides having degrees $\neq 0$, must be *spin*.[12]

This was generalised by Goette and Semmelmann [25] who proved

area extremality in $\{\mathcal{SPIN}, \mathcal{DEG} \neq 0\}$ of compact (here it means closed) Kähler manifolds with $Ricci \geq 0$, rigidity for $Ricci \geq 0$.

Moreover, they establish

area rigidity in $\{\mathcal{SPIN}, \mathcal{DEG} \neq 0\}$ of certain (non-Hermitian) compact symmetric spaces including those with non-vanishing Euler characteristics and also of Riemannian metrics on S^{2m} with positive curvature operators.

These extremality and rigidity theorems are proven in the non-Kählerian cases by *sharply evaluating* the contribution from $f^*(\mathbb{S}^+(Y))$ in the Schrödinger-Lichnerowicz formula for the Dirac operator on X twisted with the f-pullback of the *spinor$^+$ bundle* $\mathbb{S}^+(Y)$ which is, in the case where $\chi(Y) \neq 0$ is confronted with the index theorem.

(The case of odd dimensional spheres S^{2m-1}, which depends on an additional argument(s) applied to maps $X \times S^1 \to S^{2m}$[13] seems to apply only to metrics on S^{2m-1} with constant sectional curvatures.)

---

[11]Since $\pi_1(SO(n)) = \mathbb{Z}_2$ for $n \geq 3$, there are at most two isomorphism classes of vector bundles with $rank \geq 3$ over connected surfaces Σ (exactly two for closed Σ), where the trivial bundle is called *spin* and where bundles of $rank < 3$ are spin if their Whitney sums with trivial bundles are spin. An orientable vector bundle V of over a topological space B is *spin* if the pullbacks of V under continuous maps $\phi : \Sigma \to B$ for all surfaces Σ are spin. A manifold X is spin if its tangent bundle is spin.

The spin condition is necessary for the definition of the Dirac operator on X but some *twisted Dirac operators* make sense on non-spin manifolds.

[12]A map $f : X \to Y$ is *spin* if the pullbacks $\phi^*(T(X))$ for maps of surfaces, $\phi : \Sigma \to X$, satisfy
$$[\phi^*(T(X)) \text{ is spin}] \Leftrightarrow [(\phi \circ f)^*(T(Y)) \text{ is spin}]$$
for all Σ and f. Equivalently, a map f between orientable manifolds is spin if the Whitney sum $T(X) \oplus f^*(T(Y))$ is spin.

Obviously, the identity map $id : Y \to Y$ is spin and if Y is spin, e.g. $Y = S^n$, then
$$[f : X \to Y \text{ is spin}] \Leftrightarrow [X \text{ is spin}].$$

[13]Llarull uses the product metric on $X \times S^1$, where his calculation applies even though the scalar curvature $Sc(X \times S^1)$, which is $\geq Sc(S^{2m-1})$, may be smaller than $Sc(S^{2m})$.

And in the Kähler case, this is done with the *"virtual square root" of the canonical* (complex) *line bundle* on Y instead of $\mathbb{S}^+(Y)$.

Spin or non-Spin? In all of the above cases one can replace the spin condition for $f : X \to Y$ by this condition for the corresponding map between the universal coverings, $\tilde{f} : \tilde{X} \to \tilde{Y}$, where a version of Atiyah's L_2-index theorem applies.

Probably,

"spin" can be removed all together in these theorems

but this seems beyond reach of the present day methods.[14]

On the other hand, the spin condition is essential for the extremality in the class $\{\mathcal{SPIN}, \mathcal{DEG}_{\hat{A}} \neq 0\}$ where the dimension of X can be greater than $n = dim(Y)$ and where the condition $deg(f) \neq 0$ is replaced by $deg_{\hat{A}}(f) \neq 0$, where the *$\hat{A}$-degree* $deg_{\hat{A}}(f)$ stands for the $\hat{A}$-genus of the f-pull back of a generic point $y \in Y$,

$$deg_{\hat{A}}(f) = \hat{A}(f^{-1}(y)).$$

(Here, strictly speaking, f must be smooth; if f is just continuous, this applies to a smooth approximation of f, where the so defined $\hat{A}$-degree does not depend on a choice of approximation.).

This implies for instance, that

the products of the above Y, e.g. of $Y = S^n$ by the Calabi-Yau manifolds with $\hat{A} \neq 0$, e.g with Z from the above CY-example are area extremal in the class $\{\mathcal{SPIN}, \mathcal{DEG}_{\hat{A}} \neq 0\}$ as well s in the class $\{\widetilde{\mathcal{SPIN}}, \mathcal{DEG}_{\hat{A}} \neq 0\}$ where spin condition is delegated to $\tilde{f} : \tilde{X} \to \tilde{Y}$.

Notice, however, that *neither* simply connected Calabi-Yau manifolds Z themselves *nor* their products by Y are extremal in the class $\{\mathcal{SPIN}, \mathcal{DEG} \neq 0\}$, at least if $dim(Z) \geq 5$.

Indeed the connected sums $X = Z\#(-Z)$, where "$-$" stands for the reversal of orientation and where the obvious map $Z\#(-Z) \to Z$ has degree 1, admit metrics with $Sc > 0$ by Stolz' theorem mentioned in Sect. 7. sleeker

It seems that there are two divergent, yet interconnected by bridges, branches in the tree of $Sc(X) \geq 0$, where a smoother and sleeker one involves differential structure and depends on spin, while the other one is made of rougher staff such as the homotopy classes of X.[15]

Alternatively, one can use the *spherical suspension metric* g^S (of g on X) on (the bulk of) $X \times S^1$, which has $Sc(g^S) \geq Sc(S^{2m})$ and thus allows a formal reduction of the $2m - 1$ case to that of $2m$.

[14] Apparently, no single case of extremality of a closed simply connected manifold X of dimension $n \geq 3$ is amenable to the the minimal hypersurface techniques, except, may be (?) for $X = S^3$.

[15] The smooth branch is manifested by $\hat{A}$ and the mod 2 α-invariant in the index formula while the rough branch is represented by the Chern character and supported by minimal hypersurfaces.

Probably, the second branch can be transplanted to a harsh world inhabited by singular spaces but fully cleaning off spin from this branch is by no means easy even for smooth X.

Extremality and Rigidity of Products. It seems not hard to show[16] that the Riemannian products of the area extremal/rigid manifolds in the above examples are area extremal/rigid which suggests to the following.

C$_2$. Question. Are the Riemannian products of all area extremal/rigid manifolds area extremal/rigid

Smoothing Lipschitz Maps. The length extremal/rigid manifolds in some *homotopy* class of smooth maps remain extremal/rigid in the corresponding class of Lipschitz maps f.

This can be proven by a smooth approximation of these f with a minor change of their length dilations.

But

this is unclear for the *area* extremality and/or area rigidity,

since, conceivably(?) all smooth approximation f' of a Lipschitz map $f : X \to Y$ may have $area(f'(\Sigma)) >> area(f(\Sigma))$ for some Σ.

Normalisation by Scalar Curvature: Extremality/Sc and Rigidity/Sc. A map $f : X \to Y$ between Riemannian manifolds $X = (X, g)$ and $Y = (X, g_0)$ with positive scalar curvatures, $Sc(g)$, $Sc(g_0) > 0$, is called *length decreasing/Sc* if it decreases the length of the curves measured in the metrics $Sc(X)^{-1}g$ and $Sc(g_0)^{-1}g_0$, i.e. if it decreases the integrals of $\sqrt{Sc}$ over all curves in X. Similarly one understand *decrease/Sc of areas* of surfaces $\Sigma \subset X$ under maps $X \to Y$, etc.[17]

Accordingly, one defines *length/area extremality/Sc* of a Y as non existence of strictly length/area decreasing/Sc maps $X \to Y$ in a given class of manifolds and maps, while the rigidity/Sc signifies that all length/area non-increasing/Sc maps $f : X \to Y$ are homotheties (similarities) with respect to the original metrics, i.e. $f^*(g_0) = const \cdot g$.

Since the "contribution of the twist" to the Schroedinger-Lichnerowicz formula for the twisted Dirac opertor on X scales as $Sc(X)^{-1}$, the arguments from the above cited papers based on this formula deliver the corresponding extremality/Sc and rigidity/Sc results. (This was pointed out in [26])

Category $\mathcal{R}_+/_{Sc}$. Let this be the category of Riemannian manifolds with $Sc > 0$ and length (alternatively, area) non-increasing/Sc maps.

C$_3$. Question. *How much of the geometry of spaces with $Sc > 0$ can be reconstructed in the category theoretic language of $\mathcal{R}_+/_{Sc}$?*

Extremality beyond $Sc \geq 0$. The condition $Sc(g) \geq 0$ may be not indispensable for extremality of g.

---

[16] I have not verified the proof in detail at this point.

[17] It may (or may not) be worthwhile to normalise by $g \rightsquigarrow n(n-1)Sc(X)^{-1}g$, $n = dim(X)$, and see what happens for $n \to \infty$.

For instance, the double of the unit hyperbolic disk is (kind of) extremal for the natural C^0-continuous metric on it and there are similar high dimensional examples. But it is unclear if such metrics are ever smooth.

Relativisation of Non-existence Theorems for $Sc > 0$. Let Y be a closed length or area *extremal or rigid* manifold in some class of smooth manifolds X and smooth maps $f : X \to Y$, where this class is invariant under homotopies of maps.

Then, most (all?) known *Dirac operator obstructions* to the existence of metrics with $Sc > 0$ on closed manifolds X_0 **naturally extend** to *similar obstructions* to the existence of (strict) *area decreasing/Sc* maps in certain homotopy invariant classes of maps $X \to Y$, including $X = X_0 \times Y \to Y$ for $(x_0, y) \mapsto y$.

For instance, one knows that (co)*homologically symplectic manifolds* X_0 with $\pi_2(X_0) = 0$ admit no metrics with $Sc > 0$ and the proof of this (see [15] cited in the previous section) also implies that words

if Y is the above area-extremal manifold, e.g. $Y = S^n$, then no homologically symplectic[18] *map $f : X \to Y$, which, moreover, induces an isomorphism $\pi_2(X) \to \pi_2(Y)$*, can be *strictly area decreasing/Sc.*

This suggests the following.

C₄. Conjecture. Let g be a metric on X and $f_0 : X \to Y$ be a (smooth?) strictly length (area?) decreasing/Sc map in this class.

Then there exists a smooth map f homotopic to f_0 transversal to a point $y_0 \in Y$, such that the f-pullback submanifold $f^{-1}(y_0) \in X$ admits a metric with $Sc > 0$.

Also other properties, e.g. extremality, of *manifolds* X with $Sc(X) > 0$ may have counterparts for length and area decreasing/Sc *maps* $X \to Y$ and, furthermore, for *foliations* on X.

C₅. Question. *Are infinite dimensional counterparts of compact symmetric spaces, e.g. the Hilbert sphere S^∞, extremal/rigid in some class(es) of perturbations of their metrics?*

---

[18] A smooth proper map between orientable manifold, $f : X \to Y$, is *homologically symplectic* if the difference of the dimensions $n_0 = n - m$ for $n = dim(X)$ and $m = dim(Y)$ is *even* and if there exists a closed 2-form ω on X such that the integrals of $\omega^{\frac{n_0}{2}}$ over the f-pullbacks of generic points $y \in Y$ do not vanish.

In other words, the real fundamental cohomology class $[X]^\circ \in H^n_{comp}(X; \mathbb{R})$ with compact support is equal to the $\smile$ product of the f-pullback of $[Y]^\circ \in H^m_{comp}(Y, \mathbb{R})$ and the $\frac{n_0}{2}$th$\smile$power of the class $[\omega] \in H^2(X; \mathbb{R})$,

$$[X]^\circ = f^*([Y]^\circ) \smile [\omega]^{\frac{n_0}{2}}.$$

5 Extremality and Gap Extremality of Open Manifolds.

Let $U \subset Y$ be an open subset in a extremal or rigid Riemannian manifold Y where the extremality/rigidity for this Y follows by the twisted Dirac operator argument from the previous section. Then the same argument yields the following.

$\bigstar$ *If the complement $Z = Y \setminus U$ is non-empty, yet* LC-negligible (explained below) *then* **no complete** *orientable Riemannian manifold admits a smooth area non-increasing/Sc map $f : X \to U$, which has non-zero degree*[19] *and the lift of which to the universal coverings, $\tilde{f} : \tilde{X} \to \tilde{U}$, is* **spin**.

LC-negligible Sets . A piecewise smooth polyhedral subset Z in a Riemannian manifold Y is called LC-negligible if the Levi-Civita connection on the tangent bundle of X restricted to Z is *split trivial.* For instance,

• finite subsets in Y are LC-negligible;

• piecewise smooth graphs $Z \subset Y$ with trivial monodromies around the cycles, e.g. disjoint unions of trees, are LC-negligible;

• simply connected *isotropic* (e.g.*Lagrangian*) submanifolds in Kähler manifolds are LC-negligible.

This definition extends to general closed subsets Z, such as *Cantor sets*, for instance, by requiring that the monodromies along smooth curves C in the ε-neighbourhoods of Y are $o(\varepsilon\cdot$ length $(C))$ as $\varepsilon \to 0$ but the geometry behind this definition needs to be clarified.

D$_1$. Problem. *Study essential properties, such as the Hausdorff dimensions, of these subsets $Z \subset Y$ and find cases (if there are any) where $\bigstar$ remains valid for small, yet non-LC-negligible $Z \subset Y$, e.g. for (generic) smooth curves Z in Y.*

Notice in this regard that a simple surgery type argument (see Stolz' paper [5] cited in Sect. 2 and references therein) shows that

• if Z is equal to the k-skeleton $\mathcal{T}^k$ of a smooth triangulation $\mathcal{T}$ of a compact Riemannian manifold (Y, g_0), for $k \geq 2$, then $U = Y \setminus Z$ admits a complete metric $g \geq g_0$ with $Sc(g) \geq \sigma_0 = \sigma_0(Y, Z) > 0$.

Moreover, it is easy to show that

the complements $U_\varepsilon = Y \setminus \mathcal{T}^k_\varepsilon$ of the k-skeleta of the "standard fat" ε-refinements[20] *of $\mathcal{T}$ admit complete Riemannian metrics $g_\varepsilon \geq g$ the scalar curvatures of which for $k \geq 2$ satisfy*

$$Sc(g_\varepsilon) \geq const \frac{1}{\varepsilon^2}$$

for some constant $const = const(Y, \mathcal{T}) > 0$.

Thus $\bigstar$ fails to be true, for $Z = \mathcal{T}^k_\varepsilon$, $k \geq 2$, and small (how small?) ε.

---

[19]Maps $f : X \to Y$ of non-zero degree, by definition, must be equidimensional and proper.

[20]It is more practical to start with a *cubilation* $\mathcal{T}$ of Y which can be canonically ε-refined for $\varepsilon = \frac{1}{i}$, $i = 2, 3, \ldots$, by subdividing each m-cube into i^m-sub-cubes in an obvious way.

On the other hand, the *torical band width inequality* from the next section shows that if, for instance, Z is a codimension two torus in Y, e.g. $Z = \mathbb{T}^2 \subset S^4$, then the complement $U = Y \setminus Z$ admits no complete metrics with $Sc \geq \sigma > 0$ whatsoever and the same applies to a large (how large) class of codimension two polyhedra $Z \subset Y$ with contractible universal coverings.

Non-existence of *complete* metrics $g \geq g_0$ with $Sc > \sigma_0$ on the above $U = (U, g_0)$ with $Sc(g_0) = \sigma_0$ may be interesting in its own right but this can't be regarded as extremality of g_0, since a comparison of the manifolds (U, g_0), which have *bounded diameters* with their competitors (U, g) of *infinite size* is patently unfair. The true extremity issue for these U, thus, remains unresolved.

D$_2$. Question. Do there ever exist length extremal domains $U \subset Y$, $U \neq Y$, in closed connected Riemannian manifolds Y of dimensions ≥ 3?

For instance, is the the sphere S^3 minus a point (or the 3-torus minus a point) extremal?

We still do not know the answer but, on the other hand, the following *warped product* construction sometimes delivers examples of both complete and non-complete extremal and rigid manifolds (compare §12 in [14] cited in Sect. 3 and [27] cited below).

Let $Y_0 = (Y_0, g_0)$ be a Riemannian manifold with constant scalar curvature σ_0 and let $g_1 = \varphi^2 g_0 + dt^2$ be a Riemannian metric on $Y_1 = Y \times (l_-, l_+)$ for $-\infty \leq l_- < l_+ \leq \infty$, for some smooth function $\varphi = \varphi(t) > 0$ for $l_- < t < l_+$.

Then, by elementary calculation,

$$◑ \qquad Sc(g) = \frac{\sigma_0}{\varphi^2} - 2n\frac{\varphi''}{\varphi} - n(n-1)\frac{\varphi'^2}{\varphi^2}, \text{ where } n = dim(Y_0).$$

Now, let g have *constant scalar curvature*, say $Sc(g_1) = \sigma_1$ for a given $\sigma_1 \geq 0$, and prescribe: $\varphi(0) = 1$ and $\varphi'(0) = 0$.

Then ◑, regarded as an ODE and rewritten as

$$f'' = -\frac{1}{2}(n+1)f'^2 + \frac{\sigma_0}{2ne^{2f}} - \frac{\sigma_1}{2n} \text{ for } f = \log\varphi,$$

admits a unique solution f on some maximal (extremal) open interval (l_-^{ext}, l_+^{ext}) beyond which the solution does not extend.

Examples. (a) If $Y_0 = S^n$ and $\sigma_1 = n(n+1)$, then Y_1 is equal to S^{n+1} minus two opposite points.

(b) If $Y_0 = \mathbb{R}^n$ and $\sigma_1 = 0$, then $Y_1 = \mathbb{R}^{n+1}$.

(c) If $Y_0 = \mathbb{R}^n$, $\sigma_1 = n(n+1) = Sc(S^{n+1})$ and $n = 1$, then Y_1 is equal the universal covering of S^2 minus two opposite points.

In general, the manifold (Y_1, g_1) is uniquely characterised by the following three properties.

[$\bigcirc_{n(n+1)}$] The scalar curvature of Y_1 is everywhere equal to $n(n+1)$ for $n = dim(Y_1) - 1$.

[$\bigcirc_{O(n)\rtimes\mathbb{R}^n}$] The isometry group of Y_1 is $Iso(\mathbb{R}^n) = O(n) \rtimes \mathbb{R}^n$ times $\mathbb{Z}_2$. (This $\mathbb{Z}_2$ corresponds to the involution $t \leftrightarrow -t$.)

[$\bigcirc_{2\pi/n+1}$] The *band width* of Y_1 is $\frac{2\pi}{n+1}$, where this width is understood in the present case as the distance between the two (one point) boundary components of Y_1 in the metric completion $\bar{Y}_1 \supset Y_1$.

(The band-like shape of Y_1 is best seen for $dim(Y_1) = 2$, where this Y_1 is equal to the universal covering of the doubly punctured sphere S^2.)

Alternatively, one might say that the in-radius of Y_i is equal to $\frac{\pi}{n+1}$:

there are closed compact balls in Y_1 of all radii $R < \frac{\pi}{n+1}$ but no ball of radius $\geq \frac{\pi}{n+1}$ is compact.

Gap Extremality. We do not know if the above spheres minus pairs of points are extremal for $n \geq 2$ but the Euclidean spaces $\mathbb{R}^m$ are definitely *not length extremal* starting from $m = 2$.

In fact, there are (obvious, $O(m)$-invariant) metrics $g \geq g_{Eucl}$ on $\mathbb{R}^m$ with $Sc(g_1) > 0$ for all $m \geq 2$.

On the other hand,

(∗) *no metric $g \geq g_{Eucl}$ on $\mathbb{R}^m$ may have $Sc(g) \geq \varepsilon > 0$.* (See [15] cited in Sect. 3.)

This suggests the following weaker version of extremality for non-compact manifolds which we call *gap extremality.*

A metric g_0 on Y is *ε-gap length extremal* if *no* $g \geq g_0$ on Y satisfies

$$Sc(g) - Sc(g_0) > \varepsilon.$$

Then g_0 is called *gap length extremal* if it is ε-gap length extremal for all $\varepsilon > 0$ (0-gap extremal=extremal).

Similarly one defines *area gap extremality* and *gap extremality for classes of maps $f : X \to Y$*. (But I am not certain what a *workable* definition of normalized gap extremality/Sc should be.)

Whenever the twisted Dirac operator argument from the previous section yields area extremality of a closed manifold Y, e.g. if $Y = S^n$ or $Y = \mathbb{C}P^n$, this argument, combined with that from [15] (cited in Sect. 3) for $\mathbb{R}^m$, also delivers

(∗∗) *gap area extremality of $Y_m = Y \times \mathbb{R}^m$ for all $m = 1, 2, \ldots$, as well as this extremality for smooth proper spin maps $f : X \to Y_m$ of non-zero degrees.*

If a smooth proper spin map $f : X \to Y_m$ of non-zero degree decreases the areas of all surfaces $\Sigma \subset X$, then, given $\varepsilon > 0$, there exists a point $x \in X$, such that

$$Sc(X)(x) - Sc(Y')(f(x)) < \varepsilon.$$

D$_3$. Question. Does gap extremality is always stable under $Y \rightsquigarrow Y \times \mathbb{R}^m$? (Beware of $dim(Y) = 4$.)

One can't discard of ε for $m \geq 2$ but the true area (or, at least length) extremality of $Y' = Y \times \mathbb{R}$ (that allows $\varepsilon = 0$) may be provable by some twisted Dirac operator argument. For instance, if $Y = \mathbb{T}^n$ this follows from theorem 6.12 in [14] (cited in Sect. 3). Alternatively, one might use minimal hypersurfaces and soap bubble in X the f-images of which separate the two ends in $Y' = Y \times \mathbb{R}$ but then one would face a possibility of non-compact minimal hyper surfaces in X and would be obliged to resort to imposing extra assumptions on X, e.g. uniform two sided bounds on the sectional curvatures of X.

Finally, let us look at the manifold Y_1, which has the band width $\frac{2\pi}{n+1}$, in the above Example (c).

It is plausible that this Y_1 is length gap extremal but not length extremal starting from $D = dim(Y_1) = 3$.

And what we definitely know is that

the quotient space $Y_1/\mathbb{Z}^n = \mathbb{T}^n \times (-\frac{\pi}{n+1}, \frac{\pi}{n+1})$, $n + 1 = dim(Y_1)$, *is length extremal.*

We shall see the reason for this in the next section, where we shall also explain the current status of the rigidity problem for these manifolds.

6 Bounds on Widths of Bands with Positive Curvatures

Let us start with the following question which, on the face of it, has nothing to do with scalar curvature.

Given a smooth n-dimensional manifold X immersed[21] into a complete Riemannian manifold Y denote by $rad^{\perp}(X \hookrightarrow Y)$ the maximal R, such that the normal exponential map

$$exp^{\perp} : T^{\perp}(X) = T(Y)_{|X} \ominus T(X) \to Y,$$

is locally injective on the subbundle $B^{\perp}(R)(X) \subset T^{\perp}(X)$ of open normal R-balls $B_x^{N-n}(R) \subset T^{\perp}(X)$, $x \in X$.

(If the ambient space $Y = \mathbb{R}^n$, then $rad^{\perp}(X; \mathbb{R}^n)$ is equal to the reciprocal of the supremum of the principal curvatures of X.)

Take the supremum of these radii over all immersions $f : X \hookrightarrow Y$, set

$$suprad^{\perp}(X; Y) = \sup_f rad^{\perp}(X \underset{f}{\hookrightarrow} Y)$$

and let

[21] A smooth map $X \to Y$ is an immersion if it is a diffeomorphism of small neighbourhoods in X to smooth submanifolds in Y.

$$suprad^{\perp}_{N}(X) = \sup_{f_\circ} rad^{\perp}(X \underset{f_\circ}{\hookrightarrow} \mathbb{R}^N),$$

where the latter "sup" is taken over all immersion $f_\circ$ from X to the *unit ball* $B^N(1) \subset \mathbb{R}^N$.

(The notation $suprad^{\perp}(X; B^N(1))$ would be unjustified, since the image of the exponential map may be not contained in $B^N(1)$.)

E$_1$. **Problem.** Evaluate $suprad^{\perp}_{N}(X)$ in terms of the topology of X.

Examples. (a) It is obvious that $suprad^{\perp}_{N}(X) \leq 1$ for all closed manifolds X, where the equality holds if and only if X is diffeomorphic to S^n and $N > n$.
(b) Let X_k is diffeomorphic to the product of k spheres,

$$X_k = S^{n_1} \times \ldots \times S^{n_k},\ n_k \geq 1.$$

Then

$$suprad^{\perp}_{N}(X) \geq \frac{1}{\sqrt{k}} \text{ for all } N \geq (n_1 + 1) + \ldots + (n_k + 1).$$

But we do not know, for instance, whether

$$suprad^{\perp}_{N}(X_k) \to 0 \text{ for } N = dim(X_k) + 1 \text{ and } k \to \infty.$$

or, on the contrary, if

$$suprad^{\perp}_{N}(X) \geq \rho_0$$

for *all* manifolds X, (e.g. for all X_k) all sufficiently large $N \geq N(X)$ and some universal constant $\rho_0 > 0$, say $\rho_0 = 0.001$.

All known upper bounds on $suprad^{\perp}_{N}(X)$—*am I missing something obvious?*

exclusively apply to manifolds X which admit no metrics with $Sc > 0$.

A simple way to obtain such a bound is as follows.

1.Scale $B^N(1) \to B^N(\frac{1}{2})$, project $B^N(\frac{1}{2})$ to S^N from the south pole of S^N and observe that this distorts the curvatures of submanifolds X in the ball $B^N(1)$ by a finite amount independent of X and N.

2. Apply the Gauss formula to $X \hookrightarrow S^N$ and thus show that the supremum of the principal curvatures of X in S^N satisfies

$$supcurv(X \hookrightarrow S^N) \geq \frac{\sqrt{n-1}}{N-n}$$

and therefore,

$$suprad^{\perp}_{N}(X) \leq const \cdot \frac{N-n}{\sqrt{n-1}}$$

for all n-dimensional manifolds X which admit no metrics with $Sc > 0$ and for some constant $const \leq 100$. (See [27] cited below for details.)

It follows, for instance, that there are

exotic spheres Σ^n of dimensions $n = 9, 17, 25, 33, \ldots$, such that

$$suprad^{\perp}_{n+1}(\Sigma^n) \leq \frac{100}{\sqrt{n-1}},$$

but one has no idea how sharp this inequality is and if there are similar inequalities for exotic spheres which admit metrics with $Sc > 0$.

The above also applies to tori $\mathbb{T}^n$, since these admit no metrics with $Sc > 0$ either, but here the following better (but, probably, still very far from being sharp) inequality is available.

$$suprad^{\perp}_{n+1}(\mathbb{T}^n) \leq \frac{2\pi}{n+1}.$$

This is proven again by passing to S^{n+1}, where all we use of the geometry of S^{n+1} is the inequality $Sc(S^{n+1}) \geq n(n+1)$. (Isn't it amazing that there is no apparent *direct* proof of a *much stronger* bound on $rad^{\perp}(\mathbb{T}^n \subset B^{n+1}(1))$.)

Namely, the above bound on $suprad^{\perp}_{n+1}(\mathbb{T}^n)$ trivially follows from the following.

Torical Band Width Inequality. *Let g* be a metric with $Sc(g) \geq n(n+1) = Sc(S^{n+1})$ on the torical band (cylinder) $\mathbb{T}^n \times [-1, 1]$. Then the distance between the two boundary components of this band satisfies

$$[\circ_{\pm} < \tfrac{2\pi}{n+1}] \qquad dist_g(\mathbb{T}^n \times \{-1\}, \mathbb{T}^n \times \{1\}) < \frac{2\pi}{n+1}.$$

This is proven in [27] with a relative version of the Schoen-Yau minimal hypersurface method.

Besides a bound on $suprad^{\perp}_{n+1}(\mathbb{T}^n)$, the inequality $[\circ_{\pm} < \frac{2\pi}{n+1}]$ (trivially) implies that

the warped product metric $\varphi^2(t)g_{\mathbb{T}^n} + dt^2$ on $\mathbb{T}^n \times (-\frac{\pi}{n+1}, \frac{\pi}{n+1})$ with $Sc = n(n+1)$, which was introduced in the previous section, is length extremal.

Also, the argument in [27] yields *length rigidity* of this metric for $n \leq 6$, while the general case needs an elaboration on recent results on "irrelevance of singularities" of minimal hypersurfaces proved in the papers [12] and/or [13] cited in Sect. 3.

7 Extremality and Rigidity of Convex Polyhedra

Let $P \subset \mathbb{R}^n$ be a compact convex polyhedron with non-empty interior, let $Q_i \subset P$, $i \in I$, denote its $(n-1)$-faces and let

$$\angle_{ij}(P) = \angle(Q_i, Q_j)$$

denote its dihedral angles.

Say that P is *extremal* if all convex polyhedra P' which are combinatorially equivalent to P and which have

$$\angle_{ij}(P') \leq \angle_{ij}(P) \text{ for all } i, j \in I,$$

necessarily satisfy

$$\angle_{ij}(P') = \angle_{ij}(P).$$

It is known—the proof is elementary—that

the simplices and the rectangular solids are extremal and also all P with $\angle_{ij}(P) \leq \frac{\pi}{2}$, are extremal.

But it is unclear (at least to the present author) what are (if any) non-extremal P.

What we are truly interested in, however, is extremality (and rigidity) of P under transformations which keep the faces Q_i convex (rather than flat) or, even better, *mean convex*, i.e. keeping their mean curvatures non-negative.

Thus, we say that P is *mean convexly extremal* if *there is no $P' \subset \mathbb{R}^n$ diffeomorphic to P* and such that

• the faces $Q'_i \subset P'$ corresponding to all $Q_i \subset P$ have $mean.curv(Q'_i) \geq 0$,

• the dihedral angles of P', that are the angles between the tangent spaces $T_{p'}(Q'_i)$ and $T_{p'}(Q'_j)$ at the points p' on the $(n-2)$-faces $Q'_{ij} = Q'_i \cap Q_j'$, satisfy

$$\angle_{ij}(P') \leq \angle_{ij}(P),$$

• this angle inequality is strict at some point, i.e. there exits $p_0' \in Q'_{ij}$ in some Q'_{ij}, such that

$$\angle(T_{p_0'}(Q'_i), T_{p_0'}(Q'_j)) < \angle_{ij}(P).$$

F$_1$. Question. *Are all extremal convex polyhedra P are mean convexly extremal?*

It is not even known if the regular 3-simplex is mean convexly extremal, but

the mean convex extremality of the n-cube

follows by developing the cube P into a complete (orbi-covering) manifold $\hat{P}$ homeomorphic to $\mathbb{R}^n$ by reflecting P in the faces, approximating the natural continuous Riemannian metric metric on $\hat{P}$ by a smooth one with $Sc \geq \varepsilon > 0$ (see [28]) and appealing to gap extremality of $\mathbb{R}^n$ stated in Sect. 5.

And the same argument yields (see [28]) the following

[$*$] *Let a Riemannian metric g on the n-cube P satisfy:*

$*_0$ $Sc(g) \geq 0$.

$*_1$ $mean.curv_g(Q_i) \geq 0$.

$*_2$ $\angle_{ij}(P, g) \leq \frac{\pi}{2}$.

Then, necessarily, $Sc(g) = 0$, $mean.curv_g(Q_i) = 0$ and $\angle_{ij}(P, g) = \frac{\pi}{2}$.

Probably, these equalities imply that P is *isometric to a Euclidean rectangular solid* but the approximation/smoothing is no good for proving this kind of rigidity.

The main merit of [$*$] is that it provides a test for $Sc \geq 0$ in all Riemannian manifolds X:

$Sc(X) \geq 0$ *if and only if no cubical domain* $P \subset X$ *satisfies*

$$[mean.curv_g(Q_i) > 0]\&[\angle_{ij}(P, g) \leq \frac{\pi}{2}].$$

This suggests a possibility of defining $Sc(X) \geq 0$ for some singular spaces, X. e.g. for *Alexandrov spaces* with sectional curvatures bounded from below.

F_2. Conjecture *All known (and expected) properties of Riemannian manifolds with* $Sc \geq 0$, *which have no "spin" attached to their formulations, generalise to Alexandrov's spaces.*

For instance, most probably,

if an n-dimensional Alexandrov space X with curvatures bounded from below has $Sc > 0$ at all regular points $x \in X$, (or if the volumes of all infinitesimally small balls in X are bounded by the volumes of such Euclidean balls) then

every continuous map from X to a space Y with $CAT(0)$ universal covering (i.e. an Alexandrov's space with non-positive sectional curvatures) contracts to an $(n-1)$-dimensional subset in Y.

If true, this would imply that (suitably defined) *harmonic maps* $X \to Y$ must necessarily have $(n-1)$*-dimensional* images, which suggests a (non-local?) Weitzenboeck-Bochner type formula in this context and a definition of $Sc > 0$ via spectral properties of small (large?) balls (cubes?) in X.

Acknowledgements For the invitation to contribute to this volume and for his helpful comments I thank J. Kouneiher.

References

1. S. Alexander, V. Kapovitch, A. Petrunin, Alexandrov geometry. http://www.math.psu.edu/petrunin/
2. D. Hilbert, *The Foundations of Physics*, (1915)
3. A. Lichnerowicz, Spineurs harmoniques. C. R. Acad. Sci. Paris, Série A **257**, 7–9 (1963)
4. N. Hitchin, Harmonic Spinors. Adv. Math. **14**, 1–55 (1974)
5. S. Stolz, Simply connected manifolds of positive scalar curvature. Ann. of Math. (2) **136**, 511–540 (1992)
6. J. Lohkamp, Metrics of negative Ricci curvature. Ann. Math. **140**, 655–683 (1994)
7. R. Schoen, S.-T. Yau, On the proof of the positive mass conjecture in general relativity. Commun. Math. Phys. **65**, 45–76 (1979)
8. M. Min-Oo, Scalar curvature rigidity of asymptotically hyperbolic spin manifolds. Math. Ann. 285, 527–539 (1989)

9. L. Guth, Notes on Gromov's systolic estimate. Geom. Dedicata **123**, 113–129 (2006)
10. J. Kazdan, F. Warner, Existence and conformal deformation of metrics with prescribed Gaussian and Scalar curvatures. Ann. Math. **101**(2), 317–331 (1975)
11. S.T. Yau, R. Schoen, On the structure of manifolds with positive scalar curvature. Manuscripta mathematica **28**, 159–184 (1979)
12. J. Lohkamp, The Higher Dimensional Positive Mass Theorem II (2016). arXiv:1612.07505
13. R. Schoen, S.T. Yau, Positive Scalar Curvature and Minimal Hypersurface Singularities (2017), arXiv:1704.05490
14. M. Gromov, H.B. Lawson Jr., Positive scalar curvature and the Dirac operator on complete Riemannian manifolds. Publ. Math. IHS **58**, 295–408 (1983)
15. M. Gromov, Positive curvature, macroscopic dimension, spectral gaps and higher signatures, in *Proc of 1993 Conf. in Honor of of the Eightieth Birthday of I. M. Gelfand, Functional Analysis on the Eve of the 21st Century: Volume I Progress in Mathematics*, vol. 132 (1996), pp. 1–213
16. S. Markvorsen, M. Min-Oo, *Global Riemannian Geometry: Curvature and Topology*, Birkhäuser, (2012)
17. L. Guth, Metaphors in systolic geometry. in *Proceedings of the International Congress of Mathematicians*, vol. II (2010), pp. 745–768
18. L. Guth, Volumes of balls in Riemannian manifolds and Uryson width. J. Topology Anal. **09**(02), 195–219 (2017)
19. D. Bolotov, A. Dranishnikov, On Gromov's conjecture for totally non-spin manifolds (2015). arXiv:1402.4510v6
20. M. Marcinkowski, Gromov positive scalar curvature conjecture and rationally inessential macroscopically large manifolds. J. Topology 9(1), 105–116 (2016). Oxford University Press
21. J. Rosenberg, Manifolds of positive scalar curvature: a progress report. in *Surveys on Differential Geometry, vol. XI: Metric and Comparison Geometry*, International Press 2007
22. M. Gromov, Morse Spectra, Homology Measures, Spaces of Cycles and Parametric Packing Problems. www.ihes.fr/gromov/PDF/Morse-Spectra-April16-2015-.pdf
23. M. Llarull, Sharp estimates and the Dirac operator. Math. Ann. **310**, 55–71 (1998)
24. M. Min-Oo, Scalar curvature rigidity, of certain symmetric spaces, geometry, topology and dynamics (Montreal, PQ, 1995) CRM Proc. Lecture Notes, 15, Amer. Math. Soc. Providence, RI, 127–136 (1998)
25. S. Goette, U. Semmelmann, Scalar curvature estimates for compact symmetric spaces. Differ. Geom. Appl. **16**(1), 65–78 (2002)
26. M. Listing, The Scalar curvature on compact symmetric spaces. arXiv:1007.1832, 2010 - arxiv.org
27. M. Gromov, Metric Inequalities with Scalar Curvature. http://www.ihes.fr/~gromov/PDF/Inequalities-July%202017.pdf
28. M. Gromov, Dirac and Plateau Billiards in Domains with Corners. Cent. Eur. J. Math. **12**(8), 1109–1156 (2014)

Geometry and the Quantum

Alain Connes

1 Introduction

The ideas of noncommutative geometry are deeply rooted in both physics, with the predominant influence of the discovery of Quantum Mechanics, and in mathematics where it emerged from the great variety of examples of "noncommutative spaces" i.e. of geometric spaces which are best encoded algebraically by a noncommutative algebra.

It is an honor to present an overview of the state of the art of the interplay of noncommutative geometry with physics on the occasion of the celebration of the centenary of Hilbert's work on the foundations of physics. Indeed, the ideas which I will explain, those of noncommutative geometry (NCG) in relation to our model of space-time, owe a lot to Hilbert and this is so in two respects. First of course by the fundamental role of Hilbert space in the formalism of Quantum Mechanics as formalized by von Neumann, see Sect. 1.1. But also because, as explained in details in [1, 2], one can consider Hilbert to be the first person to have speculated about a unified theory of electromagnetism and gravitation, we come to this point soon in Sect. 1.2.

1.1 The Spectral Point of View

At the beginning of the eighties, motivated by the exploration of the many new spaces whose algebraic incarnation is noncommutative, I introduced a new paradigm, of spectral nature, for geometric spaces. It is based on the Hilbert space formalism of Quantum Mechanics and on mathematical ideas coming from K-theory and index

A. Connes (✉)
Mathematics, College de France, Institut des Hautes Etudes Scientifiques, OSU University, IHES 35 Route de Chartres 91310, Bures Sur Yvette, France
e-mail: alain@connes.org

J. Kouneiher (ed.), *Foundations of Mathematics and Physics One Century After Hilbert*,
https://doi.org/10.1007/978-3-319-64813-2_7

theory. A geometry is given by a "spectral triple" $(\mathcal{A}, \mathcal{H}, D)$ which consists of an involutive algebra $\mathcal{A}$ concretely represented as an algebra of operators in a Hilbert space $\mathcal{H}$ and of a (generally unbounded) self-adjoint operator D acting on the same Hilbert space $\mathcal{H}$. The main conceptual motivation came from the work of Atiyah and Singer on the index theorem and their realization that the Hilbert space formalism was the proper setting for "abstract elliptic operators" [3].

To fix ideas: a compact spin Riemannian manifold is encoded as a spectral triple by letting the algebra of functions act in the Hilbert space of spinors while the Dirac operator D plays the role of the inverse line element, as we shall amply explain below. But the key examples that showed, very early on, that the relevance of this new paradigm went far beyond the framework of Riemannian geometry comprised duals of discrete groups, leaf spaces of foliations and deformations of ordinary spaces such as the noncommutative tori which were themselves a prime example of noncommutative geometric spaces as shown in [4].

In the middle of the eighties it became clear that the new paradigm of geometry, because of its flexibility, provided a new perspective on the geometric interpretation of the detailed structure of the Standard model and of the Brout-Englert-Higgs mechanism. Over the years this new point of view has been considerably refined and is now able to account for the extremely complicated Lagrangian of Einstein gravity coupled to the standard model of particle physics. It is obtained from the spectral action developed in our joint work with Chamseddine in [5]. The spectral action is the only natural additive spectral invariant of a noncommutative geometry.

The noncommutative geometry dictated by physics is the product of the ordinary 4-dimensional continuum by a finite noncommutative geometry which appears naturally from the classification of finite geometries of *KO*-dimension equal to 6 modulo 8 (cf. [6, 7]). The compatibility of the model with the measured value of the Higgs mass was demonstrated in [8] due to the role in the renormalization of the scalar field already present in [9]. In [10, 11], with Chamseddine and Mukhanov, we gave the conceptual explanation of the finite noncommutative geometry from Clifford algebras and obtained a higher form of the Heisenberg commutation relations between p and q, whose irreducible Hilbert space representations correspond to 4-dimensional spin geometries. The role of p is played by the Dirac operator and the role of q by the Feynman slash of coordinates using Clifford algebras. The proof that all spin geometries are obtained relies on deep results of immersion theory and ramified coverings of the sphere. The volume of the 4-dimensional geometry is automatically quantized by the index theorem; and the spectral model, taking into account the inner automorphisms due to the noncommutative nature of the Clifford algebras, gives Einstein gravity coupled with a slight extension of the standard model, which is a Pati-Salam model. This model was shown in our joint work with Chamseddine and van Suijlekom [12, 13] to yield unification of coupling constants.

1.2 Gravity Coupled with Matter

As explained in detail in [1], one can consider Hilbert as the first to have fancied a unified theory of electromagnetism and gravitation. According to [1], in the course of pursuing this agenda, Hilbert reversed his original idea of founding all of physics on electrodynamics, instead treating the gravitational field equations as more fundamental. We have, in our investigations with Ali Chamseddine of the fine structure of space-time which is revealed by the Brout-Englert-Higgs mechanism, followed a parallel path: the starting point was that the NCG framework for geometry, by allowing to treat the discrete and the continuum on the same footing gives a clear geometric meaning to the Brout-Englert-Higgs sector of the Standard Model, as the signal of a discrete (but finite) component of the geometry of space-time appearing as a fine structure which refines the usual 4-dimensional continuum.

The action principle however was at the beginning of the theory still of traditional form (see [14]). In our joint work with Chamseddine [5] we understood that instead of imitating the traditional form of the Yang-Mills action, one could obtain the full package of the Einstein-Hilbert action[1] of gravity coupled with matter by a fundamental spectral principle. In the language of NCG this principle asserts that the action only depends upon the "line element" i.e. the inverse[2] of the operator D. It follows then from elementary considerations of additivity for disjoint unions of spaces that it must be of the form $\mathrm{Tr}(f(D/\Lambda))$ where f is a function and Λ is a parameter having the same dimension (that of an energy) as the inverse line element D.

1.3 Possible Relevance for Quantum Gravity

It will by now be clear to the reader that the point of view adopted in this essay is to try to understand from a mathematical perspective, how the perplexing combination of the Einstein-Hilbert action coupled with matter, with all the subtleties such as the Brout-Englert-Higgs sector, the V-A and the see-saw mechanisms etc. can emerge from a simple geometric model. The new tool is the spectral paradigm and the new outcome is that geometry does emerge on the stage where Quantum Mechanics happens, i.e. Hilbert space and linear operators.

The idea that group representations as operators in Hilbert space are relevant to physics is of course very familiar to every particle theorist since the work of Wigner and Bargmann. That the formalism of operators in Hilbert space encompasses the variable geometries which underly gravity is the *leitmotiv* of our approach.

---

[1]There is a well-known "priority episode" between Hilbert and Einstein which is discussed in great detail in [1, 2] and whose outcome, called the Einstein-Hilbert action, plays a key role in our approach.

[2]In the orthogonal complement of its kernel.

In order to estimate the potential relevance of this approach to Quantum Gravity, one first needs to understand the physics underlying the problem of Quantum Gravity. There is an excellent article for this purpose: the paper [15] explains how the problem arises when one tries to apply the perturbative method (which is so successful in quantum field theory) to the Lagrangian of gravity coupled with matter. Quoting from [15]: "Quantization of gravity is inevitable because part of the metric depends upon the other fields whose quantum nature has been well established".

Two main points are that the presence of the other fields forces one, due to renormalization, to add higher derivative terms of the metric to the Lagrangian and this in turns introduces at the quantum level an inherent instability that would make the universe blow up. This instability is instantly fatal to an interacting quantum field theory. Moreover primordial inflation prevents one from fixing the problem by discretizing space at a very small length scale. What our approach permits is to develop a "particle picture" for geometry; and a careful reading of the present paper should hopefully convince the reader that this particle picture stays very close to the inner workings of the Standard Model coupled to gravity. For now the picture is limited to the "one-particle" description and there are deep purely mathematical reasons to develop the many particles picture. The main one is that the root of the one-particle picture, described by spectral triples, is *KO*-homology and the dual topological *KO*-theory (see Sect. 3.4). The duality between the two theories is the origin of the quanta of geometry given by irreducible representations of the higher Heisenberg relation described in Sect. 4 below. As already mentioned in [16], algebraic K-theory, which is a vast refinement of the topological theory, is begging for the development of a dual theory and one should expect profound relations between this dual theory and the theory of interacting quanta of geometry. As a concrete point of departure, note that the deepest results on the topology of diffeomorphism groups of manifolds are given by the Waldhausen algebraic K-theory of spaces and we refer to [17] for a unifying picture of algebraic K-theory. For this paper, we now we discuss in depth the problem of the co-existence of the discrete and the continuum in geometry.

2 Prelude: The Discrete and the Continuum

In this preliminary section we shall discuss two solutions of the mathematical problem of treating the continuous and the discrete in a unified manner. We first briefly present Grothendieck's solution: the notion of a topos which allowed him to treat in a unified manner ordinary topological spaces and the combinatorial structures arising in the world of arithmetic. We continue with a text of Grothendieck on Riemann as a prelude for a re-reading of Riemann's inaugural lecture. We then explain how the quantum formalism provides another solution to the coexistence of discrete and continuous variables.

2.1 *Grothendieck's Solution: Topos*

Grothendieck's solution to the problem of treating the continuous and the discrete in a unified manner is the notion of a Topos. It does reconcile the usual idea of a topological space with that of a discrete combinatorial diagram. One does not concentrate on the space X itself, with its points etc. but rather on the ability of X to define a variable set Z_x depending on $x \in X$. When X is an ordinary topological space such a "variable set" indexed by X is simply a sheaf of sets on X. But this continues to make sense starting from an abstract combinatorial diagram! In short the key idea here is the idea of replacing X by its role as a parameter space.

"Space $X \to$ Category of variable sets with parameter in X"

The abstract categories of such "sets depending on parameters" fulfill almost all properties of the category of sets, except the axiom of the excluded middle, and encode in a faithful manner a topological space X through the category of sheaves of sets on X. This new idea is amazing in its simplicity, its connection with logics and the richness of the new class of spaces that it uncovers. In Grothendieck's own words (see "Récoltes et Semailles" [18, 19]) one can sense his amazement:

> Le "principe nouveau" qui restait à trouver, pour consommer les épousailles promises par des fées propices, ce n'était autre aussi que ce "lit" spacieux qui manquait aux futurs époux, sans que personne jusque-là s'en soit seulement aperçu...
>
> Ce "lit à deux places" est apparu (comme par un coup de baguette magique ...) avec l'idée du topos. Cette idée englobe, dans une intuition topologique commune, aussi bien les traditionnels espaces (topologiques), incarnant le monde de la grandeur continue, que les (soi-disant) "espaces" (ou "variétés") des géomètres algébristes abstraits impénitents, ainsi que d'innombrables autres types de structures, qui jusque-là avaient semblé rivées irrémédiablement au "monde arithmétique" des agrégats "discontinus" ou "discrets".

I would like to stress a key point of Grothendieck's idea of topos by using a metaphor. From his point of view, one understands a geometric space not by directly staring at it: no, the space remains at the back of the stage as a hidden schemer which governs the variability of every object at the front of the stage which is occupied by the usual suspects such as "abelian groups" for instance. But once one studies these usual suspects in their new environment one finds that their fine properties reveal, from their relations with ordinary abelian groups, the cohomology of the hidden parameter space. Here the word "ordinary" means "independent of the parameter" and thus ordinary sets form part of the new set theory. This makes sense because a Grothendieck topos admits a unique morphism to the topos of sets.

2.2 *Riemann*

In the prelude of "Récoltes et Semailles" [18], Alexandre Grothendieck makes the following points on the search for relevant geometric models for physics and on

Riemann's lecture on the foundations of geometry: (see Appendix for the English translation)

Il doit y avoir déjà quinze ou vingt ans, en feuilletant le modeste volume constituant l'œuvre complète de Riemann, j'avais été frappé par une remarque de lui "en passant". Il y fait observer qu'il se pourrait bien que la structure ultime de l'espace soit "discrète", et que les représentations "continues" que nous nous en faisons constituent peut-être une simplification (excessive peut-être, à la longue ...) d'une réalité plus complexe; que pour l'esprit humain, "le continu" était plus aisé à saisir que "le discontinu", et qu'il nous sert, par suite, comme une "approximation" pour appréhender le discontinu.

C'est là une remarque d'une pénétration surprenante dans la bouche d'un mathématicien, à un moment où le modèle euclidien de l'espace physique n'avait jamais encore été mis en cause ; au sens strictement logique, c'est plutôt le discontinu qui, traditionnellement, a servi comme mode d'approche technique vers le continu.

Les développements en mathématique des dernières décennies ont d'ailleurs montré une symbiose bien plus intime entre structures continues et discontinues, qu'on ne l'imaginait encore dans la première moitié de ce siècle. Toujours est-il que de trouver un modèle "satisfaisant" (ou, au besoin, un ensemble de tels modèles, se "raccordant" de façon aussi satisfaisante que possible...), que celui-ci soit "continu", "discret" ou de nature "mixte" – un tel travail mettra en jeu sûrement une grande imagination conceptuelle, et un flair consommé pour appréhender et mettre à jour des structures mathématiques de type nouveau.

Ce genre d'imagination ou de "flair" me semble chose rare, non seulement parmi les physiciens (où Einstein et Schrödinger semblent avoir été parmi les rares exceptions), mais même parmi les mathématiciens (et là je parle en pleine connaissance de cause).

Pour résumer, je prévois que le renouvellement attendu (s'il doit encore venir...) viendra plutôt d'un mathématicien dans l'âme, bien informé des grands problèmes de la physique, que d'un physicien. Mais surtout, il y faudra un homme ayant "l'ouverture philosophique" pour saisir le nœud du problème. Celui-ci n'est nullement de nature technique, mais bien un problème fondamental de "philosophie de la nature".

After reading the above text of Grothendieck, let us go to the relevant part of Riemann's Habilitation lecture on the foundations of geometry and explain why his great insight is, together with the advent of quantum mechanics, the best prelude to the new paradigm of spectral triples, the basic geometric concept in NCG.

Wenn aber eine solche Unabhängigkeit der Körper vom Ort nicht stattfindet, so kann man aus den Massverhältnissen im Grossen nicht auf die im Unendlichkleinen schliessen; es kann dann in jedem Punkte das Krümmungsmass in drei Richtungen einen beliebigen Werth haben, wenn nur die ganze Krümmung jedes messbaren Raumtheils nicht merklich von Null verschieden ist; noch complicirtere Verhältnisse können eintreten, wenn die vorausgesetzte Darstellbarkeit eines Linienelements durch die Quadratwurzel aus einem Differentialausdruck zweiten Grades nicht stattfindet. Nun scheinen aber die empirischen Begriffe, in welchen die räumlichen Massbestimmungen gegründet sind, der Begriff des festen Körpers und des Lichtstrahls, im Unendlichkleinen ihre Gültigkeit zu verlieren; es ist also sehr wohl denkbar, dass die Massverhältnisse des Raumes im Unendlichkleinen den Voraussetzungen der Geometrie nicht gemäss sind, und dies würde man in der That annehmen müssen, sobald sich dadurch die Erscheinungen auf einfachere Weise erklären liessen.

Die Frage über die Gültigkeit der Voraussetzungen der Geometrie im Unendlichkleinen hängt zusammen mit der Frage nach dem innern Grunde der Massverhältnisse des Raumes. Bei dieser Frage, welche wohl noch zur Lehre vom Raume gerechnet werden darf, kommt die obige Bemerkung zur Anwendung, dass bei einer discreten Mannigfaltigkeit das Princip der Massverhältnisse schon in dem Begriffe dieser Mannigfaltigkeit enthalten ist, bei einer

> stetigen aber anders woher hinzukommen muss. Es muss also entweder das dem Raume zu Grunde liegende Wirkliche eine discrete Mannigfaltigkeit bilden, oder der Grund der Massverhältnisse ausserhalb, in darauf wirkenden bindenen Kräften, gesucht werden.
>
> Die Entscheidung dieser Fragen kann nur gefunden werden, indem man von der bisherigen durch die Erfahrung bewährten Auffassung der Erscheinungen, wozu *Newton* den Grund gelegt, ausgeht und diese durch Thatsachen, die sich aus ihr nicht erklären lassen, getrieben allmählich umarbeitet; solche Untersuchungen, welche, wie die hier geführte, von allgemeinen Begriffen ausgehen, können nur dazu dienen, dass diese Arbeit nicht durch die Beschränktheit der Begriffe gehindert und der Fortschritt im Erkennen des Zusammenhangs der Dinge nicht durch überlieferte Vorurtheile gehemmt wird.
>
> Es führt dies hinüber in das Gebiet einer andern Wissenschaft, in das Gebiet der Physik, welches wohl die Natur der heutigen Veranlassung nicht zu betreten erlaubt.

This can be translated as follows:

"But if the independence of bodies from position is not fulfilled, we cannot draw conclusions from metric relations of the large, to those of the infinitely small; in that case the curvature at each point may have an arbitrary value in three directions, provided that the total curvature of every measurable portion of space does not differ sensibly from zero. Still more complicated relations may exist if we no longer assume that the line element is expressible as the square root of a quadratic differential. Now it seems that the empirical notions on which the metric determinations of space are based, the notion of solid body and of ray of light, cease to be valid for the infinitely small. We are therefore quite free to assume that the metric relations of space in the infinitely small do not comply with the hypotheses of geometry; and we ought in fact to do this, if we can thereby obtain a simpler explanation of phenomena.

The question of the validity of the hypotheses of geometry in the infinitely small is tied up with the question of the origin of the metric relations of space. In this last question, which we may still regard as belonging to the doctrine of space, is found the application of the remark made above; that in a discrete manifold, the origin of its metric relations is given intrinsically, while in a continuous manifold, this origin must come from outside. Either therefore the reality which underlies space must form a discrete manifold, or we must seek the origin of its metric relations outside it, in the binding forces which act upon it.

The answer to these questions can only be obtained by starting from the conception of phenomena which has hitherto been justified by experiments, and which Newton assumed as a foundation, and by making in this conception the successive changes required by facts which it cannot explain. Researches starting from general notions, like the investigation we have just made, can only be useful in preventing this work from being hampered by too narrow views, and progress in knowledge of the interdependence of things from being prevented by traditional prejudices.

This leads us into the domain of another science, of physics, into which the object of this work does not allow us to enter today."

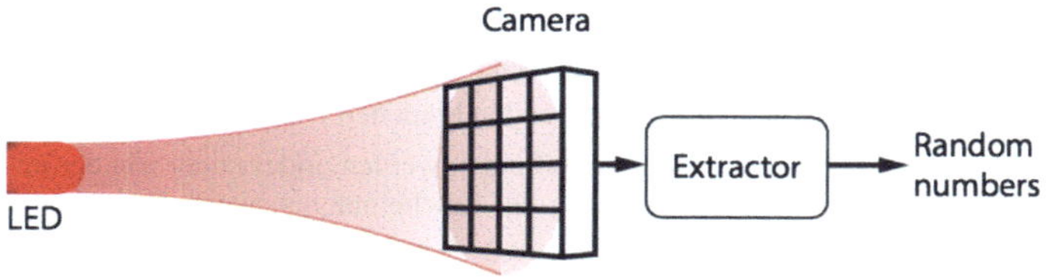

Fig. 1 The device uses the light emitted by a LED to produce a random number based on the quantum randomness of which cells of the camera are reached by emitted photons. This figure is taken from [20]

2.3 The Quantum and Variability

The originality of the quantum world (in which we actually live) as compared to its classical approximation, is already manifest at the experimental level by the "imaginative randomness" of the results of experiments in the microscopic world. In order to appreciate this point consider the problem of manufacturing a random number generator in such a way that even if an attacker happens to know the full details of the system the chance of reproducing the outcome is zero. This problem was solved concretely by Bruno Sanguinetti, Anthony Martin, Hugo Zbinden, and Nicolas Gisin from the Group of Applied Physics, University of Geneva (see [20]). They invented: "A generator of random numbers of quantum origin using technology compatible with consumer and portable electronics and whose simplicity and performance will make the widespread use of quantum random numbers a reality, with an important impact on information security (Fig. 1)."

This inherent randomness of the quantum world is not totally arbitrary since when the observable quantities that one measures happen to commute the usual classical intuition does apply. We owe to Werner Heisenberg the discovery[3] that the order of terms does matter when one deals with physical quantities which pertain to microscopic systems. We shall come back later in Sect. 3.3 to the meaning of this fact but for now we retain that when manipulating the observables quantities for a microscopic system, the order of terms in a product plays a crucial role.

The commutativity of Cartesian coordinates does not hold in the algebra of coordinates on the phase space of a microscopic system. What Heisenberg discovered was that quantum observables obey the rules of matrix mechanics and this led von Neumann to formalize quantum mechanics in terms of operators on Hilbert space. Let us explain now why this formalism actually provides a mathematical notion of "real variable" which allows for the coexistence of continuous and discrete variables. Let us first display the defect of the classical notion. In the classical formulation of real variables as maps from a set X to the real numbers $\mathbb{R}$, the set X has to be uncountable if some variable has continuous range. But then for any other variable

[3]Which he did while he was in the Island of Helgoland trying to recover from hay fever away from pollen sources (Fig. 2).

Fig. 2 Birdseye view, Helgoland, Germany, between 1890 and 1900. Image available from the United States Library of Congress's Prints and Photographs division, digital ID ppmsca.00573

with countable range some of the multiplicities are infinite. This means that discrete and continuous variables cannot coexist in this classical formalism.

Fortunately everything is fine and this problem of treating continuous and discrete variables on the same footing is completely solved using the formalism of quantum mechanics which provides another solution and treats directly the notion of real variable. The key replacement is

"Real Variable $\rightarrow$ Self Adjoint Operator in Hilbert space"

All the usual attributes of real variables such as their range, the number of times a real number is reached as a value of the variable etc. have a perfect analogue in the quantum mechanical setting. The range is the spectrum of the operator, and the spectral multiplicity gives the number of times a real number is reached. It is very comforting for instance that one can compose any measurable (Borel) map $h : \mathbb{R} \rightarrow \mathbb{R}$ with any self-adjoint operator H so that $h(H)$ makes sense and has the expected property of the composed real variable. In the early times of quantum mechanics, physicists had a clear intuition of this analogy between operators in Hilbert space (which they called q-numbers) and variables. Note that the choice of Hilbert space is irrelevant here since all separable infinite dimensional Hilbert spaces are isomorphic.

Classical	Quantum
Real variable $f : X \rightarrow \mathbb{R}$	Self-adjoint operator in Hilbert space
Possible values of the variable	Spectrum of the operator
Algebraic operations on functions	Algebra of operators in Hilbert space

In fact it is the uniqueness of the separable infinite dimensional Hilbert space that cures the above problem of coexistence of discrete and continuous variables: $L^2[0, 1]$ is the same as $\ell^2(\mathbb{N})$, and variables with continuous range (such as the operator of multiplication by $x \in [0, 1]$) coexist happily with variables with countable range (such as the operator of multiplication by $1/n$, $n \in \mathbb{N}$), but they do not commute!

It is only because one drops commutativity that variables with continuous range can coexist with variables with countable range. The only new fact is that they do not commute, and the real subtlety is in their algebraic relations.

What is surprising is that the new set-up immediately provides a natural home for the "infinitesimal variables": and here the distinction between "variables" and numbers (in many ways this is where the point of view of Newton is more efficient than that of Leibniz) is essential. It is worth quoting Newton's definition of variables and of infinitesimals, as opposed to Leibniz:

> In a certain problem, a variable is the quantity that takes an infinite number of values which are quite determined by this problem and are arranged in a definite order

> A variable is called infinitesimal if among its particular values one can be found such that this value itself and all following it are smaller in absolute value than an arbitrary given number

Indeed it is perfectly possible for an operator to be "smaller than epsilon for any epsilon" without being zero. This happens when the norm of the restriction of the operator to subspaces of finite codimension tends to zero when these subspaces decrease (under the natural filtration by inclusion). The corresponding operators are called "compact" and they share with naive infinitesimals all the expected algebraic properties.

Classical	Quantum
Infinitesimal variable	Compact operator in Hilbert space
Infinitesimal of order α	$\mu_n(T)$ of size $n^{-\alpha}$ when $n \to \infty$
Integral of function $\int f(x)dx$	$⨍ T$ = coefficient of $\log(\Lambda)$ in $\mathrm{Tr}_\Lambda(T)$

Indeed they form a two-sided ideal of the algebra of bounded operators in Hilbert space and the only property of the naive infinitesimal calculus that needs to be dropped is the commutativity.

The calculus of infinitesimals fits perfectly into the operator formalism of quantum mechanics where compact operators play the role of infinitesimals, with order governed by the rate of decay of the characteristic values, and where the logarithmic divergences familiar in physics give the substitute for integration of infinitesimals of order one, in the form of the Dixmier trace and Wodzicki's residue. We refer to [14] for a detailed description of the new integral $⨍$.

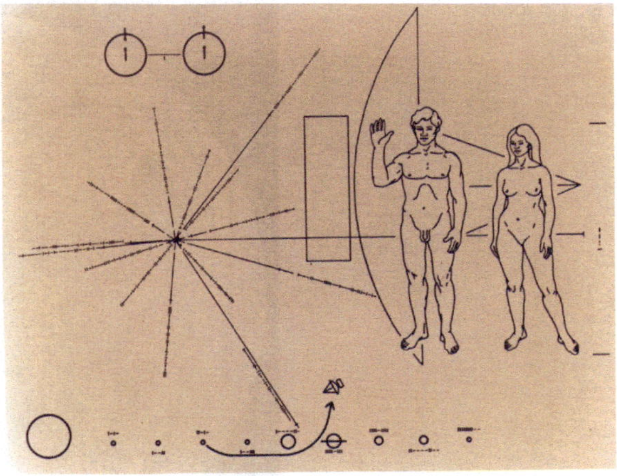

Fig. 3 The pioneer 4 probe, picture from NASA: 668774 main pioneer plaque

3 The Spectral Paradigm

Before we start the "inward bound" trip [21] to very small distances, it is worth explaining how the spectral point of view helps also when dealing with issues connected to large astronomical distances.

The simple question "Where are we?" does not have such a simple answer since giving our coordinates in a specific chart is not an invariant manner of describing our position. We refer to Fig. 3 for one attempt at an approximate answer.

In fact it is not obvious how to solve two mathematical questions which naturally arise in this context:

1. Can one specify a geometric space in an invariant manner?
2. Can one specify a point of a geometric space in an invariant manner?

3.1 Why Spectral

Given a compact Riemannian space one obtains a slew of geometric invariants of the space by considering the spectrum of natural operators such as the Laplacian. The obtained list of numbers is a bit like a scale associated to the space as made clear by Mark Kac in his famous paper[4] "Can one hear the shape of a drum?". It is well known however since a famous one page paper[5] of John Milnor that the spectrum of operators, such as the Laplacian, does not suffice to characterize a compact Riemannian space. But it turns out that the missing information is encoded by the relative position of two abelian algebras of operators in Hilbert space. Due to a theorem of von

[4]Kac [22].

[5]Milnor [23].

Fig. 4 Geodesic

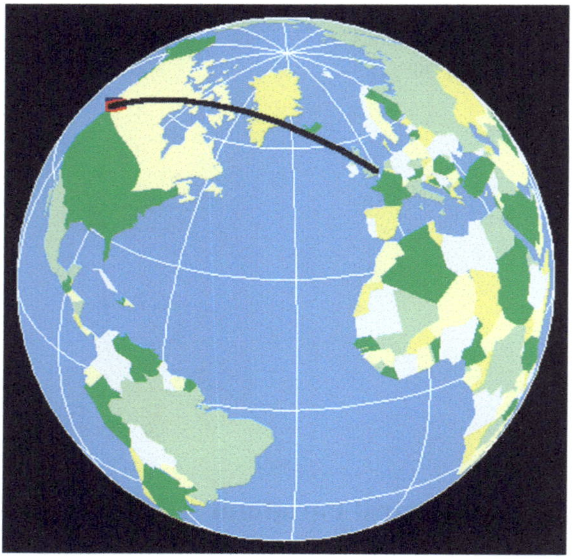

Neumann, the algebra of multiplication by all measurable bounded functions acts in Hilbert space in a unique manner, independent of the geometry one starts with. Its relative position with respect to the other abelian algebra given by all functions of the Laplacian suffices to recover the full geometry, provided one knows the spectrum of the Laplacian. For some reason which has to do with the inverse problem, it is better to work with the Dirac operator; and as we shall explain now, this gives a guess for a new incarnation of the "line element". The Riemannian paradigm is based on the Taylor expansion in local coordinates of the square of the line element, and in order to measure the distance between two points one minimizes the length of a path joining the two points as in Fig. 4

$$d(a, b) = \text{Inf} \int_\gamma \sqrt{g_{\mu\nu}\, dx^\mu\, dx^\nu} \tag{1}$$

Great efforts were done at the time of the French revolution in order to obtain a sensible unification of the various units of length that were in use across the country. It was decided (by Louis XVI, under the advice of Lavoisier) to take, as a unit, the length L such that $4 \times 10^6\, L$ would be the circumference of the earth. After using as a preliminary reduction the computation of angles from astronomical observations to reduce the actual measurement to a smaller portion of meridian, a team was sent out in 1792 to make the precise measurement of the distance between Dunkerque in the north of France and Barcelona in Spain; see Fig. 5. This measurement[6] resulted in

[6] I refer the reader to [24] for a very interesting and more detailed account of the story of the measurement performed by Delambre and Méchain.

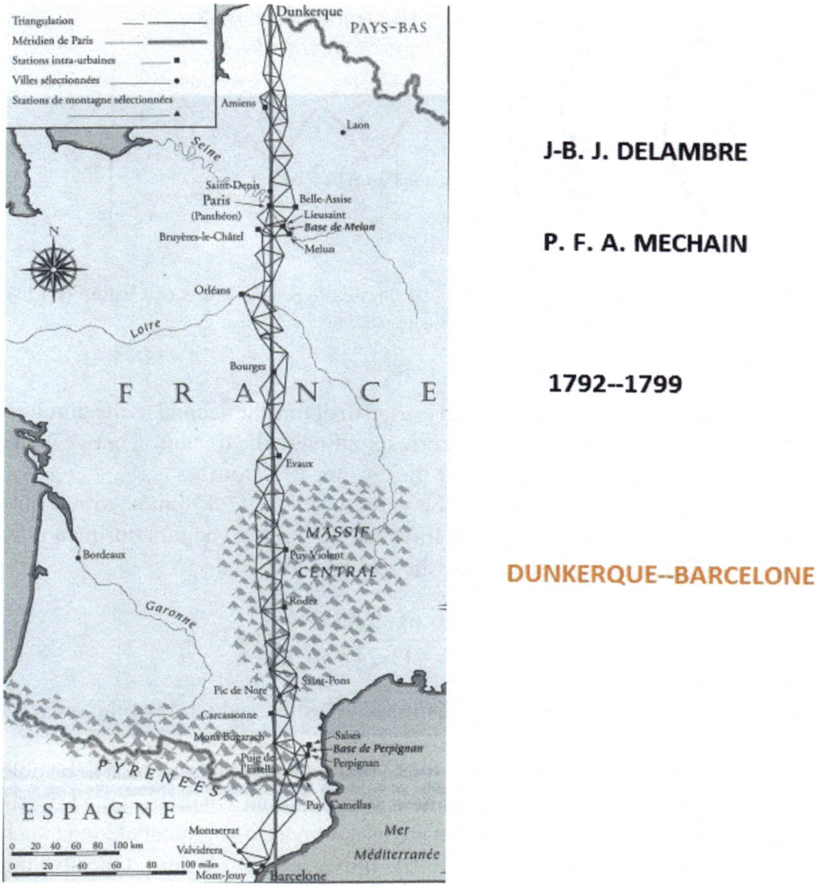

Fig. 5 Delambre and Mechain

an incarnation of L as a concrete platinum bar that was kept in Pavillon de Breuteuil near Paris. I remember learning in school this definition of the "meter".

However it turned out that in the 1930s, physicists were able to decide that the above choice of L was no good. Not only because it would seem totally unpractical if we would for instance try to transmit its definition to a far distant star, but for a more pragmatic reason, they observed that the concrete platinum bar defining L actually had a non-constant length! This observation was done by comparing it with a specific wave length of Krypton.

Then it took some time until they decided to take the obvious step: to replace L by the wavelength of a specific atomic transition (the chosen one is called $2S_{1/2}$ of Cesium 133), as was done in 1967 (Fig. 6).

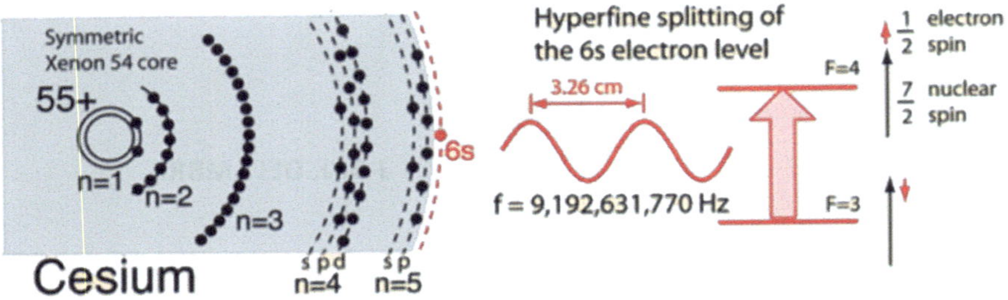

Fig. 6 Meter → Wave length, the 13th CGPM (1967) uses hyperfine levels of Cesium (C133). Adaptation of original found at hyperphysics.phy-astr.gsu.edu

More precisely, this hyperfine transition is used to define the second as the duration of 9,192,631,770 periods of the radiation corresponding to the transition between the two hyperfine levels of the ground state of the cesium 133 atom.

Moreover the speed of light is set to the value of 299,792,458 m/s, which thus defines the meter as the length of the path travelled by light in vacuum during a time interval of 1/299,792,458 of a second, i.e. the meter is

$$\frac{9,192,631,770}{299,792,458} = \frac{656,616,555}{21,413,747} \sim 30.6633$$

times the wavelength of the hyperfine transition of Cesium (which is of the order of 3.26 cm).

What is manifest with this new choice of L is that one now has a chance to be able to communicate our "unit of length" with aliens without telling them to come to Paris etc. Probably in fact this issue should motivate us to choose a chemical element such as hydrogen which is far more common in the universe than Cesium. One striking advantage of the new choice of L is that it is no longer "localized" (as it was before near Paris) and is available anywhere using the constancy of the spectral properties of atoms. It will serve us as a motivation for our spectral paradigm.

3.2 *The Line Element*

The presence of the square root in (1) is the witness of Riemann's prescription for the square of the line element as $ds^2 = g_{\mu\nu}\, dx^\mu\, dx^\nu$. In the spectral framework the extraction of the square root of the Laplacian goes back to Hamilton who already wrote, using his quaternions, the key combination

$$D = i\partial_x + j\partial_y + k\partial_z$$

Fig. 7 Hamilton [25], Clifford, Dirac

The conceptual algebraic device for extracting the square root of sums of squares such as $X^2 + Y^2$ is provided by the Clifford algebra where the anti-commutation $XY = -YX$ provides the simplification $(X + Y)^2 = X^2 + Y^2$ (Fig. 7).

P. Dirac showed how to extract the square root of the Laplacian in order to obtain a relativistic form of the Schrödinger equation. For curved spaces Atiyah and Singer devised a general formula for the Dirac operator on a spin Riemannian manifold and this provides us with our prescription: the line element is the propagator

$$ds = D^{-1}$$

(where one takes the value 0 on the kernel). This allows us to measure distances and (1) becomes

$$d(a,b) = \text{Sup}\, |f(a) - f(b)|, f \text{ such that } \|[D, f]\| \leq 1. \tag{2}$$

which gives the same answer as (1) and is a "Kantorovich dual" of the usual formula. But we now have the possibility to define and measure distances without the need of paths joining two points as in (1). And indeed one finds plenty of examples of totally disconnected spaces in which the new formula (2) makes sense and gives sensible results while (1) would not, due to the absence of connected arcs.

The link of this new definition of distances (and hence of geometry) with the quantum world appears in many ways: first the "line element" ought to be an "infinitesimal". This indeed fits since in a compact Riemannian spin manifold the above operator ds is compact i.e. infinitesimal as explained in Sect. 2.3. But there are two more facts which help us to appreciate the relevance of the new concept: both are displayed in Fig. 8. In the upper part the directed line is a common ingredient of Feynman diagrams, it represents the internal legs of fermionic diagrams and is called the "fermion propagator". Physically it represents a very tiny interval in which the interaction takes place. Mathematically it is our "ds" (modulo a bit of agility in understanding the physics language and in particular the need to pass from the Minkowski signature to the Euclidean one). The lower part of Fig. 8 displays

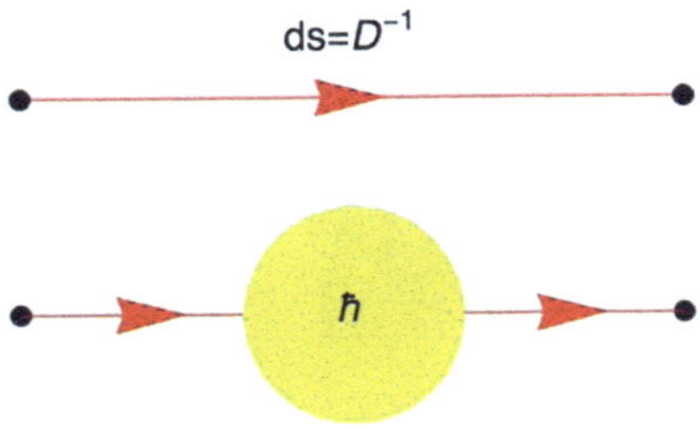

Fig. 8 Line element

an even more important feature: the above fermionic propagator undergoes quantum corrections due to its role in quantum field theory and we can interpret these corrections as quantum corrections to the geometry!

3.3 *The Bonus from Non-commutativity*

In algebra the commutativity assumption often appears as a welcome simplification which makes many algebraic manipulations much easier. But in fact we should realize that our use of the written language makes us perfectly familiar with non-commutativity. The advantage, as far as meaning is concerned, of paying attention to the order of terms, becomes clear when considering anagrams i.e. writings which become equal when "abelianized" but nevertheless have quite different meanings when the order of terms is respected. Here is a recent anagram which can be found in "Anagrammes pour lire dans les pensées" by Raphael Enthoven and Jacques Perry-Salkow,

"*ondes gravitationnelles*"

"*le vent d'orages lointains*"

When we permit ourselves to commute the various letters involved in each of these phrases we find the same result:

$$a^2de^3gi^2l^2n^3o^2rs^2t^2v$$

This shows that in projecting a phrase in the commutative world one looses an enormous amount of information encoded by non-commutativity. Natural languages respect non-commutativity and a phrase is a much more informative datum than its commutative algebraic shadow.

Here are two more key features of the noncommutative world:

1. Non-commuting discrete variables of the simplest kind generate continuous variables.
2. A noncommutative algebra possesses inner automorphisms.

We always think of variables through their representations as operators in Hilbert space as explained in Sect. 2.3 and since the product of two self-adjoint operators is not self-adjoint unless they commute, one deals with algebras $\mathcal{A}$ which are $*$-algebras i.e. which are endowed with an antilinear involution which obeys the rule $(xy)^* = y^* x^*$ for any $x, y \in \mathcal{A}$. The simplest noncommutative algebra of this kind is $M_2(\mathbb{C})$ the algebra of 2×2 matrices

$$a = \begin{pmatrix} a_{11} & a_{12} \\ a_{21} & a_{22} \end{pmatrix}, \quad b = \begin{pmatrix} b_{11} & b_{12} \\ b_{21} & b_{22} \end{pmatrix}, \quad ab = \begin{pmatrix} a_{11}b_{11} + a_{12}b_{21} & a_{11}b_{12} + a_{12}b_{22} \\ a_{21}b_{11} + a_{22}b_{21} & a_{21}b_{12} + a_{22}b_{22} \end{pmatrix}$$

and the antilinear involution is given using the complex conjugation $z \mapsto \bar{z}$ by the conjugate transpose, i.e.

$$a^* = \begin{pmatrix} \bar{a}_{11} & \bar{a}_{21} \\ \bar{a}_{12} & \bar{a}_{22} \end{pmatrix}$$

This algebra $M_2(\mathbb{C})$ only represents discrete variables taking at most two values but as soon as one adjoins another non-commuting variable Y, such that $Y = Y^*$ and $Y^2 = 1$ one generates all matrix valued functions on the two-sphere.

To be more precise, write the above generic matrix in the form $a = a_{11}e_{11} + a_{12}e_{12} + a_{21}e_{21} + a_{22}e_{22}$ where the $e_{ij} \in M_2(\mathbb{C})$, and the coefficients $a_{ij} \in \mathbb{C}$ are complex numbers. Then using algebra one can write $Y = y_{11}e_{11} + y_{12}e_{12} + y_{21}e_{21} + y_{22}e_{22}$ where the y_{ij} are no longer complex numbers but commute with $M_2(\mathbb{C})$. For instance $y_{11} = e_{11}Ye_{11} + e_{21}Ye_{12}$. One imposes the additional condition that the trace of Y is zero, i.e. that $y_{11} + y_{22} = 0$. It is then an exercise using the relations $Y = Y^*$ and $Y^2 = 1$, to show that the C^*-algebra generated by the y_{ij} is the algebra $C(S^2)$ of continuous functions on the two sphere S^2. It contains of course plenty of "continuous variables" and the traditional sup norm of complex valued functions is

$$\mathrm{Sup}_{x \in S^2} |f(x)| = \mathrm{Sup}_\pi \|\pi(f)\|$$

where in the right hand side π runs through all Hilbert space representations (compatible with the involution $*$) of the above relations. One obtains all continuous functions by completion and thus one keeps inside the algebra $C(S^2)$ the nicer smooth functions such as those algebraically obtained from the y_{ij}. The sphere itself is recovered as the **Spectrum** of the algebra, and the points of the sphere are the characters i.e. the morphisms of involutive algebras to $\mathbb{C}$.

This is a prototype example of how a connected space (here the two sphere S^2) can spring out of the discrete (here $M_2(\mathbb{C})$ and the two valued variable Y) due to noncommutativity. Note also the compatibility of the two notions of spectrum. Indeed for f in the commutative algebra generated by the y_{ij}, the spectrum of the operator $\pi(f)$ is the image by the corresponding function on S^2 of the support of the representation π which is a closed subset of the spectrum of the algebra. To put this in a suggestive manner: what happens is that the geometric space S^2 appeared in a spectral manner and from familiar players of the quantum world: the algebra $M_2(\mathbb{C})$, for instance, is familiar from spin systems.

There is another great bonus from non-commutativity: the natural algebra which springs out of the non-commuting $M_2(\mathbb{C})$ and Y discussed above is not the algebra generated by the y_{ij} but the algebra generated by $M_2(\mathbb{C})$ and Y. It contains the former but is larger and gives the algebra $C(S^2, M_2(\mathbb{C}))$ of matrix valued continuous functions on the two sphere. If we take the subalgebra of smooth functions $\mathcal{A} = C^\infty(S^2, M_2(\mathbb{C}))$ (which is canonically obtained inside $C(S^2, M_2(\mathbb{C}))$ by applying the smooth functional calculus to the generators) and one looks at its automorphism group,[7] one finds that it fits in an exact sequence

$$1 \to \mathrm{Int}(\mathcal{A}) \to \mathrm{Aut}(\mathcal{A}) \to \mathrm{Out}(\mathcal{A}) \to 1.$$

Such an exact sequence exists for any non-commutative $*$-algebra, the inner automorphisms $\mathrm{Int}(\mathcal{A})$ are those of the form $x \mapsto uxu^*$ where $u \in \mathcal{A}$ is a unitary element i.e. fulfills $uu^* = u^*u = 1$. The nice general fact is that these automorphisms always form a normal subgroup of the group $\mathrm{Aut}(\mathcal{A})$ and the quotient group $\mathrm{Out}(\mathcal{A})$ is called the group of outer automorphisms of $\mathcal{A}$. Now when one computes these groups in our example i.e. for $\mathcal{A} = C^\infty(S^2, M_2(\mathbb{C}))$, one finds that the group $\mathrm{Out}(\mathcal{A})$ is the group of diffeomorphisms $\mathrm{Diff}(S^2)$ while $\mathrm{Int}(\mathcal{A})$ is the group of smooth maps from S^2 to the Lie group $PSU(2)$ whose Lie algebra is $su(2)$. Thus we witness in this example the marriage of the gauge group of gravity i.e. the diffeomorphism group, with the gauge group of matter i.e. here of an $su(2)$-gauge theory.

3.4 *The Notion of Manifold*

The notion of spectral geometry has deep roots in pure mathematics. They have to do with the understanding of the notion of (smooth) manifold. While this notion is simple to define in terms of local charts i.e. by glueing together open pieces of finite dimensional vector spaces, it is much more difficult and instructive to arrive at a global understanding. To be specific we now discuss the notion of a compact oriented smooth manifold.

What one does is to detect global properties of the underlying space with the goal of characterizing manifolds. At first one only looks at the space up to homotopy. The broader category of "manifolds" that one first obtains is that of "Poincaré complexes" i.e. of CW complexes X which satisfy Poincaré duality with respect to the fundamental homology class with coefficients in $\mathbb{Z}$. It is important to take into account the fundamental group $\pi_1(X)$, and to assume Poincaré duality with arbitrary local coefficients. In the simply connected case, a result of Spivak [26] shows the existence (and uniqueness up to stable fiber homotopy equivalence) of a spherical fibration, called the Spivak normal bundle $p : E \to X$. Such a fibration satisfies the covering homotopy property and each fiber $p^{-1}(x)$ has the homotopy type of a sphere. At this point one is still very far from dealing with a manifold and the obstruction

[7]Compatible with the $*$-operation.

to obtain a smooth manifold in the given homotopy type is roughly the same as that of finding a vector bundle whose associated spherical fibration is $p : E \to X$. This follows from the work of Novikov and Browder at the beginning of the 1960s. There are important nuances between piecewise linear (PL) and smooth but they do not affect the 4-dimensional case in which we are interested.

The first key root of the notion of "spectral geometry" is a result of D. Sullivan (see [27], epilogue) that a PL-bundle is the same thing (modulo the usual "small-print" qualifications at the prime 2, [28]) as a spherical fibration together with a *KO*-orientation. What we retain is that the key property of a "manifold" is not Poincaré duality in ordinary homology but is Poincaré duality in the finer theory called *KO*-homology. To understand how much finer that theory is, it is enough to state that the fundamental class $[X] \in KO_*(X)$ contains all the information about the Pontrjagin classes of the manifold and these are not at all determined by its homotopy type: in the simply connected case only the signature class is fixed by the homotopy type.

Here comes now the second crucial root of the notion of spectral geometry from pure mathematics. In their work on the index theorem, Atiyah and Singer understood that operators in Hilbert space provide the right realization for *KO*-homology cycles [3, 29]. Their original idea was developed by Brown-Douglas-Fillmore, Voiculescu, Mischenko and acquired its definitive form in the work of Kasparov at the end of the 1970s. The great new tool is bivariant Kasparov theory, but as far as K-homology cycles are concerned[8] the right notion is already in Atiyah's paper [3]: A K-homology cycle on a compact space X is given by a representation of the algebra $C(X)$ (of continuous functions on X) in a Hilbert space $\mathcal{H}$, together with a Fredholm operator F acting in the same Hilbert space fulfilling some simple compatibility condition (of commutation modulo compact operators) with the action of $C(X)$. One striking feature of this representation of K-homology cycles is that the definition does not make any use of the commutativity of the algebra $C(X)$.

At the beginning of the 1980s, motivated by numerous examples of noncommutative spaces arising naturally in geometry from foliations or in physics from the Brillouin zone in the work of Bellissard on the quantum Hall effect, I realized that specifying an unbounded representative of the Fredholm operator gave the right framework for spectral geometry. The corresponding K-homology cycle only retains the stable information and is insensitive to deformations while the unbounded representative encodes the metric aspect. These are the deep mathematical reasons which are the roots of the notion of spectral triple.

3.5 Real Structure

The additional structure on a K-homology cycle that upgrades it into a *KO*-homology cycle is given by requiring a *real structure* [30], i.e. an antilinear unitary operator J

[8]The nuance between K and *KO* is important and gives rise to the real structure discussed in the next section.

acting in $\mathcal{H}$ which plays the same role and has the same algebraic properties as the charge conjugation operator in physics.

- In physics J is the charge conjugation operator.
- It is deeply related to Tomita's (Fig. 9) operator which conjugates the algebra with its commutant. The basic relation always satisfied is Tomita's relation:

$$[a, b^{\text{op}}] = 0\,, \quad \forall a, b \in \mathcal{A},\, b^{\text{op}} := Jb^* J^{-1}.$$

- In *KO*-homology, one obtains a *KO*-homology cycle for the algebra $\mathcal{A} \otimes \mathcal{A}^{\text{op}}$ and an intersection form:

$$K(\mathcal{A}) \otimes K(\mathcal{A}) \to \mathbb{Z}, \ \text{Index}(D_{e \otimes f})$$

In the even case, the chirality operator γ plays an important role, both γ and J are decorations of the spectral triple.

The following further relations hold for D, J and γ

$$J^2 = \varepsilon\,, \ DJ = \varepsilon' JD, \quad J\,\gamma = \varepsilon'' \gamma J\,, \quad D\gamma = -\gamma D$$

Fig. 9 Minoru Tomita

The values of the three signs $\varepsilon, \varepsilon', \varepsilon''$ depend only, in the classical case of spin manifolds, upon the value of the dimension n modulo 8 and are given in the following table:

n	0	1	2	3	4	5	6	7
ε	1	1	−1	−1	−1	−1	1	1
ε'	1	−1	1	1	1	−1	1	1
ε''	1		−1		1		−1	

In the classical case of spin manifolds there is a relation between the metric (or spectral) dimension given by the rate of growth of the spectrum of D and the integer modulo 8 which appears in the above table. For more general spaces, however, the two notions of dimension (the dimension modulo 8 is called the "*KO*-dimension" because of its origin in K-theory) become independent, since there are spaces F of metric dimension 0 but of arbitrary *KO*-dimension.

The search to identify the structure of the noncommutative space followed the bottom-up approach where the known spectrum of the fermionic particles was used to determine the geometric data that defines the space.

This bottom-up approach involved an interesting interplay with experiments. While at first the experimental evidence of neutrino oscillations contradicted the first attempt, it was realized several years later[9] in 2006 (see [7]), that the obstruction to getting neutrino oscillations was naturally eliminated by dropping the equality between the metric dimension of space-time (which is equal to 4 as far as we know) and its *KO*-dimension which is only defined modulo 8. When the latter is set equal to 2 modulo 8 (using the freedom to adjust the geometry of the finite space encoding the fine structure of space-time) everything works fine: the neutrino oscillations are there as well as the see-saw mechanism which appears for free as an unexpected bonus. Incidentally, this also solved the fermion doubling problem by allowing a simultaneous Weyl-Majorana condition on the fermions to halve the degrees of freedom.

3.6 *The Inner Fluctuations of the Metric*

In our joint work with Chamseddine and van Suijlekom [31], we obtained a conceptual understanding of the role of the gauge bosons in physics as the inner fluctuations of the metric. I will describe this result here in a non-technical manner.

In order to comply with Riemann's requirement that the inverse line element D embodies the forces of nature, it is evidently important that we do not separate artificially the gravitational part from the gauge part, and that D encapsulates both forces in a unified manner. In the traditional geometrization of physics, the gravitational part specifies the metric while the gauge part corresponds to a connection on a principal

[9]This crucial step was taken independently by John Barrett.

bundle. In the NCG framework, D encapsulates both forces in a unified manner and the gauge bosons appear as inner fluctuations of the metric but form an inseparable part of the latter. Ignoring at first the important nuance coming from the real structure J, the inner fluctuations of the metric were first defined as the transformation

$$D \mapsto D + A, \;\; A = \sum a_j[D, b_j], \;\; a_j, b_j \in \mathcal{A}, \;\; A = A^*$$

which imitates the way classical gauge bosons appear as matrix-valued one-forms in the usual framework. The really important facts were that the spectral action applied to $D + A$ delivers the Einstein-Yang-Mills action which combines gravity with matter in a natural manner, and that the gauge invariance becomes transparent at this level since an inner fluctuation coming from a gauge potential of the form $A = u[D, u^*]$ where u is a unitary element (i.e. $uu^* = u^*u = 1$) simply results in a unitary conjugation $D \mapsto uDu^*$ which does not change the spectral action.

An equally important fact which emerged very early on, is that as soon as one considers the product of an ordinary geometric space by a finite space of the simplest nature, such as two points, the inner fluctuations generate the Higgs field and the spectral action gives the desired quartic potential underlying the Brout-Englert-Higgs mechanism. The inverse line element D_F for the finite space F is given by the Yukawa coupling matrix which thus acquires geometric meaning as encoding the geometry of F.

What we discovered in our joint work with Chamseddine and van Suijlekom [31] is that the inner fluctuations arise in fact from the action on metrics (i.e. the D) of a canonical *semigroup* $\mathrm{Pert}(\mathcal{A})$ which only depends upon the algebra $\mathcal{A}$ and extends the unitary group. The semigroup is defined as the self-conjugate elements:

$$\mathrm{Pert}(\mathcal{A}) := \{A = \sum a_j \otimes b_j^{\mathrm{op}} \in \mathcal{A} \otimes \mathcal{A}^{\mathrm{op}} \mid \sum a_j b_j = 1, \;\; \theta(A) = A\}$$

where θ is the antilinear automorphism of the algebra $\mathcal{A} \otimes \mathcal{A}^{\mathrm{op}}$ given by

$$\theta : \sum a_j \otimes b_j^{\mathrm{op}} \mapsto \sum b_j^* \otimes a_j^{*\mathrm{op}}.$$

The composition law in $\mathrm{Pert}(\mathcal{A})$ is the product in the algebra $\mathcal{A} \otimes \mathcal{A}^{\mathrm{op}}$. The action of this semigroup $\mathrm{Pert}(\mathcal{A})$ on the metrics is given, for $A = \sum a_j \otimes b_j^{\mathrm{op}}$ by

$$D \mapsto D' = {}^A D = \sum a_j D b_j.$$

Moreover, the transitivity of inner fluctuations results from

$${}^{A'}({}^A D) = {}^{(A'A)} D.$$

What is remarkable is that it allows one to obtain the inner fluctuations in the real case (see Sect. 3.5), i.e. in the presence of the anti-unitary involution J, without having to make the "order one" hypothesis. To do this one uses instead of the algebra $\mathcal{A}$ the finer one given by $\mathcal{B} = \mathcal{A} \otimes \hat{\mathcal{A}}$ where the conjugate algebra $\hat{\mathcal{A}}$ acts in Hilbert space using JaJ^{-1} for $a \in \mathcal{A}$. The commutation of the actions of $\mathcal{A}$ and of $\hat{\mathcal{A}}$ in Hilbert space ensure that $\mathcal{B}$ acts. One then simply defines a semigroup homomorphism $\mu : \text{Pert}(\mathcal{A}) \to \text{Pert}(\mathcal{B})$ by

$$A \in \mathcal{A} \otimes \mathcal{A}^{\text{op}} \mapsto \mu(A) = A \otimes \hat{A} \in \left(\mathcal{A} \otimes \hat{\mathcal{A}}\right) \otimes \left(\mathcal{A} \otimes \hat{\mathcal{A}}\right)^{\text{op}}.$$

This gives the inner fluctuations in the real case and they take the form

$$D \mapsto D' := D + A_{(1)} + \widetilde{A}_{(1)} + A_{(2)}$$

where, with $A = \sum a_j \otimes b_j^{\text{op}}$ as above

$$A_{(1)} = \sum_i a_i \, [D, b_i]$$

$$\widetilde{A}_{(1)} = \sum_i \hat{a}_i \left[D, \hat{b}_i\right], \qquad \hat{a}_i = Ja_iJ^{-1}, \quad \hat{b}_i = Jb_iJ^{-1}$$

$$A_{(2)} = \sum_{i,j} \hat{a}_i a_j \left[[D, b_j], \hat{b}_i\right] = \sum_{i,j} \hat{a}_i \left[A_{(1)}, \hat{b}_i\right].$$

The new quadratic term $A_{(2)}$ vanishes when the order 1 condition is fulfilled but not in general. This conceptual understanding of the inner fluctuations allowed us, with Chamseddine and van Suijlekom [12, 13, 31] to determine the inner fluctuations for the natural extension of the Standard Model obtained from the classification of irreducible finite geometries of *KO*-dimension 6 of [6, 32]. This gives a Pati-Salam extension of the Standard model and we showed in [12, 13] that it yields a natural unification of couplings.

4 Quanta of Geometry

The above extension of the Standard Model obtained from the classification of irreducible finite geometries of *KO*-dimension 6 is based on the finite dimensional algebra $M_2(\mathbb{H}) \oplus M_4(\mathbb{C})$. While this algebra occurred as one of the simplest in the classification of [6, 32], its choice remained motivated by the bottom-up approach that we had followed all along up to that point. For instance there was no conceptual explanation for the difference of the real dimensions: 16 for $M_2(\mathbb{H})$ and 32 for $M_4(\mathbb{C})$.

This state of the theory changed drastically in our joint work with Chamseddine and Mukhanov [10, 11] where the above finite dimensional algebra $M_2(\mathbb{H}) \oplus M_4(\mathbb{C})$ appeared unexpectedly from a completely different motivation. The framework is the same, "spectral geometries" and the question is how to encode all spin Riemannian

4-manifolds in an operator theoretic manner. The key new idea is that since spectral triples only quantize the fundamental *KO*-homology class one should look at the same time for the quantization of the dual *KO*-theory class.

A hint of this idea can be understood easily in the one dimensional case, i.e. for the geometry of the circle. It is an exercise to prove that for unitary representations of the relations

$$UU^* = U^*U = 1, \; D = D^*, \;\; U^*[D, U] = 1 \tag{3}$$

with D unbounded self-adjoint playing as above the role of the inverse line element, one has[10]

1. ds infinitesimal $\Rightarrow \displaystyle{-\!\!\!\!\!\!\int} |ds| \in \mathbb{N}$.
2. The formula $d(a, b) = \operatorname{Sup}\{|f(a) - f(b)| \mid \|[D, f]\| \leq 1\}$ gives the standard distance on the spectrum of U which is the unit circle in $\mathbb{C}$.
3. Let M be a dimension 1 compact Riemannian manifold, $(\mathcal{A}, \mathcal{H}, D)$ the associated spectral triple. Then a solution $U \in \mathcal{A}$ of the equation $U^*[D, U] = 1$ exists if and only if the length $|M| \in 2\pi\mathbb{N}$.

One may understand the relations (3) as representations of a group which is a close relative of the Heisenberg group and this would lead one to group representations: but this theme would stay far away from our goal which is 4-dimensional geometries – and which was achieved in [10, 11]. What we have discovered is a higher geometric analogue of the Heisenberg commutation relations $[p, q] = i\hbar$. The role of the momentum p is played by the Dirac operator, as amply discussed above. The role of the position variable q in the higher analogue of $[p, q] = i\hbar$ was the most difficult to uncover, and another hint was given in Sect. 3.3 where the 2-sphere appeared from very simple non-commuting discrete variables. The general idea of [10, 11] is to encode the analogue of the position variable q in the same way as the Dirac operator encodes the components of the momenta, just using the Feynman slash. As explained below there are two levels. In the first, which is discussed in Sect. 4.1, the quantization is done for the K-theory class, and this justifies the terminology of K-theory higher Heisenberg equation. However, geometrically, the only solutions are disjoint unions of spheres of unit volume. To reach arbitrary compact oriented spin 4-manifolds, one needs the *KO*-theory refinement. This is treated in Sect. 4.2.

4.1 The K-Theory Higher Heisenberg Equation; Spheres

Let us first rewrite the description of the algebra of Sect. 3.3, which was presented as

$$M_2(\mathbb{C}) \star Y, \; Y = Y^*, \; Y^2 = 1, \; < Y > = 0.$$

---

[10]We refer to [14] for the meaning of the integral symbol.

As explained in Sect. 3.3 one can represent its elements as matrices with entries in the commutant of $M_2(\mathbb{C})$

$$Y = \begin{pmatrix} y_{11} & y_{12} \\ y_{21} & y_{22} \end{pmatrix}, \quad Y = \begin{pmatrix} t & z \\ z^* & -t \end{pmatrix}$$

where the second form is deduced from the relations. We can rewrite the result in terms of 3 gamma matrices Γ_A, $0 \leq A \leq 2$,

$$\Gamma_0 = \begin{pmatrix} 1 & 0 \\ 0 & -1 \end{pmatrix}, \quad \Gamma_1 = \begin{pmatrix} 0 & 1 \\ 1 & 0 \end{pmatrix}, \quad \Gamma_2 = \begin{pmatrix} 0 & i \\ -i & 0 \end{pmatrix}.$$

which fulfill:

$$\{\Gamma_A, \Gamma_B\} = 2\,\delta_{AB}, \quad (\Gamma_A)^* = \Gamma_A$$

and Y now takes the simple form:

$$Y = Y^A \Gamma_A, \quad Y^2 = 1, \quad Y^* = Y.$$

4.1.1 One-Sided Higher Heisenberg Equation

This suggests the following extension for arbitrary even n. We let $Y \in \mathcal{A} \otimes C_+$ be of the Feynman slashed form:

$$Y = Y^A \Gamma_A, \ Y^A \in \mathcal{A}, \ Y^2 = 1, \ Y^* = Y. \tag{4}$$

Here $C_+ \subset M_s(\mathbb{C})$, $s = 2^{n/2}$, is an irreducible representation of the Clifford algebra on $n + 1$ gamma matrices Γ_A, $0 \leq A \leq n$

$$\Gamma_A \in C_+, \quad \{\Gamma_A, \Gamma_B\} = 2\,\delta_{AB}, \ (\Gamma_A)^* = \Gamma_A.$$

The one-sided higher analogue of the Heisenberg commutation relations is

$$\frac{1}{n!}\left\langle Y\,[D, Y]^n \right\rangle = \gamma \tag{5}$$

where the notation $\langle T \rangle$ means the *normalized* trace of $T = T_{ij}$ with respect to the above matrix algebra $M_s(\mathbb{C})$ ($1/s$ times the sum of the s diagonal terms T_{ii}).

4.1.2 Quantization of Volume

For even n, Eq. (5), together with the hypothesis that the eigenvalues of D grow as in dimension n (i.e. that ds is an infinitesimal of order $1/n$) imply that the volume, expressed as the leading term in the Weyl asymptotic formula for counting eigenvalues of the operator D, is *quantized* by being equal to the index pairing of the operator D with the K-theory class of $\mathcal{A}$ defined by (note that s is even)

$$[e - 1/2] := [e] - s/2[1_{\mathcal{A}}] \in K_0(\mathcal{A}), \quad e = (1 + Y)/2.$$

To understand this result, we need to recall that the integral pairing between K-homology and K-theory is computed by the pairing of the Chern characters in cyclic theory according to the diagram:

$$\begin{array}{ccc} K - \text{Theory} & \longleftrightarrow & K - \text{Homology} \\ \Big\downarrow {\scriptstyle Ch_*} & & \Big\downarrow {\scriptstyle Ch^*} \\ HC_* & \longleftrightarrow & HC^* \end{array}$$

While the Chern character from K-homology to cyclic cohomology is difficult, its counterpart from K-theory to cyclic homology can be explained succinctly as follows. Given a unital (not assumed commutative) algebra $\mathcal{A}$, the (b, B)-bicomplex is obtained from the (b, B) bicomplex:

$$\underline{\mathcal{A}} := \mathcal{A}/\mathbb{C}1, \; \underline{C}_n(\mathcal{A}) := \mathcal{A} \otimes \underline{\mathcal{A}} \otimes \cdots \otimes \underline{\mathcal{A}}$$

$$b(a_0 \otimes \cdots \otimes a_n) := a_0 a_1 \otimes \cdots \otimes a_n - a_0 \otimes a_1 a_2 \otimes \cdots \otimes a_n + \cdots +$$

$$(-1)^{n-1} a_0 \otimes \cdots \otimes a_{n-1} a_n + (-1)^n a_n a_0 \otimes \cdots \otimes a_{n-1}$$

$$B(a_0 \otimes \cdots \otimes a_n) := \sum_0^n (-1)^{nj} 1 \otimes a_j \otimes a_{j+1} \otimes \cdots \otimes a_{j-1}.$$

The operations (b, B) fulfill

$$b^2 = 0, \; B^2 = 0, \; bB = -Bb$$

and an even (resp. odd) cycle $c = (c_n)$ is given by its components $c_n \in \underline{C}_n(\mathcal{A})$ for n even (resp. odd) which fulfill

$$Bc_n + bc_{n+2} = 0\,, \quad \forall n \text{ even (resp. odd).} \tag{6}$$

The Chern character of an idempotent $e \in \mathcal{A}$, $e^2 = e$, is then given by the cycle with components $\mathrm{Ch}_0(e) = e \in \mathcal{A} = \underline{C}_0(\mathcal{A})$ and for $k > 0$

$$\mathrm{Ch}_{2k}(e) = \lambda_k \times (e - \frac{1}{2}) \otimes e \otimes e \otimes \cdots \otimes e \in \underline{C}_{2k}(\mathcal{A}).$$

One has

$$b(e - \frac{1}{2}) \otimes e \otimes e \otimes \cdots \otimes e = \frac{1}{2}(1 \otimes e \otimes e \otimes \cdots \otimes e)$$

$$B(e - \frac{1}{2}) \otimes e \otimes e \otimes \cdots \otimes e = B(e \otimes e \otimes e \otimes \cdots \otimes e)$$

$$= (2k+1)\,(1 \otimes e \otimes e \otimes \cdots \otimes e)\,.$$

Thus one can choose the λ_k so that $(2k+1)\lambda_k + \frac{1}{2}\lambda_{k+1} = 0$

$$B\mathrm{Ch}_{2k}(e) + b\mathrm{Ch}_{2k+2}(e) = 0$$

and one gets a cycle $\mathrm{Ch}_*(e)$ in the (b, B)-bicomplex which gives the Chen character in K-theory.

In general the idempotent e does not belong to $\mathcal{A}$ but to matrices $M_s(\mathcal{A})$ and the next step is to pass to matrices. To do this one considers partial trace maps

$$\mathrm{tr} : \underline{C}_n(M_s(\mathcal{A})) \to \underline{C}_n(\mathcal{A}).$$

One defines

$$\mathrm{tr} : M_s(\mathcal{A}) \otimes M_s(\mathcal{A}) \otimes \cdots \otimes M_s(\mathcal{A}) \to \mathcal{A} \otimes \mathcal{A} \otimes \cdots \otimes \mathcal{A}$$

as the linear map such that:

$$\mathrm{tr}\,((a_0 \otimes \mu_0) \otimes (a_1 \otimes \mu_1) \otimes \cdots \otimes (a_m \otimes \mu_m)) = \mathrm{Trace}(\mu_0 \cdots \mu_m)\; a_0 \otimes a_1 \otimes \cdots \otimes a_m$$

where Trace is the ordinary trace of matrices. Let us denote by ι_k the operation which inserts a 1 in a tensor at the k-th place. So for instance

$$\iota_0(a_0 \otimes a_1 \otimes \cdots \otimes a_m) = 1 \otimes a_0 \otimes a_1 \otimes \cdots \otimes a_m$$

One has $\mathrm{tr} \circ \iota_k = \iota_k \circ \mathrm{tr}$ since (taking $k = 0$ for instance)

$$\mathrm{tr} \circ \iota_0\,((a_0 \otimes \mu_0) \otimes (a_1 \otimes \mu_1) \otimes \cdots \otimes (a_m \otimes \mu_m)) =$$

$$= \mathrm{tr}\,((1 \otimes 1) \otimes (a_0 \otimes \mu_0) \otimes (a_1 \otimes \mu_1) \otimes \cdots \otimes (a_m \otimes \mu_m))$$

$$= \mathrm{Trace}(1\mu_0 \cdots \mu_m) 1 \otimes a_0 \otimes a_1 \otimes \cdots \otimes a_m =$$

$$= \iota_0\,(\mathrm{tr}\,((a_0 \otimes \mu_0) \otimes (a_1 \otimes \mu_1) \otimes \cdots \otimes (a_m \otimes \mu_m)))\,.$$

Thus the map tr induces a map tr : $\underline{C}_n(M_s(\mathcal{A})) \to \underline{C}_n(\mathcal{A})$ and one checks that this map is compatible with the operations (b, B). For an idempotent $e \in M_s(\mathcal{A})$ the components of its Chern character in $\underline{C}_*(\mathcal{A})$ are given by $\mathrm{tr}(\mathrm{Ch}_{2k}(e))$. Thus they are

$$\mathrm{Ch}_{2k}(e) = \lambda_k \times \mathrm{tr}\left((e - \frac{1}{2}) \otimes e \otimes e \otimes \cdots \otimes e\right), \quad k > 0.$$

Moreover this formula still holds for $k = 0$ when replacing e by $[e - 1/2]$:

$$\mathrm{Ch}_{2k}([e - 1/2]) = \lambda_k \times \mathrm{tr}\left((e - \frac{1}{2}) \otimes e \otimes e \otimes \cdots \otimes e\right), \quad \forall k \geq 0.$$

Using $Y = 2e - 1$ and the construction of the $\underline{C}_m(\mathcal{A})$ one thus gets, for $m = 2k$ even,

$$\mathrm{Ch}_m([e - 1/2]) = 2^{-(m+1)}\lambda_k \mathrm{tr}\,(Y \otimes Y \otimes Y \otimes \cdots \otimes Y) \in \underline{C}_m(\mathcal{A}). \tag{7}$$

The fundamental fact which is behind the quantization of the volume is, for Y fulfilling (4), the vanishing of all the lower components

$$\mathrm{Ch}_m([e - 1/2]) = 0, \ \forall m < n. \tag{8}$$

This follows because for a product P of an odd number $2k + 1 < n + 1$ of Γ_A, the trace of P vanishes since one can still find a $\Gamma = \Gamma_X$ which anti-commutes with the Γ_A's involved in P and thus

$$P = \Gamma^2 P = -\Gamma P \Gamma \Rightarrow \mathrm{Trace}(P) = 0.$$

It follows from (8) that the component $\mathrm{Ch}_n([e - 1/2])$ is a Hochschild cycle and that for any cyclic n-cocycle ϕ_n the pairing $< \phi_n, e >$ is the same as $< I(\phi_n), \mathrm{Ch}_n(e) >$ where $I(\phi_n)$ is the Hochschild class of ϕ_n. This applies to the cyclic n-cocycle ϕ_n which is the Chern character ϕ_n in K-homology of the spectral triple $(\mathcal{A}, \mathcal{H}, D)$ with grading γ where $\mathcal{A}$ is the algebra generated by the components Y^A of Y. One then uses the following formula for the Hochschild class τ of the Chern character ϕ_n in K-homology of the spectral triple $(\mathcal{A}, \mathcal{H}, D)$, up to normalization[11]:

$$\tau(a_0, a_1, \ldots, a_n) = \fint \gamma a_0 [D, a_1] \cdots [D, a_n] D^{-n}, \ \ \forall a_j \in \mathcal{A}.$$

This follows from the local index formula of Connes-Moscovici [33]. But in fact, one does not need the technical hypothesis since, when the lower components of the operator theoretic Chern character all vanish, one can use the non-local index formula in cyclic cohomology and the determination in the book [14] of the Hochschild class of the index cyclic cocycle. We refer to [34] for an optimal formulation of the result.

[11] We refer to [14] for the meaning of the integral symbol.

Moreover since D commutes with the algebra $M_s(\mathbb{C})$ one has

$$\tau \circ \mathrm{tr}(y_0, y_1, \ldots, y_n) = s \,⨍\, \gamma \, \langle y_0[D, y_1] \cdots [D, y_n]\rangle \, D^{-n}, \quad \forall y_j \in M_s(\mathcal{A})$$

so that with $Y = 2e - 1$, one gets

$$< \tau, \mathrm{Ch}_n([e - 1/2]) >= s \,⨍\, \gamma \left\langle Y \, [D, Y]^n \right\rangle D^{-n}$$

and, up to normalization, Eq. (5) thus implies

$$⨍\, D^{-n} = ⨍\, \gamma^2 D^{-n} = \frac{1}{n!} ⨍\, \gamma \left\langle Y \, [D, Y]^n \right\rangle D^{-n} = \frac{1}{n!s} < \tau, \mathrm{Ch}_n([e - 1/2]) > \in \frac{1}{n!s} \mathbb{Z} :$$

which is the quantization of the volume.

4.1.3 Disjoint Quanta

We recall that given a smooth compact oriented spin manifold M, the associated spectral triple $(\mathcal{A}, \mathcal{H}, D)$ is given by the action in the Hilbert space $\mathcal{H} = L^2(M, S)$ of L^2-spinors of the algebra $\mathcal{A} = C^\infty(M)$ of smooth functions on M, and the Dirac operator D which in local coordinates is of the form

$$D = \gamma^\mu \left(\frac{\partial}{\partial x^\mu} + \omega_\mu \right)$$

where $\gamma^\mu = e^\mu_a \gamma^a$ and ω_μ is the spin-connection.

Theorem 4.1 *Let M be a spin Riemannian manifold of even dimension n and $(\mathcal{A}, \mathcal{H}, D)$ the associated spectral triple. Then a solution of the one-sided equation* (5) *exists if and only if M decomposes as the disjoint sum of spheres of unit volume. On each of these irreducible components the unit volume condition is the only constraint on the Riemannian metric which is otherwise arbitrary for each component (Fig. 10).*

Equation (4) shows that a solution Y gives a map $Y : M \to S^n$ from the manifold M to the n-sphere. Let us compute the left hand side of (5). The normalized trace of the product of $n + 1$ Gamma matrices is the totally antisymmetric tensor

$$\langle \Gamma_A \Gamma_B \cdots \Gamma_L \rangle = i^{n/2} \epsilon_{AB\ldots L}, \quad A, B, \ldots, L \in \{1, \ldots, n + 1\}.$$

One has

$$[D, Y] = \gamma^\mu \frac{\partial Y^A}{\partial x^\mu} \Gamma_A = \nabla Y^A \Gamma_A$$

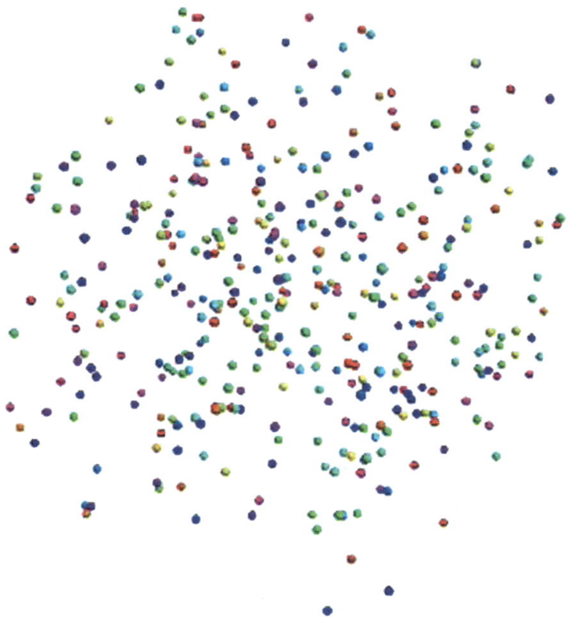

Fig. 10 Collection of tiny spheres

where we let ∇f be the Clifford multiplication by the gradient of f. Thus one gets at any $x \in M$ the equality

$$\langle Y\,[D,Y]\cdots[D,Y]\rangle = i^{n/2}\epsilon_{AB\ldots L}Y^A\nabla Y^B\cdots\nabla Y^L.$$

Given n operators $T_j \in \mathcal{C}$ in an algebra $\mathcal{C}$ the multiple commutator

$$[T_1,\ldots,T_n] := \sum \epsilon(\sigma)T_{\sigma(1)}\cdots T_{\sigma(n)}$$

(where σ runs through all permutations of $\{1,\ldots,n\}$) is a multilinear totally antisymmetric function of the $T_j \in \mathcal{C}$. In particular, if the $T_i = a_i^j S_j$ are linear combinations of n elements $S_j \in \mathcal{C}$ one gets

$$[T_1,\ldots,T_n] = \mathrm{Det}(a_i^j)[S_1,\ldots,S_n]. \qquad (9)$$

For fixed A, and $x \in M$ the sum over the other indices

$$\epsilon_{AB\ldots L}Y^A\nabla Y^B\cdots\nabla Y^L = (-1)^A Y^A[\nabla Y^1,\nabla Y^2,\ldots,\nabla Y^{n+1}]$$

where all other indices are $\neq A$. At $x \in M$ one has $\nabla Y^j = \gamma^\mu\partial_\mu Y^j$ and by (9) the multi-commutator (with ∇Y^A missing) gives

$$[\nabla Y^1, \nabla Y^2, \ldots, \nabla Y^{n+1}] = \epsilon^{\mu\nu\ldots\lambda}\partial_\mu Y^1 \cdots \partial_\lambda Y^{n+1}[\gamma^1, \ldots, \gamma^n].$$

Since $\gamma^\mu = e^\mu_a \gamma_a$ and $i^{n/2}[\gamma_1, \ldots, \gamma_n] = n!\gamma$ one thus gets

$$\langle Y\,[D, Y] \cdots [D, Y] \rangle = n!\gamma \mathrm{Det}(e^\alpha_a)\,\omega$$

where

$$\omega = \epsilon_{AB\ldots L} Y^A \partial_1 Y^B \cdots \partial_n Y^L$$

so that $\omega dx_1 \wedge \cdots \wedge dx_n$ is the pullback $Y^\#(\rho)$ by the map $Y : M \to S^n$ of the rotation invariant volume form ρ on the unit sphere S^n given by

$$\rho = \frac{1}{n!}\epsilon_{AB\ldots L} Y^A dY^B \wedge \cdots \wedge dY^L.$$

Thus, using the inverse vierbein, the one-sided equation (5) is equivalent to

$$\det\left(e^a_\mu\right) dx_1 \wedge \cdots \wedge dx_n = Y^\#(\rho).$$

This equation implies that the Jacobian of the map $Y : M \to S^n$ cannot vanish anywhere, and hence that the map Y is a covering.

It would seem at this point that only disconnected geometries fit in this framework. But this would be to ignore an essential piece of structure of the NCG framework, which allows one to refine (5). Namely: the real structure J, an antilinear isometry in the Hilbert space $\mathcal{H}$ which is the algebraic counterpart of charge conjugation.

4.2 *The KO-Theory Higher Heisenberg Equation*

We now take into account the real structure J and this gives the refinement from K to KO. One replaces (4) by (with summation on indices A and $\kappa \in \{\pm 1\}$)

$$Y = Y^A_\kappa \Gamma_{A,\kappa}, \; Y^4 = 1, \; Y^* Y = 1, \tag{10}$$

The Hilbert space splits according to the spectrum of Y^2 as a direct sum $\mathcal{H} = \mathcal{H}(+) \oplus \mathcal{H}(-)$

$$Y = Y_+ \oplus Y_-, \;\; Y_\pm^2 = \pm 1, \;\; Y_\pm^* = \pm Y_\pm, \;\; Y_\pm = Y^A_\pm \Gamma_{A,\pm}.$$

For $\kappa \in \{\pm 1\}$ the $\Gamma_{A,\kappa}$ fulfill in $\mathcal{H}(\kappa)$ the Clifford relations

$$\left\{\Gamma_{A,\kappa}, \Gamma_{B,\kappa}\right\} = 2\kappa\,\delta_{AB}, \;\; (\Gamma_{A,\kappa})^* = \kappa\Gamma_{A,\kappa}.$$

The compatibility with J is given by the relations:

$$JY^2 = -Y^2 J, \; [Y, JYJ^{-1}] = 0$$

Let $C_{\pm} = C(\Gamma_{A,\pm})$ be the algebra generated over $\mathbb{R}$ by the $\Gamma_{A,\kappa}$. In $\mathcal{H}(+)$, C_+ commutes with $C'_- = JC_-J^{-1}$ to take into account the relation $[Y, JYJ^{-1}] = 0$. We thus view Y as a smooth section:

$$Y = Y_+ \oplus Y_- \in C^\infty(M, C_+ \oplus C_-)$$

This leads us to refine the quantization condition by taking J into account as the two-sided equation

$$\frac{1}{n!}\langle Z\,[D, Z]^n\rangle = \gamma, \quad Z = 2EJEJ^{-1} - 1, \ [D, Y^2] = 0, \tag{11}$$

where E is the spectral projection $E_+ \oplus E_-$ of the unitary $Y = Y_+ \oplus Y_- \in C^\infty(M, C_+ \oplus C_-)$.

$$E = E_+ \oplus E_- = \frac{1}{2}(1 + Y_+) \oplus \frac{1}{2}(1 + iY_-)$$

It turns out that in dimension $n = 4$, the irreducible pieces give :

$$C_+ = M_2(\mathbb{H}), \quad C_- = M_4(\mathbb{C})$$

which give the algebraic constituents of the Standard Model exactly in the form of our previous work. This can be seen using the following table:

$p - q$ mod 8	$\mathrm{Cliff}_{p,q}(\mathbb{R})$ $n = p + q$	$p - q$ mod 8	$\mathrm{Cliff}_{p,q}(\mathbb{R})$ $n = p + q$
0	$M(2^{n/2}, \mathbb{R})$	1	$M_u(\mathbb{R}) \oplus M_u(\mathbb{R})$ $u = 2^{(n-1)/2}$
2	$M(2^{n/2}, \mathbb{R})$	3	$M(2^{(n-1)/2}, \mathbb{C})$
4	$M(2^{(n-2)/2}, \mathbb{H})$	5	$M_v(\mathbb{H}) \oplus M_v(\mathbb{H})$ $v = 2^{(n-3)/2}$
6	$M(2^{(n-2)/2}, \mathbb{H})$	7	$M(2^{(n-1)/2}, \mathbb{C})$

Indeed in dimension $n = 4$ one needs $n + 1 = 5$ gamma matrices, and the irreducible pieces of the Clifford algebras $\mathrm{Cliff}_{p,q}(\mathbb{R})$ are $M_2(\mathbb{H})$ for $(p, q) = (5, 0)$ and $M_4(\mathbb{C})$ for $(p, q) = (0, 5)$. Moreover in the 4-dimensional case one has, in the Hilbert space $\mathcal{H}(+)$, by the detailed calculation of [11],

$$\langle Z\,[D, Z]^4\rangle_+ = \frac{1}{2}\left\langle Y_+\left[D, Y_+\right]^4\right\rangle + \frac{1}{2}\left\langle Y'_-\left[D, Y'_-\right]^4\right\rangle,$$

where $Y'_- = iJY_-J^{-1}$. One now gets two maps $Y_\pm : M \to S^n$ while (11) becomes, up to normalization

$$\det\left(e^a_\mu\right) dx_1 \wedge \cdots \wedge dx_n = Y^\#_+(\rho) + Y^\#_-(\rho), \tag{12}$$

where $Y^\#_\pm(\rho)$ is the pull back of the volume form ρ of the sphere.

For an n-dimensional smooth compact manifold we let $D(M)$ be the set of pairs of smooth maps $\phi_\pm : M \to S^n$ such that the differential form

$$\phi_+^\#(\rho) + \phi_-^\#(\rho) = \omega$$

does not vanish anywhere on M (ρ is the standard volume form on the sphere S^n).

Definition 4.2 Let M be an n-dimensional oriented smooth compact manifold

$$q_M := \{\deg(\phi_+) + \deg(\phi_-) \mid (\phi_+, \phi_-) \in D(M)\}$$

where $\deg(\phi)$ is the topological degree of ϕ.

Theorem 4.3 (*i*) *Let M be a compact oriented spin Riemannian manifold of dimension* 4. *Then a solution of* (12) *exists if and only if the volume of M is quantized to belong to the invariant $q_M \subset \mathbb{Z}$.*

(*ii*) *Let M be a smooth connected oriented compact spin* 4*-manifold. Then q_M contains all integers $m \geq 5$.*

The invariant q_M makes sense in any dimension. For $n = 2, 3$, and any M, it contains all sufficiently large integers. The case $n = 4$ is more difficult; but we showed in [11] that for any Spin manifold it contains all integers $m > 4$. This uses fine results on the existence of ramified covers of the sphere and on immersion theory going back to Smale, Milnor and Poenaru. By a result of Iori and Piergallini [35], any orientable closed (connected) smooth 4-manifold is a simple 5-fold cover of S^4 branched over a smooth surface (meaning that the covering map can be assumed to be smooth) (Fig. 11). The key lemma[12] which allows one to then rely on immersion theory and apply the fundamental result of Poenaru [36] (on the existence of an immersion in $\mathbb{R}^n$ of any open parallelizable n-manifold) is the following:

Lemma 4.4 *Let $\phi : M \to S^4$ be a smooth map such that $\phi^\#(\alpha)(x) \geq 0\, \forall x \in M$ and let $R = \{x \in M \mid \phi^\#(\alpha)(x) = 0\}$. Then there exists a map ϕ' such that $\phi^\#(\alpha) + \phi'^\#(\alpha)$ does not vanish anywhere if and only if there exists an immersion $f : V \to \mathbb{R}^4$ of a neighborhood V of R. Moreover if this condition is fulfilled one can choose ϕ' to be of degree* 0.

The spin condition on the 4-manifold allows one to prove that the neighborhood V is parallelizable. By a result of A. Haefliger, the spin condition is equivalent to the vanishing of the second Stiefel-Whitney class w_2 of the tangent bundle. In the converse direction, Jean-Claude Sikorav and Bruno Sevennec found the following obstruction which implies for instance that $D(\mathbb{C}P^2) = \emptyset$. Let M be an oriented compact smooth 4-dimensional manifold, then, with w_2 the second Stiefel-Whitney class of the tangent bundle,

$$D(M) \neq \emptyset \implies w_2^2 = 0$$

---

[12] I am indebted to Simon Donaldson for his generous help in finding this key result.

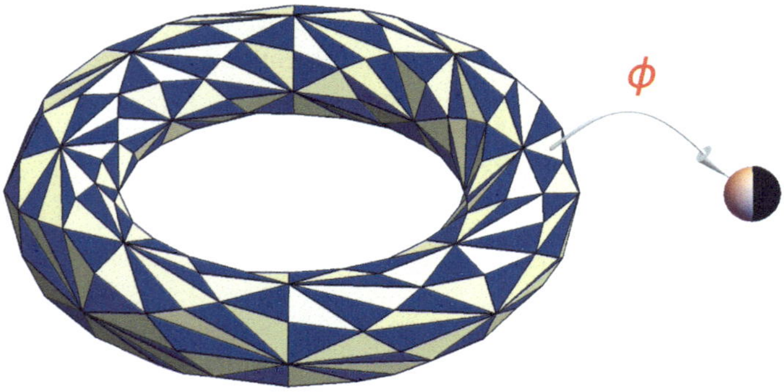

Fig. 11 Ramified cover of the sphere

Indeed if $D(M) \neq \emptyset$ one has a cover of M by two open sets on which the tangent bundle is stably trivialized. Thus the above product of two Stiefel-Whitney classes vanishes.

4.3 Emerging Geometry

Theorem 4.3 shows how 4-dimensional spin geometries arise from irreducible representations of simple algebraic relations. There is no restriction to fix the Hilbert space $\mathcal{H}$ as well as the actions of the Clifford algebras $C_\pm$ and of J and γ. The remaining indeterminate operators are D and Y. They fulfill Eq. (11). The geometry appears from the joint spectrum of the $Y_\pm^A$ and is a 4-dimensional immersed submanifold in the 8-dimensional product $S^4 \times S^4$. Thus this suggests taking the operators Y, D as being the correct variables for a first shot at a theory of quantum gravity. In the sequel the algebraic relations between $Y_\pm$, D, J, $C_\pm$, γ are assumed to hold. As we have seen above a compact spin 4-dimensional manifold M appears as immersed by a map $(Y_+, Y_-) : M \to S^4 \times S^4$. An interesting question which comes in this respect is whether, given a compact spin 4-dimensional manifold M, one can find a map $(Y_+, Y_-) : M \to S^4 \times S^4$ which embeds M as a submanifold of $S^4 \times S^4$. One has the strong Whitney embedding theorem: $M^4 \subset \mathbb{R}^4 \times \mathbb{R}^4 \subset S^4 \times S^4$ so there is no a-priori obstruction to expect an embedding rather than an immersion. It is worthwhile to mention that a generic immersion would in fact suffice to reconstruct the manifold. Next, in general, if one starts from a representation of the algebraic relations, there are two natural questions:

(A): Is it true that the joint spectrum of the Y_+^A and Y_-^B is of dimension 4 while one has 8 variables?

(B): Is it true that the volume $⨍ D^{-4}$ remains quantized?

4.3.1 Dimension

The reason why (A) holds in the case of classical manifolds is that in that case the joint spectrum of the Y^A and Y'^B is the subset of $S^4 \times S^4$ which is the image of the manifold M by the map $x \in M \mapsto (Y(x), Y'(x))$ and thus its dimension is at most 4.

The reason why (A) holds in general is because of the assumed boundedness of the commutators $[D, Y]$ and $[D, Y']$ together with the commutativity $[Y, Y'] = 0$ (order zero condition) and the fact that the spectrum of D grows like in dimension 4.

4.3.2 Quantization of Volume

The reason why (B) holds in the general case is that the results of Sect. 4.1.2 apply separately to Y_+ and Y'_-. This gives, up to a normalization constant $c_4 \neq 0$, the integrality

$$\int\!\!\!\!\!- \gamma \left\langle Y_+ [D, Y_+]^4 \right\rangle D^{-4} = c_4 < [D], [e - 1/2] > \in c_4 \mathbb{Z}$$

$$\int\!\!\!\!\!- \gamma \left\langle Y'_- [D, Y'_-]^4 \right\rangle D^{-4} = c_4 < [D], [e' - 1/2] > \in c_4 \mathbb{Z}.$$

Thus the equality

$$\left\langle Z [D, Z]^4 \right\rangle_+ = \frac{1}{2} \left\langle Y_+ [D, Y_+]^4 \right\rangle + \frac{1}{2} \left\langle Y'_- [D, Y'_-]^4 \right\rangle.$$

together with Eq. (11) gives in $\mathcal{H}(+)$,

$$\frac{1}{2} \left\langle Y_+ [D, Y_+]^4 \right\rangle + \frac{1}{2} \left\langle Y'_- [D, Y'_-]^4 \right\rangle = 4!\gamma$$

and one gets from $\gamma^2 = 1$:

Theorem 4.5 *In any operator representation of the two sided equation* (11) *in which the spectrum of D grows as in dimension* 4 *the volume (the leading term of the Weyl asymptotic formula) is quantized, (up to a normalization constant $c > 0$)*

$$\int\!\!\!\!\!- D^{-4} \in c\,\mathbb{N}.$$

This quantization of the volume implies that the bothersome cosmological leading term of the spectral action is now quantized; and thus it no longer appears in the differential variation of the spectral action. Thus and provided one understands better how to reinstate all the fine details of the finite geometry (the one encoded by the Clifford algebras) the variation of the spectral action will reproduce the Einstein equations coupled with matter.

4.4 Final Remarks

Finally, we briefly discuss a few important points which would require more work of clarification if one wants to get a bit closer to the goal of unification at the "pre-quantum" level, best described in Einstein's words (see H. Nicolai, Cern Courier, January 2017) as follows:

"Roughly but truthfully, one might say: we not only want to understand how nature works, but we are also after the perhaps utopian and presumptuous goal of understanding why nature is the way it is and not otherwise."

1. All our discussion of geometry takes place in the Euclidean signature. Physics takes place in the Minkowski signature. The Wick rotation plays a key role in giving a mathematical meaning to the Feynman integral in QFT for flat space-time but becomes problematic for curved space-time. But following Hawking and Gibbons one can investigate the Euclidean Feynman integral over compact 4-manifolds implementing a cobordism between two fixed 3-geometries. Two interesting points occur if one uses the above spectral approach. First the new boundary terms, involving the extrinsic curvature of the boundary, which Hawking and Gibbons had to add to the Einstein action, pop up automatically from the spectral action: as shown in [37]. Second, in the functional integral, the kinetic term of the Weyl term (i.e. the "dilaton") has the wrong sign. In our formalism the higher Heisenberg equation fixes the volume form and automatically freezes the dilaton.
2. The number of generations is not predicted by the above theory. The need to have this multiplicity in the representation of the finite algebra $\mathcal{A}_F$ might be related to the discussion of Sect. 3.4 in the following way. For non-simply connected spaces the Poincaré duality *KO*-fundamental class should take into account the fundamental group. We skipped over this point in Sect. 3.4; and in the non-simply connected case one needs to twist the fundamental *KO*-homology class by flat bundles. It is conceivable that the generations appear from such a twist by a 3-dimensional representation. This could be a good motivation to extend the classical treatment of flat bundles (i.e. of representations of the fundamental group) to the general case of noncommutative spaces.

Acknowledgements I am grateful to Joseph Kouneiher and Jeremy Butterfield for their help in the elaboration of this paper.

Appendix

Here is a possible translation of the second quote of Grothendieck:

> It must be already fifteen or twenty years ago that, leafing through the modest volume constituting the complete works of Riemann, I was struck by a remark of his "in passing". He pointed out that it could well be that the ultimate structure of space is discrete, while the

continuous representations that we make of it constitute perhaps a simplification (perhaps excessive, in the long run ...) of a more complex reality; That for the human mind, "the continuous" was easier to grasp than "the discontinuous", and that it serves us, therefore, as an "approximation" to apprehend the discontinuous.

This is a remark of a surprising penetration in the mouth of a mathematician, at a time when the Euclidean model of physical space had never yet been questioned; in the strictly logical sense, it is rather the discontinuous which traditionally served as a mode of technical approach to the continuous.

Mathematical developments of recent decades have, moreover, shown a much more intimate symbiosis between continuous and discontinuous structures than was imagined, even in the first half of this century.

In any case finding a "satisfactory" model (or, if necessary, a set of such models, "satisfactorily connecting" to each other) of "continuous", "discrete" or of "mixed" nature - such work will surely involve a great conceptual imagination, and a consummate flair for apprehending and unveiling new type mathematical structures.

This kind of imagination or "flair" seems rare to me, not only among physicists (where Einstein and Schrödinger seem to have been among the rare exceptions), but even among mathematicians (and here I speak with full knowledge).

To summarize I predict that the expected renewal (if it must yet come) will come from a mathematician in soul well informed about the great problems of physics, rather than from a physicist. But above all, it will take a man with "philosophical openness" to grasp the crux of the problem. This is by no means a technical one but rather a fundamental problem of natural philosophy.

References

1. L. Corry *David Hilbert and the axiomatization of physics*. Springer-Science + Business-Media B.V. (2004)
2. J. Kouneiher, J. Stachel, *Einstein and Hilbert*. In this volume
3. M. Atiyah, Global theory of elliptic operators, in *Proceedings of the International Conference on Functional Analysis and Related Topics (Tokyo, 1969)* (University of Tokyo Press, Tokyo, 1970), p. 2130
4. A. Connes, *C*-algèbres et géométrie différentielle*. C.R. Acad. Sci. Paris Sér. A-B **290**, A599–A604 (1980)
5. A. Chamseddine, A. Connes, The spectral action principle. Commun. Math. Phys. **186**, 731–750 (1997)
6. A. Chamseddine, A. Connes, *Why the Standard Model?* hep-th 0706.3688
7. A. Chamseddine, A. Connes, M. Marcolli, *Gravity and the Standard Model with Neutrino Mixing*, hep-th/0610241
8. A. Chamseddine, A. Connes, Resilience of the spectral standard model. JHEP **1209**, 104 (2012)
9. A. Chamseddine, A. Connes, Noncommutative geometry as a framework for unification of all fundamental interactions including gravity. Fortschr. Phys. **58**, 553 (2010)
10. A. Chamseddine, A. Connes, V. Mukhanov, Quanta of geometry: noncommutative aspects. Phys. Rev. Lett. **114**, 091302 (2015)
11. A. Chamseddine, A. Connes, V. Mukhanov, Geometry and the quantum: basics. JHEP **12**, 098 (2014)
12. A. Chamseddine, A. Connes, W.D. van Suijlekom, Beyond the spectral standard model: emergence of Pati-Salam unification. JHEP **11**, 132 (2013)
13. A. Chamseddine, A. Connes, W.D. van Suijlekom, Grand unification in the spectral Pati-Salam models. JHEP **2511**, 011 (2015)

14. A. Connes, *Noncommutative geometry* (Academic Press, 1994)
15. R.P. Woodard, How far are we from the quantum theory of gravity? Rep. Progr. Phys. **72**(12), 126002 (2009), 42 pp. 81V17 (83C45)
16. A. Connes, *Leçon inaugurale au Collège de France*, 11 Janvier 1985. http://www.alainconnes.org/docs/lecollege.pdf
17. B. Dundas, T. Goodwillie, R. McCarthy, *The local structure of algebraic K-theory*. Algebra and Applications, vol. 18. (Springer, London, 2013)
18. A. Grothendieck, *Récoltes et Semailles*
19. C. MacLarty, *Grothendieck on Foundations for the Rebirth of Geometry*. In this volume
20. N. Gisin, A. Martin, B. Sanguinetti, H. Zbinden, Quantum random number generation on a mobile phone. Phys. Rev. X **4**, 031056 (2014)
21. A. Pais, *Inward Bound: Of Matter and Forces in the Physical World* (Oxford University Press, Oxford, 1986)
22. M. Kac, Can one hear the shape of a drum? Am. Math. Mon. **73**((4, part 2)), 1–23 (1966)
23. J. Milnor, Eigenvalues of the Laplace operator on certain manifolds. Proc. Natl. Acad. Sci. U.S.A. **51**, 542 (1964)
24. http://acces.ens-lyon.fr/clea/lunap/Triangulation/TriangCompl1.html
25. T. Hankins, *Sir William Rowan Hamilton* (Johns Hopkins University Press, Baltimore, MD, 1980)
26. M. Spivak, Spaces satisfying Poincaré duality. Topology **6**, 77–101 (1967)
27. J. Milnor, J. Stasheff, *Characteristic Classes*. Annals of Mathematics Studies, vol. 76 (Princeton University Press, Princeton, N.J.; University of Tokyo Press, Tokyo, 1974)
28. P.H. Siegel, Witt spaces: a geometric cycle theory for KO-homology at odd primes. Am. J. Math. **105**(5), 1067–1105 (1983)
29. I. Singer, Future extensions of index theory and elliptic operators. Prosp. Math. Ann. Math. Stud. **70**, 171–185 (1971). (Princeton University Press)
30. A. Connes, Noncommutative geometry and reality. J. Math. Phys. **36**, 6194 (1998)
31. A. Chamseddine, A. Connes, W.D. van Suijlekom, Inner fluctuations in noncommutative geometry without the first order condition. J. Geom. Phys. **73**, 222 (2013)
32. A. Chamseddine, A. Connes, *A dress for SM the beggar*, hep-th 0706.3690
33. A. Connes, H. Moscovici, The local index formula in noncommutative geometry. GAFA **5**, 174–243 (1995)
34. A. Carey, A. Rennie, F. Sukochev, D. Zanin, Universal measurability and the Hochschild class of the Chern character. J. Spectral Theory **6**(1), 1–41 (2016)
35. M. Iori, R. Piergallini, 4-manifolds as covers of the 4-sphere branched over non-singular surfaces. Geom. Topology **6**, 393–401 (2002)
36. V. Poenaru, Sur la théorie des immersions. Topology **1**, 81–100 (1962)
37. A. Chamseddine, A. Connes, Universal formula for noncommutative geometry actions: unification of gravity and the Standard Model. Phys. Rev. Lett. **77**, 486804871 (1996)

What Every Physicist Should Know About String Theory

Edward Witten

The aim of this article[1] is to describe the minimum that any physicist might want to know about string theory, focusing on a few basic questions. How does string theory generalize standard quantum field theory? Why does string theory force us to unify General Relativity with the other forces of nature, while standard quantum field theory makes it so difficult to incorporate General Relativity? Why are there no ultraviolet divergences in string theory? And what happens to Einstein's conception of spacetime?

Anyone who has studied physics is aware that although physics—like history—does not precisely repeat itself, it does rhyme, with similar structures appearing in different areas. For example, Einstein's gravitational waves are analogous to electromagnetic waves, or to the water waves at the surface of a pond. We will begin with one of nature's rhymes—an analogy between the problem of quantum gravity and the theory of a single particle.

Even though we do not really understand it, quantum gravity is supposed to be some sort of theory in which, at least from a macroscopic point of view, we average, in a quantum mechanical sense, over all possible spacetime geometries. (We do not know to what extent this description is valid microscopically.) The averaging is performed, in the simplest case, with a weight factor $\exp(iI/\hbar)$, where I is the Einstein-Hilbert action:

$$I = \frac{1}{16\pi G}\int \mathrm{d}^4x\sqrt{g}(R - 2\Lambda).$$

[1]To be published in *Foundations of Mathematics and Physics, one century after Hilbert*, ed. Joseph Kouneiher, Mathematical physics Studies, Springer 2017. Adapted with permission from an article that appeared in the November, 2015 issue of Physics Today.

E. Witten (✉)
Institute for Advanced Study, School of Natural Sciences, Princeton, USA
e-mail: witten@ias.edu

J. Kouneiher (ed.), *Foundations of Mathematics and Physics One Century After Hilbert*,
https://doi.org/10.1007/978-3-319-64813-2_8

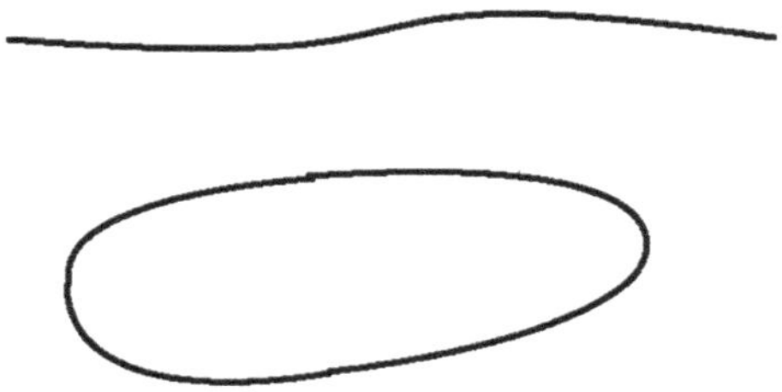

Fig. 1 Manifolds of dimension 1

Here G is Newton's constant, g is the determinant of the metric tensor, R is the curvature scalar, Λ is a cosmological constant, and x^i are spacetime coordinates. We *could* add matter fields, but we do not seem to need them.

Let us try to make a theory like this with 1 spacetime dimension instead of 4. As indicated in Fig. 1, there are not many choices for a one-manifold. Moreover, the curvature scalar is identically zero in 1 dimension and there is no Einstein-Hilbert action. However, Einstein's fundamental insight was not the specific Einstein-Hilbert action but the broader idea that the spacetime geometry can vary dynamically and the laws of nature are generally covariant (invariant under arbitrary diffeomorphisms of spacetime). In this more general sense, we can make a nontrivial quantum gravity theory in dimension 1 provided we include matter fields.

The simplest matter fields are scalar fields $X_i, i = 1, \ldots, D$. The standard General Relativistic action for scalar fields is

$$I = \int \mathrm{d}t \sqrt{g} \left(\frac{1}{2} \sum_{i=1}^{D} g^{tt} \left(\frac{dX_i}{\mathrm{d}t} \right)^2 - \frac{1}{2} m^2 \right),$$

where $g = (g_{tt})$ is a 1×1 metric tensor, and Λ has been replaced with $m^2/2$.

Let us introduce the "canonical momentum" $P_i = \mathrm{d}X_i/\mathrm{d}t$. The "Einstein field equation"—which is the equation of motion obtained by varying the action I with respect to g—is just

$$g^{tt} \sum_{i=1}^{D} P_i^2 + m^2 = 0.$$

We pick the gauge $g_{tt} = 1$, so the equation is $P^2 + m^2 = 0$, with $P^2 = \sum_i P_i^2$. Quantum mechanically (in units with $\hbar = 1$), $P_i = -i \frac{\partial}{\partial X_i}$ and the meaning of the equation $P^2 + m^2 = 0$ is that the wavefunction $\Psi(X)$ must be annihilated by the differential operator that corresponds to $P^2 + m^2$:

$$\left(-\sum_{i=1}^{D} \frac{\partial^2}{\partial X_i^2} + m^2 \right) \Psi(X) = 0.$$

This is a familiar equation—the relativistic Klein-Gordon equation in D dimensions—but in Euclidean signature. To give this fact a sensible physical interpretation, we should reverse the kinetic energy of one of the scalar fields X_i so that the action becomes

$$I = \int \mathrm{d}t \sqrt{g} \left(\frac{1}{2} g^{tt} \left(- \left(\frac{\mathrm{d}X_0}{\mathrm{d}t} \right)^2 + \sum_{i=1}^{D-1} \left(\frac{\mathrm{d}X_i}{\mathrm{d}t} \right)^2 \right) - m^2 \right).$$

Now the wavefunction obeys a Klein-Gordon equation in *Lorentz* signature:

$$\left(\frac{\partial^2}{\partial X_0^2} - \sum_{i=1}^{D-1} \frac{\partial^2}{\partial X_i^2} + m^2 \right) \Psi(X) = 0.$$

So we have found an exactly soluble theory of quantum gravity in one dimension that describes a spin 0 particle of mass m propagating in D-dimensional Minkowski spacetime. Actually, we can replace Minkowski spacetime by any D-dimensional spacetime M with a Lorentz (or Euclidean) signature metric G_{IJ}, the action being then

$$I = \int \mathrm{d}t \sqrt{g} \left(\frac{1}{2} \sum_{i=1}^{D} g^{tt} G_{IJ} \frac{dX^I}{\mathrm{d}t} \frac{\mathrm{d}X^J}{\mathrm{d}t} - m^2 \right).$$

The equation obeyed by the wavefunction is now a Klein-Gordon equation on M:

$$\left(-G^{IJ} \frac{D}{DX^I} \frac{D}{DX^J} + m^2 \right) \Psi(X) = 0.$$

This is the massive Klein-Gordon equation in curved spacetime.

Just to make things more familiar, let us go back to the case of flat spacetime (we work in Euclidean signature to avoid having to keep track of some factors of i). Let us calculate the amplitude for a particle to start at one point x in spacetime and end at another point y (Fig. 2). We do this by evaluating a Feynman path integral in

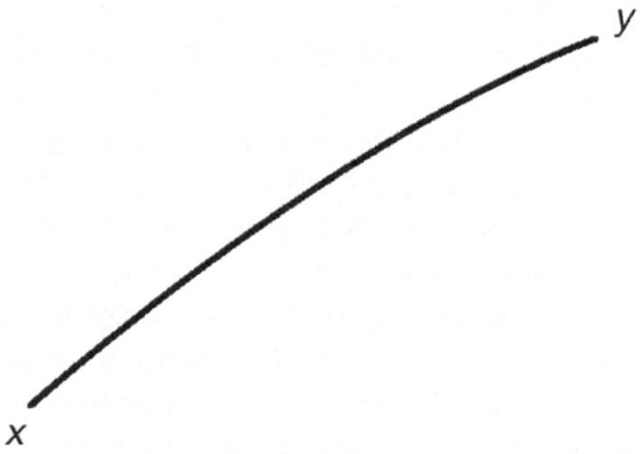

Fig. 2 Propagation of a particle between specified spacetime points x and y

our quantum gravity model. The path integral is performed over all metrics $g(t)$ and scalar fields $X_i(t)$ on the one-manifold sketched in the figure, with the condition that $X_i(t)$ is equal to x at one end and to y at the other.

Part of the process of evaluating the path integral in our quantum gravity model is to integrate over the metric on the one-manifold, modulo diffeomorphisms. But up to diffeomorphism, this one-manifold has only one invariant, its total length τ, which we will interpret as the elapsed proper time. For a given τ, we can take the one-metric to be just $g_{tt} = 1$ where $0 \leq t \leq \tau$. Now on this one-manifold, we have to integrate over all paths $X(t)$ that start at x at $t = 0$ and end at y at $t = \tau$. This is the basic Feynman integral of quantum mechanics with the Hamiltonian being $H = \frac{1}{2}\left(P^2 + m^2\right)$. According to Feynman, the result is the matrix element of $\exp(-\tau H)$:

$$G(x, y; \tau) = \int \frac{\mathrm{d}^D p}{(2\pi)^D} e^{ip\cdot(y-x)} \exp\left(-\frac{\tau}{2}(p^2 + m^2)\right).$$

But we have to remember to do the "gravitational" part of the path integral, which in the present context means to integrate over τ.

The integral over τ gives our final answer:

$$G(x, y) = \int_0^\infty \mathrm{d}\tau\, G(x, y; \tau) = \int \frac{\mathrm{d}^D p}{(2\pi)^D} e^{ip\cdot(y-x)} \frac{2}{p^2 + m^2}.$$

This is the output of the complete path integral—an integral over metrics $g(t)$ and paths $X(t)$ with the given endpoints, modulo diffeomorphisms—in our quantum gravity model.

But the function $G(x, y)$ is the standard Feynman propagator in Euclidean signature, apart from a convention-dependent normalization factor. Moreover, an analogous derivation in Lorentz signature (for both the spacetime M and the particle worldline) gives the correct Lorentz signature Feynman propagator, with the $i\varepsilon$.

So we have interpreted a free particle in D-dimensional spacetime in terms of one-dimensional quantum gravity. How can we include interactions? There is actually a perfectly natural way to do this. There are not a lot of smooth one-manifolds, but there is a large supply of singular one-manifolds in the form of graphs (Fig. 3). Our "quantum gravity" action makes sense on such a graph. We simply take the same action that we used before, summed over all of the line segments that make up the graph.

Now to do the quantum gravity path integral, we have to integrate over all metrics on the graph, up to diffeomorphism. The only invariants are the total lengths or "proper times" of each of the segments. Some of the lines in Fig. 3 have been labeled by length or proper time variables τ_i.

The natural amplitude to compute is one in which we hold fixed the positions $x_1, \ldots, x_4$ of the external particles and integrate over all the τ's and over the paths the particle follow on the line segments. To evaluate such an integral, it is convenient to first perform a computation in which we also hold fixed the positions $y_1, \ldots, y_4$

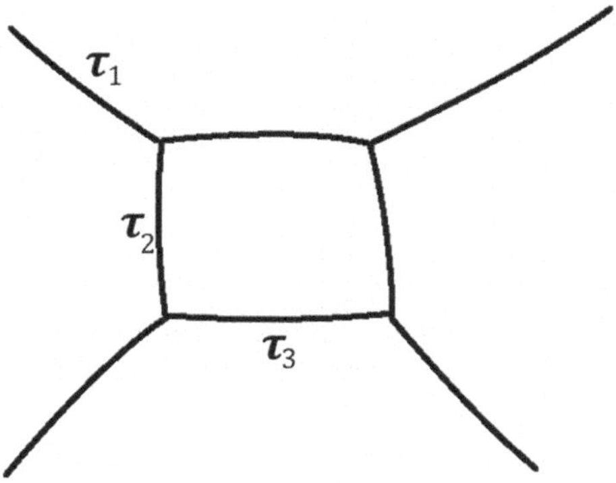

Fig. 3 A graph with trivalent vertices. Some lines have been labeled with proper time variables τ_1, τ_2, τ_3

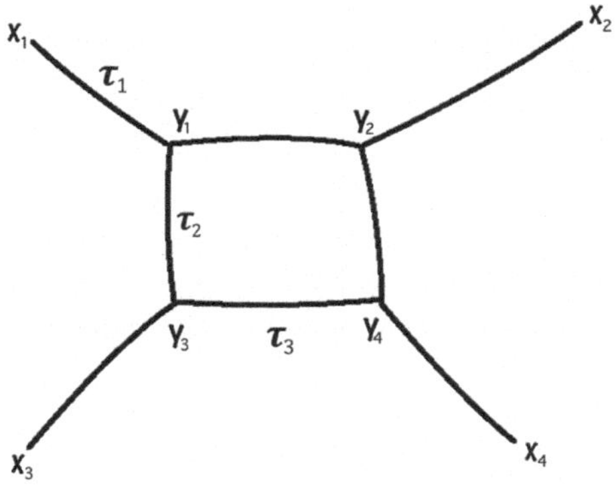

Fig. 4 The natural path integral to consider is one in which the positions $x_1, \ldots, x_4$ of the external particles are fixed, and one integrates over everything else. To perform such a path integral, it is convenient to first evaluate an integral in which the positions $y_1, \ldots, y_4$ of vertices are also fixed

of the vertices in the graph. This means that all endpoints of all segments are labeled (Fig. 4). The computation that we have to perform on each segment is the same as before and gives the Feynman propagator. The final integration over $y_1, \ldots, y_4$ imposes momentum conservation at each vertex. Thus we arrive at Feynman's recipe to compute the amplitude associated to a Feynman graph: a Feynman propagator for each line, and an integration over all momenta subject to momentum conservation.

We have arrived at one of nature's rhymes. If we imitate in one dimension what we would expect to do in $D = 4$ dimensions to describe quantum gravity, we arrive at something that is certainly important in physics, namely ordinary quantum field theory in a possibly curved spacetime. In the example that we have considered, the "ordinary quantum field theory" is scalar ϕ^3 theory, because of the particular matter system we started with and assuming we take the graphs to have cubic vertices. Quartic vertices (for instance) would give ϕ^4 theory, and a different matter system would give fields of different spins. So many or maybe all quantum field theories in D dimensions can be derived in this sense from quantum gravity in 1 dimension.

There is actually a much more perfect rhyme if we repeat this in two dimensions, that is for a string instead of a particle. We immediately run into the fact that a two-manifold Σ can be curved (Fig. 5). Related to this, two-dimensional metrics are not

Fig. 5 Generically, a two-dimensional surface has nonzero Ricci curvature

all locally equivalent under diffeomorphisms. A two-dimensional metric in general is a 2×2 symmetric matrix constructed from 3 functions

$$g_{ij} = \begin{pmatrix} g_{11} & g_{12} \\ g_{21} & g_{22} \end{pmatrix}, \quad g_{21} = g_{12}.$$

But a transformation of the two-dimensional coordinates, generated by

$$\sigma^i \to \sigma^i + h^i(\sigma), \quad i = 1, 2,$$

can only remove two functions, leaving the Ricci curvature scalar as an invariant.

All this suggests that the integral over two-dimensional metrics will not much resemble what we found in the one-dimensional case. But now we notice that the obvious analog of the action function that we used for the particle, namely

$$I = \int_\Sigma \mathrm{d}^2\sigma \sqrt{g}\, g^{ij} G_{IJ} \frac{\partial X^I}{\partial \sigma^i} \frac{\partial X^J}{\partial \sigma^j},$$

is *conformally-invariant*, that is, it is invariant under a Weyl transformation of the metric

$$g_{ij} \to e^{\phi} g_{ij}$$

for any real function ϕ on Σ. This is true precisely in two dimensions (and only[2] if there is no "cosmological constant"). If we require Weyl invariance as well as diffeomorphism invariance, then this is enough to make any metric g_{ij} on Σ locally trivial (locally equivalent to δ_{ij}), similarly to what we said for one-manifolds.

Some very pretty 19th century mathematics now comes into play. A two-manifold whose metric is given up to a Weyl transformation is called a Riemann surface. As in the one-dimensional case, a Riemann surface can be characterized up to diffeomorphism and Weyl transformation by finitely many parameters. There are two big differences: the parameters are now complex rather than real, and their range is

[2]While maintaining conformal invariance, we can add to the action the usual Einstein-Hilbert term, the integral of the scalar curvature R. This plays no role in one dimension because a one-manifold has no intrinsic curvature. In two dimensions, there is a curvature scalar but its integral $\frac{1}{4\pi}\int_\Sigma \mathrm{d}^2x\sqrt{g}R$ is a topological invariant, the Euler characteristic of Σ. We can and should include this term in the action. It turns out that the coefficient with which it appears determines the "string coupling constant," the strength with which strings interact.

restricted in a way that leaves no room for an ultraviolet divergence. We will return to that last point later.

We give an example of the relation between the one-dimensional parameters and the two-dimensional ones in Fig. 6. A metric on the indicated graph depends up to diffeomorphism on three real length or proper time parameters τ_1, τ_2, τ_3. If the graph is "thickened" to a two-manifold, then a metric on this two-manifold depends up to diffeomorphism and Weyl transformation on three complex parameters $\widehat{\tau}_1, \widehat{\tau}_2, \widehat{\tau}_3$. For another illustration of the relation between a Feynman graph and a corresponding Riemann surface, see Fig. 7.

Now we come to a deeper rhyme. We used one-dimensional quantum gravity to describe quantum field theory in a possibly curved spacetime, *but not to describe quantum gravity in spacetime*. The reason that we did not get quantum gravity in spacetime is that there is no correspondence between operators and states in quantum mechanics. We considered the one-dimensional quantum mechanics with action

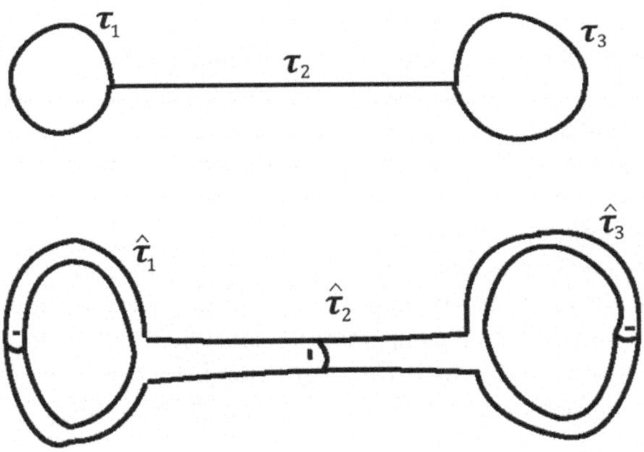

Fig. 6 A Feynman diagram with proper time parameters τ_1, τ_2, τ_3, and a corresponding Riemann surface obtained by slightly thickening all the lines in the Feynman diagram into tubes that join together smoothly. The Riemann surface is parametrized up to diffeomorphism and Weyl transformation by complex variables $\widehat{\tau}_1, \widehat{\tau}_2, \widehat{\tau}_3$

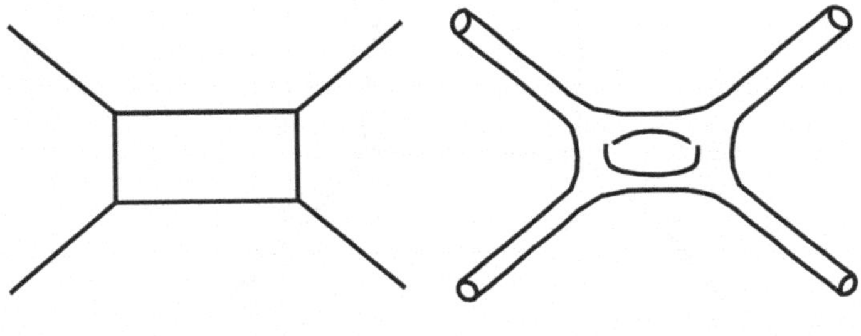

Fig. 7 A one-loop Feynman diagram for 2 → 2 scattering (left), and its string theory analog (right), which is obtained by thickening all of the lines in the Feynman diagram into tubes, and replacing the vertices in the Feynman diagram by smooth junctions between tubes

$$I = \int \mathrm{d}t \sqrt{g} \left(\frac{1}{2} g^{tt} G_{IJ} \frac{\mathrm{d}X^I}{\mathrm{d}t} \frac{\mathrm{d}X^J}{\mathrm{d}t} - \frac{1}{2} m^2 \right).$$

What turned out to be the external states in a Feynman diagram were just the states in this quantum mechanics. A deformation of the spacetime metric is represented not by a state in this quantum mechanics, but by an operator. When we make a change δG_{IJ} in the spacetime metric G_{IJ}, the action changes by $I \to I + \int \mathrm{d}t \sqrt{g} \mathcal{O}$, where $\mathcal{O} = \frac{1}{2} g^{tt} \delta G_{IJ} \partial_t X^I \partial_t X^J$ is the operator that encodes a change in the spacetime metric.

A state would appear at the end of an *external* line in the Feynman graph. But an operator $\mathcal{O}$ such as the one describing a perturbation in the spacetime metric appears at an *interior* point in the graph, as in Fig. 8. Technically, to compute the effect of the perturbation, we include in the path integral a factor $\delta I = \int \mathrm{d}t \sqrt{g} \mathcal{O}$, integrating over the position at which the operator $\mathcal{O}$ is inserted. Just one possible insertion point is shown in the figure. Since states enter at ends of external lines and operators are inserted at internal points, there is in general no simple relation between operators and states.

But in conformal field theory, there *is* a correspondence between states and operators; this correspondence actually is important in some areas of statistical mechanics and condensed matter physics, as well as in string theory. And hence the operator $\mathcal{O} = g^{ij} \delta G_{IJ} \partial_i X^I \partial_j X^J$ that represents a fluctuation in the spacetime metric automatically represents a state in the quantum mechanics. That is why the theory describes quantum gravity in spacetime.

The operator-state correspondence arises from a 19th century relation between two pictures that are conformally equivalent. In Fig. 9, we show a two-manifold Σ with a marked point p at which an operator $\mathcal{O}$ is inserted. In Fig. 10, the point p has been removed from Σ, and a Weyl transformation has been made of the metric of Σ, converting what used to be a small neighborhood of the point p to a semi-infinite

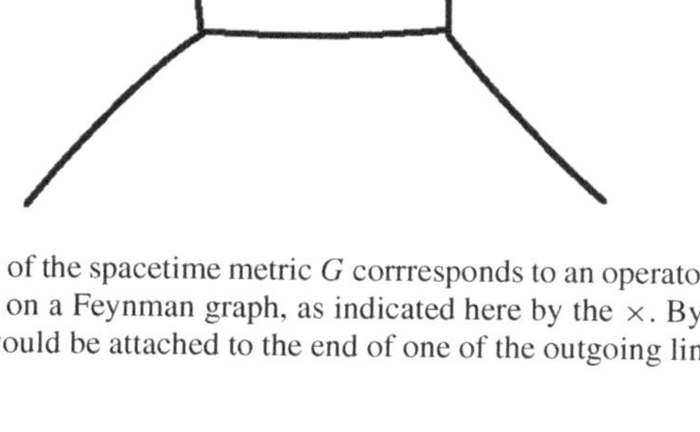

Fig. 8 A deformation of the spacetime metric G corrresponds to an operator $\mathcal{O}$ that can be inserted at some internal point on a Feynman graph, as indicated here by the $\times$. By contrast, a state in the quantum mechanics would be attached to the end of one of the outgoing lines of the graph

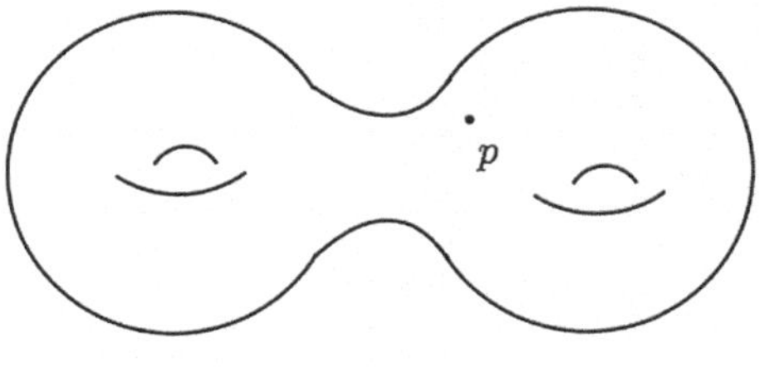

Fig. 9 A Riemann surface with a marked point labeled p at which an operator $\mathcal{O}$ is inserted

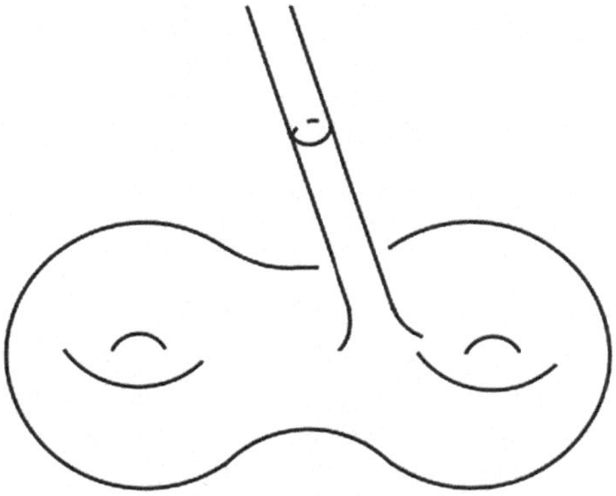

Fig. 10 After deleting the marked point, the Riemann surface of Fig. 9 is conformally equivalent to this one, with an outgoing "tube" that is analogous to an external line of a Feynman graph. The operator $\mathcal{O}$ that was inserted at the marked point is converted to a quantum state of the string that propagates on the tube

tube. This is analogous to an external line of a Feynman graph, and what would be inserted at the end of the tube is a quantum string state. The relation between the two pictures is the correspondence between operators and states.

To understand the Weyl transformation between the two pictures, consider the metric of the plane in polar coordinates:

$$ds^2 = dr^2 + r^2 d\theta^2.$$

We think of inserting an operator at the point $r = 0$. Now remove this point, and make a Weyl transformation, multiplying ds^2 by $1/r^2$. This gives a new metric

$$(ds')^2 = \frac{1}{r^2}dr^2 + d\phi^2.$$

In terms of $w = \log r$, $-\infty < w < \infty$, the new metric is

$$(ds')^2 = dw^2 + d\phi^2,$$

which describes a cylinder. The point $r = 0$ in one description corresponds in the other description to the end $w \to -\infty$ of the cylinder. What is interpreted in one description as an operator inserted at $r = 0$ is interpreted in the other description as a quantum state flowing in from $w = -\infty$ (Fig. 11).

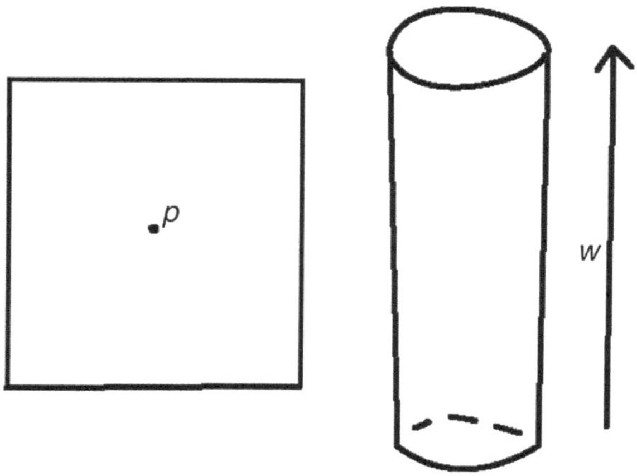

Fig. 11 Left: A plane $\mathbb{R}^2$ with a point p labeled. Right: The plane with p omitted is equivalent via a Weyl transformation to a cylinder with flat metric. The point p is mapped to the "bottom" end of the cylinder

Thus, string theory describes quantum gravity in spacetime. But it does not describe quantum gravity only. It describes quantum gravity unified with a variety of particles and forces in spacetime. These other particles and forces correspond to other operators in the conformal field theory of the string—apart from the operator $\mathcal{O}$ that is related to a fluctuation in the spacetime geometry—or equivalently to other quantum states of the string.

The next step is to explain why this type of theory does not have ultraviolet divergences. This contrasts sharply with what happens if we simply apply textbook recipes of quantization to the Einstein-Hilbert action for gravity. If we do that, we encounter intractable ultraviolet divergences that were first found in the 1930s. Back then, it was not entirely clear that this type of problem is special to gravity, because there were also troublesome ultraviolet divergences when other particle forces were studied in the framework of relativistic quantum theory. However, as the problems were overcome for the other forces—most completely with the emergence of the Standard Model of particle physics in the 1970s—it became clear that the problems for gravity are serious.

To understand why there are no ultraviolet divergences in string theory, we should begin by asking how ultraviolet divergences arise in ordinary quantum field theory. They arise when all the proper time variables in a loop go simultaneously to zero. So in the example of Fig. 12, there can be an ultraviolet divergence when $\tau_1, \tau_2, \tau_3, \tau_4$ simultaneously vanish.

It is true that, as stated earlier, a Riemann surface can be characterized by complex parameters that roughly parallel the proper time parameters of a Feynman graph (Fig. 7). But there is one very important difference, which is the reason there are no ultraviolet divergences in string theory. The proper time variables τ_i of a Feynman graph cover the whole range $0 \leq \tau_i \leq \infty$. By contrast, the corresponding Riemann surface parameters $\widehat{\tau}_i$ are bounded away from 0. Given a Feynman diagram, one can make a corresponding Riemann surface, but only if the proper time variables τ_i are

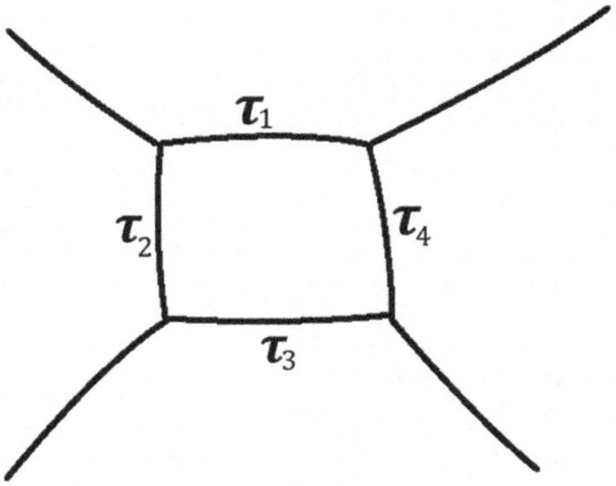

Fig. 12 This Feynman diagram can generate an ultraviolet divergence in the limit that the proper time parameters τ_1, τ_2, τ_3, τ_4 in the loop all vanish

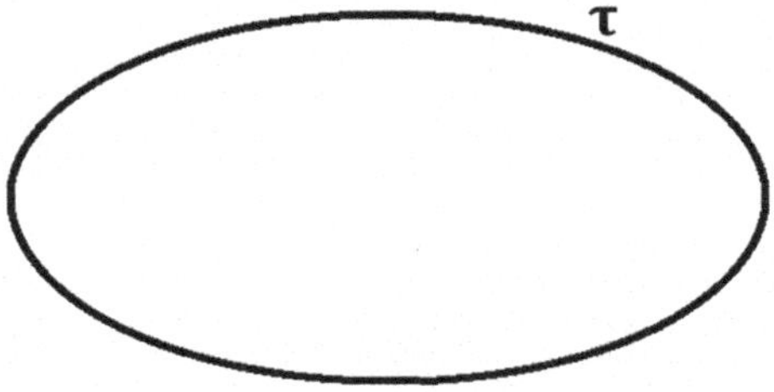

Fig. 13 This simple Feynman diagram, with a single proper time parameter τ, underlies the one-loop cosmological constant in quantum field theory

not too small. The region of the parameter space where ultraviolet divergences occur in field theory simply has no counterpart in string theory.

Instead of giving a general explanation of this, we will just explain how it works in the case of the one-loop cosmological constant. The Feynman diagram is a simple circle (Fig. 13), with a single proper time parameter τ. The resulting expression for the one-loop cosmological constant is

$$\Gamma_1 = \frac{1}{2}\int_0^\infty \frac{\mathrm{d}\tau}{\tau} \mathrm{Tr}\, \exp(-\tau H)$$

where H is the particle Hamiltonian. This integral diverges at $\tau = 0$, and the divergence is more severe than it looks because of the momentum integration that is part of the trace.

Going to string theory means replacing the classical one-loop diagram with its stringy counterpart, which is a torus (Fig. 14). Nineteenth century mathematicians showed that every torus is conformally equivalent to a parallelogram in the plane with opposite sides identified (Fig. 15). But to explain the idea without any extraneous technicalities, we will consider only rectangles instead of parallelograms. We label the height and base of the rectangle as s and s', respectively (Fig. 16). Only the ratio

$$u = \frac{s'}{s}$$

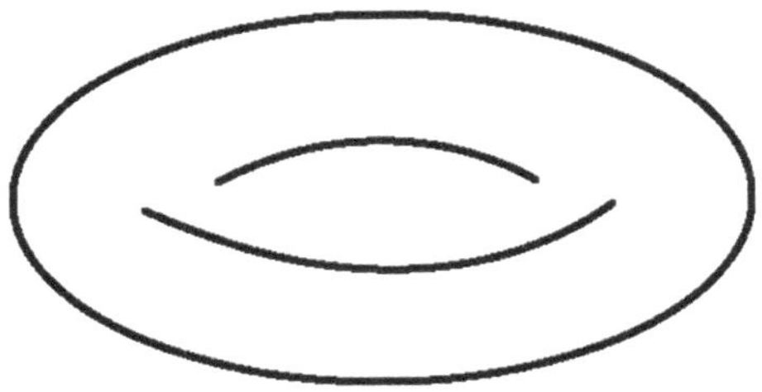

Fig. 14 A torus, which underlies the one-loop cosmological constant in string theory

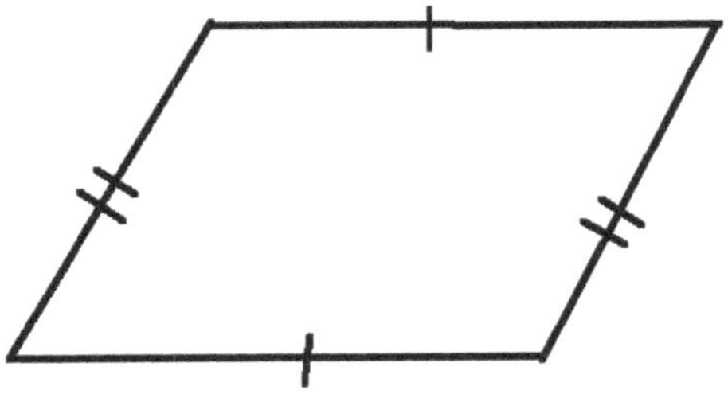

Fig. 15 A parallelogram with opposite sides identified to make a torus

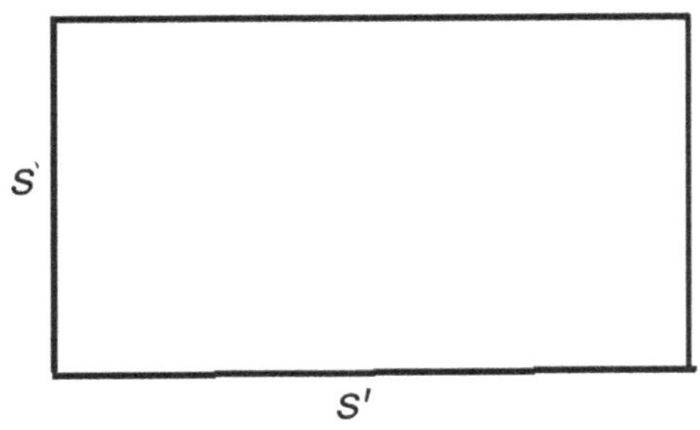

Fig. 16 A rectangle with height and base labeled as s and s'

is conformally-invariant. Also since it is arbitrary what we call the "height" as opposed to the "base" of a rectangle, we are free to exchange $s \leftrightarrow s'$, which corresponds to

$$u \leftrightarrow \frac{1}{u}.$$

So we can restrict to $s' \geq s$, and thus the range of u is

$$1 \leq u < \infty.$$

The proper time parameter τ of the particle corresponds to u in string theory, with the key difference that $0 \leq \tau < \infty$ but $1 \leq u < \infty$. So the one-loop cosmological constant in field theory is

$$\Gamma_1 = \frac{1}{2}\int_0^\infty \frac{\mathrm{d}\tau}{\tau} \mathrm{Tr}\, \exp(-\tau H)$$

but (in the approximation of considering only rectangles and not parallelograms), the one-loop cosmological constant in string theory is

$$\Gamma_1 = \frac{1}{2}\int_1^\infty \frac{\mathrm{d}u}{u}\mathrm{Tr}\,\exp(-\tau H).$$

There is no ultraviolet divergence, because the lower limit on the integral is 1 instead of 0. (A more complete analysis with parallelograms shifts the lower bound on u from 1 to $\sqrt{3}/2$.)

We have described a special case, but this is a general story. The stringy formulas generalize the field theory formulas, but without the region that can give ultraviolet divergences in field theory. The infrared region ($\tau \to \infty$ or $u \to \infty$) lines up properly between field theory and string theory and this is why a string theory can imitate field theory in its predictions for the behavior at low energies or long times and distances.

Our final goal here is to explain, at least partly, in what sense spacetime "emerges" from something deeper if string theory is correct. Let us focus on the following fact. The spacetime M with its metric tensor $G_{IJ}(X)$ was encoded as the data that enabled us to define a two-dimensional conformal field theory that we used in this construction. Moreover, that is the only way that spacetime entered the story.

We could have used in this construction a different two-dimensional conformal field theory (subject to a few general rules that we will omit). Now if $G_{IJ}(X)$ is slowly varying (the radius of curvature is everywhere large), the Lagrangian by which we described the two-dimensional conformal field theory is weakly coupled and useful. This is the situation in which string theory matches to ordinary physics that we are familiar with. We may say that in this situation, the theory has a semiclassical interpretation in terms of strings in spacetime (and this will reduce at low energies to an interpretation in terms of particles and fields in spacetime).

When we get away from a semiclassical limit, the Lagrangian is not so useful and the theory does not have any particular interpretation in terms of strings in spacetime. This leads to many nonclassical consequences, such as the ability to make continuous transitions from one spacetime manifold to another, or the fact that certain types of singularities in classical General Relativity (but not black hole singularities) turn out to represent perfectly smooth and harmless situations in string theory. An example of the nonclassical behavior of string theory is sketched in Fig. 17.

We can say that from this point of view, spacetime "emerges" from a seemingly more fundamental concept of two-dimensional conformal field theory. In general a string theory comes with no particular spacetime interpretation, but such an interpretation can emerge in a suitable limit, somewhat as classical mechanics sometimes arises as a limit of quantum mechanics.

This is not a complete explanation of the sense in which, in the context of string theory, spacetime emerges from something deeper. A completely different side of the story, beyond the scope of the present article, involves quantum mechanics and the duality between gauge theory and gravity. However, what we have described is

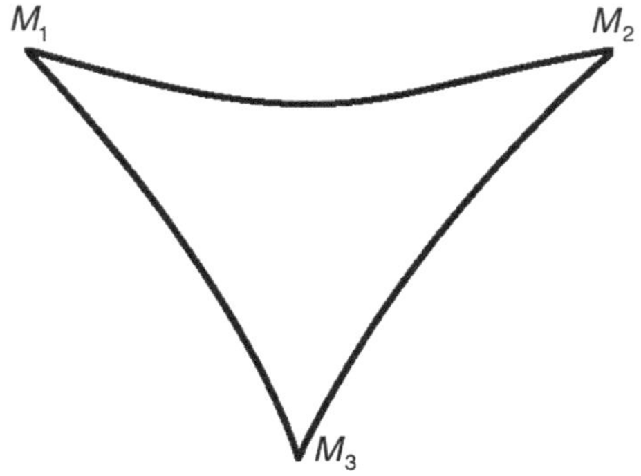

Fig. 17 This picture schematically represents a family of two-dimensional conformal field theories that depends on two parameters. In different limits, these theories have semiclassical interpretations in terms of strings propagating in a spacetime M_1, M_2, or M_3. Generically there is no such interpretation. One can make a continuous transition between the different possible classical spacetimes that appear at corners of the picture by passing through the "stringy" region in the interior

certainly one important and relatively well-understood piece of the puzzle. It is at least a partial insight about how spacetime as conceived by Einstein can emerge from something deeper.

Quanta of Space-Time and Axiomatization of Physics

Ali H. Chamseddine

Abstract We consider Hilbert's sixth problem on the axiomatization of physics starting with a higher degree Heisenberg commutation relation involving the Dirac operator and the Feynman slash of scalar fields. The two sided version of the commutation relation in dimension 4 implies volume quantization and determines a noncommutative space which is a tensor product of continuous and discrete spaces. This noncommutative space predicts the full structure of a unified model of all particle interactions based on Pati-Salam symmetries or, as a special case, the Standard Model. We study implications of this quantization condition on Particle Physics, General Relativity, the cosmological constant and dark matter. We demonstrate that, with little input, noncommutative geometry gives a compelling and attractive picture about the nature and structure of space-time.

1 Introduction

David Hilbert research on the axiomatization of geometry led him to suggest the sixth problem on his list for the axiomatization of Physics which have received the least attention [1]. Hilbert contributed prominently to the formulation of the gravitational equations in the General Theory of Relativity which was presented in November 1915, almost simultaneously with Einstein [2, 3]. Weyl has asserted that during the period 1910–1922 Hilbert has devoted considerable time to research in Physics which was an integral part of his mathematical world. Indeed, in 1915 Hilbert has presented a unified theory of electromagnetism and gravitation based on the use of the variational principle derived in an axiomatic fashion from the two principles of general invariance and "Mie's axiom of the world function". This attempt can be considered as the seed that motivated much work on ideas on unification of all fundamental interactions such as in Kaluza-Klein theory, supersymmetry, superstring

A. H. Chamseddine (✉)
Physics Department, American University of Beirut, Beirut, Lebanon
e-mail: chams@aub.edu.lb

A. H. Chamseddine
Institut des Hautes Etudes Scientifique (I.H.E.S.), 91440 Bures-sur-Yvette, France

J. Kouneiher (ed.), *Foundations of Mathematics and Physics One Century After Hilbert*,
https://doi.org/10.1007/978-3-319-64813-2_9

theory and noncommutative geometry. In this article I will follow up on the contribution of Alain Connes to this volume and show that starting with the axioms of noncommutative geometry supplemented by a minimal number of physical assumptions would result, unambiguously, in a unified theory of all fundamental interactions and matter content of space-time [4, 5]. We will be able to establish a link between the quantization of volume of space at Planck energy and the constituents of matter and their symmetries. In addition we uncover the origin of the Higgs fields and symmetry breaking, and indicate possible solutions to long standing problems such as resolving the singularities in GR, dark matter and dark energy.

All the material covered in this review is a result of a long time collaboration with Alain Connes which started in 1996 and continues until now. More recently our collaboration included Walter van Suijlekom and, in separate publications, Slava Mukhanov. An excellent introduction to the material covered in this review is the accompanying article by Alain Connes in this volume. However, an attempt is made to make this article self-contained.

The Planck scale is the scale at which all rescaled curvature invariants of a Riemannian manifold are of the same order. The volume of any manifold at scales below the Planck scale, will be many orders of magnitude larger than that scale. To avoid the problem of infinities, which are expected to arise in a quantized theory of gravity, it is a natural proposition to assume that the volume of a physical space is an integer multiple of a unit volume of Planckian size and thus provide a cutoff scale. It is well known that the degree of a smooth map Y from a connected, compact, oriented n-manifold to the sphere S^n is an integer

$$Y : M_n \to S^n, \tag{1}$$

where Y is $\mathbb{R}^{n+1}$ valued on M_n. This map is normalized by $\langle Y(x), Y(x)\rangle = 1$ where $x \in M_n$ and if we let Δ be the positive normed determinant function in $\mathbb{R}^{n+1}$, then the degree of the map is given by [6]

$$\deg(Y) \equiv \frac{1}{\kappa_n} \int_{M_n} \left\langle \Delta, Y(dY)^n \right\rangle \in \mathbb{Z} \tag{2}$$

where κ^n is the volume of the n-sphere:

$$\kappa_{2m} = \frac{2^{m+1}}{(2m-1)!!}\pi^m, \qquad \kappa_{2m+1} = \frac{2}{m!}\pi^{m+1}, \quad m = 1, \ldots, \infty. \tag{3}$$

We propose to identify the integrand in (2), which is an n-form over an n-dimensional connected, compact oriented manifold, with the volume form:

$$w_n = \frac{1}{\kappa_n} \left\langle \Delta, Y(dY)^n \right\rangle, \tag{4}$$

then the volume of M_n will be an integer multiple of the unit Planckian n-sphere. From this we deduce that the pullback $Y^* (w_n)$ is a differential form that does not vanish anywhere. This in turn implies that the Jacobian of the map Y does not vanish anywhere, and that Y is a covering of the sphere. The sphere is simply connected, and on each connected component $M_j \subset M_n$, the restriction of the map Y to M_j is a diffeomorphism, implying that the manifold must be disconnected, with each piece having the topology of a sphere [7]. We will show how to avoid this unsatisfactory conclusion and how the attractive idea of volume quantization works in a convincing way within the formulation of noncommutative geometry.

Extensive research over the last two decades have shown that there are many advantages to work with noncommutative geometry instead of Riemannian geometry [4]. The approach is spectral in nature and its concepts are modeled after quantum mechanics where geometry is defined in terms of spectral data. These are specified in terms of spectral triple $(\mathcal{A}, \mathcal{H}, D)$ where $\mathcal{A}$ is an associative algebra with unit 1 and involution $*$, $\mathcal{H}$ a complex Hilbert space carrying a faithful representation of the algebra $\mathcal{A}$ and D is a slef-adjoint operator on $\mathcal{H}$ with the resolvent $(D - \lambda 1)^{-1}$, where $\lambda \notin \mathbb{R}$ of D, compact. The operator D plays the role of inverse line element. In addition the real structure J is an anti-unitary operator that sends the algebra $\mathcal{A}$ to its commutant $\mathcal{A}^o$ such that [8]

$$\left[a, b^o\right] = 0, \qquad a, b \in \mathcal{A}, \qquad b^o = J b^* J^{-1} \in \mathcal{A}^o. \tag{5}$$

The chirality operator γ is a unitary operator in $\mathcal{H}$ defined in even dimensions such that $\gamma^2 = 1$ and commutes with $\mathcal{A}$

$$\left[\gamma, a\right] = 0 \qquad \forall a \in \mathcal{A}. \tag{6}$$

There are commutativity or anti-commutativity relations between D, J, and γ :

$$J^2 = \epsilon, \qquad JD = \epsilon' DJ, \qquad J\gamma = \epsilon'' \gamma J, \qquad D\gamma = -\gamma D, \tag{7}$$

where $\epsilon, \epsilon', \epsilon'' \in \{-1, 1\}$. The operators γ and J are similar to the chirality and charge conjugation operators and to every fixed value of $\epsilon, \epsilon', \epsilon''$ is associated a KO dimension, which may be non-metric, and thus is defined only modulo 8. It is then evident that the generalized Heisenberg relation must be modified to include not only the mapping Y from M_n to S^n but also the effects of the operator J which requires two mappings Y and Y'. We have shown that using the two mappings Y and Y' to set the volume quantization condition would avoid limiting the topology of the manifold to be that of a sphere in dimensions two and four [7, 9]. We shall elaborate on the form of the generalized Heisenberg relation and show that this leads, unambiguously, to the construction of a noncommutative space whose geometry gives naturally a unified model of all particle interactions based on Pati-Salam symmetry group which also includes the Standard Model as a special case.

This article is organized as follows. In Sect. 2 the conjectured Heisenberg quantization two sided relation is constructed in such a way as to give the volume of the underlying manifold to be given by the sum of two integers times the volume of a unit Planckian sphere. In Sect. 3 the algebra of the finite noncommutative space is derived to be the sum of two algebras, which in dimension four, is given by the sum $M_2(\mathbb{H})$ and $M_4(\mathbb{C})$ [10–12]. In Sect. 4 we determine the noncommutative space and make contact with our previous work on noncommutative geometry [13–15]. In Sect. 5 we show the the unified model associated with this noncommutative space is of the Pati-Salam type and in Sect. 6 we give the Standard Model obtained as a limiting case [15]. Section 7 is a summary of the minimal Pati-Salam model [12, 16]. In Sect. 8 we present the spectral action principle and calculate the spectral action of the Standard Model. In Sect. 9 we study consequences of volume quantization on the equations of motion in both instances when the fields Y and Y' are with or without kinetic terms. In Sect. 10 we give the solitonic solutions and show that these are identical to the $O(5)$ non-linear gravitational sigma model. In Sect. 11 we consider the case of a Riemannian manifold with Lorentzian signature where the four-dimensional manifold is viewed as a $3+1$ space formed from the motion of three dimensional hypersurfaces. We show that it is possible to impose quantization of the three dimensional compact space provided that the field mapping the one-dimensional non-compact space satisfies a length preserving relation. In Sect. 12 we further discuss the conditions under which a quantization of a two dimensional hypersurface is possible. In Sect. 13 we study the equations of motion for the cases of three dimensional volume and two dimensional surface quantization. We discuss quantization on the special spaces $\mathbb{R}\times S^3$ and $\mathbb{R}^2\times S^2$. Section 14 contains a discussion and the conclusion.

2 Heisenberg Volume Quantization in Dimensions 2 and 4

For a Riemannian manifold of dimension n the algebra $\mathcal{A}$ is taken to be $C^\infty(M)$, the algebra of continuously differentiable functions, while the operator D is identified with the Dirac operator given by

$$D_M=\gamma^\mu\left(\frac{\partial}{\partial x^\mu}+\omega_\mu\right), \tag{8}$$

where $\gamma^\mu=e_a^\mu\gamma^a$ and $\omega_\mu=\frac{1}{4}\omega_{\mu bc}\gamma^{bc}$ is the $SO(n)$ Lie-algebra valued spin-connection with the (inverse) vielbein e_a^μ being the square root of the (inverse) metric $g^{\mu\nu}=e_a^\mu\delta^{ab}e_b^\nu$. The gamma matrices γ^a are anti-hermitian $(\gamma^a)^*=-\gamma^a$ and define the Clifford algebra $\{\gamma^a,\gamma^b\}=-2\delta^{ab}$. The Hilbert space $\mathcal{H}$ is the space of square integrable spinors $L^2(M,S)$. The Dirac operator is Hermitian with respect to the inner product

$$(\psi,D_M\psi)=(D_M\psi,\psi)=\int d^nxe\psi^*D_M\psi, \tag{9}$$

where $e=\det\left(e_\mu^a\right)$ with e_μ^a being the inverse of e_a^μ. The chirality operator γ in even dimensions is then given by

$$\gamma = (i)^{\frac{n}{2}} \gamma^1 \gamma^2 \ldots \gamma^n \tag{10}$$

From the above discussion, it is very suggestive to associate with the map fields Y^A, $A = 1, 2, \ldots, n+1$ a Clifford algebra valued field $Y = Y^A \Gamma_A$ where [17]

$$\Gamma_A \in C_\kappa, \quad \{\Gamma_A, \Gamma_B\} = 2\kappa\, \delta_{AB}, \; (\Gamma_A)^* = \kappa \Gamma_A. \tag{11}$$

Here $\kappa = \pm 1$ and $C_\kappa \subset M_s(\mathbb{C})$ is the algebra of $s \times s$ matrices, where $s = 2^{n/2}$. A generalization of the Heisenberg commutation relation $[p, q] = -i\hbar$ is conjectured to be given by [7]

$$\langle Y\,[D, Y] \ldots [D, Y] \rangle = \sqrt{\kappa}\, \gamma \quad (n \text{ terms } [D, Y]), \tag{12}$$

where $Y \in C^\infty(M) \otimes C_\kappa$ is of the Feynman slashed form $Y = Y^A \Gamma_A$, and fulfill the equations

$$Y^2 = \kappa, \qquad Y^* = \kappa Y. \tag{13}$$

The notation $\langle T \rangle$ means the trace of T with respect to the above matrix algebra $M_s(\mathbb{C})$. In a coordinate basis Eq. (12) takes the form [7]

$$\frac{1}{n!} \epsilon^{\mu_1 \mu_2 \ldots \mu_n} \epsilon_{A_1 A_2 \ldots A_{n+1}} Y^{A_{n+1}} \partial_{\mu_1} Y^{A_1} \partial_{\mu_2} Y^{A_2} \ldots \partial_{\mu_n} Y^{A_n} = \det\left(e^a_\mu\right), \tag{14}$$

which is a constraint on the volume form. This can be thought of as a generalization of the coordinate-momenta $[p, q] = -i\hbar$ phase space quantization where p is replaced with the Dirac operators D and q is replaced with the Feynman slash coordinates Y. We have seen, however, that this quantization condition implies that the n-manifold decomposes into a set of bubbles. The difference now is that the quantization condition is given in terms of the noncommutative data. One cannot fail to notice that the operator J is missing from Eq. (12) which suggests that this equation must be modified to take this operator into account. We first define the projection operator $e = \frac{1}{2}(1 + Y)$ satisfying $e^2 = e$ [18] but now there are two possibilities, Y corresponding to the case $\kappa = 1$ and Y' to the case $\kappa = -1$. Thus let $Y = Y^A \Gamma_A \equiv Y$ and let $Y' = iJYJ^{-1}$ and $\Gamma'_A = iJ\Gamma_A J^{-1}$ so that we can write

$$Y = Y^A \Gamma_A, \qquad Y' = Y'^A \Gamma'_A, \tag{15}$$

satisfying $Y^2 = 1$ and $Y'^2 = 1$. The projection operators $e = \frac{1}{2}(1 + Y)$ and $e' = \frac{1}{2}\left(1 + Y'\right)$ satisfy $e^2 = e$, $e'^2 = e'$ with e and e' commuting. This allows to define the projection operator $E = ee'$ and the associated field

$$Z = 2E - 1, \tag{16}$$

satisfying $Z^2 = 1$. The conjectured quantization condition takes the elegant form of a two-sided relation [7, 9]

$$\left\langle Z\left[D,Z\right]^{n}\right\rangle =\gamma. \tag{17}$$

Our proposal is that this quantization condition is valid for all noncommutative geometries defined by the spectral data where the metric dimension of the operator D as determined from the Weyl asymptotic formula is less than or equal to four. The presence of the chirality operator γ indicates that the dimension n should be even, and this would limit us to the two cases $n=2$ and $n=4$. For odd dimensional n the form of the quantization condition should be modified, but will not be considered here. We have shown that for both $n=2$ and $n=4$ Eq. (17) splits as the sum of two pieces [7]

$$\left\langle Z\left[D,Z\right]^{n}\right\rangle =\left\langle Y\left[D,Y\right]^{n}\right\rangle +\left\langle Y'\left[D,Y'\right]^{n}\right\rangle . \tag{18}$$

This implies that the volume form of the n-dimensional Riemannian manifold is the sum of two n-forms and thus

$$\det\left(e_{\mu}^{a}\right)=\frac{1}{n!}\epsilon^{\mu_1\mu_2\ldots\mu_n}\epsilon_{A_1A_2\ldots A_{n+1}}Y^{A_{n+1}}\partial_{\mu_1}Y^{A_1}\partial_{\mu_2}Y^{A_2}\ldots\partial_{\mu_n}Y^{A_n}+ \tag{19}$$

$$+\frac{1}{n!}\epsilon^{\mu_1\mu_2\ldots\mu_n}\epsilon_{A_1A_2\ldots A_{n+1}}Y^{'A_{n+1}}\partial_{\mu_1}Y^{'A_1}\partial_{\mu_2}Y^{A_2}\ldots\partial_{\mu_n}Y^{'A_n}. \tag{20}$$

Consider the smooth maps $\phi_{\pm}:M_n\rightarrow S^n$ then their pullbacks $\phi_{\pm}^{\#}$ would satisfy

$$\phi_{+}^{\#}\left(\alpha\right)+\phi_{-}^{\#}\left(\alpha\right)=\omega, \tag{21}$$

where α is the volume form on the unit sphere S^n [19] and $\omega\left(x\right)$ is an $n-$form that does not vanish anywhere on M_n. We stress that the quantization condition does not split as the sum of two terms except for $n=2,4$, however, if one starts with the conjecture that the volume form is the sum of the two traces in terms of the coordinates Y and Y' then Eq. (21) would follow and would then not be limited to the two values for n. We have shown that for a compact connected smooth oriented manifold with $n<4$ one can find two maps $\phi_{+}^{\#}\left(\alpha\right)$ and $\phi_{-}^{\#}\left(\alpha\right)$ whose sum does not vanish anywhere, satisfying Eq. (21) such that $\int_M\omega\in\mathbb{Z}$. The proof for $n=4$ is more difficult and there is an obstruction unless the second Stieffel-Whitney class w_2 vanishes, which is satisfied if M is required to be a spin-manifold and the volume to be larger than or equal to five units. The key idea in the proof is to note that the kernel of the map Y is a hypersurface Σ of co-dimension 2 and therefore [7]

$$\dim\Sigma=n-2. \tag{22}$$

We can then construct a map $Y'=Y\circ\psi$ where ψ is a diffeomorphism on M such that the sum of the pullbacks of Y and Y' does not vanish anywhere. The important point to stress here is that the conjectured two sided relation (17) is taken to hold for arbitrary noncommutative spaces where $n\leq4$ where n is the dimension as determined in the Weyl asymptotic formula for the growth of eigenvalues of the

Dirac operator, and is not restricted for Riemannian manifolds. In other words, one can seek solutions for this equation in general and find the noncommutative space satisfying this equation.

3 Clifford Algebras and Feynman Slash

We have seen that the coordinates Y are defined over a Clifford algebra C_+ spanned by $\{\Gamma_A, \Gamma_B\} = 2\delta_{AB}$. For $n = 2$, $C_+ = M_2(\mathbb{C})$ while for $n = 4$, $C_+ = M_2(\mathbb{H}) \oplus M_2(\mathbb{H})$ where $\mathbb{H}$ is the field of quaternions [17]. However, for $n = 4$, since we will be dealing with irreducible representations we take $C_+ = M_2(\mathbb{H})$. Similarly the coordinates Y' are defined over the Clifford algebra C_- spanned by $\{\Gamma'_A, \Gamma'_B\} = -2\delta_{AB}$ and for $n = 2$, $C_- = \mathbb{H} \oplus \mathbb{H}$ and for $n = 4$, $C_- = M_4(\mathbb{C})$. The operator J acts on the two algebras $C_+ \oplus C_-$ in the form $J(x, y) = (y^*, x^*)$ (i.e. it exchanges the two algebras and takes the Hermitian conjugate). The coordinates $Z = \frac{1}{2}(Y + 1)(Y' + 1) - 1$, then define the matrix algebras [10]

$$\mathcal{A}_F = M_2(\mathbb{C}) \oplus \mathbb{H}, \qquad n = 2 \tag{23}$$

$$\mathcal{A}_F = M_2(\mathbb{H}) \oplus M_4(\mathbb{C}), \qquad n = 4. \tag{24}$$

One, however, must remember that the maps Y and Y' are functions of the coordinates of the manifold M and therefore the algebra associated with this space must be

$$\mathcal{A} = C^\infty(M, \mathcal{A}_F) \tag{25}$$

$$= C^\infty(M) \otimes \mathcal{A}_F. \tag{26}$$

To see this consider, for simplicity, the $n = 2$ case with only the map Y. The Clifford algebra $C_- = \mathbb{H}$ is spanned by the set $\{1, \Gamma^A\}$, $A = 1, 2, 3$, where $\{\Gamma^A, \Gamma^B\} = -2\delta^{AB}$. We then consider functions which are made out of words of the variable Y formed with the use of constant elements of the algebra [18]

$$\sum_{i=1}^{\infty} a_1 Y a_2 Y \ldots a_i Y, \qquad a_i \in \mathbb{H},$$

which will generate arbitrary functions over the manifold, which is the most general form since $Y^2 = 1$. One can easily see that these combinations generate all the spherical harmonics. This result could be easily generalized by considering functions of the fields

$$Z = \frac{1}{2}(Y + 1)(Y' + 1) - 1, \qquad Y \in \mathbb{H}, \quad Y' \in M_2(\mathbb{C}),$$

showing that the noncommutative algebra generated by the constant matrices and the Feynman slash coordinates Z is given by [18]

$$\mathcal{A} = C^{\infty}(M_2) \otimes (\mathbb{H} + M_2(\mathbb{C})) .$$

4 Finite Noncommutative Space

Having explained the simple case $n = 2$, for the remainder of this paper we restrict ourselves to the physical case of $n = 4$. Here the algebra is given by

$$\mathcal{A} = C^{\infty}(M_4) \otimes (M_2(\mathbb{H}) + M_4(\mathbb{C})) . \tag{27}$$

The associated Hilbert space is

$$\mathcal{H} = L^2(M_4, S) \otimes \mathcal{H}_F . \tag{28}$$

The Dirac operator mixes the finite space and the continuous manifold non-trivially

$$D = D_M \otimes 1 + \gamma_5 \otimes D_F , \tag{29}$$

where D_F is a self adjoint operator in the finite space. The chirality operator is

$$\gamma = \gamma_5 \otimes \gamma_F , \tag{30}$$

and the anti-unitary operator J is given by

$$J = J_M \gamma_5 \otimes J_F , \tag{31}$$

where J_M is the charge-conjugation operator C on M and J_F the anti-unitary operator for the finite space. Thus an element $\Psi \in \mathcal{H}$ is of the form $\Psi = \begin{pmatrix} \psi_A \\ \psi_{A'} \end{pmatrix}$ where ψ_A is a 16 component $L^2(M, S)$ spinor in the fundamental representation of $\mathcal{A}_F$ of the form $\psi_A = \psi_{\alpha I}$ where $\alpha = 1, \ldots, 4$ with respect to $M_2(\mathbb{H})$ and $I = 1, \ldots, 4$ with respect to $M_4(\mathbb{C})$ and where $\psi_{A'} = C\psi_A^*$ is the charge conjugate spinor to ψ_A [15]. The chirality operator γ must commute with elements of $\mathcal{A}$ which implies that γ_F must commute with elements in $\mathcal{A}_F$. Commutativity of the chirality operator γ_F with the algebra $\mathcal{A}_F$ and that this $\mathbb{Z}/2$ grading acts non-trivially reduces the algebra $M_2(\mathbb{H})$ to $\mathbb{H}_R \oplus \mathbb{H}_L$ [10]. Thus the γ_F is identified with $\gamma_F = \Gamma^5 = \Gamma^1\Gamma^2\Gamma^3\Gamma^4$ and the finite space algebra reduces to

$$\mathcal{A}_F = \mathbb{H}_R \oplus \mathbb{H}_L \oplus M_4(\mathbb{C}) . \tag{32}$$

This can be easily seen by noting that an element of $M_2(\mathbb{H})$ takes the form $\begin{pmatrix} q_1 & q_2 \\ q_3 & q_4 \end{pmatrix}$ where each q_i, $i = 1, \ldots, 4$, is a 2×2 matrix representing a quaternion. Taking the representation of $\Gamma^5 = \begin{pmatrix} 1_2 & 0 \\ 0 & -1_2 \end{pmatrix}$ to commute with $M_2(\mathbb{H})$ implies that $q_2 = 0 = q_3$, thus reducing the algebra to $\mathbb{H}_R \oplus \mathbb{H}_L$. Therefore the index $\alpha = 1, \ldots, 4$ splits into two parts, $\dot{a} = \dot{1}, \dot{2}$ which is a doublet under $\mathbb{H}_R$ and $a = 1, 2$ which is a doublet under $\mathbb{H}_L$. The spinor Ψ further satisfies the chirality condition $\gamma\Psi = \Psi$ which implies that the spinors $\psi_{\dot{a}I}$ are in the $(2_R, 1_L, 4)$ with respect to the algebra $\mathbb{H}_R \oplus \mathbb{H}_L \oplus M_4(\mathbb{C})$ while ψ_{aI} are in the $(1_R, 2_L, 4)$ representation.[1] The finite space Dirac operator D_F is then a 32×32 Hermitian matrix acting on the 32 component spinors Ψ. In addition we take three copies of each spinor to account for the three families, but will omit writing an index for the families. At present we have no explanation for why the number of generations should be three. The Dirac operator for the finite space is then a 96×96 Hermitian matrix. The Dirac action is then given by [14]

$$(J\Psi, D\Psi) . \tag{33}$$

We note that we are considering compact spaces with Euclidean signature and thus the condition $J\Psi = \Psi$ could not be imposed. It could, however, be imposed if the four dimensional space is Lorentzian [20].The reason is that the KO dimension of the finite space is 6 because the operators D_F, γ_F and J_F satisfy

$$J_F^2 = 1, \qquad J_F D_F = D_F J_F, \qquad J_F \gamma_F = -\gamma_F J_F. \tag{34}$$

The operators D_M, $\gamma_M = \gamma_5$, and $J_M = C$ for a compact manifold of dimension 4 satisfy

$$J_M^2 = -1, \qquad J_M D_M = D_M J_M, \qquad J_M \gamma_5 = \gamma_5 J_M. \tag{35}$$

Thus the *KO* dimension of the full noncommutative space $(\mathcal{A}, \mathcal{H}, D)$ with the decorations J and γ included is 10 and satisfies

$$J^2 = -1, \qquad JD = DJ, \qquad J\gamma = -\gamma J. \tag{36}$$

We have shown in [14] that the path integral of the Dirac action, thanks to the relations $J^2 = -1$ and $J\gamma = -\gamma J$, yields a Pfaffian of the operator D instead of its determinant and thus eliminates half the degrees of freedom of Ψ and have the same effect as imposing the condition $J\Psi = \Psi$.

We have also seen that the operator J sends the algebra $\mathcal{A}$ to its commutant, and thus the full algebra acting on the Hilbert space $\mathcal{H}$ is $\mathcal{A} \otimes \mathcal{A}^o$. Under automorphisms of the algebra

$$\Psi \to U\Psi, \tag{37}$$

[1] Due to a typographical error in the abstract of [12] the fermionic representation was listed incorrectly as $(2_R, 2_L, 4)$ while in the body of the paper the correct representation appears.

where $U = u\hat{u}$ with $u \in \mathcal{A}$, $\hat{u} \in \mathcal{A}^o$ with $\left[u, \hat{u}\right] = 0$, it is clear that Dirac action is not invariant. This is similar to the situation in electrodynamics where the Dirac action is not invariant under local phase transformations but the invariance is easily restored by introducing the vector potential A_μ through the transformation

$$\gamma^\mu \partial_\mu \to \gamma^\mu \left(\partial_\mu + ieA_\mu\right). \tag{38}$$

In our case, the Dirac operator D is replaced with

$$D_A = D + A, \tag{39}$$

where the connection A is given by [11]

$$A = \sum a\hat{a}\left[D, b\hat{b}\right]. \tag{40}$$

It can be shown that under automorphisms U of the algebra we have

$$D_A \to UD_AU^*. \tag{41}$$

The connection A splits into three pieces

$$A = A_{(1)} + JA_{(1)}J^{-1} + A_{(2)}, \tag{42}$$

where

$$A_{(1)} = \sum a\left[D, b\right] \tag{43}$$

$$A_{(2)} = \sum \hat{a}\left[A_{(1)}, \hat{b}\right], \tag{44}$$

which satisfies $JA_{(2)}J^{-1} = A_{(2)}$. At this point we have to distinguish few possibilities.

5 Pati-Salam Models

In the first possibility we assume that the double commutator

$$\left[a, \left[D, \hat{b}\right]\right] \neq 0, \tag{45}$$

which implies that $A_{(2)} \neq 0$. The fluctuations A of the inner automorphisms were computed in [12]. The calculation is straightforward and could be easily done using symbolic manipulation programs such as Mathematica or Maple. We shall content ourselves in this paper by collecting some of the important results. Starting with

$a \in M_4(\mathbb{C}) \oplus M_4(\mathbb{C})$ we write

$$a = \begin{pmatrix} X_\alpha^\beta \delta_I^J & 0 \\ 0 & \delta_{\alpha'}^{\beta'} Y_{I'}^{J'} \end{pmatrix}, \tag{46}$$

where $X_\alpha^\beta \in \mathbb{H}_R \oplus \mathbb{H}_L$ and $Y_I^J \in M_4(\mathbb{C})$. Thus we now have

$$X_\alpha^\beta = \begin{pmatrix} X_{\dot a}^{\dot b} & 0 \\ 0 & X_a^b \end{pmatrix}, \qquad X_a^b = \begin{pmatrix} X_1^1 & X_1^2 \\ -\overline{X}_1^2 & \overline{X}_1^1 \end{pmatrix} \in \mathbb{H}_L, \tag{47}$$

and similarly for $X_{\dot a}^{\dot b} \in \mathbb{H}_R$. The anti-linear isometry $J = C\gamma_5 \otimes J_F$ is represented by

$$J_F = \begin{pmatrix} 0 & \delta_\alpha^{\beta'} \delta_I^{J'} \\ \delta_{\alpha'}^\beta \delta_{I'}^J & 0 \end{pmatrix} \times \text{complex conjugation}, \tag{48}$$

and satisfies $J_F^2 = 1$ which implies that $J^2 = -1$. In this form

$$\hat{a} = J a^* J^{-1} = \begin{pmatrix} \delta_\alpha^\beta Y_I^{tJ} & 0 \\ 0 & X_{\alpha'}^{t\beta'} \delta_{I'}^{J'} \end{pmatrix} \tag{49}$$

where the superscript t denotes the transpose matrix. This clearly satisfies the commutation relation

$$\left[a, \hat{b}\right] = 0, \tag{50}$$

which is simply the statement that the right action and left action commute. We shall now show that the relations that D must satisfy greatly constrain its form. The (finite) Dirac operator can be written in matrix form

$$D_F = \begin{pmatrix} D_A^B & D_A^{B'} \\ D_{A'}^B & D_{A'}^{B'} \end{pmatrix}, \tag{51}$$

and must satisfy the properties

$$\gamma_F D_F = -D_F \gamma_F \qquad J_F D_F = D_F J_F, \tag{52}$$

where $J_F^2 = 1$. We also adopt the notation $D_{A'}{}^B = D^{AB}$.

A matrix realization of γ_F and J_F is given by

$$\gamma_F = \begin{pmatrix} G_F & 0 \\ 0 & -\overline{G}_F \end{pmatrix}, \quad G_F = \begin{pmatrix} 1_2 & 0 \\ 0 & -1_2 \end{pmatrix}, \quad J_F = \begin{pmatrix} 0_4 & 1_4 \\ 1_4 & 0_4 \end{pmatrix} \circ \text{cc}. \tag{53}$$

These relations, together with the hermiticity of D imply the relations

$$(D_F)^{B'}_{A'} = \left(\overline{D}_F\right)^{B}_{A} \qquad (D_F)^{B}_{A'} = \left(\overline{D}_F\right)^{A'}_{B}, \tag{54}$$

with the bar denoting complex conjugation. The operator D_F have the following zero components [15]

$$(D_F)^{bJ}_{aI} = 0 = (D_F)^{bJ}_{\dot{a}I} \tag{55}$$

$$(D_F)^{\dot{b}'J'}_{aI} = 0 = (D_F)^{b'J'}_{\dot{a}I}, \tag{56}$$

leaving the components $(D_F)^{\dot{b}J}_{aI}$, $(D_F)^{b'J'}_{aI}$ and $(D_F)^{\dot{b}'J'}_{\dot{a}I}$ arbitrary. These restrictions lead to important constraints on the structure of the connection that appears in the inner fluctuations of the Dirac operator.

We have shown, using elementary algebra, that the components of the connection A which is tensored with the Clifford gamma matrices γ^{μ} are the gauge fields of the Pati-Salam model with the symmetry of $SU(2)_R \times SU(2)_L \times SU(4)$. On the other hand, the non-vanishing components of the connection which is tensored with the gamma matrix γ_5 are given by

$$(A)^{\dot{b}J}_{aI} \equiv \gamma_5 (\Sigma)^{\dot{b}J}_{aI}, \qquad (A)^{b'J'}_{aI} = \gamma_5 H_{aIbJ}, \qquad (A)^{\dot{b}'J'}_{\dot{a}I} \equiv \gamma_5 H_{\dot{a}I\dot{b}J}, \tag{57}$$

where $H_{aIbJ} = H_{bJaI}$ and $H_{\dot{a}I\dot{b}J} = H_{\dot{b}J\dot{a}I}$, which is the most general Higgs structure possible. These correspond to the representations with respect to $SU(2)_R \times SU(2)_L \times SU(4)$ [12]

$$\Sigma^{\dot{b}J}_{aI} = \left(\bar{2}_R, 2_L, 1\right) + \left(\bar{2}_R, 2_L, 15\right) \tag{58}$$

$$H_{aIbJ} = (1_R, 1_L, 6) + (1_R, 3_L, 10) \tag{59}$$

$$H_{\dot{a}I\dot{b}J} = (1_R, 1_L, 6) + (3_R, 1_L, 10). \tag{60}$$

We note, however, that the inner fluctuations form a semi-group and if a component $(D_F)^{\dot{b}J}_{aI}$ or $(D_F)^{b'J'}_{aI}$ or $(D_F)^{\dot{b}'J'}_{\dot{a}I}$ vanish, then the corresponding A field will also vanish. We distinguish three cases: (1) Left-right symmetric Pati–Salam model with fundamental Higgs fields $\Sigma^{\dot{b}J}_{aI}$, H_{aIbJ} and $H_{\dot{a}I\dot{b}J}$. In this model the field H_{aIbJ} should have a zero vev. (2) A Pati-Salam model where the Higgs field H_{aIbJ} that couples to the left sector is set to zero (and then remain zero under fluctuations) which is desirable because there is no symmetry between the left and right sectors at low energies. (3) The initial values for $(D_F)^{\dot{b}J}_{aI}$, $(D_F)^{b'J'}_{aI}$ and $(D_F)^{\dot{b}'J'}_{\dot{a}I}$ before fluctuations are given by those that are determined for the Standard Model, where order one condition is satisfied for the subalgebra, then the Higgs fields $\Sigma^{\dot{b}J}_{aI}$, H_{aIbJ} and $H_{\dot{a}I\dot{b}J}$ will become dependent fields and expressible in terms of more fundamental fields (as will be shown in the next section).

In matrix form the operator D_F has the sub-matrices [15]

$$(D_F)_{\alpha I}^{\beta J} = \begin{pmatrix} 0 & D_{aI}^{\dot{b}J} \\ D_{\dot{a}I}^{bJ} & 0 \end{pmatrix}, \qquad D_{aI}^{\dot{b}J} = \left(D_{\dot{a}J}^{bI}\right)^{*}. \tag{61}$$

Then the components of the Dirac operator tensored with γ^{μ}, including inner fluctuations, is given by [12]

$$(D_A)_{\dot{a}I}^{\dot{b}J} = \gamma^{\mu}\left(D_{\mu}\delta_{\dot{a}}^{\dot{b}}\delta_I^J - \frac{i}{2}g_R W_{\mu R}^{\alpha}\left(\sigma^{\alpha}\right)_{\dot{a}}^{\dot{b}}\delta_I^J - \delta_{\dot{a}}^{\dot{b}}\left(\frac{i}{2}g V_{\mu}^{m}\left(\lambda^{m}\right)_I^{J} + \frac{i}{2}g V_{\mu}\delta_I^J\right)\right) \tag{62}$$

$$(D_A)_{aI}^{bJ} = \gamma^{\mu}\left(D_{\mu}\delta_a^b\delta_I^J - \frac{i}{2}g_L W_{\mu L}^{\alpha}\left(\sigma^{\alpha}\right)_a^b\delta_I^J - \delta_a^b\left(\frac{i}{2}g V_{\mu}^{m}\left(\lambda^{m}\right)_I^{J} + \frac{i}{2}g V_{\mu}\delta_I^J\right)\right), \tag{63}$$

where the fifteen 4×4 matrices $\left(\lambda^{m}\right)_I^{J}$ are traceless and generate the group $SU(4)$ and $W_{\mu R}^{\alpha}$, $W_{\mu L}^{\alpha}$, V_{μ}^{m} are the gauge fields of $SU(2)_R$, $SU(2)_L$, and $SU(4)$. The requirement that A is unimodular implies that

$$\mathrm{Tr}\,(A) = 0, \tag{64}$$

which gives the condition

$$V_{\mu} = 0. \tag{65}$$

This shows that the resulting gauge group is $SU(2)_R \times SU(2)_L \times SU(4)$, which is the Pati-Salam gauge symmetry. In addition we have for the components of the Dirac operator tensored with γ_5,

$$(D_A)_{\dot{a}I}^{bJ} = \gamma_5\Sigma_{\dot{a}I}^{bJ} \tag{66}$$

$$(D_A)_{\dot{a}I}^{\dot{b}'J'} = \gamma_5 H_{\dot{a}I\dot{b}J} \tag{67}$$

$$(D_A)_{aI}^{b'J'} = \gamma_5 H_{aIbJ}, \tag{68}$$

where $\Sigma_{\dot{a}I}^{bJ}$ is in the $(2_R, 2_L, 1+15)$ representation, $H_{\dot{a}I\dot{b}J} = H_{\dot{b}J\dot{a}I}$ is in the $(3_R, 1_L, 10) + (1_R, 1_L, 6)$ representation and H_{aIbJ} is in the $(1_R, 1_L, 6) + (1_R, 3_L, 10)$ with respect to $SU(2)_R \times SU(2)_L \times SU(4)$. To conclude, there are only three Pati-Salam models with fixed Higgs structure, where the first one is the most general case, and the other two are special cases of the first one.

6 The Standard Model

We now consider the situation when the order one condition is satisfied

$$\left[a, \left[D, \hat{b}\right]\right] = 0, \tag{69}$$

and the center of the algebra $Z(\mathcal{A})$ is non-trivial in such a way that the space is connected. Physically, this means that there is a mixing term between the fermions and their conjugates. The Dirac operator connects the spinors ψ_A and their conjugates $\psi_{A'}$ so that

$$[D, Z(\mathcal{A})] \neq 0. \tag{70}$$

In physical terms this would allow a Majorana mass term for the fermions. It was shown in [10] that the unique solution to this equation constrains the algebra $\mathcal{A}_F = \mathbb{H}_R \oplus \mathbb{H}_L \oplus M_4(\mathbb{C})$ to be restricted to a subalgebra

$$\mathbb{C} \oplus \mathbb{H}_L \oplus M_3(\mathbb{C}), \tag{71}$$

so that an element of $\mathcal{A}$ takes the form [15]

$$a = \begin{pmatrix} X \otimes 1_4 & & & & \\ & \overline{X} \otimes 1_4 & & & \\ & & q \otimes 1_4 & & \\ & & & 1_4 \otimes X & \\ & & & & 1_4 \otimes m \end{pmatrix}, \tag{72}$$

where $X \in \mathbb{C}, \quad q \in \mathbb{H}, \quad m \in M_3(\mathbb{C})$ and the operator D_F have a singlet non-zero entry in the mixing term $(D_F)_A^{A'}$

$$\begin{aligned} (D_F)_{\alpha I}^{\beta J} &= \left(\delta_\alpha^1 \delta_{\dot{1}}^\beta k^{*\nu} + \delta_\alpha^{\dot{1}} \delta_1^\beta k^\nu + \delta_\alpha^2 \delta_{\dot{2}}^\beta k^{*e} + \delta_\alpha^{\dot{2}} \delta_2^\beta k^e\right) \delta_I^1 \delta_1^J \\ &\quad + \left(\delta_\alpha^1 \delta_{\dot{1}}^\beta k^{*u} + \delta_\alpha^{\dot{1}} \delta_1^\beta k^u + \delta_\alpha^2 \delta_{\dot{2}}^\beta k^{*d} + \delta_\alpha^{\dot{2}} \delta_2^\beta k^d\right) \delta_I^i \delta_j^J \delta_i^j \end{aligned} \tag{73}$$

$$(D_F)_{\alpha I}^{\beta' K'} = \delta_\alpha^{\dot{1}} \delta_{\dot{1}'}^{\beta'} \delta_I^1 \delta_{1'}^{K'} k^{*\nu_R} \sigma, \tag{74}$$

where k^ν, k^e, k^u, k^d and k^{ν_R} are 3×3 Yukawa couplings in generation space. The field σ is a singlet (which could be complex) whose vev is responsible for the right-handed neutrino Majorana mass. The operator D must be replaced with the operator

$$D_A = D + A + JAJ^{-1}, \tag{75}$$

and

$$A_{(2)} = 0, \tag{76}$$

which greatly simplifies the Higgs structure. The various components of the Dirac operator are exactly those of the Standard Model, in addition to the Higgs fields which are the components of the connection A along discrete directions

$$
\begin{aligned}
&(D)^{\dot{1}1}_{\dot{1}1} = \gamma^{\mu} \otimes D_{\mu} \otimes 1_3, \quad D_{\mu} = \partial_{\mu} + \frac{1}{4}\omega^{cd}_{\mu}(e)\,\gamma_{cd}, \quad 1_3 = \text{generations} \\
&(D)^{a1}_{\dot{1}1} = \gamma_5 \otimes k^{*\nu} \otimes \epsilon^{ab} H_b \qquad k^{\nu} = 3 \times 3 \text{ neutrino mixing matrix} \\
&(D)^{\dot{2}1}_{\dot{2}1} = \gamma^{\mu} \otimes \left(D_{\mu} + i g_1 B_{\mu}\right) \otimes 1_3 \\
&(D)^{a1}_{\dot{2}1} = \gamma_5 \otimes k^{*e} \otimes \overline{H}^{a} \\
&(D)^{\dot{1}1}_{a1} = \gamma_5 \otimes k^{\nu} \otimes \epsilon_{ab} \overline{H}^{b} \\
&(D)^{\dot{2}1}_{a1} = \gamma_5 \otimes k^{e} \otimes H_a \\
&(D)^{b1}_{a1} = \gamma^{\mu} \otimes \left(\left(D_{\mu} + \frac{i}{2} g_1 B_{\mu}\right)\delta^{b}_{a} - \frac{i}{2} g_2 W^{\alpha}_{\mu} \left(\sigma^{\alpha}\right)^{b}_{a}\right) \otimes 1_3, \qquad \sigma^{\alpha} = \text{Pauli} \\
&(D)^{\dot{1}j}_{\dot{1}i} = \gamma^{\mu} \otimes \left(\left(D_{\mu} - \frac{2i}{3} g_1 B_{\mu}\right)\delta^{j}_{i} - \frac{i}{2} g_3 V^{m}_{\mu} \left(\lambda^{m}\right)^{j}_{i}\right) \otimes 1_3, \qquad \lambda^{i} = \text{Gell-Mann} \\
&(D)^{aj}_{\dot{1}i} = \gamma_5 \otimes k^{*u} \otimes \epsilon^{ab} H_b \delta^{j}_{i} \\
&(D)^{\dot{2}j}_{\dot{2}i} = \gamma^{\mu} \otimes \left(\left(D_{\mu} + \frac{i}{3} g_1 B_{\mu}\right)\delta^{j}_{i} - \frac{i}{2} g_3 V^{m}_{\mu} \left(\lambda^{m}\right)^{j}_{i}\right) \otimes 1_3 \\
&(D)^{aj}_{\dot{2}i} = \gamma_5 \otimes k^{*d} \otimes \overline{H}^{a} \delta^{j}_{i} \\
&(D)^{bj}_{ai} = \gamma^{\mu} \otimes \left(\left(D_{\mu} - \frac{i}{6} g_1 B_{\mu}\right)\delta^{b}_{a}\delta^{j}_{i} - \frac{i}{2} g_2 W^{\alpha}_{\mu} \left(\sigma^{\alpha}\right)^{b}_{a} \delta^{j}_{i} - \frac{i}{2} g_3 V^{m}_{\mu} \left(\lambda^{m}\right)^{j}_{i} \delta^{b}_{a}\right) \otimes 1_3 \\
&(D)^{\dot{1}j}_{ai} = \gamma_5 \otimes k^{u} \otimes \epsilon_{ab} \overline{H}^{b} \delta^{j}_{i} \\
&(D)^{\dot{2}j}_{ai} = \gamma_5 \otimes k^{d} \otimes H_a \delta^{j}_{i} \\
&(D)^{\dot{1}'1'}_{\dot{1}1} = \gamma_5 \otimes k^{*\nu_R} \sigma \qquad \text{generate scale } M_R \text{ by } \sigma \to M_R \\
&(D)^{\dot{1}1}_{\dot{1}'1'} = \gamma_5 \otimes k^{\nu_R} \sigma \\
&D^{B'}_{A'} = \overline{D}^{B}_{A}, \qquad D^{B}_{A'} = \overline{D}^{B'}_{A}, \qquad D^{B'}_{A} = \overline{D}^{B}_{A'}
\end{aligned}
$$

where in this notation the fermions are enumerated as

$$\psi_{\dot{1}1} = \nu_R \tag{77}$$

$$\psi_{\dot{2}1} = e_R \tag{78}$$

$$\psi_{a1} = l_a = \begin{pmatrix} \nu_L \\ e_L \end{pmatrix} \tag{79}$$

$$\psi_{\dot{1}i} = u_{iR} \tag{80}$$

$$\psi_{\dot{2}i} = d_{iR} \tag{81}$$

$$\psi_{ai} = q_{ia} = \begin{pmatrix} u_{iL} \\ d_{iL} \end{pmatrix}. \tag{82}$$

It is clear that the associated gauge group is $U(1) \times SU(2) \times SU(3)$ and that there is only one Higgs doublet H. We note the presence of the singlet field σ which is the field whose vev will give a Majorana mass to the right-handed neutrinos. This field plays an essential role in stabilizing the Higgs coupling so that it does not turn negative at very high energies [21]. We note in passing that the number of generations is inserted by hand in the Dirac operator of the finite space, and at present we do not have any geometrical explanation to single out three generations.

7 A Special Pati-Salam Model

We have shown that inner fluctuations resulting from the action on operators in Hilbert space form a semi-group Pert$(\mathcal{A})$. There exists configurations for which the inverse transformation to the perturbation does not exist. One such Dirac operator D_A corresponds to the case where the initial operator D is taken to be the one deduced for the Standard Model as given in (73) and (74), but not restricting its action to the subalgebra $\mathbb{C} \oplus \mathbb{H}_L \oplus M_3(\mathbb{C})$ but to the full algebra $\mathbb{H}_R \oplus \mathbb{H}_L \oplus M_4(\mathbb{C})$. In this case one finds out that the resultant vector fields are the same as in the case of Pati-Salam models, but where the Higgs fields $\Sigma_{\dot{a}I}^{bJ}$ and $H_{\dot{a}I\dot{b}J}$ become composite fields determined in function of fundamental Higgs fields while H_{aIbJ} vanishes. These are given by [12]

$$\Sigma_{\dot{a}I}^{bJ} = \left(\left(k^{\nu}\phi_{\dot{a}}^{b} + k^{e}\tilde{\phi}_{\dot{a}}^{b}\right)\Sigma_{I}^{J} + \left(k^{u}\phi_{\dot{a}}^{b} + k^{d}\tilde{\phi}_{\dot{a}}^{b}\right)\left(\delta_{I}^{J} - \Sigma_{I}^{J}\right)\right) \tag{83}$$

$$H_{\dot{a}I\dot{b}J} = k^{*\nu_R}\Delta_{\dot{a}J}\Delta_{\dot{b}I} \tag{84}$$

$$H_{aIbJ} = 0, \tag{85}$$

where the Higgs field $\phi_{\dot{a}}^{b}$ is in the $\left(2_R, \bar{2}_L, 1\right)$ of the product gauge group $SU(2)_R \times SU(2)_L \times SU(4)$, $\tilde{\phi}_{\dot{a}}^{b} = \tau_2\bar{\phi}_{\dot{a}}^{b}\tau_2$ and $\Delta_{\dot{a}J}$ is in the $\left(2_{R,}, 1_L, 4\right)$ representation while Σ_I^J is in the $(1_R, 1_L, 1+15)$ representation. The fact that one gets a simpler Higgs representations in this case makes it more attractive. It is certainly an interesting question to determine all Dirac operators which lead to singular transformations where the resultant Higgs fields are composites of more fundamental ones. The scalar potential which contains quartic interactions in the bosonic fields, which because of compositness, are of order 8. All terms of orders higher than four will be suppressed by the cut-off scale and could be truncated. Similarly the coupling of such terms to the fermionic fields will be suppressed by the cut-off scale. To conclude this section, it is remarkable that starting with the simple quantization condition which represents the Chern-character of the noncommutative space and is a special case of the orientability condition, fixes uniquely the structure of space-time as well as the matter content in the form of a very specific Pati-Salam unification model, or three of its truncations, including the Standard Model. This enables us to track gravitational and matter interactions, starting from the Planck scale where the starting point is few spheres of Planck size, and ending up with the present scale. This compelling picture could represent a valid framework for the realization of Hilbert's program for axiomatization of physics.

8 Spectral Action

The coordinates $Y^A(x)$ are topological fields, and apart from being coordinates of a sphere and satisfying the volume quantization condition, are not constrained. They do play a role serving as coordinates conjugate to the momentum represented by the Dirac operator. In particular, since now D and Y play the role of momenta and coordinates, it is natural to consider the spectral action to be of the form [10]

$$\mathrm{Tr} f\left(D_A, Y\right),$$

which, because $Y^2 = 1$, implies the dependence on terms of the form $[D, Y]$. The lowest order contribution of such terms come from $[D, Y]^2$ which corresponds to adding the following term to the action

$$\frac{1}{2}\int_M d^4x\sqrt{g}g^{\mu\nu}\partial_\mu Y^A \partial_\nu Y^A. \tag{86}$$

It is also clear that in the case of the two sided quantization with the field Z the contribution of the term $[D, Z]^2$ gives the sum of two contributions without interference terms

$$\frac{1}{2}\int_M d^4x\sqrt{g}g^{\mu\nu}\left(\partial_\mu Y^A \partial_\nu Y^A + \partial_\mu Y'^A \partial_\nu Y'^A\right).$$

We have shown that the spectral action for the part dependent on D_A gives the bosonic action for all dynamical fields appearing in the connection A. In particular, in the case of the Standard Model the bosonic action for the part independent of the fields Y^A and Y'^A is given by [13–15, 22]

$$S = 2f_4\Lambda^4 a_0 + 2f_2\Lambda^2 a_2 + f_0 a_4 + \cdots \tag{87}$$

and

$$\begin{aligned} S_b = {} & \frac{48}{\pi^2} f_4 \Lambda^4 \int d^4x\sqrt{g} \\ & - \frac{4}{\pi^2} f_2\Lambda^2 \int d^4x \sqrt{g}\left(R + \frac{1}{2}a\overline{H}H + \frac{1}{4}c\sigma^2\right) \end{aligned} \tag{88}$$

$$+\frac{1}{2\pi^2}f_0\int d^4x\sqrt{g}\left[\frac{1}{30}\left(-18C^2_{\mu\nu\rho\sigma}+11R^*R^*\right)\right.$$
$$+\frac{5}{3}g_1^2B^2_{\mu\nu}+g_2^2\left(W^\alpha_{\mu\nu}\right)^2+g_3^2\left(V^m_{\mu\nu}\right)^2$$
$$+\frac{1}{6}aR\overline{H}H+b\left(\overline{H}H\right)^2+a\left|\nabla_\mu H_a\right|^2$$
$$\left.+2e\overline{H}H\,\sigma^2+\frac{1}{2}d\,\sigma^4+\frac{1}{12}cR\sigma^2+\frac{1}{2}c\left(\partial_\mu\sigma\right)^2\right]$$
$$+\cdots$$

where a, c, d, e are defined in terms of the Yukawa couplings, $f_0 = f(0)$ and f_k are the Mellin transforms of the function f

$$f_k = \int_0^\infty f(v)\, v^{k-1}dv, \qquad k > 0. \tag{89}$$

This action is calculated using heat kernel methods and was shown to contain unification of gravity with gauge symmetries and Higgs field and the scalar singlet. All couplings are related at unification scale. The zeroth order term in the expansion gives the cosmological constant, the first order gives the Einstein-Hilbert action and the scalar masses, and the second order gives the Yang-Mills and scalar kinetic terms as well as the second order in curvature terms. The presence of the singlet field σ whose vev gives mass to the right-handed neutrino plays an important role in stabilizing the Higgs coupling which will not become negative at very high energies as well as being consistent with a low Higgs mass of 126 Gev [21]. The form of the gauge and Higgs kinetic terms and potential implies unification of the gauge couplings and the Higgs coupling. In addition there is a relation between the fermion masses and the gauge field masses. A study of the RGE showed that these relations are consistent with present experimental data and predicts the top quark mass to be around 170 Gev. However, gauge coupling unification is off by 4% indicating that the Standard Model is an excellent approximation to a Pati-Salam model listed above. We have shown [16] that gauge coupling unification is indeed possible for Pati-Salam models at a unification scale of the order of 10^{16} Gev.

It is also worthwhile to summarize the fermionic action

$$S_f = \int d^4x\sqrt{g}\left(\nu_R^*\gamma^\mu D_\mu\nu_R\right. \tag{90}$$
$$+\,e_R^*\gamma^\mu\left(D_\mu+ig_1B_\mu\right)e_R$$
$$+\,l_L^{a*}\gamma^\mu\left(\left(D_\mu+\frac{i}{2}g_1B_\mu\right)\delta_a^b-\frac{i}{2}g_2W_\mu^\alpha\left(\sigma^\alpha\right)_a^b\right)l_{bL}$$

$$+ u_R^{i*}\gamma^\mu\left(\left(D_\mu - \frac{2i}{3}g_1 B_\mu\right)\delta_i^j - \frac{i}{2}g_3 V_\mu^m\left(\lambda^m\right)_i^j\right)u_{jR}$$
$$+ d_R^{i*}\gamma^\mu\left(\left(D_\mu + \frac{i}{3}g_1 B_\mu\right)\delta_i^j - \frac{i}{2}g_3 V_\mu^m\left(\lambda^m\right)_i^j\right)d_{jR}$$
$$+ q_L^{ia*}\gamma^\mu\left(\left(D_\mu - \frac{i}{6}g_1 B_\mu\right)\delta_a^b\delta_i^j - \frac{i}{2}g_2 W_\mu^\alpha\left(\sigma^\alpha\right)_a^b\delta_i^j - \frac{i}{2}g_3 V_\mu^m\left(\lambda^m\right)_i^j\delta_a^b\right)q_{jbL}$$
$$+ \nu_R^*\gamma_5 k^{*\nu}\epsilon^{ab}H_b l_{aL} + e_R^*\gamma_5 k^{*e}\overline{H}^a l_{aL}$$
$$+u_R^{i*}\gamma_5 k^{*u}\epsilon^{ab}H_b\delta_i^j q_{jaL} + d_R^{i*}\gamma_5 k^{*d}\overline{H}^a\delta_i^j q_{jaL} + \nu_R^*\gamma_5 k^{*\nu_R}\sigma\left(\nu_R^*\right)^c + \text{h.c}\Big).$$

Note that the singlet field σ after getting a vev from the minima of its potential, will give a Majorana mass to the right-handed neutrino and implies that the left handed neutrino will have a small mass through a see-saw mechanism.

9 Consequences of Volume Quantization

Having established the importance of the volume quantization condition, which in turn implies that the two sets of fields Y and Y' mapping the four dimensional manifold to four spheres must be taken into consideration when studying the dynamical content of the resulting model. In particular, the Einstein equations of motion will be modified. The volume constraint, imposed through a Lagrange multiplier, will result in traceless Einstein equations, with the trace part equated to the Lagrange multiplier. We will show that Bianchi identities give rise to a cosmological constant as an integration constant. We now study the implications of the presence of the fields Y and Y' on the structure of the model.

For simplicity and to avoid cluttering of fields and indices, in what follows we shall consider only one set of fields Y^A and not two sets Y^A and Y'^A as required by the reality condition. The effects on the equations of motion will be minimal. Here we take $Y \in M_2(\mathbb{H})$ a 2×2 matrix whose elements are quaternions. This can be written as

$$Y = Y^A\Gamma_A, \qquad A = 1,\ldots,5, \tag{91}$$

where Γ_A are Hermitian gamma matrices satisfying $\left\{\Gamma^A,\Gamma^B\right\} = 2\delta^{AB}$ where Cliff $(+,+,+,+,+) = M_2(\mathbb{H})\oplus M_2(\mathbb{H})$ and we take one of the irreducible representations $M_2(\mathbb{H})$. The condition $Y^2 = 1$ implies

$$Y^AY^A = 1, \tag{92}$$

which defines coordinates on the four dimensional sphere S^4. We can check that

$$\frac{1}{2^2(4!)}\left\langle Y[D,Y][D,Y][D,Y][D,Y]\right\rangle = \gamma, \tag{93}$$

implies the relation

$$\det\left(e_{\mu}^{a}\right)=\frac{1}{4!}\epsilon^{\mu\nu\kappa\lambda}\epsilon_{ABCDE}Y^{A}\partial_{\mu}Y^{B}\partial_{\nu}Y^{C}\partial_{\kappa}Y^{D}\partial_{\lambda}Y^{E}, \tag{94}$$

which fixes the volume density and whose integral quantizes the volume. This last condition can be imposed through a Lagrange multiplier. To do this consider the action

$$\begin{aligned}I=&-\frac{1}{2\kappa^{2}}\int d^{4}x\sqrt{g}R+\frac{1}{2}\int d^{4}x\lambda\left(\frac{1}{\kappa^{4}}\sqrt{g}-\frac{1}{4!}\epsilon^{\mu\nu\kappa\lambda}\epsilon_{ABCDE}Y^{A}\partial_{\mu}Y^{B}\partial_{\nu}Y^{C}\partial_{\kappa}Y^{D}\partial_{\lambda}Y^{E}\right)\\&+\frac{1}{2\kappa^{4}}\int d^{4}x\sqrt{g}\lambda'\left(Y^{A}Y^{A}-1\right),\end{aligned} \tag{95}$$

where $\kappa^{2}=8\pi G$ which will be set to 1. Notice that the third term is a four-form and represents the volume element of a unit four-sphere. It can be written in terms of differential forms without any tensor indices

$$-\frac{1}{2(4!)}\int\lambda\epsilon_{ABCDE}Y^{A}dY^{B}\wedge dY^{C}\wedge dY^{D}\wedge dY^{E} \tag{96}$$

$$=-\frac{1}{8(4!)}\int\lambda\mathrm{Tr}\left(YdY\wedge dY\wedge dY\wedge dY\right), \tag{97}$$

and is independent of the variation of the metric. Varying the action with respect to the metric, after imposing the two Lagrange multipliers constraints

$$Y^{A}Y^{A}=1 \tag{98}$$

$$\sqrt{g}=\frac{1}{4!}\epsilon^{\mu\nu\kappa\lambda}\epsilon_{ABCDE}Y^{A}\partial_{\mu}Y^{B}\partial_{\nu}Y^{C}\partial_{\kappa}Y^{D}\partial_{\lambda}Y^{E}, \tag{99}$$

gives

$$G_{\mu\nu}+\frac{1}{2}g_{\mu\nu}\lambda=0. \tag{100}$$

Tracing it with $g^{\mu\nu}$ then gives

$$\lambda=-\frac{1}{2}G, \tag{101}$$

which when substituted back yields the traceless Einstein equation

$$G_{\mu\nu}-\frac{1}{4}g_{\mu\nu}G=0. \tag{102}$$

Applying the Bianchi identity to this equation implies

$$\partial_{\mu}G=0=\partial_{\mu}\lambda, \tag{103}$$

and thus

$$\lambda = -4\Lambda \tag{104}$$

$$G = 4\Lambda, \tag{105}$$

where Λ is the cosmological constant arising as an integrating constant [23]. Therefore we see that an added benefit of having the quantization condition is that the cosmological constant now appears as an integrating constant in the equations of motion and is not necessary to be present in the action. This result is similar to the one encountered in unimodular gravity, with a major difference that in our case the diffeomorphism symmetry is not restricted but only the volume is quantized with all symmetries being intact.

Next, varying the fields Y^A gives (using $\partial_\mu \lambda = 0$)

$$-\frac{5}{2(4!)}\lambda \epsilon^{\mu\nu\kappa\lambda}\epsilon_{ABCDE}\partial_\mu Y^B \partial_\nu Y^C \partial_\kappa Y^D \partial_\lambda Y^E + \lambda' Y_A \sqrt{g} = 0. \tag{106}$$

Tracing this equation with Y^A gives

$$\lambda' = \frac{5}{2}\lambda = -\frac{5}{4}G. \tag{107}$$

Assuming that $G \neq 0$ (the case $G = 0$ recovers the full set of Einstein equations without cosmological constant), we further have

$$Y_A = \frac{1}{4!}\frac{1}{\sqrt{g}}\epsilon^{\mu\nu\kappa\lambda}\epsilon_{ABCDE}\partial_\mu Y^B \partial_\nu Y^C \partial_\kappa Y^D \partial_\lambda Y^E, \tag{108}$$

which implies the equation

$$\epsilon^{\mu\nu\kappa\lambda}\epsilon_{A'BCDE}\partial_\mu Y^B \partial_\nu Y^C \partial_\kappa Y^D \partial_\lambda Y^E \left(\delta_A^{A'} - Y_A Y^{A'}\right) = 0. \tag{109}$$

Note that the expression

$$\frac{3}{8\pi^2}\frac{1}{4!}\int_{S^4} d^4x \epsilon^{\mu\nu\kappa\lambda}\epsilon_{ABCDE} Y^A \partial_\mu Y^B \partial_\nu Y^C \partial_\kappa Y^D \partial_\lambda Y^E = \pi_4\left(S^4\right) \tag{110}$$

$$= \mathbb{Z}, \tag{111}$$

is the winding of the sphere S^4 ($\pi_n\left(S^n\right) = \mathbb{Z}$ [6, 24]). Thus

$$\int_M \sqrt{g} d^4x = N\left(\frac{8\pi^2}{3}\right), \tag{112}$$

where N is the winding number of the mapping $M_4 \to S^4$ [25, 26]. We can easily see that the Y^A equation of motion (108) follows from Eq. (99) and does not give any new information because it appears through a topological term. To see this use the identity resulting from anti-symmetrizing six indices taking five values,

$$\begin{aligned} 0 &= Y_{[A}\,\epsilon_{A'BCDE]}\epsilon^{\mu\nu\kappa\lambda}\partial_\mu Y^B\partial_\nu Y^C\partial_\kappa Y^D\partial_\lambda Y^E \\ &= \left(Y_A\epsilon_{A'BCDE} - Y_{A'}\epsilon_{ABCDE} - 4Y_B\epsilon_{AA'CDE}\right)\epsilon^{\mu\nu\kappa\lambda}\partial_\mu Y^B\partial_\nu Y^C\partial_\kappa Y^D\partial_\lambda Y^E, \end{aligned} \tag{113}$$

which, after using the property $Y_B\partial_\mu Y^B = 0$ and Eq. (99), implies Eq. (108).

10 Solitonic Solution

We have seen that if we consider the spectral action to be of the form $\mathrm{Tr}\, f\,(D, Y)$, it will then contain the kinetic term

$$\frac{1}{2}\int_M d^4x\sqrt{g}g^{\mu\nu}\partial_\mu Y^A\partial_\nu Y^A. \tag{114}$$

Including this term in the action gives the modified Einstein equations

$$G_{\mu\nu} + \frac{1}{2}g_{\mu\nu}\lambda = \partial_\mu Y^A\partial_\nu Y^A - \frac{1}{2}g_{\mu\nu}\left(\partial Y\cdot\partial Y\right), \tag{115}$$

where we have denoted by $\partial Y\cdot\partial Y = g^{\kappa\lambda}\partial_\kappa Y^A\partial_\lambda Y^A$. Taking the trace of this equation determines λ :

$$\lambda = -\frac{1}{2}\left(G + \partial Y\cdot\partial Y\right), \tag{116}$$

and when this is plugged back into Eq. (115) it gives two equations, the first of which is traceless

$$G_{\mu\nu} - \frac{1}{4}g_{\mu\nu}G = \partial_\mu Y^A\partial_\nu Y^A - \frac{1}{4}g_{\mu\nu}\left(\partial Y\cdot\partial Y\right). \tag{117}$$

Taking covariant derivative of Eq. (115) using Bianchi identity, gives

$$\frac{1}{2}\partial_\mu\lambda = \partial_\mu Y^A\Box Y^A, \tag{118}$$

where $\Box Y^A = g^{\mu\nu}\nabla_\mu\partial_\nu Y^A$ and after making use of the identity $Y^A\Box Y^A = -\partial Y\cdot\partial Y$ that follows by differentiating $Y^A\partial_\mu Y^A = 0$. We now examine the Y^A equation

$$- \frac{5}{2(4!)} \lambda \epsilon^{\mu\nu\kappa\lambda} \epsilon_{ABCDE} \partial_\mu Y^B \partial_\nu Y^C \partial_\kappa Y^D \partial_\lambda Y^E + \lambda' Y_A \sqrt{g}$$
$$= \sqrt{g} \Box Y^A + \frac{1}{12} \epsilon^{\mu\nu\kappa\lambda} \epsilon_{ABCDE} \partial_\mu \lambda Y^B \partial_\nu Y^C \partial_\kappa Y^D \partial_\lambda Y^E. \quad (119)$$

Tracing with Y^A gives

$$\lambda' = \frac{5}{2} \lambda + Y^A \Box Y^A. \quad (120)$$

Plugging this back and using Eq. (108) gives

$$\Box Y^A - Y^A \left(Y^B \Box Y^B \right) = - \frac{1}{12\sqrt{g}} \epsilon^{\mu\nu\kappa\lambda} \epsilon_{ABCDE} \partial_\mu \lambda Y^B \partial_\nu Y^C \partial_\kappa Y^D \partial_\lambda Y^E. \quad (121)$$

The left-hand side of Eq. (118) is a total derivative, while the right-hand side is not. The general solution of Eqs. (118) and (121) is not easy to find. We shall restrict ourselves to the subspace where

$$\partial_\mu \lambda = 0,$$

so that

$$G + g^{\mu\nu} \partial_\mu Y^A \partial_\nu Y^A = 4\Lambda.$$

Equation (121) then simplifies to

$$\Box Y^A - Y^A \left(Y^B \Box Y^B \right) = 0. \quad (122)$$

This equation, being traceless, could be recast in terms of the dependent variables Y^a, $a = 1, \ldots 4$, substituting the relation $Y^5 = \sqrt{1 - Y^a Y^a}$ so that the kinetic term $g^{\mu\nu} \partial_\mu Y^A \partial_\nu Y^A$ takes the form

$$g^{\mu\nu} \partial_\mu Y^a \partial_\nu Y^b h_{ab}, \quad (123)$$

where

$$h_{ab} = \left(\delta_{ab} + \frac{Y_a Y_b}{1 - Y^c Y^c} \right). \quad (124)$$

The Eq. (122) then takes the form [27]

$$g^{\mu\nu} \left(\nabla_\mu \partial_\nu Y^a + \partial_\mu Y^b \partial_\nu Y^c \Gamma^a_{bc} \right) = 0, \quad (125)$$

where Γ^a_{bc} is the Christoffel connection of the metric h_{ab} on the sphere S^4 which is given by

$$\Gamma^a_{bc} = h_{bc} Y^a. \quad (126)$$

This shows that the fields Y^a are harmonic maps which shows that maps from the four-manifolds M_4 to S^4 satisfying the equations of motion are harmonic. We conclude that the equations of motion are identical to those of the $O(5)$ non-linear sigma model, which is also equivalent to the Projective quaternionic model HP^1 [28, 29]. These works have derived the instanton solution (for a conformally flat metric) with $N = 1$ and the multi-instanton solution $N = n$.

First for the $N = 1$ instanton solution we have

$$g_{\mu\nu} = \delta_{\mu\nu}\frac{1}{\left(1+x^2\right)^2}, \quad x^2 = x^a x^a, \qquad a = 1, \ldots, 4 \tag{127}$$

$$Y^a = \frac{2x^a}{1+x^2}, \quad Y^5 = \frac{x^2-1}{1+x^2}, \tag{128}$$

which satisfies

$$R_{\mu\nu} = \frac{1}{4}g_{\mu\nu}R, \quad R = 48. \tag{129}$$

The multi-instanton solution is given by

$$g_{\mu\nu} = 2\left(\partial_\mu x^n \partial_\nu \overline{x}^n + \partial_\nu x^n \partial_\mu \overline{x}^n\right)\frac{1}{\left(1+x^n\overline{x}^n\right)^2}, \tag{130}$$

where x is a quaternionic coordinate

$$x = x^4 1 + e_i x^i, \tag{131}$$

where e_i, $i = 1, 2, 3$ are the three quaternionic complex structures $e_1^2 = e_2^2 = e_3^2 = -1$ and $e_1 e_2 = -e_2 e_1 = e_3$. We also have

$$Y = \frac{2x^n}{\left(1+x^n\overline{x}^n\right)} = Y^4 1 + e^i Y^i \tag{132}$$

$$Y^5 = \frac{\left(x^n\overline{x}^n - 1\right)}{\left(x^n\overline{x}^n + 1\right)}. \tag{133}$$

This solution gives a winding number n.

11 Three Dimensional Volume Quantization

Up to this point we have been dealing with compact manifolds. Physical space-time has a Lorentzian signature, and is thus topologically equivalent to $\mathbb{R} \times M_3$.

Alternatively, we can envision the following picture. Consider as a starting point any three dimensional hypersurface Σ_3 whose normals at any point has time-like directions and with a family of geodesic lines normal to the hypersurface. Let these lines be time coordinates and set t to be the distance as measured from the initial hypersurface. Denote by y^i, $i = 1, 2, 3$ as the coordinates on the hypersurface Σ. There will still be arbitrary coordinate transformations $x^\alpha = x^\alpha\left(y^i\right)$, $\alpha = 1, 2, 3$. Denote the four coordinates by $x^\mu = (t, x^\alpha)$ and define the functions [30]

$$e_i^\mu = \frac{\partial x^\mu}{\partial y^i}, \tag{134}$$

and the corresponding normal vectors n_μ such that

$$n_\mu e_i^\mu = 0. \tag{135}$$

The inverse functions e_μ^i are defined with the aid of the vectors n_μ so that

$$e_i^\mu e_\mu^j = \delta_i^j, \qquad e_i^\mu e_\nu^i = \delta_\nu^\mu - n^\mu n_\nu, \tag{136}$$

where the vectors n_μ satisfy

$$n_\mu n^\mu = \varepsilon, \tag{137}$$

where $\varepsilon = 1$ for metric with signature $(+, +, +, +)$ and $\varepsilon = -1$ for signature $(-, +, +, +)$. The metric on the four-dimensional manifold generated due to the motion of the three dimensional hypersurface is then given by

$$g_{\mu\nu} = e_\mu^i h_{ij} e_\nu^j + \varepsilon n_\mu n_\nu, \tag{138}$$

where h_{ij} is the metric on the three dimensional hypersurface Σ. The inverse metric is given by

$$g^{\mu\nu} = e_i^\mu h^{ij} e_j^\nu + \varepsilon n^\mu n^\nu, \tag{139}$$

where h^{ij} is the inverse metric of h_{ij} which implies that

$$n^\mu = g^{\mu\nu} n_\nu, \qquad n^\mu e_\mu^i = 0. \tag{140}$$

For simplicity we can chose the gauge where

$$e_i^t = \frac{\partial t}{\partial y^i} = 0, \tag{141}$$

which implies that

$$n_\alpha = 0. \tag{142}$$

Denoting

$$n_t = N, \qquad e_t^i = N^i, \tag{143}$$

the components of the metric $g_{\mu\nu}$ will be given by

$$g_{tt} = \varepsilon N^2 + N^\alpha h_{\alpha\beta} N^\beta, \qquad g_{t\alpha} = N_\alpha, \qquad g_{\alpha\beta} = h_{\alpha\beta}, \tag{144}$$

where

$$h_{\alpha\beta} = e^i_\alpha h_{ij} e^j_\beta, \qquad N_\alpha = e^i_\alpha h_{ij} N^j. \tag{145}$$

In particular, the vector n^μ is given by

$$n^\mu = \left(\frac{1}{N}, -\frac{N^\alpha}{N} \right). \tag{146}$$

This gives the familiar 3 + 1 ADM splitting of the metric [31]

$$ds^2 = h_{\alpha\beta} \left(dx^\alpha + N^\alpha dt \right) \left(dx^\beta + N^\beta dt \right) + \varepsilon N^2 dt^2. \tag{147}$$

At this point we note that for the three dimensional hypersurface Σ_3 we will utilize the two maps Y and Y' from Σ to the three sphere S^3, which are defined with respect to the Clifford algebras Cliff $(+,+,+,+) = M_2(\mathbb{H})$ and Cliff $(-,-,-,-) = M_2(\mathbb{H})$ where

$$Y = Y^a \Gamma_a, \qquad Y' = iY'^a \Gamma'_a, \qquad a = 1, \ldots, 4, \tag{148}$$

where

$$\{\Gamma_a, \Gamma_b\} = 2\delta_{ab}, \qquad \left\{\Gamma'_a, \Gamma'_b\right\} = -2\delta_{ab}, \tag{149}$$

and $Y^2 = 1,\ Y'^2 = 1$. In reality, we can consider the mappings from the moving hypersurfaces Σ_3 which generate the four dimensional manifold and thus we have $Y^a(x^\mu)$ and $Y'^a(x^\mu)$. These could be extended by the field $X(x^\mu)$ which maps the geodesics normal to Σ_3 into $\mathbb{R}$. We can then consider the field X to be measure of the distance

$$X = \sqrt{g_{\mu\nu} dx^\mu dx^\nu}, \tag{150}$$

which according to the Hamilton-Jacobi equation will then satisfy [32]

$$g^{\mu\nu} \frac{\partial X}{\partial x^\mu} \frac{\partial X}{\partial x^\nu} = \varepsilon, \tag{151}$$

and this is a requirement that the mapping function X preserves the length of a curve on M_4. This relation could be viewed as a condition to minimize the distance between two points in noncommutative geometry

$$[D, X]^2 = -1. \tag{152}$$

Thus, in contrast to the four dimensional case where the mapping is from M_4 to $S^4 \times S^4$, the mapping now is from $\mathbb{R} \times \Sigma_3$ to $\mathbb{R} \times S^3 \times S^3$. The Feynman slashed fields $Y^5\Gamma_5$ and $Y'^5\Gamma'_5$ must now be replaced with the field X slashed with some combination of 1, Γ_5, Γ'_5 and $\Gamma_5\Gamma'_5$. To find out the correct procedure, we make the following observation. In the four-dimensional case, we used the Feynman slashed coordinates $Y = Y^A\Gamma_A$, $A = 1, \ldots, 5$. The matrices $\frac{1}{4}\Gamma_{AB} = \frac{1}{8}\left(\Gamma_A\Gamma_B - \Gamma_B\Gamma_A\right)$ are generators of the Lie Algebra $SO(5)$. Denoting these by J_{AB}, they have the commutation relations

$$[J_{AB}, J_{CD}] = -\left(\delta_{AC}J_{BD} - \delta_{BC}J_{AD} - \delta_{AD}J_{BC} + \delta_{BD}J_{AC}\right). \tag{153}$$

Denoting $A = a, 5$ where $a = 1, \ldots, 4$ and $J_{a5} = RP_a$ we then have

$$[P_a, P_b] = -\frac{1}{R^2}J_{ab}. \tag{154}$$

In the limit $R = \frac{1}{\eta} \to \infty$, the generators P_a become, locally, the translation generators and J_{ab} will correspond to $SO(4)$ Lorentz generators. This is the procedure we will follow to decompose one of the coordinates, say Y^5 by writing

$$Y^5 = \eta X, \tag{155}$$

and simultaneously rescale one of the coordinates, say x^4

$$x^4 \to \eta t, \tag{156}$$

then taking the limit $\eta \to 0$. We will obtain the volume quantization condition by compactifying the four-dimensional two sided relation to $3 + 1$ in the above limit, where the fields Y^5 and Y'^5 are not coordinates on the fours sphere, but independent fields. To this end, let

$$\begin{aligned}
Z &= 2EE' - 1 &(157)\\
&= \frac{1}{2}\left(Y^a\Gamma_a + \eta X\Gamma_5 + 1\right)\left(Y'^a\Gamma_a + \eta X\Gamma'_5 + 1\right) - 1 &(158)\\
&= 2ee' - 1 + \eta X\left(\Gamma_5 e' + \Gamma'_5 e\right) + O\left(\eta^2\right) &(159)\\
&= z + \eta X\left(\Gamma_5 e' + \Gamma'_5 e\right) + O\left(\eta^2\right), &(160)
\end{aligned}$$

where

$$e = \frac{1}{2}\left(Y^a\Gamma_a + 1\right), \qquad e' = \frac{1}{2}\left(Y'^a\Gamma'_a + 1\right), \qquad z = 2ee' - 1. \tag{161}$$

Notice that we have identified the fields Y^5 and Y'^5 with the same field X because this is the field corresponding to the motion of the hypersurface. The correct quantization condition of the $3 + 1$ dimensional space, which also results from compactification of the four dimensional quantization condition is given by

$$\lim_{\eta\to 0}\frac{1}{\eta}\left\langle\left(z+\eta X\left(\Gamma_5 e'+\Gamma_5' e\right)\right)\left([D,z]+\eta\left[D,X\left(\Gamma_5 e'+\Gamma_5' e\right)\right]\right)^4\right\rangle=\gamma, \quad (162)$$

where γ is the chirality operator of the generated $3+1$ dimensional manifold. For consistency, one must first show that all terms of order $\frac{1}{\eta}$ are zero. For example

$$\left\langle z\,[D,z]^4\right\rangle=0, \quad (163)$$

as this would involve terms like $\langle\Gamma_a\Gamma_b\Gamma_c\Gamma_d\Gamma_e\rangle=0$ because this is the trace of an odd number of Γ matrices. Therefor we have to worry only about terms independent of η as the terms of order η vanish in the limit. Terms which are linear in X (and not its derivative) also vanish because terms of the form

$$X\left\langle\left(\Gamma_5 e'+e\Gamma_5'\right)[D,z]^4\right\rangle, \quad (164)$$

will give the terms

$$X\epsilon^{\mu\nu\kappa\lambda}\epsilon_{abcd}\partial_\mu Y^a\partial_\nu Y^b\partial_\kappa Y^c\partial_\lambda Y^d=X\det\left|\partial_\mu Y^a\right|=0, \quad (165)$$

as the Jacobian $\left|\partial_\mu Y^a\right|$ vanishes because the four Y^a are not independent. After some algebra, one can check that the only non-vanishing terms are

$$\sum_{p=0}^{3}\left\langle z\,[D,z]^p\,([D,X])\left(\Gamma_5 e'+e\Gamma_5'\right)[D,z]^{3-p}\right\rangle=\gamma. \quad (166)$$

There is no need to repeat the calculation done in the $d=4$ case as the result holds in general, and in particular in the limit $\eta\to 0$ and this is a smooth limit as terms of order $\frac{1}{\eta}$ vanish identically. We thus conclude that this condition implies

$$\frac{1}{3!}\epsilon^{\mu\nu\kappa\lambda}\epsilon_{abcd}\partial_\mu X\left(Y^a\partial_\nu Y^b\partial_\kappa Y^c\partial_\lambda Y^d+Y'^a\partial_\nu Y'^b\partial_\kappa Y'^c\partial_\lambda Y'^d\right)=\det\left|e_\mu^a\right|. \quad (167)$$

The field X could be identified with the time coordinate in a certain gauge. For example, in the synchronous gauge we have $g^{tt}=1$, $g^{ti}=0$ which implies that $X=t$ is a solution of the above constraint. If we define the three-dimensional hypersurface Σ_3 by $t=$constant, then the lapse function N could be defined by $\partial_t X=N$ with the boundary condition

$$\partial_i X|_\Sigma=0. \quad (168)$$

We could have obtained the $3+1$ quantization condition, directly by compactifying the four-dimensional condition of the mapping from $M_4\to S^4$. Let $Y^5=\eta X=Y'^5$ and simultaneously rescale one of the coordinates, say x^4

$$x^4\to\eta x^0, \quad (169)$$

so that the constraint in the limit $\eta \to 0$, becomes (written covariantly)

$$\sqrt{g} = \lim_{\eta \to 0} \left(\frac{1}{4!} \epsilon^{\mu\nu\kappa\lambda} \epsilon_{ABCDE} \left(Y^A \partial_\mu Y^B \partial_\nu Y^C \partial_\kappa Y^D \partial_\lambda Y^E \right. \right. \tag{170}$$

$$\left. \left. + Y'^A \partial_\mu Y'^B \partial_\nu Y'^C \partial_\kappa Y'^D \partial_\lambda Y'^E \right) \right) \tag{171}$$

$$= \frac{1}{3!} \epsilon^{\mu\nu\kappa\lambda} \epsilon_{abcd} \left(\partial_\mu X \right) \left(Y^a \partial_\nu Y^b \partial_\kappa Y^c \partial_\lambda Y^d + Y'^a \partial_\nu Y'^b \partial_\kappa Y'^c \partial_\lambda Y'^d \right), \tag{172}$$

where the X field is unconstrained, while the fields Y^a and Y'^a satisfy

$$Y^a Y^a = 1, \qquad Y'^a Y'^a = 1, \qquad a = 1, \ldots, 4. \tag{173}$$

Notice that the term

$$\frac{1}{4!} \epsilon^{\mu\nu\kappa\lambda} X \partial_\mu Y^a \partial_\nu Y^b \partial_\kappa Y^c \partial_\lambda Y^d \epsilon_{abcd} = X dY^1 \wedge dY^2 \wedge dY^3 \wedge dY^4, \tag{174}$$

is equal to zero because dY^4 depends on a linear combination of $dY^1, \ldots, dY^3$.

The Clifford algebra $M_2(\mathbb{H}) \oplus M_2(\mathbb{H})$ spanned by $Y^a \Gamma_a$ and $Y'^a \Gamma'_a$ will be extended by the generators $X\Gamma_5$ and $X\Gamma'_5$. The first $M_2(\mathbb{H})$ corresponding to the Clifford algebra Cliff$(+, +, +, +)$ is not effected by the addition of Γ_5. The second $M_2(\mathbb{H})$ corresponding to the Clifford algebra Cliff$(-, -, -, -)$ changes to $M_4(\mathbb{C})$ when extended by Γ'_5. Thus the algebra associated with the two sided relation (166) for the $3 + 1$ manifold is the same as the four dimensional case and is given by

$$M_2(\mathbb{H}) \oplus M_4(\mathbb{C}). \tag{175}$$

Thus, this compactification corresponds to a mapping from $\mathbb{R} \times \Sigma_3 \to \mathbb{R} \times S^3$ where Σ_3 is a three dimensional hypersurface. Although imposing this condition could be made and leads to the mimetic matter phenomena [33, 34], it is worth noting that we need to impose this condition only on the hypersurface Σ_3 to be defined below:

$$g^{\mu\nu} \partial_\mu X \partial_\nu X |_\Sigma = 1. \tag{176}$$

To get acquainted with this condition, we first consider the situation where we have a three dimensional hypersurface in space-time, a case dealt with in the ADM decomposition [31]. Consider the $3 + 1$ splitting of space-time so that (for Lorentzian signature)

$$ds^2 = h_{ij} \left(dx^i + N^i dt \right) \left(dx^j + N^j dt \right) - N^2 dt^2, \tag{177}$$

where $N\left(x^i, t\right)$ and $N^i\left(x^i, t\right)$ are the lapse and shift functions. Then

$$\sqrt{-g} = N\sqrt{h}. \tag{178}$$

We, therefore, supplement the volume quantization condition

$$\sqrt{g} = \frac{1}{3!}\epsilon^{\mu\nu\kappa\lambda}\epsilon_{abcd}\partial_\mu X\left(Y^a\partial_\nu Y^b\partial_\kappa Y^c\partial_\lambda Y^d + Y'^a\partial_\nu Y'^b\partial_\kappa Y'^c\partial_\lambda Y'^d\right), \quad (179)$$

by adding the constraints (151) to hold on the hypersurface

$$\partial_i X|_\Sigma = 0, \qquad \partial_t X|_\Sigma = N|_\Sigma, \quad (180)$$

from which we deduce that the constraint (179), when restricted to the hypersurface Σ_3, gives

$$\left(N\sqrt{h}\right)_\Sigma = \frac{1}{3!}N\epsilon^{ijk}\epsilon_{abcd}\left(Y^a\partial_i Y^b\partial_j Y^c\partial_k Y^d + Y'^a\partial_i Y'^b\partial_j Y'^c\partial_k Y'^d\right), \quad (181)$$

and we finally have

$$\int_\Sigma \sqrt{h}d^3x = \frac{1}{3!}\int_\Sigma \epsilon^{ijk}\epsilon_{abcd}\left(Y^a\partial_i Y^b\partial_j Y^c\partial_k Y^d + Y'^a\partial_i Y'^b\partial_j Y'^c\partial_k Y'^d\right)d^3x \quad (182)$$

$$= \frac{1}{3!}\int_\Sigma \epsilon_{abcd}\left(Y^a dY^b dY^c dY^d + Y'^a dY'^b dY'^c dY'^d\right) \quad (183)$$

$$= \frac{4}{3}\pi^2\left(w + w'\right) \quad (184)$$

where w and w' are integers given by the winding numbers on S^3. One can check that an exact solitonic solution with winding number one, is given by

$$X = t, \qquad Y^m = \frac{2x^m}{1 + x^m x^m}, \qquad Y^4 = \frac{x^m x^m - 1}{1 + x^m x^m}, \quad (185)$$

with the metric

$$g_{tt} = 1, \qquad g_{t\alpha} = 0, \qquad g_{\alpha\beta} = \frac{\delta_{\alpha\beta}}{(1 + x^m x^m)^2}, \quad (186)$$

and this corresponds to a quantized three dimensional volume.

To understand the condition $g^{\mu\nu}\partial_\mu X\partial_\nu X|_\Sigma = 1$ we notice that in the synchronous gauge [32] we can take $X = \tau$, $g^{tt} = \frac{1}{N^2}$ so that $\left|\frac{\partial\tau}{\partial t}\right| = N$ and thus the line measure $Ndt \to N\frac{\partial t}{\partial\tau}d\tau = d\tau$ which is consistent with $g^{\tau\tau} = 1$. Thus this condition amounts to length preserving transformation. We deduce that in a Lorentzian space-time volume quantization is possible, provided that the field corresponding to the non-compact transformation satisfy a length preserving condition. For the two sided equation where we have both Y^A and Y'^A it is important to truncate both Y^5 and Y'^5 to the same field X

$$Y^5 = \eta X, \qquad Y'^5 = \eta X,$$

which avoids imposing further unnatural conditions. There are many advantages to impose the condition (151) locally as this constraint modifies Einstein gravity only in the longitudinal sector as the field X is not dynamical. In the synchronous gauge, this field is identified with the time coordinate and modifies Einstein equations by giving an energy-momentum tensor in the absence of matter, giving rise to mimetic cold matter. We have shown that this field, which arises naturally from the three space quantization condition can be used to construct realistic cosmological models such as inflation without the need to introduce additional scalar fields. By including terms in the action of the form $f(\Box X)$ which do occur in the spectral action as can be seen from considerations of the scale invariance, it is possible to avoid singularities in Friedmann, Kasner [35] or Black hole solutions [36]. This is possible because the contributions of the field X to the energy-momentum tensor would allow, for special functions $f(\Box X)$ to limit the curvature, preventing the singularities from occurring.

12 Area Quantization

Next consider the compactification of two fields, keeping only three compact fields Y^m, $m = 1, 2, 3$, and rescale the two fields

$$Y^4 = \eta X^1, \qquad Y^5 = \eta X^2 \tag{187}$$

$$Y'^4 = \eta X^1, \qquad Y'^5 = \eta X^2, \tag{188}$$

and simultaneously rescale the coordinates

$$x^\alpha \to \eta x^\alpha, \qquad \alpha = 1, 2, \tag{189}$$

where x^α are coordinates along directions transverse to the two dimensional hypersurface, so that

$$\sqrt{g} = \lim_{\eta\to 0}\left(\frac{1}{4!}\epsilon^{\mu\nu\kappa\lambda}\epsilon_{ABCDE}\left(Y^A\partial_\mu Y^B\partial_\nu Y^C\partial_\kappa Y^D\partial_\lambda Y^E\right.\right. \tag{190}$$

$$\left.\left.+Y'^A\partial_\mu Y'^B\partial_\nu Y'^C\partial_\kappa Y'^D\partial_\lambda Y'^E\right)\right) \tag{191}$$

$$= \frac{1}{2}\epsilon^{\mu\nu\kappa\lambda}\epsilon_{ab}\partial_\mu X^a\partial_\nu X^b\epsilon_{mnp}\left(Y^m\partial_\kappa Y^n\partial_\lambda Y^p + Y'^m\partial_\kappa Y'^n\partial_\lambda Y'^p\right), \tag{192}$$

where $X^a(x^\mu)$, $a = 1, 2$, while the $Y^m(x^\mu)$ and $Y'^m(x^\mu)$ are subject to the constraints

$$Y^pY^p = 1, \qquad Y'^pY'^p = 1, \qquad p = 1, 2, 3. \tag{193}$$

Again, since the functions X^a are unconstrained to be coordinates on a sphere, normalization conditions must be imposed

$$\det\left(g^{\mu\nu}\partial_\mu X^a\partial_\nu X^b\right)_\Sigma = 1. \tag{194}$$

In case of Minkowski signature we must replace 1 with -1. It is known that this condition is the area preserving transformation on the two dimensional surface from the original surface with coordinates x^α to the surface with coordinates X^a. We note that in order to completely characterize this transformation we still have the option of specifying the *trace* of the matrix $g^{\mu\nu}\partial_\mu X^a\partial_\nu X^b$ which turns out to determine the stability of the map under linear perturbations [37].

Thus this compactification corresponds to the mapping $M_4 \to \mathbb{R}^2\times S^2$. We assume that there is a hypersurface Σ_2 endowed with an induced metric and with coordinates x^i so that the four dimensional metric can be written in the form

$$ds^2 = h_{ij}\left(dx^i + h^{ik}N_{i\alpha}dx^\alpha\right)\left(dx^j + h^{jl}N_{l\beta}dx^\beta\right) + k_{\alpha\beta}dx^\alpha dx^\beta, \tag{195}$$

where h^{ij} is the inverse of h_{ij}, the metric on Σ_2 with $i, j = 1, 2$ and $\alpha, \beta = 3, 4$. In matrix form, the four-metric is

$$\begin{pmatrix} k_{\alpha\beta} + N_{i\alpha}N_{j\beta}h^{ij} & N_{i\alpha} \\ N_{i\alpha} & h_{ij} \end{pmatrix}. \tag{196}$$

The inverse of this metric is given by

$$\begin{pmatrix} k^{\alpha\beta} & -N^{j\alpha} \\ -N^{j\alpha} & h^{ij} + N^{i\alpha}N^{j\beta}k_{\alpha\beta} \end{pmatrix}, \tag{197}$$

where $k^{\alpha\beta}$ is the inverse of $k_{\alpha\beta}$ and $N^{i\alpha}$ is obtained from $N_{i\alpha}$ by raising indices with the metrics h^{ij} and $k^{\alpha\beta}$. The hypersurface Σ_2 is then defined by the equations

$$x^\alpha = \text{const}, \qquad \alpha = 1, 2, \tag{198}$$

parametrized by the coordinates x^i, $i = 3, 4$. In this form we have

$$\sqrt{g} = \sqrt{h}\sqrt{k}. \tag{199}$$

The constraint (194) is then solved by

$$\partial_i X^a|_\Sigma = 0, \tag{200}$$

so that

$$\det\left(k^{\alpha\beta}\partial_\alpha X^a\partial_\beta X^b\right)_\Sigma = 1, \tag{201}$$

which implies

$$(\det k)_\Sigma = \left(\det\left|\partial_\alpha X^a\right|_\Sigma\right)^2 . \tag{202}$$

Using

$$\left(\epsilon^{ij}\epsilon_{ab}\epsilon^{\alpha\beta}\partial_\alpha X^a\partial_\beta X^b\epsilon_{mnp}Y^m\partial_i Y^n\partial_j Y^p\right)_\Sigma = \det\left|\partial_\alpha X^a\right|_\Sigma \left(\epsilon^{ij}\epsilon_{mnp}Y^m\partial_i Y^n\partial_j Y^p\right)_\Sigma \tag{203}$$

$$= \left(\sqrt{k}\epsilon^{ij}\epsilon_{mnp}Y^m\partial_i Y^n\partial_j Y^p\right)_\Sigma . \tag{204}$$

The volume constraint becomes

$$\left(\sqrt{h}\sqrt{k}\right)_\Sigma = \frac{1}{2}\left(\sqrt{k}\epsilon^{ij}\epsilon_{mnp}\left(Y^m\partial_i Y^n\partial_j Y^p + Y'^m\partial_i Y'^n\partial_j Y'^p\right)\right)_\Sigma . \tag{205}$$

One important point to realize is that the fundamental constraint equation is (192), and that we can integrate this equation over any hypersurface we like, and not only over the full space. In particular, let us choose to integrate over a two dimensional hypersurface Σ_2 with coordinates x^α, then this implies that

$$\int_{\Sigma_2} d^2x\sqrt{h} = \frac{1}{2}\int_\Sigma \epsilon^{ij}\epsilon_{mnp}\left(Y^m\partial_i Y^n\partial_j Y^p + Y'^m\partial_i Y'^n\partial_j Y'^p\right)dx^i dx^j \tag{206}$$

$$= \frac{1}{2}\int_\Sigma \epsilon_{mnp}\left(Y^m dY^n dY^p + Y'^m dY'^n dY'^p\right) \tag{207}$$

$$= 4\pi\left(w + w'\right), \tag{208}$$

where w and $w'$are integers and equal to the winding numbers of the two maps.

13 Equations of Motion for $\mathbb{R}\times S^3$ and $\mathbb{R}^2\times S^2$

13.1 $\mathbb{R}\times S^3$ Case

Start by taking the action

$$I = -\frac{1}{2}\int d^4x\sqrt{g}R + \frac{1}{2}\int d^4x\lambda\left(\sqrt{g} - \frac{1}{3!}\epsilon^{\mu\nu\kappa\lambda}\partial_\mu X\epsilon_{abcd}Y^a\partial_\nu Y^b\partial_\kappa Y^c\partial_\lambda Y^d\right)$$
$$+\frac{1}{2}\int d^4x\sqrt{g}\lambda'\left(Y^aY^a - 1\right) + \frac{1}{2}\int d^4x\sqrt{g}\lambda''\left(g^{\mu\nu}\partial_\mu X\partial_\nu X - 1\right). \tag{209}$$

We have included a constraint on the X field, which is known to have the effect of replacing the scale factor in gravity by the field X which mimics dark matter [33, 34]. We also have the option of not including this field, and in that case the effects of the field X will only be topological providing only the joining of the disconnected pieces. For simplicity, we have included only the coordinates of one of the maps Y^a. First, we have the λ'' and $g^{\mu\nu}$ equations

$$g^{\mu\nu}\partial_\mu X\partial_\nu X = 1 \tag{210}$$

$$G_{\mu\nu} + \frac{1}{2}\lambda g_{\mu\nu} - \lambda''\partial_\mu X\partial_\nu X = 0. \tag{211}$$

Taking the trace of Einstein equation gives

$$\lambda'' = G + 2\lambda, \tag{212}$$

resulting in the traceless equation

$$G_{\mu\nu} - G\partial_\mu X\partial_\nu X + \frac{1}{2}\lambda\left(g_{\mu\nu} - 4\partial_\mu X\partial_\nu X\right) = 0. \tag{213}$$

Next the variation of the field X gives

$$\partial_\mu\left(\sqrt{g}g^{\mu\nu}\partial_\nu X\left(G + 2\lambda\right)\right) = \frac{1}{2}\partial_\mu\lambda V^\mu, \tag{214}$$

where we have denoted

$$V^\mu = \frac{1}{3!}\epsilon^{\mu\nu\kappa\lambda}\epsilon_{abcd}Y^a\partial_\nu Y^b\partial_\kappa Y^c\partial_\lambda Y^d, \tag{215}$$

and used the property

$$\partial_\mu V^\mu = 0. \tag{216}$$

This last equation is a consequence of the identity $dY^1 \wedge dY^2 \wedge dY^3 \wedge dY^4 = 0$ which follows from

$$dY^4 = -\frac{1}{Y^4}\left(Y^1dY^1 + Y^2dY^2 + Y^3dY^3\right). \tag{217}$$

The Y^a equation gives

$$\sqrt{g}\lambda' Y_a - \frac{1}{2}\lambda\partial_\mu X\frac{1}{3!}\epsilon^{\mu\nu\kappa\lambda}\epsilon_{abcd}\partial_\nu Y^b\partial_\kappa Y^c\partial_\lambda Y^d \tag{218}$$

$$= \partial_\mu X\partial_\nu\left(\lambda\frac{1}{3!}\epsilon^{\mu\nu\kappa\lambda}\epsilon_{abcd}Y^b\partial_\kappa Y^c\partial_\lambda Y^d\right). \tag{219}$$

Contracting this equation with Y^a gives

$$\lambda' = \frac{3}{2}\lambda. \tag{220}$$

The Bianchi identity gives

$$\frac{1}{2}\partial_\mu\lambda = \nabla^\nu\left((G+2\lambda)\,\partial_\mu X\partial_\nu X\right). \tag{221}$$

Using the property $\left(\nabla^\nu\partial_\mu X\right)\partial_\nu X = 0$, obtained by differentiating Eq. (210) this simplifies to

$$\frac{1}{2}\partial_\mu\lambda = \frac{1}{\sqrt{g}}\partial_\rho\left(\sqrt{g}g^{\rho\nu}\left(G+2\lambda\right)\right)\partial_\mu X. \tag{222}$$

For example, in the synchronous gauge where $g_{tt}=1$ and $X=t$, we find $\partial_i\lambda=0$ and

$$\frac{\partial}{\partial t}\left(G+\frac{3}{2}\lambda\right)+\frac{1}{2}\frac{\partial}{\partial t}\ln g = 0. \tag{223}$$

For Friedmann type universe this condition simplifies to

$$\frac{\partial}{\partial t}\left(G+3\frac{\dot{a}}{a}+\frac{3}{2}\lambda\right)=0, \tag{224}$$

which is the Einstein equation allowing mimetic dark matter and cosmological constant arising as integration constants.

One can easily verify that the Bianchi identity (221) upon contracting by V^μ gives

$$\frac{1}{2}\partial_\mu\lambda V^\mu = \partial_\mu X V^\mu\,\nabla^\nu\left((G+2\lambda)\,\partial_\nu X\right) \tag{225}$$

$$= \partial_\mu\left(\sqrt{g}g^{\mu\nu}\left(G+2\lambda\right)\partial_\nu X\right), \tag{226}$$

which coincides with the X equation after contracting with V^μ.

Note that if the constraint $\left(g^{\mu\nu}\partial_\mu X\partial_\nu X\right)_\Sigma = 1$ is only imposed on the boundary, then there will be no need for a Lagrange multiplier and the equations do simplify to give

$$G_{\mu\nu}-\frac{1}{4}g_{\mu\nu}G=0 \tag{227}$$

$$G+2\lambda=0 \tag{228}$$

$$\partial_\mu\lambda=0 \tag{229}$$

$$\lambda'=\frac{3}{2}\lambda. \tag{230}$$

without any new information from the Y^a and X equations.

13.2 $\mathbb{R}^2 \times S^2$ Case

We start with the action

$$I = -\frac{1}{2\kappa^2}\int d^4x\sqrt{g}R + \frac{1}{2}\int d^4x\lambda\left(\frac{1}{\kappa^3}\sqrt{g} - \frac{1}{2!}\epsilon^{\mu\nu\kappa\lambda}\epsilon_{ab}\partial_\mu X^a\partial_\nu X^b\epsilon_{mnp}Y^m\partial_\kappa Y^n\partial_\lambda Y^p\right)$$
$$+\frac{1}{2\kappa^4}\int d^4x\sqrt{g}\lambda'\left(Y^mY^m - 1\right). \tag{231}$$

Varying $g^{\mu\nu}$ and setting $\kappa^2 = 1$, gives

$$G_{\mu\nu} + \frac{1}{2}\lambda g_{\mu\nu} = 0, \tag{232}$$

and by tracing this equation we get

$$G + 2\lambda = 0. \tag{233}$$

After substituting back we get the traceless equation

$$G_{\mu\nu} - G\partial_\mu X\partial_\nu X + \frac{1}{2}\lambda g_{\mu\nu} = 0. \tag{234}$$

The Bianchi identity gives

$$\frac{1}{2}\partial_\mu\lambda = 0.$$

Next, we have the X^a equation

$$-\partial_\mu\left(\epsilon^{\mu\nu\kappa\lambda}\epsilon_{ab}\lambda\partial_\nu X^b\epsilon_{mnp}Y^m\partial_\kappa Y^n\partial_\lambda Y^p\right)$$
$$= \partial_\mu\lambda\epsilon^{\mu\nu\kappa\lambda}\epsilon_{ab}\partial_\nu X^b\epsilon_{mnp}Y^m\partial_\kappa Y^n\partial_\lambda Y^p, \tag{235}$$

and finally the Y^m equation gives

$$\sqrt{g}\lambda' Y_m - \frac{1}{2}\lambda\epsilon_{ab}\partial_\mu X^a\partial_\nu X^b\frac{1}{2}\epsilon^{\mu\nu\kappa\lambda}\epsilon_{mnp}\partial_\kappa Y^n\partial_\lambda Y^p \tag{236}$$
$$= \epsilon_{ab}\partial_\mu X^a\partial_\nu X^b\partial_\kappa\left(\lambda\frac{1}{2}\epsilon^{\mu\nu\kappa\lambda}\epsilon_{mnp}Y^n\partial_\lambda Y^p\right). \tag{237}$$

Contracting this equation with Y^m gives

$$\lambda' = \frac{3}{2}\lambda, \tag{238}$$

and thus

$$\frac{3}{2}\sqrt{g}Y_m - \frac{3}{2}\epsilon_{ab}\partial_\mu X^a\partial_\nu X^b \frac{1}{2}\epsilon^{\mu\nu\kappa\lambda}\epsilon_{mnp}\partial_\kappa Y^n\partial_\lambda Y^p = 0, \tag{239}$$

together with

$$\sqrt{g} = \frac{1}{2!}\epsilon^{\mu\nu\kappa\lambda}\epsilon_{ab}\partial_\mu X^a\partial_\nu X^b\epsilon_{mnp}Y^m\partial_\kappa Y^n\partial_\lambda Y^p. \tag{240}$$

This implies

$$Y_m\epsilon_{pqr}Y^p\partial_{[\kappa} Y^q\partial_{\lambda]}Y^r = \epsilon_{mnp}\partial_{[\kappa} Y^n\partial_{\lambda]}Y^p. \tag{241}$$

This relation is an identity which follows from the vanishing of a rank four antisymmetric tensor $[mpqr]$ taking three values

$$0 = Y_{[m}\,\epsilon_{pq\,r]}\partial_{[\kappa} Y^q\partial_{\lambda]}Y^r \tag{242}$$
$$= Y_m\epsilon_{pqr}Y^p\partial_{[\kappa} Y^q\partial_{\lambda]}Y^r - Y_p\epsilon_{mqr}\partial_{[\kappa} Y^q\partial_{\lambda]}Y^r + 2Y_q\epsilon_{pmr}Y^p\partial_{[\kappa} Y^q\partial_{\lambda]}Y^r, \tag{243}$$

the last term being zero because $Y_q\partial_\kappa Y^q = 0$. Thus, as expected, no new information comes from the Y equation, except for its trace.

The X_a equation reduces to

$$-\partial_\mu\left(\epsilon^{\mu\nu\kappa\lambda}\epsilon_{ab}\lambda\partial_\nu X^b\epsilon_{mnp}Y^m\partial_\kappa Y^n\partial_\lambda Y^p\right)$$
$$= \partial_\mu\lambda\epsilon^{\mu\nu\kappa\lambda}\epsilon_{ab}\partial_\nu X^b\epsilon_{mnp}Y^m\partial_\kappa Y^n\partial_\lambda Y^p, \tag{244}$$

which is identically satisfied since $\partial_\mu\lambda = 0$. This shows that the resulting system is that of gravity plus mimetic dark matter, with the topological fields $Y^m\,(x)$ connecting the different unit spheres, constituting the building fabric of space-time.

Finally we comment on the possibility of adding mimetic matter to the system corresponding to the quantization of $\mathbb{R}^2 \times \Sigma_2$ where Σ_2 is a two dimensional surface. Looking at the induced metric

$$h^{ab} = g^{\mu\nu}\partial_\mu X^a\partial_\nu X^b, \qquad a, b = 1, 2, \tag{245}$$

we notice that we had to impose, on the boundary, the constraint

$$\det h^{ab} = 1, \tag{246}$$

which is the area preserving condition for the two dimensional surfaces. These maps will be characterized by the value of the trace of h^{ab} and their stability will depend on the value of $t = \operatorname{tr} h^{ab}$. These are stable and of the elliptic type when $-2 < t < 2$. Unfortunately, the resulting system of equations is not easy to solve, and it is not clear whether such system can lead to realistic models. It is therefore doubtful whether using more than one scalar field associated with imposing one or more constraints is useful. We conclude that for our purposes, it is enough to characterize the conditions for area quantization is to have an area preserving conditions on the mapping defined by the two fields X^1 and X^2 taken as boundary conditions.

14 Discussion and Conclusions

It is an ambitious goal to initiate a program of axiomatization of physics as suggested by Hilbert. Our proposal is to start from an analogue of the Heisenberg commutation relation to quantize the geometry. The Dirac operator plays the role of momentum while the Feynman slash of scalar fields plays the role of coordinates. When the dimension of the noncommutative space, as determined by the growth of eigenvalues, is 2 or 4 there are two possible Clifford algebras with which the scalar fields are contracted with the corresponding gamma matrices. These two Clifford algebras are related to each other through the reality operator J which is an anti-unitary operator that is part of the data defining the noncommutative space. In four dimensions the sum of the two Clifford algebras is $M_2(\mathbb{H}) \oplus M_4(\mathbb{C})$ which is the algebra of the finite space that is tensored with the continuous Riemannian space. The quantization condition implies that the volume of the continuous part of the space is quantized in terms of the winding numbers of the two mappings Y and Y' from M_4 to S^4. The presence of two maps instead of one allows for the representation of a spin-manifold M_4, with arbitrary topology and large volume as the pullback of the two maps which yields four coordinates given on local charts. This construction determines, in a unique way, the noncommutative space that defines our space-time. Inner fluctuations of the Dirac operator by automorphisms of the algebra extends it to include a connection, which is a one form defined over the noncommutative space. Components of the connection along the continuous directions are the gauge fields of the resulting gauge group, and the components along the discrete directions are the Higgs fields. The connection then includes all the bosonic fields of a unified field theory, which is a Pati-Salam model with a definite Higgs structure. There are two special cases when these Higgs fields are either truncated or are in composite representations of more fundamental fields. The Standard Model with neutrinos (and a singlet) is a special case of the Pati-Salam model which satisfies an order one condition where the connection becomes restricted to the algebra $\mathcal{A}$ but not its opposite. Elements of the Hilbert space define the fermions which are 16 in the representation $(2,1,4)+(1,2,4)$ with respect to the symmetry $S(2)_R \times SU(2)_L \times SU(3)$. Thus all bosonic fields in the -form of gravity, gauge and Higgs fields are unified in the Dirac operator and all fermion fields are unified in the fundamental representation in the Hilbert space. The dynamics is governed by the spectral action principle where the spectral action is an arbitrary positive function of the Dirac operator valid up to a cutoff scale, which is taken to be near the Planck scale. In other words, by starting from a quantization condition on the volume of the noncommutative space, all fields and their interactions are predicted and given by a Pati-Salam model which has three special cases one of which is the Standard Model with neutrino masses and a singlet field. The spectral Standard Model predicts unification of gauge couplings and the correct mass for the top quark and is consistent with a low Higgs mass of 125 Gev. The unification model is assumed to hold at the unification scale and when the gauge, Yukawa and Higgs couplings relations are taken as initial conditions on the RGE, one finds complete agreement with experiment, except for the meeting of the gauge couplings which are off by 4%. This suggests that a Pati-Salam model defines the physics beyond the Standard Model, and where we have shown [16] that it allows for unification of gauge couplings, consistent with experimental data.

The assumption of volume quantization has consequences on the structure of General Relativity. Equations of motion agree with Einstein equations except for the trace condition, which now determines the Lagrange multiplier enforcing volume quantization. The cosmological constant, although not included in the action, is now an integration constant. The two mapping fields Y and Y' from the four-manifold to S^4 can be considered to be be solutions of instanton equations and give the physical picture that coordinates of a point are represented as the localization of instantons with finite energy. To have a physical picture of time we have also considered a four-manifold formed with the topology of $R \times \Sigma_3$, where Σ_3 is a three dimensional hypersurface, to allow for space-times with Lorentzian signature. The quantization condition is modified to have two mappings from $\Sigma_3 \to S^3$ and a mapping $X : \mathbb{R} \to \mathbb{R}$. The resulting algebra of the noncommutative space is unchanged, and the three dimensional volume is quantized provided that the mapping field X is constrained to have unit gradient. This field X modifies only the longitudinal part of the graviton and plays the role of mimetic dust. It thus solves, without extra cost, the dark matter problem [33]. Recently, we have shown that this field X can be used to build realistic cosmological models [34]. In addition, and under certain conditions, could be used to avoid singularities in General relativity for Friedmann, Kasner [35] and Black hole solutions [36]. This is possible because this scalar field modifies the longitudinal sector in GR. We have presented various implications of the quantization condition such as the absence of the cosmological constant from the action, quantizing volumes and areas of maps of M_4 to S^4, $\mathbb{R} \times S^3$ and $\mathbb{R}^2 \times S^2$.

We have presented enough evidence that a framework where space-time assumed to be governed by noncommutative geometry results in a unified picture of all particles and their interactions. The axioms could be minimized by starting with a volume quantization condition, which is the Chern character formula of the noncommutative space and a special case of the orientability condition. This condition determines uniquely the structure of the noncommutative space. Remarkably, the same structure was also derived, in slightly less unique way, by classifying all finite noncommutative spaces [10]. The picture is very compelling, in contrast to other constructions, such as grand unification, supersymmetry or string theory, where there is no limit on the number of possible models that could be constructed. The picture, however, is still incomplete as there are still many unanswered questions and we now list few of them. Further studies are needed to determine the structure and hierarchy of the Yukawa couplings, the number of generations, the form of the spectral function and the physics at unification scale, quantizing the fields appearing in the spectral action and in particular the gravitational field. To conclude, noncommutative geometry as a basis for unification, is a predictive and exciting field with very appealing features and many promising new directions for research.

Acknowledgements I would like to thank Alain Connes for a fruitful and pleasant collaboration on the topic of noncommutative geometry for the last twenty years. I would also like to thank Walter van Suijlekom and Slava Mukhanov for essential contributions to this program of research. This research is supported in part by the National Science Foundation under Grant No. Phys-1518371.

References

1. L. Corry, *David Hilbert and the Axiomatization of Physics (1898–1918)* (Springer, Dordrecht, 2004)
2. D. Hilbert, Die Grundlagen der Physik, Konigl. Gesell. d. Wiss. Gottingen, Nachr. Math. Phys. Kl. 395–407 (1915)
3. D. Hilbert, Die Grundlagen der Physik, (Zweite Mitteilung), Konigl. Gesell.d. Wiss. Gottingen, Nachr. Math. Phys. Kl. 53–76 (1917)
4. A. Connes, *Noncommutative Geometry* (Academic Press, New York, 1994)
5. J. Gracia-Bondia, J. Varilly, H. Figueroa, *Elements of Noncommutative Geometry, Birkhauser*
6. W. Greub, S. Halperin, R. Vanstone, *Connections, Curvature and Cohomology*, vols. 1–3, vol. 2 (sphere maps) (Academic Press, 1973), pp. 347–351
7. A.H. Chamseddine, A. Connes, V. Mukhanov, Geometry and the quantum: basics. JHEP **12**, 098 (2014). arXiv:1411.0977
8. A. Connes, Noncommutative geometry and reality. J. Math. Phys. **36**, 6194 (1995)
9. A.H. Chamseddine, A. Connes, V. Mukhanov, Quanta of geometry: noncommutative aspects. Phys. Rev. Lett. **114**, 091302 (2015)
10. A.H. Chamseddine, A. Connes, Why the standard model. J. Geom. Phys. **58**, 38 (2008)
11. A.H. Chamseddine, A. Connes, W. van Suijlekom, Inner fluctuations in noncommutative geometry without the first order condition. J. Geom. Phys. **73**, 222 (2013)
12. A.H. Chamseddine, A. Connes, W. van Suijlekom, Beyond the spectral standard model: emergence of Pati-Salam unification. JHEP **11**, 132 (2013)
13. A.H. Chamseddine, A. Connes, The spectral action principle. Commun. Math. Phys. **186**, 731 (1997)
14. A.H. Chamseddine, A. Connes, M. Marcolli, Gravity and the standard model with neutrino mixing. Adv. Theor. Math. Phys. **11**, 991–1089 (2007)
15. A.H. Chamseddine, A. Connes, Noncommutative geometry as a framework for unification of all fundamental interactions including gravity. Fortsch. Phys. **58**, 553 (2010)
16. A.H. Chamseddine, A. Connes, W. van Suijlekom, Grand unification in the spectral Pati-Salam model. JHEP **1511**, 011 (2015)
17. H. Lawson, M. Michelson, *Spin Geometry* (Princeton University Press, Princeton, 1989)
18. A. Connes, A short survey of noncommutative geometry. arXiv: hep-th/0003006
19. J. Moser, On the volume elements on a manifold. Trans. Am. Math. Soc. **120**, 286–294 (1965)
20. J. Barrett, A Lorentzian version of the noncommutative geometry of the standard model of particle physics. J. Math. Phys. **48**, 012303 (2007)
21. A.H. Chamseddine, A. Connes, Reselience of the spectral standard model. JHEP **09**, 104 (2009)
22. A.H. Chamseddine, A. Connes, The uncanny precision of the spectral action. Commun. Math. Phys. **293**, 867–897 (2010)
23. M. Henneaux, C. Teitelboim, The cosmological constant and general covariance. Phys. Lett. B **222**, 195 (1989)
24. R. Bott, L. Tu, *Differential Forms in Algebraic Topology* (Springer, New York, 1982), p. 215
25. R. Rajaraman, *Solitons and Instantons, An introduction* (North Holland, Amsterdam, 1989)

26. S. Coleman, *Aspects of Symmetry*, Chapter 6 (Cambridge University Press, 1985)
27. Y. Xin, *Geometry of Harmonic Maps* (Birkhauser, Boston, 1996)
28. E. Gava, R. Jengo, C. Omero, The $O(5)$ non-linear sigma model as a $SU(2)$ gauge theory. Phys. Lett. **81B**, 187 (1979)
29. F. Gursey, M. Jafarizadeh, H. Tze, Quaternionic $S^4 \approx HP(1)$ gravitational and chiral instantons. Phys. Lett. **88B**, 282 (1979)
30. K. Kuchar, Geometry of hypersurface. 1. J. Math. Phys. **17**, 777 (1976)
31. C. Misner, K. Thorne, J. Wheeler, *Gravitation (W* (Freeman and Company, San Francisco, 1973)
32. L. Landau, E. Lifshitz, *The Classical Theory of Fields*, 4th edn. (Butterworth Heinemann, Oxford)
33. A.H. Chamseddine, V. Mukhanov, Mimetic dark matter. JHEP **1311**, 135 (2013)
34. A.H. Chamseddine, V. Mukhanov, A. Vikman, Cosmology with mimetic matter. JCAP **1406**, 017 (2014)
35. A.H. Chamseddine, V. Mukhanov, *Resolving cosmological singularities*. JCAP **1703**, 009 (2017)
36. A.H. Chamseddine, V. Mukhanov, *Nonsingular black hole*. Eur. Phys. J. C **77**, 183 (2017)
37. R. MacKay, *Renormalization in Area Preserving Maps* (World Scientific, Singapore, 1992), pp. 30–32

Twistor Theory as an Approach to Fundamental Physics

Roger Penrose

Abstract The original motivations underlying the introduction of twistor theory are described, demanding a (3+1)-dimensional space-time theory dependent upon complex analysis and geometry. Space-time points are relegated to a secondary role, light rays, with a twisting aspect to them, being taken as more fundamental. The twistor treatment of wavefunctions for massless fields leads to a representation in terms of holomorphic sheaf cohomology. This, in turn, leads to a description of anti-self-dual (left-handed) gravitational (and Yang-mills) feeds. Failed attempts to remove this anti-self-dual restriction (the googly problem) led to a 40-year blockage to the development of twistor theory as a possible overall approach to fundamental physics. In recent years, a hopeful approach to deal with this problem—palatial twistor theory—has arisen, but the detailed development of these ideas has so far proved technically difficult.

1 Underlying Motivations

1.1 Twistor Aspirations

Twistor theory is a body of unusual ideas, initiated in 1963, that was intended eventually to provide a coherent overall approach to fundamental physics [1]. Despite various successes over the years, in which it provided significant inputs into certain areas of pure mathematics, most particularly differential geometry, representation theory and integrable systems (see, in particular, [2, 3]), twistor theory had, for many years, practically no impact at all on the theoretical physics community itself. This remained true for some four decades until around 2005, when (following certain earlier publications of others [4–11]), Witten [12] introduced several novel ideas that then stimulated further work, e.g. [13–20], which allowed twistor methods to help in powerfully simplifying the calculations of scattering amplitudes at very high

R. Penrose (✉)
Mathematical Institute, Oxford, UK
e-mail: rpenroad@gmail.com

J. Kouneiher (ed.), *Foundations of Mathematics and Physics One Century After Hilbert*,
https://doi.org/10.1007/978-3-319-64813-2_10

energies (where all particles involved could be considered to be massless). Nonetheless, as a candidate for an overarching general theory of physics, the proposal remained vastly short of its original aims.

In my view there has been one key obstruction to progress towards this overall objective, which is connected with the *chiral* nature of the formalism. One sees this most strikingly in the way that the curved space-times of *general relativity* began to be addressed by the theory, whereas the original scheme was restricted to describing special-relativistic physics of flat Minkowskian space-time. From 1975 (see [21, 22]), by means of what became known as the "non-linear graviton construction", it became possible to use twistor theory to generate general (complex) curved-space solutions of the Einstein vacuum equations (with or without a cosmological constant Λ), but only for "gravitons" of *left*-handed helicity. To describe the right-handed ones using the same formalism, and to combine the two in a satisfactory way, was fundamentally elusive, and this became known as the "googly problem" (a term borrowed from the game of cricket). Although, for around 40 years, no satisfactory solution to this conundrum was found, it is my opinion that a relatively new approach to this issue, referred to as "palatial twistor theory" may well hold the key [23, 24], and this scheme will be outlined here in §C7 and §C8. However, genuine progress in adopting these ideas in an effective way has not yet been achieved.

1.2 Space-Time Dimensionality: Two Roles for the Riemann Sphere

As for twistor theory's actual origins, that effectively occurred in early December 1963, when I was on a 9-month appointment at the University of Texas. Various motivational ideas had been troubling me for a number of years previously, concerning what I had felt to be a need for a novel approach to foundational physics, and these had then largely come together in my mind at that time. This was the initial stage of the proposal that, a little later, I indeed referred to as "twistor theory", owing to a key role that the twisted configuration of interlocking circles shown in Fig. 1 (a stereographically projected family of the Clifford parallels on a 3-sphere) had played for me. I shall come to the role of this configuration in §B1).

One of my main motivations had arisen from my feeling that there was a need for a formalism that was geared to that specific dimensionality of space-time structure that we directly perceive around us. This line of thinking was very unlike that of various other ideas for an underlying physics of the world that later became popular, e.g. string theory [25]. I had earlier become convinced that what was needed would be a formalism that should indeed be very specific to the number of space and time dimensions, namely 3 and 1, respectively, that macroscopically present themselves to us, and I took the view that this should be central to the scheme. This goes very much in opposition to the role of space-time dimensionality underlying many of the current trends, most particularly string theory, where extra space dimensions (and

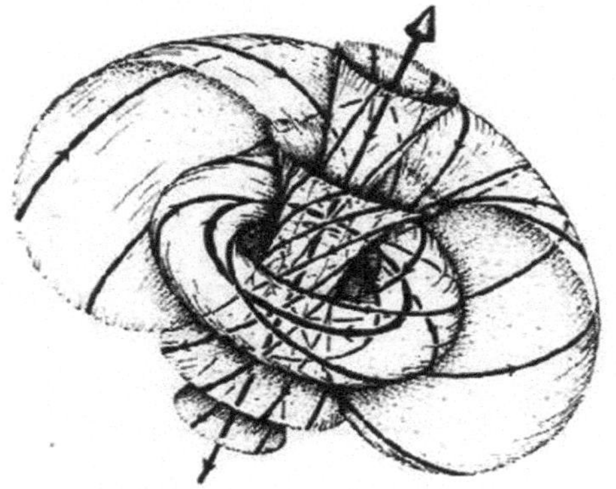

Fig. 1 A picture representing a non-null twistor: stereographic projection of Clifford parallels on a 3-sphere to Euclidean 3-space E. The tangent directions to the circles point in the direction (projected into E) of the rays of a Robinson congruence. By continuingly reassembling itself the entire configuration travels with the speed of light, as E evolves in time, in the direction of the large arrow at the top right

even an extra time dimension, in the case of "F-theory") are put forward [25] as serious proposals for the overall space-time geometry of the physical world that we inhabit. It also goes against the very natural and commendable desire, in *pure* mathematics, for formalisms that can be applied, generally, to any spatial dimensionality whatever, but the aims of theoretical physics are very different from those of pure mathematics, even though much of theoretical physics depends vitally on the latter.

Another of my basic motivations had been for a formalism that was essentially *complex* in the sense that it would be able to take advantage of what I had regarded, ever since my days as a mathematics undergraduate, as the "magic" of complex analysis and holomorphic (i.e. complex analytic) geometry. I had learnt that the complex number system had not only a profoundly deep power and elegance, but that it had also found a basic realization in its underlying role in the formalism of quantum theory. When I had begun to study quantum mechanics in a serious way, and particularly following the superb course of lectures given by Paul Dirac when I was a graduate student (in algebraic geometry) at Cambridge, I became fascinated by the quantum description of spin, and how the complex numbers of quantum mechanics were directly related to the 3-dimensiality of physical space, via the 2-sphere of spatial directions being appropriately identified as a Riemann (or Bloch) sphere of the ratios of pairs of complex numbers (quantum amplitudes) where, in the case of a massive particle of spin $^1/_2$ such as an electron (see Fig. 2), we can think of these as being the complex components of a 2-spinor. Moreover, I had realized that in the relativistic context, there was another role for the Riemann sphere, this time as the celestial sphere that an astronaut in space would observe. The transformation of this celestial sphere to that of a second astronaut, moving at a relativistic speed while passing nearby the first would be one that preserves the complex structure of the Riemann sphere (i.e. conformal without reflection). The special (i.e. non-reflective)

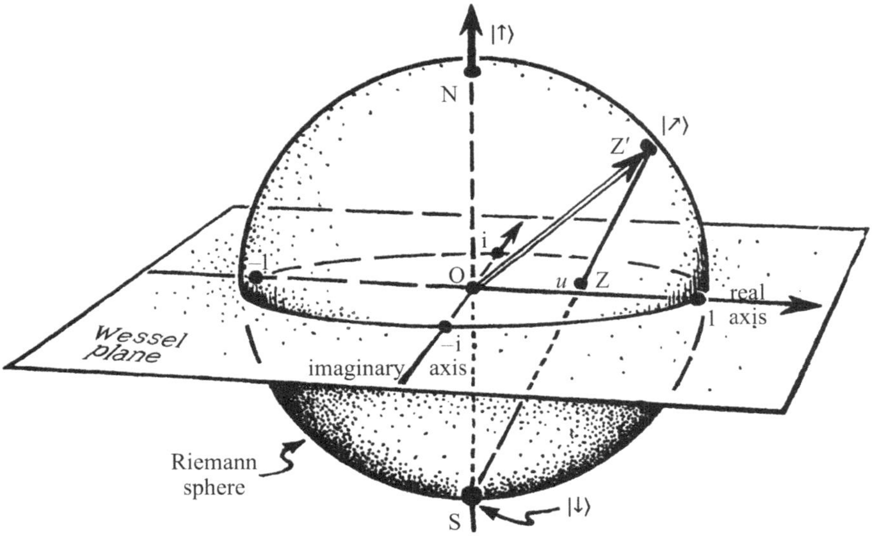

Fig. 2 The Riemann sphere (here in its role as a Bloch sphere) projects stereographically from its south pole S to the complex (Wessel) plane, whose unit circle coincides with the equator of the sphere. A general spin state $|\nearrow\rangle = w|\uparrow\rangle + z|\downarrow\rangle$, of a spin-1/2 massive particle is represented by the pint Z on the Wessel plane denoting the complex number $u = z/w$, which is the stereographic image of Z′ on the sphere (so S, Z, and Z′ are collinear). The spin direction $\nearrow$ is then OZ, where O is the sphere's center

Lorentz group is thus seen to be identical with these holomorphic transformations of this Riemann sphere (Mobius transformations). Again this was clear from the 2-spinor formalism, this time in the relativistic context (see [26]).

1.3 The 2-Spinor Formalism

This dual role for the Riemann sphere, one fundamentally to do with quantum mechanics in the case of 3 spatial dimensions, and the other fundamentally to do with macroscopic relativity, in (3+1)-dimensional space-time, struck me as being no accident, but something that linked together these two great revolutions of 20th century physics—of the small and of the large—via the magic of complex numbers. I felt that this might represent a definite clue to a deep unifying relation between the two. Both could be seen as a feature of the 2-spinor calculus, as introduced be Cartan [27] and van der Waerden [28], and which I had learnt how to use [29] from Dirac, in an unusual deviation from his normal Cambridge course on quantum mechanics.

I liked to think of a 2-spinor (often referred to by physicists as a "Weyl spinor") in a very geometrical way, and I realized that, up to an overall sign, a non-zero

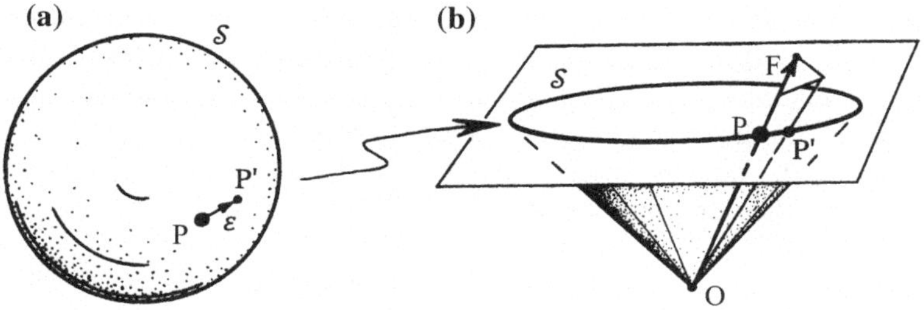

Fig. 3 **a** The space of null directions at some space-time point O is represented as a Riemann 2-sphere $\mathcal{S}$. The flagpole direction of a 2-spinor is represented, on $\mathcal{S}$, as the point P. Infinitesimally near to P is P′, where the direction $\overrightarrow{PP'}$ provides the 2-spinor's flag plane. **b** In space-time terms, the 2-spnor's flagpole is shown as the null 4-vector $\overrightarrow{OF}$, where we realize $\mathcal{S}$ as a particular 3-plane intersection of the future null cone of O (all this taken in O's tangent 4-space), so that P lies on the line OF. The 2-spinor's flag plane is now seen as the null half-2-plane extending away from the line OF in the direction of P′

2-spinor can be represented as a future-pointing null vector (a vector pointing along the future null cone), referred to as the "flagpole", together with a "flag plane" direction through that flagpole [30, 31]. The flag plane would be a null half-plane bounded by the flagpole. This flag geometry can be thought of in the following way. Imagine the Riemann sphere $\mathcal{S}$ of null (i.e. lightlike) directions at some point O in space-time. (See Fig. 3.) We are thinking of the geometry in the tangent 4-space of the point O. The flagpole direction is represented by some point P on a sphere of cross-section of the future null cone of O, which we identify with $\mathcal{S}$, and we choose a point P′ on $\mathcal{S}$ infinitesimally separated from P. The straight line extended out from P in the direction of P′, when joined to O, defines the required flag half-plane. We note that as the point P′ rotates about P, the flag plane rotates about the flagpole. The spinor itself is defined only up to sign by this geometry, but we must take note that if P′ rotates continuously around P through 2π, the spinor becomes replaced by its *negative*. To reach the original 2-spinor by this procedure, the rotation of the flag plane would have to be through 4π.

I had found that 2-spinor methods were surprisingly valuable in giving us insights into the formalism of *general relativity* that were different from those that the standard Lorentzian tensor framework readily provides. Most immediately striking was the very simple-looking 2-spinor expression for Weyl's conformal curvature [29] (see also [32]). Whereas the usual Weyl-tensor quantity C_{abcd}, has a somewhat complicated collection of symmetry and trace-free conditions, the corresponding 2-spinor is simply a *totally symmetric* complex 2-spinor quantity Ψ_{ABCD}.

Some comments concerning the 2-spinor index notation being used here are appropriate. Capital italic Latin index letters $A, B, C, \ldots$ refer to the (2-complex dimensional) spin space if they are upper indices, and to the *dual* of this space if lower ones; *primed* such letters $A', B', C', \ldots$ refer to the complex-conjugate spin space. The tensor product of the spin space with its complex conjugate is identified with the (complexified) tangent space to the space-time, at each of its points. In general, I shall take these as *abstract* indices, in the sense described in my book with Wolfgang Rindler, *Spinors and Space-Time*, volume 1 [31], so that no coordinate system is implied, either for the space-time or for a basis for the spin-space. This is notationally very handy, because the space-time indices $a, b, c, \ldots$ can then be thought of as "shorthand" for the spinor index pairs:

$$a = AA', \quad b = BB', \quad c = CC', \ldots$$

The spin-space (and hence also its dual and complex conjugate) has a symplectic structure defined by the skew-symmetric quantities

$$\varepsilon_{AB}, \quad \varepsilon^{AB}, \quad \varepsilon_{A'B'}, \quad \varepsilon^{A'B'},$$

these being used for lowering or raising indices, (where we must be a little careful about signs and index orderings):

$$\kappa_B = \kappa^A \varepsilon_{AB}, \quad \kappa^A = \kappa_B \varepsilon^{AB}, \qquad \eta_{B'} = \eta^{A'} \varepsilon_{A'B'}, \quad \eta^{A'} = \eta_{B'} \varepsilon^{A'B'}$$

so that on terms of components,

$$\kappa_1 = \kappa^0, \quad \kappa_0 = -\kappa^1, \quad \eta_{1'} = \eta^{0'}, \quad \eta_{0'} = -\eta^{1'},$$

where the component form of each of the epsilons is

$$\begin{pmatrix} 0 & 1 \\ -1 & 0 \end{pmatrix}.$$

The metric tensor, in abstract-index form is

$$g_{ab} = \varepsilon_{AB} \varepsilon_{A'B'},$$

and the abstract-index form of the Weyl conformal curvature tensor is

$$C_{abcd} = \Psi_{ABCD} \varepsilon_{A'B'} \varepsilon_{C'D'} + \varepsilon_{AB} \varepsilon_{CD} \widetilde{\Psi}_{A'B'C'D'}.$$

Here, I have allowed for the case of a complex metric g_{ab}, both Ψ_{ABCD} and $\widetilde{\Psi}_{A'B'C'D'}$ being totally symmetric, where Ψ_{ABCD} describes the anti-self-dual (left-handed) Weyl curvature and $\widetilde{\Psi}_{A'B'C'D}$, the self-dual (right-handed) part. In the case of a real Lorenzian space-time metric ($\bar{\varepsilon}_{AB} = \varepsilon_{AB}$) and $\widetilde{\Psi}_{A'B'C'D'}$ is the complex conjugate of Ψ_{ABCD}:

$$\widetilde{\Psi}_{A'B'C'D'} = \overline{\Psi}_{A'B'C'D'},$$

but it will be important for what follows that we consider the complex case also, as we shall be concerned with self-dual (complex vacuum) space-times, for which

$\Psi_{ABCD} = 0$ and anti-self-dual ones, for which $\widetilde{\Psi}_{A'B'C'D} = 0$, later (these complex fields being regarded as wavefunctions).

1.4 Zero Rest-Mass Fields

We find that in the case of a (real Lorentzian) vacuum metric (with or without cosmological constant), the Bianchi identities become

$$\nabla^{AA'}\Psi_{ABCD} = 0$$

which may be compared with the Maxwell equations in charge-free space-time

$$\nabla^{AA'}\varphi_{AB} = 0,$$

where φ_{AB} relates to a (complex) Maxwell field tensor F_{ab} in the same way as Ψ_{ABCD} relates to C_{abcd}, namely

$$F_{ab} = \varphi_{AB}\varepsilon_{A'B'} + \varepsilon_{AB}\tilde{\varphi}_{A'B'},$$

where φ_{AB} describes the anti-self-dual (left-handed) part of the field and $\tilde{\varphi}_{A'B'}$, the right-handed (self-dual) part. For a real Maxwell field, they are complex conjugates of each other:

$$\tilde{\varphi}_{A'B'} = \overline{\varphi}_{A'B'}.$$

I had become interested in the issue of finding solutions of the general equation

$$\nabla^{AA'}\phi_{ABc....E} = 0$$

in (conformally) flat space-time, $\phi_{ABc....E}$ being symmetric in its n spinor indices, the equation being the (conformally invariant) free-field equation for a massless field of spin $n/2$ [33–35]. This equation (together with the wave equation in suitably conformally invariant form) had a particular importance for me, and I believed it to have a rather basic status in relativistic physics. For I had come to the view that nature might have a "massless' structure at its roots, mass itself being a secondary phenomenon. In around 1961 (see [36]) I had found a formula for obtaining the solution of this field equation from general data freely specified on a null initial hypersurface. I had formed the view that this formula had a certain kinship with the Cauchy integral formula for obtaining the value of a holomorphic function at some point of the complex plane in terms of the function's values along a closed contour surrounding that point. I had felt that, in some sense, this massless field equation might be akin to the Cauchy-Riemann equations. There had to be an unusual "complex" way of looking at Minkowski space, I had surmised, in which the massless field equations were simply a statement of *holomorphicity*—but in what sense could this be true?

There was one remaining feature that I felt sure must be represented, as part of this mysterious "complex" way of looking at space-tine. This arose from a discussion that I had had with Engelbert Schücking when I shared an office with him in the spring of 1961 at Syracuse University in New York State. Engelbert had persuaded me of the key importance to quantum field theory of the splitting of field amplitudes into positive and negative frequency parts. I was not happy with the standard procedure of first resolving these amplitudes into Fourier components and then selecting the positive ones, as not only did this strike me as too "top-heavy", but also the Fourier analysis is not conformally invariant—and I had come to believe that this conformal invariance, being a feature of massless fields, was important (again, something that had been stressed to me by Engelbert).

I had become aware that for complex functions defined on a line (thought of as the time line) we may understand their splitting into positive- and negative-frequency parts in the following way. We view this time line as being the equator of real numbers in a *Riemann sphere* which, as before, is the complex plane compactified by the single point labelled by "∞", but where the sphere is now being oriented somewhat differently from that of Fig. 2, with the real numbers now featuring as the equator (increasing as we proceed in an anti-clockwise sense un the horizontal plane), rather than the unit circle. Functions defied on this equatorial circle which extend holomorphically into the southern hemisphere (with usual conventions) are the functions of positive frequency, and those which extend holomorphically into the northern hemisphere are those of negative frequency. An arbitrary complex function defined on this circle can be split into a function extending globally into the southern hemisphere and one globally into the northern hemisphere—uniquely except for an ambiguity with regard to the constant part—and this provides us with the required positive/negative frequency split, without any resort to Fourier analysis. I wanted to extend this picture into something more global, with regard to space-time, and I had in mind that my sought-for "complex" way of looking at Minkowski space should exhibit something strongly analogous to this division into two halves, where the boundary between the two could be interpreted in "real" terms, in some direct way. This had then set the stage for the emergence of twistor theory!

2 The Emergence of Twistor Theory

2.1 Robinson Congruences

A colleague of mine, Ivor Robinson, who had taken up a position at what later became the University of Texas at Dallas, had been working on finding global non-singular *null* solutions of Maxwell's free-field equations in Minkowski space-time

$\mathbb{M}$, where "null" in this context means that the invariants of the field tensor F_{ab} vanish, i.e. $F_{ab}F^{ab} = 0 = {}^*F_{ab}F^{ab}$ where ${}^*F_{ab}$ is the Hodge dual of F_{ab}. Equivalently, in 2-spinor terms, $\varphi_{AB}\varphi^{AB} = 0$, which tells us that

$$\varphi^{AB} = \kappa^A\kappa^B,$$

for some κ^A. It is not hard to show that the Maxwell equations then imply that the flagpole direction of κ^A points along a 3-parrameter family—a *congruence*—of null straight lines, which turn out to be what is called "shear-free", which means that although the lines may diverge, converge, or rotate, locally, there is no shear (or distortion) as we follow along the lines.

Although, not relevant to the discussion at the moment, it is worth noting that the study of shear-free congruences of rays in *curved* space-times has a considerable historical significance—where I use the term "ray" simply to mean a null (i.e. light-like) geodesic in space-time. In particular, the well-known Kerr solution [37, 38] of the Einstein vacuum equations for a rotating black hole possesses a shear-free ray congruence, and this played a key role in its discovery, as it did also in Newman's generalization to an electrically charged black hole [39], and also in the Robinson-Trautman gravitationally radiating exact solutions [40], among other examples. As in the case of Minkowski space $\mathbb{M}$, as described above, it is also true that for any null solution φ^{AB} of Maxwell's equations in curved space-times, the flagpole directions of the κ^A-spinors point along a shear-free family of rays.

A simple example of a shear-free null congruence in $\mathbb{M}$ is obtained from any fixed choice of a ray L in $\mathbb{M}$, where the family of all rays that meet L provides a shear-free ray congruence. I refer to such a congruence as a*special Robinson congruence*, and this includes the limiting case when L is taken out to infinity, so our congruence becomes a family of parallel rays in $\mathbb{M}$. Ivor Robinson had developed ways of producing null solutions of the Maxwell equations, starting from any given shear-free null congruence, but when applied to the special congruences just described, he found that singularities would arise along the line L itself (except in the otherwise unsatisfactory case where L is a is at infinity). Desiring a singularity-free solution, he provided the following ingenious trick. Consider, instead, solutions of Maxwell's equations in the *complexified* Minkowski space-time $\mathbb{CM}$, and displace the line L in a complex direction, so that it lies in $\mathbb{CM}$, but entirely outside its real part $\mathbb{M}$. Complex analytic solutions of Maxwell's equations, based on the complex "special Robinson congruence" defined by the *displaced* L need not now be singular within $\mathbb{M}$, and the flagpoles of the κ^A-spinors within $\mathbb{M}$ now point along an entirely *non-singular* sear-free ray congruence in $\mathbb{M}$, which I later named a (general) *Robinson congruence*.

I became highly intrigued by the geometry of general Robinson congruences, and I soon realized that one could describe them in the following way. Consider an arbitrary spacelike 3-plane E in Minkowski 4-space $\mathbb{M}$. E will have the geometry of ordinary Euclidean 3-space, and each ray of the congruence will meet E in a single point, at which we can determine the location of that ray within $\mathbb{M}$ by specifying a unit 3-vector $\boldsymbol{n}$ at that point, pointing in the spatial direction that is the orthogonal projection into E of the null direction of L there. Thus we have a vector field of $\boldsymbol{n}s$ within E to represent the Robinson congruence. After some thought I realized what the nature of this vector field must be. The $\boldsymbol{n}$-vectors are tangents to the oriented circles (together with one oriented straight line) obtained by stereographic projection of a family of Clifford parallels on a 3-sphere. See Fig. 1 for a picture of this configuration, and Ref. [41] for a detailed derivation. The large arrow at the top right indicates the direction in which the configuration appears to move with the speed of light by continually reassembling itself in that direction, as E evolves into the future.

By examining this configuration, and counting the number of degrees of freedom that such configurations have, I realized that the space of Robinson congruences must be 6-dimensional. Moreover, it was reasonably clear to me that by its very mode of construction, this space ought to have a *complex structure*, and so must be, in a natural way, a complex 3-manifold. Within this space would lie the space of special Robinson congruences, each of which would be determined by a single ray (namely L). The space of rays in $\mathbb{M}$ is 5-real-dimensional, and it divides the space of general Robinson congruences into two halves, namely those with a right-handed twist and those with a left-handed twist. The complex 3-space of Robinson congruences, which came to be known as "projective twistor space" appeared to be just what I believed was needed, where the "real" part of the space (representing light rays in $\mathbb{M}$, or their limits at infinity) would, like the "real" equator of the Riemann sphere described at the end of §A4, divide the entire space into two halves. This, indeed appeared to be exactly the kind of thing that I was looking for!

2.2 Twistors in Terms of 2-Spinors

To be more explicit about things, and to understand precisely how the space of Robinson congruences does provide a compact complex 3-manifold divided in two by the real 5-space of special Robinson congruences, let us turn again to the relativistic 2-spinor formalism of §A3. We shall see how this allows us to provide a very neat description of individual rays in $\mathbb{M}$ and how this generalizes to describe general Robinson congruences. Consider some ray Z in $\mathbb{M}$, and let us assign a *strength* to this ray in the form of a null 4-momentum covector p_a pointing along Z at each of its points, parallel-propagated along Z. In fact, let us go a little further than this by assigning a (dual, conjugate) 2-spinor $\pi_{A'}$ parallel-propagated along Z, where

$$p_a = \bar{\pi}_A \pi_{A'}$$

so that in addition to $\pi_{A'}$'s flagpole pointing along Z, it also assigns a flag plane (and spinor sign) to Z which is parallel-propagated along it. This will be referred to as a *spinor scaling* for the ray Z.

We need to choose a space-time origin point O within $\mathbb{M}$, so that any point X of $\mathbb{M}$ can be labelled by a position vector x^a at O. Then if X is any point on the ray Z, we can define a 2-spinor ω^A by the equation,

$$\omega^A = ix^{AA'}\pi_{A'}$$

and we find that ω^A remains unchanged if X is replaced by any other point on the ray Z. The pair $(\omega^A, \pi_{A'})$, serve to identify the ray Z, together with a spinor scaling for Z.

The 2-spinors ω^A and $\pi_{A'}$ are the *spinor parts* (with respect to the origin O) of the *twistor* Z^α, which represents the spinor-scaled ray Z, and often one simply writes

$$Z^\alpha = (\omega^A, \pi_{A'}).$$

However, for a *ray*, there is a particular equation that must hold between the spinor parts, namely

$$\omega^A\bar{\pi}_A + \pi_{A'} + \bar{\omega}^{A'} = 0$$

which follows from the fact that the vector x^a is *real*, so that $x^{AB'}$ has the *Hermitian* property $\overline{x^{AB'}} = x^{BA'}$. The above equation can be rewritten as

$$Z^\alpha\bar{Z}_\alpha = 0.$$

where $\bar{Z}_\alpha$ the complex conjugate of Z^α

$$\bar{Z}_\alpha = (\bar{\pi}_A, \bar{\omega}^{A'}),$$

(and note the reverse order of the spinor parts) is a *dual* twistor. When $Z^\alpha\bar{Z}_\alpha = 0$, we refer to Z^α as a *null* twistor, so it is that the null twistors represent (spinor-scaled) rays in $\mathbb{M}$—or rays at $\mathbb{M}$'s infinity.

The above equation

$$\omega^A = \mathrm{i}x^{AA'}\pi_{A'}$$

is referred to as the *incidence relation* between the space-time point X and the twistor $Z^\alpha = (\omega^A, \pi_{A'})$. We may also be interested in this incidence relation when X is allowed to be a complex point. Likewise, for a dual twistor

$$W_\alpha = (\lambda_A, \mu^{A'}),$$

incidence with a (possibly complex) point X is expressed as

$$\mu^{A'} = -\mathrm{i}x^{AA'}\lambda_A.$$

2.3 *Minkowski Space Compactified, Complexified, and Twistor Spaces*

At this juncture It would be helpful to clarify the nature of "infinity", with regard to Minkowski space $\mathbb{M}$. We recall that when a ray L is characterized in terms of the null congruence of rays that intersect L, we were led to consider the ray congruences that consist entirely of *parallel* rays, arising when L is moved out to infinity. There is a whole 2-sphwere's-worth of such systems of parallel rays one for each null direction. Thus the family of limiting rays L at infinity constitutes a kind of "light cone at infinity". Indeed, this provides us with the picture of *compactified Minkowski space* $\mathbb{M}^{\#}$ (with topology $S^1 \times S^3$), as illustrated in Fig. 4, where Fig. 4a shows how a future and past null boundary regions can be supplied for Minkowski space, while Fig. 4b shows how these two boundaries are to be identified so as to produce the highly symmetrical compact Lorentzian-conformal manifold $\mathbb{M}^{\#}$. Every ray within $\mathbb{M}^{\#}$ is compactified by a single point to become a topological circle. The global symmetry group of $\mathbb{M}^{\#}$ is what it referred to as the 15-parameter conformal symmetry group of flat space-time.

Now let us consider how to represent a *non*-null twistor Z^{α} in a geometrical way, it is best to think in terms of the family of null twistors Y^{α} that are *orthogonal* to Z^{α} in the sense that

$$Z^{\alpha}\bar{Y}_{\alpha} = 0$$

(or, equivalently $Y^{\alpha}\bar{Z}_{\alpha} = 0$). If Z^{α} were a null twistor—where Y^{α} is *given* as a null twistor—these respectively representing rays Z and Y, then this vanishing of their

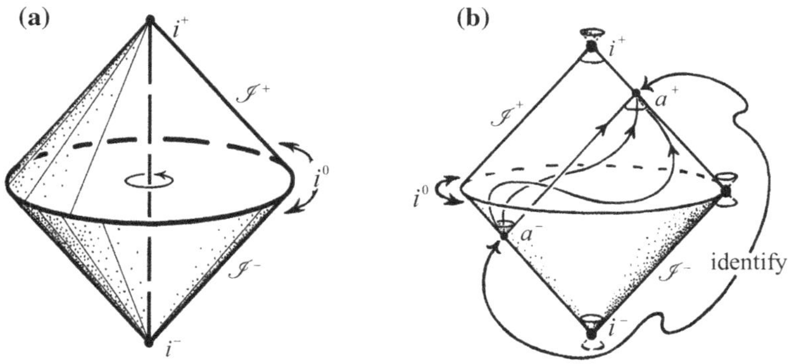

Fig. 4 **a** A conformal picture indicating how Minkowski space-time $\mathbb{M}$ acquires its future null boundary I^+, a null 3-surface supplying future end-points to rays in $\mathbb{M}$ and, similarly, a past null boundary I^- supplying past end-points to rays in $\mathbb{M}$. There are also three other conformal boundary points i^+, i^-, and i^0 denoting future, past, and spacelike infinity, respectively. **b** To complete the picture of compactified Minkowski space $\mathbb{M}^{\#}$, we must identify I^+, with I^-, so that the future end-point a^+ of any ray in $\mathbb{M}$ is identified with its past end-point a^-. Also the three points i^+, i^-, and i^0 must be identified

scalar product asserts that these rays *intersect* (perhaps at infinity). Accordingly, if Z is fixed, then this condition on Y tells us that the Y belongs to the special Robinson congruence defined by the ray Z. Now, let Z be a fixed *non*-null twistor (but where Y remains null). Then the congruence of Y-rays subject to orthogonality with Z will provide a *general* Robinson congruence. See [41] for details.

The space $\mathbb{T}$ of all twistors Z^α is a 4-dimensional complex vector space, with pseudo-Hermitian scalar product $(Z^\alpha \bar{Y}_\alpha)$ of split signature $(+\ +\ -\ -)$. Geometrical notions are often best expressed in terms of the *projective* twistor space $\mathbb{PT}$ of twistors up to proportionality, this being a complex projective 3-space $\mathbb{CP}^3$. This compact complex manifold $\mathbb{PT}$—or, more strictly, in accordance with the above discussion, the $\mathbb{CP}^3$ of *dual* projective twistors $\mathbb{PT}^*$—can indeed be identified with the space of Robinson congruences referred to above. The *dual* twistor space $\mathbb{T}^*$ is identified with the *complex conjugate* space $\bar{\mathbb{T}}$ of $\mathbb{T}$ via this pseudo-Hermitian structure. The points of the dual projective space $\mathbb{PT}^*$ represent the complex projective planes within $\mathbb{PT}$. The complex projective lines within $\mathbb{PT}$ correspond to points of the complexified compactified Minkowski space $\mathbb{CM}^\#$.

Whereas, generally speaking, it is the projective twistor space $\mathbb{PT}$ that is useful to us if we are thinking of geometrical matters, the space $\mathbb{T}$ is appropriate if we are concerned with the *algebra* of twistors. For a non-zero twistor Z^α, we can have three algebraic alternatives. These are:

$Z^\alpha \bar{Z}_\alpha > 0$, for a *positive* or *right-handed* twistor Z^α, belonging to the space $\mathbb{T}^+$,

$Z^\alpha \bar{Z}_\alpha < 0$, for a *negative* or *right-handed* twistor Z^α, belonging to the space $\mathbb{T}^-$,

$Z^\alpha \bar{Z}_\alpha = 0$, for a *null* twistor Z^α, belonging to the space $\mathbb{N}$.

The entire twistor space $\mathbb{T}$ is the disjoint union of the three parts $\mathbb{T}^+$, $\mathbb{T}^-$, and $\mathbb{N}$, as is its projective version $\mathbb{PT}$ the disjoint union of the three parts $\mathbb{PT}^+$, $\mathbb{PT}^-$, and $\mathbb{PN}$ (see Fig. 5).

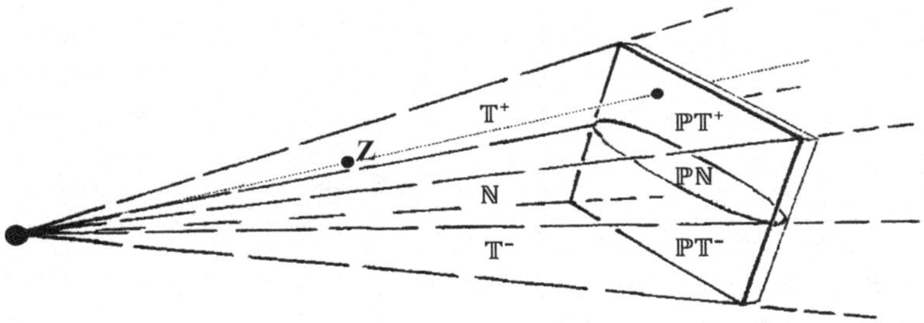

Fig. 5 The way that the various parts of twistor space $\mathbb{T}$ relate to their various projective counterparts of $\mathbb{PT}$. Each point of $\mathbb{PT}$ represents a 1-dimenional vector subspace of $\mathbb{T}$

2.4 Helicity and Relativistic Angular Momentum

It is the space $\mathbb{PN}$ that has the most direct physical interpretation, since its points correspond to world-lines of free classical massless particles, which we can think of as the classical histories of (pointlike) photons in free motion, though possibly at infinity, as a limiting case in Minkowski space-time $\mathbb{M}$; see Fig. 6. As indicated above, points of complexified Minkowski space $\mathbb{CM}$ (that are not at infinity) are represented as (complex projective) lines in $\mathbb{PT}$, but so also are *all* the points of the complexified *compactified* Minkowski space $\mathbb{CM}^\#$. Those lines that lie in $\mathbb{PN}$, represent points of the *real* space-time $\mathbb{M}$ (possibly at infinity), but since these lines are still complex projective lines, they are indeed *Riemann spheres*, in accordance with the ambitions put forward in §A2; see Fig. 6.

In Fig. 7 this picture is extended to include a physical interpretation of *non*-null twistors, where points of $\mathbb{PT}^+$ and $\mathbb{PT}^-$ are represented, in Minkowski space, as though they are light rays with a twist about them. This is schematic, but indeed these points can be regarded as representing massless particles with spin. In relativistic physics, if a massless particle has a *non-zero spin*, the "spin-axis" must be directed parallel or anti-parallel to the particle's velocity. We say that the particle has a *helicity* s, that can be positive or negative (or zero, for a spinless massless particle). If $s \neq 0$, then the particle's space-time trajectory is not precisely defined (in a relativistically invariant way) as a world-line, but can be specified in terms of its 4-momentum p_a and 6-angular momentum M^{ab} about some chosen space-time origin point $\mathbf{O}$. These must be subject to

$$p_a p^a = 0, \quad p_0 > 0, \quad M^{(ab)} = 0, \quad \tfrac{1}{2}\varepsilon_{abcd}\, p^b M^{cd} = s p_a$$

(curved or square brackets around indices respectively denoting symmetric or anti-symmetric parts), where $\varepsilon_{abcd} = \varepsilon_{[abcd]}$ is the Levi-Civita tensor fixed by its

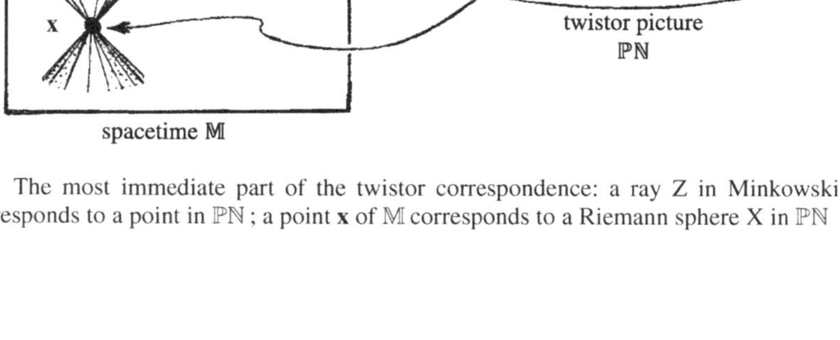

Fig. 6 The most immediate part of the twistor correspondence: a ray Z in Minkowski space $\mathbb{M}$ corresponds to a point in $\mathbb{PN}$; a point $\mathbf{x}$ of $\mathbb{M}$ corresponds to a Riemann sphere X in $\mathbb{PN}$

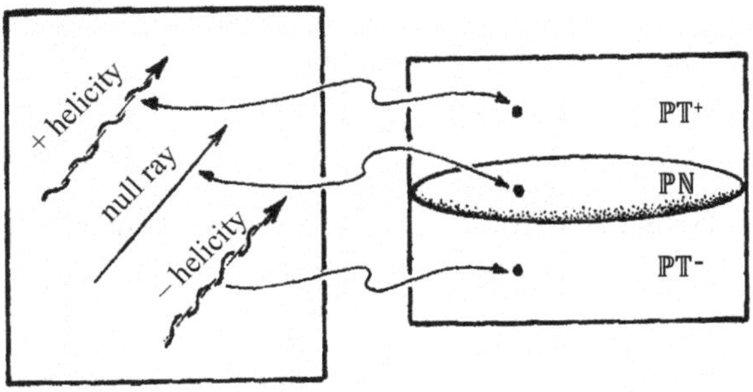

Fig. 7 Classical massless particles with positive (right-handed) helicity can be represented as points of $\mathbb{PT}^+$ and those with negative (left-handed) helicity, as points of $\mathbb{PT}^-$

component value $\varepsilon_{0123} = 1$ in a right-handed orthonormal Minkowskian frame (with time-axis basis vector δ_0^a so p_0 is the particle's energy, in units where the speed of light $c = 1$). Note that $\frac{1}{2}\varepsilon_{abcd}M^{cd} =^* M_{ab}$ is the Hodge dual of M^{ab}. The connection between these quantities and twistor theory is that if

$$Z^\alpha = (\omega^A, \pi_{A'})$$

then we can make the interpretation

$$p_{AA'} = \pi_{A'}\bar{\pi}_A, \quad M^{AA'BB'} = \mathrm{i}\omega^{(A}\bar{\pi}^{B)}\varepsilon^{A'B'} - \mathrm{i}\bar{\omega}^{(A'}\pi^{B')}\varepsilon^{AB},$$

and all the above conditions are automatically satisfied, provided that $\pi_{A'} \neq 0$. Conversely, the twistor Z^α (with $\pi_{A'} \neq 0$) is determined, uniquely up to a phase multiplier $e^{\mathrm{i}\theta}$, by p_a and M^{ab}, subject to these conditions. The helicity s finds the very simple (and fundamental) expression

$$\begin{aligned} 2s &= \omega^A\bar{\pi}_A + \bar{\pi}_{A'}\bar{\omega}^{A'} \\ &= Z^\alpha\,\bar{Z}_\alpha. \end{aligned}$$

There is, however, the subtlety referred to above that in this interpretation of a non-null twistor, when the helicity s is non-zero, there is no actual world-line that can describe the particle's location in a relativistically invariant way. This is an important undercurrent to the application of twistor ideas in general relativity as discussed in C7 and C8.

2.5 Description Under Shift of Origin

Under a displacement of the origin O to a new point Q of $\mathbb{M}$,

$$\mathrm{O} \mapsto \mathrm{Q},$$

where the position vector $\overrightarrow{OQ}$ is (in abstract-index form) q^a, the spinor parts of the twistor $Z^\alpha = (\omega^A, \pi_{A'})$ undergo

$$\omega^A \mapsto \omega^A - iq^{AA'}\pi_{A'}, \quad \pi_{A'} \mapsto {}_{A'} .$$

For a dual twistor $W_\alpha = (\lambda_A, \mu^{A'})$, we correspondingly have

$$\lambda_A \mapsto \lambda_A, \quad \mu^{A'} \quad \mapsto \mu^{A'} + iq^{AA'}\lambda_A.$$

This turns out to be consistent with the standard transformation of M^{ab} (and p_a) under origin change, where the position vector x^a of a space-time point X correspondingly undergoes

$$x^a \mapsto x^a - q^a.$$

There is a connection between the above direct *physical* interpretation of a twistor in terms of angular momentum—particularly a non-null twistor Z^α—and the *Robinson congruence* defined by Z^α. This congruence is provided by the family of rays defined by the null (dual) twistors W_α satisfying

$$Z^\alpha W_\alpha = 0.$$

To see the connection with singular momentum, let us examine this relation at an arbitrary point Q of $\mathbb{M}$, where we now take Q as our origin point. We are interested in the ray W of the congruence which passes through Q. With respect to Q, as origin, W_α then takes the form

$$W_\alpha = (\lambda_A, 0)$$

($\mu^{A'}$ being zero, since W_α is now incident with the origin point Q; see §B2), so that the relation $Z^\alpha W_\alpha = 0$ now becomes

$$\omega^A \lambda_A = 0,$$

at the point Q. This tells us that the flagpole direction of ω^A is the same as that of λ^A, namely the direction of the ray W. Tus, the angular momentum M^{ab} of the spinning massless particle determined by Z^α has a structure that is characterized by the flagpole directions of its two spinor parts with respect to Q. We may refer back to Fig. 1, to see the curious spatial geometry of all this, where the flagpole directions of ω^A twist around in this complicated (Robinson congruence) way, while that of π_A simply points in the direction of motion of the configuration. It may perhaps be mentioned, that the choice of letters "π" and "ω" come from the normal usage of "p" for momentum, and "omega" for angular momentum.

3 Fields, Quantization and Curved Space-Time

3.1 Twistor Quantization Rules

Up to this point, we have been considering twistor theory only in relation to classical physics in flat space-time geometry. *Quantum* twistor theory—and, indeed, as we shall be seeing later, space-time *curvature,* involves considering twistors (and dual twistors) as non-commuting operators, satisfying certain *commutation laws*:

$$Z^\alpha \bar{Z}_\beta - \bar{Z}_\beta Z^\alpha = \hbar\delta^\alpha_\beta$$

and, as far as our current considerations go,

$$Z^\alpha Z^\beta - Z^\beta Z^\alpha = 0, \quad \bar{Z}_\alpha \bar{Z}_\beta - \bar{Z}_\beta \bar{Z}_\alpha = 0$$

[41, 42]. Now the twistors are taken to be *linear operators* generating a non-commutative algebra $\mathbb{A}$ whose elements are taken to be acting on an appropriate quantum "ket-space" $|\ldots\rangle$ of some kind [43], but it is best not to be specific about this, just now. We could alternatively think of our operators as *dual* twistors, subject to the commutation laws

$$W_\alpha \bar{W}^\beta - \bar{W}^\beta W_\alpha = -\hbar\delta^\alpha_\beta$$

and

$$W_\alpha W_\beta - W_\beta W_\alpha = 0, \quad \bar{W}^\alpha \bar{W}_\beta - \bar{W}_\beta \bar{W}_\alpha = 0,$$

which is the same thing as before, but with $\bar{Z}_\alpha$ re-labelled as W_α.

These commutation laws are *almost* implied by the standard quantum commutators for 4-position and 4-momentum

$$p_a x^b - x^b p_a = i\hbar\delta^b_a$$

but there appears to be an additional input related to the issue of helicity. By direct calculation, we may verify that the twistor commutation laws reproduce exactly the (more complicated-looking) standard commutation laws for p_a and M^{ab} that arise from their roles as translation and Lorentz-rotation generators of the Poincaré group (see [41]). In this calculation, there is no factor-ordering ambiguity in the expressions for p_a and M^{ab} in terms of the spinor parts of Z^α and $\bar{Z}_\alpha$ (owing to the symmetry brackets). Yet, the calculation for the helicity s (writing the operator as $\mathbf{s}$) yields:

$$\mathbf{s} = \frac{1}{4}(Z^\alpha \bar{Z}_\alpha + \bar{Z}_\alpha Z^\alpha).$$

3.2 Twistor Wavefunctions

In accordance with standard quantum-mechanical procedures, in order to express wavefunctions for massless particles in twistor terms, we need functions of Z^α that are "independent of $\bar{Z}''_\beta$. This means "annihilated by $\partial/\partial\bar{Z}_\beta$" i.e. *holomorphic* in Z^α (Cauchy–Riemann equations). Thus, a twistor wavefunction (in the Z^α-description) is holomorphic in Z^α and the operators representing Z^α and $\bar{Z}_\alpha$ act:

$$Z^\alpha \rightsquigarrow Z^\alpha \times, \quad \bar{Z}_\alpha \rightsquigarrow -\hbar\frac{\partial}{\partial Z^\alpha},$$

Alternatively, we could be thinking of functions of $\bar{Z}_\alpha$ that are "independent of Z^β", i.e. *anti*-holomorphic in Z^α. Here it would be better to re-name $\bar{Z}_\alpha$ as W_α and consider functions *holomorphic* in W_α. Accordingly, in the *dual* twistor W_α-description, a wavefunction must be holomorphic in W_α and we have the operators representing $\bar{W}^\alpha$ and W_α, again satisfying the required commutation relations, but now with:

$$\bar{W}^\alpha \rightsquigarrow \hbar\frac{\partial}{\partial W_\alpha}, \quad W_\alpha \rightsquigarrow W_\alpha \times .$$

If we are asking that our wavefunction describe a (massless) particle of *definite helicity*, then we need to put it into an eigenstate of the *helicity operator* **s**, which, by the above, is

$$\mathbf{s} = -\frac{1}{2}\hbar(Z^\alpha\frac{\partial}{\partial Z^\alpha} + 2)$$

in the Z^α-description, and

$$\mathbf{s} = \frac{1}{2}\hbar(W_\alpha\frac{\partial}{\partial W_\alpha} + 2)$$

in the W_α-description. These are simply displaced *Euler homogeneity operators*

$$\Upsilon = Z^\alpha\frac{\partial}{\partial Z^\alpha} \quad \text{or} \quad \widetilde{\Upsilon} = W_\alpha\frac{\partial}{\partial W_\alpha},$$

so a helicity eigenstate, with eigenvalue s, in the Z^α-description requires a holomorphic twistor wavefunction $f(Z^\alpha)$ that is *homogeneous* of degree

$$n = -2s - 2,$$

where I henceforth adopt $\hbar = 1$. Then $2s$ is an integer (odd for a fermion and even for a boson). In the W_α-description, the dual twistor wavefunction $\tilde{f}(W_\alpha)$ is homogeneous of degree $\tilde{n}$ where

$$\tilde{n} = 2s - 2.$$

3.3 Twistor Generation of Massless Fields and Wavefunctions

In ordinary space-time terms, the position-space wavefunction of a massless particle of helicity $2s$ [33–35] satisfies a *field equation*, this being expressible in the 2-spinor form

$$\nabla^{AA'}\psi_{AB\ldots E}=0, \Box\psi=0, \text{ or } \nabla^{AA'}\tilde{\psi}_{A'B'\ldots E'}=0,$$

for the integer $2s$ satisfying $s<0$, $s=0$, or $s>0$, respectively, these equations having been already considered in §A4, but where the scalar case $s=0$ is now included also, involving the D'Alembertian

$$\Box=\nabla_a\nabla^a.$$

We have *total symmetry* for each of the $|2s|$-index quantities

$$\psi_{AB\ldots E}=\psi_{(AB\ldots E)} \text{ and } \tilde{\psi}_{A'B'\ldots E'}=\tilde{\psi}_{A'B'\ldots E')}.$$

What is the connection between the holomorphic twistor wavefunction $f(Z^\alpha)$, or dual twistor wavefunction $\tilde{f}(W_\alpha)$, with these space-time equations? In most direct terms this is given (for suitable numerical constants k, k') by contour integrals [42, 44, 45][1]:

$$\psi_{AB\ldots E}(\mathbf{x})=k\oint_{\omega=\mathrm{i}\mathbf{x}\cdot\pi}\frac{\partial}{\partial\omega^A}\frac{\partial}{\partial\omega^B}\cdots\frac{\partial}{\partial\omega^E}f(\boldsymbol{\omega},\boldsymbol{\pi})\delta\boldsymbol{\pi}, \qquad \text{if } s\le 0;$$

$$\tilde{\psi}_{A'B'\ldots E'}(\mathbf{x})=k'\oint_{\omega=\mathrm{i}\mathbf{x}\cdot\pi}\pi_{A'}\pi_{B'}\ldots\pi_{E'}f(\boldsymbol{\omega},\boldsymbol{\pi})\delta\boldsymbol{\pi}, \qquad \text{if } s\ge 0.$$

Here $AB\ldots E$ or $A'B'\ldots E'$ are $|2s|$ in number, and the 1-form $\delta\boldsymbol{\pi}$ is

$$\delta\boldsymbol{\pi}=\varepsilon^{F'G'}\pi_{F'}\mathrm{d}\pi_{G'},$$

and where k and k' are suitable constants. I have taken the liberty of writing x^a, ω^A, and $\pi_{A'}$ without their abstract indices in places here, and using bold-face upright type instead. The *contour*, for these integrals, lies within the Riemann sphere, in $\mathbb{PT}$, of twistors $Z^\alpha=(\omega^A,\pi_{A'})$ satisfying the incidence relation $\omega^A=ix^{AA'}\pi_{A'}$ (written $\boldsymbol{\omega}=\mathbf{ix}\cdot\pi$, below the integral sign), which removes the ω^A-dependence and introduces x^a-dependence, and then the contour integration itself removes the

[1] The upper one of these two integral expressions was put forward by Lane Hughston, as a complement to the lower one, which I had found earlier. The significance of having both of these was not recognized, initially, but it was later realized that both are needed for the complete picture.

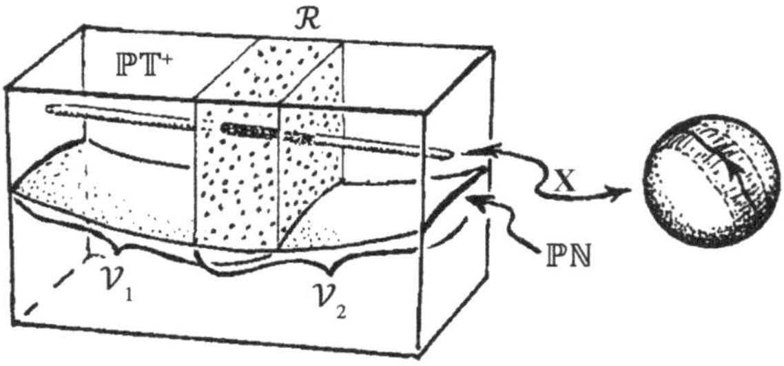

Fig. 8 The geometry relevant to the twistor contour integral for a wavefunction. The regions Q_1 and Q_2 of the text are the respective complements, within $\mathbb{PT}^+$, of the depicted regions $\mathcal{V}_2$ and $\mathcal{V}_1$. Here, the open sets $\mathcal{V}_1$, $\mathcal{V}_2$ provide a 2-set open covering of $\mathbb{PT}^+$ and $\mathcal{R}$ is their intersection

$\pi_{A'}$-dependence, leaving us with just x^a-dependence. See Fig. 8, where "X" is the Riemann sphere representing the (complex) point labelled x^a. Satisfaction of the field equations is an immediate consequence of these holomorphic expressions. The 2-form $d\pi_{0'} \wedge d\pi_1' = \frac{1}{2} d\delta\pi$ is sometimes more appropriate to use, rather than $\delta\pi$, the contour then being 2-dimensional, lying in $\mathbb{T}$ rather than $\mathbb{PT}$. In the *dual* twistor description, we have corresponding expressions.

3.4 Singularity Structure for Twistor Wavefunctions

For such expressions to provide non-zero answers, the function f must have appropriate singularities. The situation of specific interest to us here is the case of a wavefunction for a free massless particle, although these formulae can also be used under many other circumstances, such as for real solutions of Maxwell's equations in particular domains. Real solutions can clearly be obtained from the complex ones described here, by taking the real part, the equations to be satisfied being linear. Completely general solutions of the equations are obtained in this way provided that they are analytic. In fact, precursors of these equations were found long ago, for the Laplace equation by Whittaker [46] in 1903, and for the wave equation (in 1904) by Bateman [47] who later generalized it for the Maxwell equations in the 1930s, see [48].

For a wavefunction, we require complex solutions of *positive frequency*, and here is where the important early motivation for twistor theory referred to at the end of §A4 was finally satisfied. To start with, this was only in a way that seemed somewhat odd. But eventually this apparent oddness was re-interpreted as something remarkably "natural" when properly understood, with potentially deep implications.

Let us see how this works. First, we take note of the fact that the family of points of $\mathbb{CM}$ that constitute the sub-region $\mathbb{CM}^+$ known as the "forward tube"—namely the family of points of $\mathbb{CM}$ with position vectors whose imaginary parts are past-pointing timelike—corresponds to the family of lines that lie entirely in $\mathbb{PT}^+$. A complex function ψ, defined on $\mathbb{M}$, which extends smoothly to a holomorphic function throughout $\mathbb{CM}^+$ is indeed of *positive frequency* and conversely, positive frequency being a key requirement for a wavefunction [49]. Thus, for our twistor wavefunction f, we require "regularity" of an appropriate sort throughout the region $\mathbb{PT}^+$. Yet it would be far too restrictive to demand holomorphicity for f over the whole of $\mathbb{PT}^+$ and, in any case, such a function would simply give the answer *zero* when contour integrated. What we seem to need is a function with two separated regions of singularity on each Riemann sphere (complex projective line) that corresponds to a point in $\mathbb{CM}^+$, i.e. to a projective line in $\mathbb{PT}^+$, since then we could obtain a non-trivial answer to the contour integration, the contour being a closed loop on the Riemann sphere that separates the two regions of singularity on the sphere. The situation is depicted on the right-hand side of Fig. 8. This is achieved if the singularities of f are constrained to lie in two disjoint regions Q_1 and Q_2 (each closed in $\mathbb{PT}^+$ so our contour integrations can take place within the holomorphic region $\mathcal{R}$ between them (Fig. 8). Our twistor wavefunction f is thus taken to be holomorphic throughout the (open) region

$$\mathcal{R} \quad = \quad \mathbb{PT}^+ - (Q_1 \cup Q_2).$$

This, indeed, appears to be a somewhat odd requirement for the twistor description of such a fundamental thing as a massless particle's wavefunction. Moreover, the region $\mathcal{R}$ is very far from being invariant under the holomorphic motions of $\mathbb{PT}^+$, some of these representing the non-reflective Poincaré (inhomogeneous Lorentz) motions of Minkowski space $\mathbb{M}$. Any particular choice of the region $\mathcal{R}$ clearly cannot take precedence over any other such choice obtained from the original one by such a motion, so there is clearly much non-uniqueness involved in the choices of $\mathcal{R}$ and f in this description. This difficulty looms large if we try to add two twistor wavefunctions which might have incompatible singularity structures. Linearity is, after all a central feature of quantum mechanics as we currently understand that subject, so how are we to deal with this problem?

3.5 Čech Cohomology

The resolution of these puzzling features, leads us to an understanding of what kind of an entity a twistor wavefunction actually "is". This lies in the notion of *Čech sheaf cohomology*. It is not appropriate that we go into much detail, here, but some indication of the issues involved will be of importance for us. What we find is that the twistor wavefunction f is not really to be viewed as being "just a function" in the ordinary sense, but as representing an element of "1st cohomology" (actually 1st *sheaf* cohomology). I shall call such an entity a *1-function*. An ordinary function, in this terminology, would be a 0-function. There are also higher-order entities referred to as 2-functions, 3-functions, etc., but we shall not need to consider these here.

An important aspect of 1-functions (or of n-functions, where $n > 0$) is that they are *non-local* entities in an essential way (a feature of twistor theory which appears to reflect aspects of non-locality that occur in quantum mechanics). A good intuitive way of appreciating the idea of a 1-function is to contemplate the "impossible tribar" depicted in Fig. 9. Here we have a picture that for each local region, there is an interpretation provided, of a 3-dimensional structure that is unambiguous, except for an uncertainty as to its distance from the viewer's eye. As we follow around the triangular shape, our interpretation remains consistent (though with this mild-seeming ambiguity) until we return to our starting point, only to find that it has actually become *in*consistent! The element of 1st cohomology that is expressed by the picture is a measure of this *global* inconsistency [50].

How might we assign such a measure to the degree of this impossibility? I shall not go into full details here, but the idea is to regard the object under consideration—here the tribar—as being built up from a number of regions (open sets) which together cover the whole object, but which are "locally trivial" in some appropriate topological (or differential) sense. In the case of the tribar, we might have a local picture of each vertex, say $\mathcal{V}_1$, $\mathcal{V}_2$, $\mathcal{V}_3$, where the three pictures overlap pairwise in smaller open regions $\mathcal{V}_i \cap \mathcal{V}_j$, somewhere along each relevant arm of the tribar, so that taken together they provide a picture of the entire tribar. On each overlap region $\mathcal{V}_i \cap \mathcal{V}_j$, we require some numerical measure F_{ij} which describes the ratio of the displacement from the eye that needs to be made for the pictures to be considered to match, and since we need an additive measure we take F_{ij} to be te logarithm of this ratio, and accordingly the F_{ij} are anti-symmetric ($F_{ji} = -F_{ij}$). The triple (F_{12}, F_{23}, F_{31}), taken *modulo* the particular triples of the form ($H_1-H_2, H_2-H_3, H_3-H_1$) where H_i refers to the freedom in the interpretation of the particular vertex picture $\mathcal{V}_i$ provides. The resulting algebraic notion gives us the required cohomology element, describing the

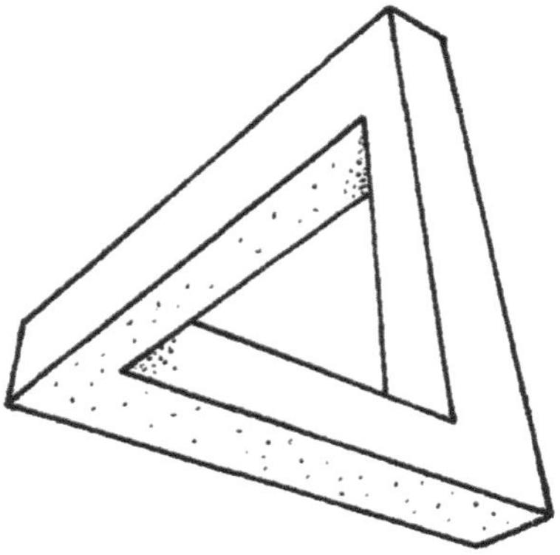

Fig. 9 An impossible "tribar", as an illustration of the notion of (1st) cohomology. There is an unambiguous interpretation of each local part, except for an ambiguity as to the distance from the viewer's eye, but globally this ambiguity leads to a non-local inconsistency. The measure of this inconsistency is an element of 1st cohomology. Twistor wavefunctions exhibit a similar feature, where the rigidity of analytic continuation replaces the rigidity of a material body

degree of impossibility in the figure. This notion is what I am calling a 1-function. For further issues see [3].

This is just to give a little flavor of what sort of an entity a 1-function actually is. More specifically, in the context of twistor theory, we are concerned with complex spaces and holomorphic functions on them. Thus, in the case of a twistor wavefunction there is the important subtlety, in that the global "impossibility" arises from the "rigidity" of *holomorphic functions* rather than that of the solid structures conjured up by the local parts of Fig. 9. But let us be a bit more general here, and imagine some complex manifold $\mathcal{K}$. We shall need a locally finite open covering $\mathcal{C} = (\mathcal{V}_1, \mathcal{V}_2, \mathcal{V}_3, \ldots)$, of $\mathcal{K}$. To define a 1-function f, with respect to $\mathcal{C}$, we assign a holomorphic function f_{ij} on each non-empty pairwise intersection:

$$f_{ij} = -f_{ji} \text{ is holomorphic on } \mathcal{V}_i \cap \mathcal{V}_j,$$

and on each non-empty triple intersection:

$$f_{ij} + f_{jk} + f_{ki} = 0 \quad \text{on} \quad \mathcal{V}_i \cap \mathcal{V}_j \cap \mathcal{V}_k,$$

where the collection $\{f_{ij}\}$ is taken *modulo* corresponding collections of the form $\{h_i - h_j\}$, where each

$$h_i \text{ is holomorphic on } \mathcal{V}_i,$$

so that two 1-functions are considered to be equal if the difference between their $\{f_{ij}\}$ representations is of the form $\{h_i - h_j\}$. This defines a 1-function *with respect to the particular covering* $\mathcal{C}$. For the full definition, we would have to take the direct limit for finer and finer coverings. Fortunately, in the case of complex manifolds, as is being considered here, we are assured that provided that the sets $\mathcal{V}_i$ are of suitable type (e.g. Stein spaces; see [51]) then we gain nothing from taking such a limit, and the 1-function concept is already with us. Nevertheless, in order to add two 1-functions defined by different coverings, we do need to take their common refinement in order to perform this operation, which can be a little complicated in practice.

In the case of main interest here, namely $\mathcal{K} = \mathbb{PT}^+$, it will be adequate for our immediate purposes here simply to take a 2-set covering of $\mathbb{PT}^+$, namely $\mathcal{C} = \{\mathcal{V}_1, \mathcal{V}_2\}$, with open sets given by the complements, within $\mathbb{PT}^+$ of the respective singularity regions Q_2 and Q_1. Then we have our required covering $\mathcal{C}$ (not actually with Stein spaces, but that is not of importance here)

$$\mathbb{PT}^+ = \mathcal{V}_1 \cup \mathcal{V}_2, \qquad \mathcal{R} = \mathcal{V}_1 \cap \mathcal{V}_2;$$

(see Fig. 8), just as we had earlier. The family $\{f_{ij}\}$ consists of the single twistor function f, which, by an abuse of notation I may identify with the 1-function it determines.

In the cases of homogeneity 0 or 2 (left-handed electromagnetism or left-handed linearized gravity, respectively), there are generalizations of the twistor-space contour-integral expressions that allow one to view the 1-function nature of a twistor function in a different light, in which non-linearities of general relativity and particle physics begin to play a significant role. To appreciate this, let us return to the*general* Čech descriptions given earlier, where a locally finite open covering $\mathcal{C} = (\mathcal{V}_1, \mathcal{V}_2, \mathcal{V}_3, \ldots)$ of a complex space $\mathcal{K}$ was considered. In the specification of a 1-function f, in relation to this covering, we required a family of holomorphic functions $\{f_{ij}\}$ defined on the non-empty overlaps $\mathcal{V}_i \cap \mathcal{V}_j$. Here, the functions are entirely *passive*, being just "painted on" the space $\mathcal{K}$. However, we can consider a somewhat more active role for such a 1-function f, such as (a) specifying the generation fa *bundle* above $\mathcal{K}$, or (b) using f to specify the generation of a *deformation* of $\mathcal{K}$ itself. In each case, the rules (see [51]) defining a 1-function are exactly what is needed to fulfil this purpose. However, in each case, this specification by a 1-function would only be as an infinitesimal generator of the bundle or deformed space [except for an Abelian group in case (a)] because of non-linearities. Nevertheless, the general idea expressed in (a) and (b) still holds true; it is just that the linear nature of a 1-function ceases to hold. In effect, we have a kind of "non-linear 1-function".

It was in 1977 that Richard Ward introduced the procedure indicated in case (a) above, first in the situation provided by the (left-handed) Maxwell equations, which allowed interactions of the field with charged particles to be considered. Almost immediately afterwards he showed how this procedure could be generalized to the (left-handed) Yang-Mills equations [52]. This turned out to have considerable importance in the theory of integrable systems (see, for example, [2, 3]). Shortly before all this, in 1976, procedure (b) had been introduced [21], to provide a twistorial representation of all conformally complex-Riemannian 4-manifolds which ae anti-self-dual (i.e. $\tilde{\Psi}_{A'B'C'D'} = 0$; see end of §A3). When an additional simple condition is imposed, this provides not only a (complex) metric but automatically generates the general anti-self-dual solution of the Einstein vacuum equations, either without [21] or with a cosmological constant Λ [22].

3.6 Infinity Twistors and Einstein's Equations

It is a fairly straight-forward procedure to generate the desired deformed twistor spaces satisfying the required conditions ensuring satisfaction of the Einstein Λ-vacuum equations. Basically, what is required is to match appropriate portions on (non-projective) twistor space, while preserving the Euler operator

$$\Upsilon = Z^{\alpha} \frac{\partial}{\partial Z^{\alpha}}$$

and the 2-form

$$\Theta = I_{\alpha\beta} \mathrm{d}Z^\alpha \wedge \mathrm{d}Z^\beta$$

where the anti-symmetrical *infinity twistor* $I_{\alpha\beta}$ (and its dual $I^{\alpha\beta}$) are given by

$$I_{\alpha\beta} = \begin{pmatrix} \frac{\Lambda}{6}\varepsilon_{AB} & 0 \\ 0 & \varepsilon^{A'B'} \end{pmatrix}, \qquad I^{\alpha\beta} = \begin{pmatrix} \varepsilon^{AB} & 0 \\ 0 & \frac{\Lambda}{6}\varepsilon_{A'B'} \end{pmatrix}.$$

We see that $I^{\alpha\beta}$ and $I_{\alpha\beta}$ are both complex conjugates and *duals* of one another:

$$I_{\alpha\beta} = \overline{I^{\alpha\beta}}, \quad I^{\alpha\beta} = \overline{I_{\alpha\beta}},$$
$$I_{\alpha\beta} = \tfrac{1}{2}\varepsilon_{\alpha\beta\rho\sigma} I^{\rho\sigma}, \quad I^{\alpha\beta} = \tfrac{1}{2}\varepsilon^{\alpha\beta\rho\sigma} I_{\rho\sigma},$$

where $\varepsilon_{\alpha\beta\rho\sigma}$ and $\varepsilon^{\alpha\beta\rho\sigma}$ are Levi–Civita twistors, fixed by their anti-symmetry and $\varepsilon_{0123} = 1 = \varepsilon^{0123}$ in standard twistor coordinates. The preservation of Υ and Θ on the overlaps where $\mathcal{V}_i$ as matched to $\mathcal{V}_j$, is ensured, if we shift infinitesimally along the vector field

$$I^{\alpha\beta} \frac{\partial f_{ij}}{\partial Z^\alpha} \frac{\partial}{\partial Z^\beta},$$

where each f_{ij} has homogeneity degree 2 (i.e. $\Upsilon f_{ij} = 2 f_{ij}$, which corresponds to helicity –2). We can imagine exponentiating these infinitesimal deformations to a finite one. In the case of a 2-set covering, we can achieve this explicitly by exponentiating the single function f_{12}, but with larger numbers of sets, we can encounter difficulties in satisfying the required condition on triple overlaps. A simpler procedure for satisfying the required condition of preserving Υ and ΘI is to use generating functions, see [21].

It is, however, not at all a direct matter to obtain the (complex) curved space-time $\mathcal{M}$ from the deformed twistor space $\mathcal{T}$, according to this construction. The points of $\mathcal{M}$ correspond to "lines" in $\mathbb{P}\mathcal{T}$, that are completed Riemann spheres, stretching across from one patch to the other, or perhaps others, if the covering involves more than two patches. These Riemann spheres are not easy to locate, in a general way, since they are determined by the global requirement that they be compact holomorphic curves within $\mathbb{P}\mathcal{T}$ of spherical topology (and belonging to the correct topological family). The very existence of these "lines", as I shall call them (provided that the deformation from $\mathbb{PT}^+$, or some other part of $\mathbb{PT}$, is not too drastic), together with the fact that they belong to a 4-parameter family, depends upon key theorems by Kodaira and Kodaira-Spencer (see [53, 54]). The space representing these lines is the required complex 4-manifold $\mathcal{M}^{\mathbb{C}}$. Its complex conformal structure comes about from the simple fact that meeting lines in $\mathbb{P}\mathcal{T}$ correspond to null separated points in $\mathcal{M}$, and the definition of its metric scaling comes about through use of the form Θ. With this complex metric, the complex 4-manifold automatically satisfies the Einstein Λ-vacuum equations, and the construction provides the general anti-self-dual solution. This procedure has become known as the "non-linear graviton construction" [21],[2] [22]. It has found numerous applications in differential geometry [2, 3].

At this point, it is worth emphasizing an essential but unusual feature of the non-linear graviton construction. This is that the "curvature" in the deformed twistor space is not local, in the sense that a small-enough neighbourhood of a point in the deformed space is identical in structure to that of ordinary flat twistor space (for given Λ). The "curvature" in the deformed space is a *non-local* feature of the space $\mathbb{P}\mathcal{T}$, but in the construction of the "space-time" manifold $\mathcal{M}^{\mathbb{C}}$, we consequently find genuine local curvature in the normal sense (Riemann curvature, Weyl curvature).

As a general approach to physics, however, there has been a fundamental obstruction to progress for some four decades, namely what has become known as the "googly problem", referred to briefly in A1. The problem is that, by the very nature of the construction. The *points* of $\mathbb{P}\mathcal{T}$ have an interpretation within $\mathcal{M}^{\mathbb{C}}$ as what are called "α-surfaces" (totally null self-dual complex 2-surfaces) [21], and there would have to be a 1-complex-parameter family of such surfaces through each point of $\mathcal{M}^{\mathbb{C}}$ (corresponding to the 1-parameter family of points on each line of $\mathbb{P}\mathcal{T}$) and this would imply $\tilde{\Psi}_{A'B'C'D'} = 0$, i.e. $\mathcal{M}^{\mathbb{C}}$ being conformally anti-self-dual. Clearly, if twistor theory is to have any hope of providing a basis for fundamental physics, there needs to be a way around this "googly problem" (see A1). Many ideas for addressing this issue have been made over the years, often resorting to examining the twistor structure at infinity, where the geometry is simpler than that at finite regions (see, for example, [57]), but none has been able to achieve much. The most successful approach has been that of ambitwistors [58, 59], complex null geodesics, modelled on twistor, dual twistor pairs (Z^{α}, W_{β}), subject to $Z^{\alpha} W_{\alpha} = 0$. This enables complex-Riemannian 4-manifolds to be studied in relation to twistor-type ideas, and the Einstein vacuum equations to be examined in this light. But it does not follow the twistor route of "non-linearizing" the 1-function description of quantum wavefunctions, where left- and right-handed helicities can be combined together to describe gravitational interactions.

3.7 The Ideas of Palatial Twistor Theory

It is fortunate, therefore, that there is a novel approach to generalizing the non-linear graviton construction, so that both helicities can be accommodated within the same general framework, and that classical conformal space-times should also come under the same umbrella. To understand the basic idea, let us first return to the procedure that we considered earlier, in the non-linear graviton construction, where to produce a suitably deformed twistor space, we "glue together" pieces of complex manifold, preserving the complex structure from patch to patch. The notion

[2] The historical point must be made here that a key input to the development of the non-linear graviton construction was the introduction, in 1976, by Ezra T. Newman, of his notion of the $\mathcal{H}$-space, for an asymptotically flat space-time, as described initially in [55]; for more detail, see [56].

of "complex structure" can be encapsulated in terms of the algebra of holomorphic functions on each patch—or, technically, the "sheaf" of such functions, where we require holomorphicity throughout each small neighbourhood of every point. Now, we saw from the above that the basic conundrum was the existence of the actual *points* in each patch, since it was the interpretation of the *points* in $\mathbb{P}\mathcal{T}$ that gave rise to the unwanted α-surfaces. Thus, it would seem, we somehow need to find a way of matching the sheaves of algebras of holomorphic functions from one patch to another, without actually having "patches" that consist of individual points! This will not do, however, because the algebras already "know" the points, so long as the algebras are, like the algebra of holomorphic functions, *commutative*.

This suggests that we take, instead, a holomorphic algebra that is *non*-commutative.[3] As we have seen in §C1, there is a natural non-commutative algebra in twistor theory, namely that generated (via complex linear combinations and products, i.e. repeated basic operations, and the taking of appropriate limits) from the operators (see C2):

$$Z^\alpha \times \qquad \text{and} \qquad -\frac{\partial}{\partial Z^\alpha}.$$

I shall refer to this algebra as $\mathbb{A}$ the basic quantum twistor algebra for Minkowski space $\mathbb{M}$. The idea is that, in some sense, we have two (or more) "sub-regions", analogous to the $\mathcal{V}_1$ and $\mathcal{V}_2$ of the non-linear graviton construction that, in an appropriate sense "cover" the entire region of interest, which is, ultimately, to represent some open portion of the complex(ified) space-time $\mathcal{M}^{\mathbb{C}}$.

The essential idea is that the algebra $\mathbb{A}$ can be thought of as a system of complex linear operators acting on holomorphic functions defined locally on some complex space which, initially, we think of as twistor space $\mathbb{T}$. In Dirac's quantum-mechanical terminology [43], $\mathbb{T}$ is a "ket" space for the algebra $\mathbb{A}$ of quantum operators. We are to think of $\mathbb{A}$ as an abstract algebra that is not dependent upon this particular realization. For example, the same $\mathbb{A}$ could also be thought of as the space of complex linear operators acting on holomorphic functions on the *dual* twistor space $\mathbb{T}^*$, where the respective operators above would (as displayed in §C2) now be

$$\frac{\partial}{\partial W_\alpha} \qquad \text{and} \qquad W_\alpha \times$$

which satisfy the same commutation rules as before. Thus, in Dirac's terminology, $\mathbb{T}^*$ would be an alternative ket space for $\mathbb{A}$. Another way of basically saying this would be to assert that the complex parameters Z^0, Z^1, Z^2, Z^3 constitute a *complete set of commuting operators* for the algebra $\mathbb{A}$, and so also do W_0, W_1, W_2, W_3, where the elements of a complete set of commuting operators would all have to commute with one another, and that, in the space that they all coordinatize, they and their partial derivatives generate the whole of $\mathbb{A}$. The general idea would be to patch

[3] I am very grateful to Michael Atiyah for making me aware of this important requirement, in a conversation in 2013.

together various different "ket patches", by analogy with the patching together of different open regions in a locally finite covering $\mathcal{C} = (\mathcal{V}_1, \mathcal{V}_2, \mathcal{V}_3, \ldots)$, in order to build up a non-trivial complex manifold. In some sense, the algebras have to "agree" on overlaps, but the algebra $\mathcal{A}$ that we end up with, would differ from $\mathbb{A}$, but agreeing with it in some local sense.

There are various issues that need to be faced, with regard to this sort of "patching". We might, for example consider some sub-region $\mathcal{X}$ of twistor space $\mathbb{T}$, which we propose to use of as a ket-space "patch". The operator $\exp(A^\alpha \partial/\partial Z^\alpha)$, for constant A^α, could be a bit of a dangerous entity, as a candidate for membership of the algebra whose ket space is indeed to be $\mathcal{X}$. A problematic issue could be that a holomorphic function f, defined on $\mathcal{X}$, whose analytic continuation from inside $\mathcal{X}$ to a point displaced by the vector A^α to a somewhere outside $\mathcal{X}$ where this analytic continuation of f becomes *singular* would exclude $\exp(A^\alpha \partial/\partial Z^\alpha)$ from membership of the algebra, since its action on f would be singular. On the other hand, the operation of multiplication by $\exp(A^\alpha W_\alpha)$, on a ket space that is any sub-region of $\mathbb{T}^*$ would be completely harmless. Such issues need to be better understood fur a properly rigorous picture of this intended procedure.

It is clear from all this that there is a considerable vagueness in this proposal, as put forward above. Most particularly, we do not have a clear notion of topological issues, such as "local" and "open set", when it comes to these algebras. These difficult issues are not properly resolved as things stand, but I think that a helpful approach is to imagine that we are staying close to the region that would be $\mathbb{P}\mathbb{N}$, in flat-space twistor theory. This is the space whose points represent light rays in Minkowski space, and this notion clearly carries over to a general (let' us say globally hyperbolic [60]) space-time $\mathcal{M}$. The 5-manifold $\mathbb{P}\mathcal{N}$ of rays (null geodesics) in $\mathcal{M}$ will be Hausdorff, by virtue of $\mathcal{M}$'s global hyperbolicity [60]. By providing also a *spinor scaling* (see §B2) to each ray (taken to be parallel-propagated along the ray), we can unambiguously define the smooth Hausdorff 7-manifold $\mathcal{N}$.

We now have the ordinary notions of locality and open coverings, as applied to $\mathbb{P}\mathcal{N}$ and this extends to $\mathcal{N}$, by virtue of the spinor scaling, as described in §B2. Thus, we can imagine that $\mathbb{P}\mathcal{N}$ has a locally finite open covering $(\mathbb{P}\mathcal{N}_1, \mathbb{P}\mathcal{N}_2, \ldots)$, and a corresponding locally finite open covering $(\mathcal{N}_1, \mathcal{N}_2, \ldots)$ of $\mathcal{N} - \{0\}$. The idea is that if the individual spaces $\mathcal{N}_k$, together with their intersections, are, in an appropriate sense, "simple", then they can be (non-canonically) assigned respective *flat* twistor quantum algebras $\mathbb{A}_1, \mathbb{A}_2 \ldots$, where "flat" refers to being some kind of "subalgebra" of $\mathbb{A}$ in some appropriate notional sense, where we must bear in mind the comments made two paragraphs above. On the various overlapped regions $\mathcal{N}_j \cup \mathcal{N}_k$ the algebras $\mathbb{A}_i$ and $\mathbb{A}_j$ need to be patched together in the same kind of appropriate sense, but the resulting "totally patched" algebra $\mathcal{A}$ for the whole space should *not* be "flat" in its total structure if $\mathcal{M}$ is conformally curved. Moreover, it would be intended that the conformal structure of $\mathcal{M}$ would be completely encoded within the algebra $\mathcal{A}$.

At least, that is the general idea, but there are numerous issues that remain unresolved, as of now. Nevertheless, there are some possible ways that one might proceed. Since we are considering the situation "close to $\mathcal{N}$", we might well work in terms of a "power series in $Z^\alpha \bar{Z}_\alpha$", which 'in the quantized version would mean a "power series in $\Upsilon = Z^\alpha \partial/\partial Z^\alpha$". This raises some intriguing issues currently under investigation. As a positive comment, In relation to all this, it may be remarked that it is perfectly legitimate for the "intersections" of the algebras $\mathbb{A}_{\mathcal{N}_i}$ and $\mathbb{A}_{\mathcal{N}_j}$ to relate to the actual *pointwise* intersections $\mathcal{N}_i \cap \mathcal{N}_j$ of the regions $\mathcal{N}_i$ and $\mathcal{N}_j$, since there is no issue of "α-surfaces" arising here. However, it is clear that much further work needed to be done in order that this proposal can become well defined.

3.8 Local Twistors and the Einstein Equations

There is a significant feature of twistor theory that appears to be necessary to incorporate into the discussion of §C7, especially if Einstein's equations are to be incorporated in an appropriate way. This is the notion of a *local twistor* and the accompanying notion of *local twistor transport*. These we come to next.

A local twistor is a quantity $Z^\alpha = (\omega^A, \pi_{A'})$, defined at any point Q of a space-time manifold $\mathcal{M}$, which transforms as

$$\omega^A \mapsto \omega^A, \qquad \pi_{A'} \mapsto_{A'} + \mathrm{i}\omega^A\, \Omega^{-1}\nabla_{AA'}\Omega,$$

under a conformal rescaling of $\mathcal{M}$'s metric, according to $g_{ab} \mapsto \Omega^2 g_{ab}$ (Ω being a smooth positive-valued function on $\mathcal{M}$). To get an exact correspondence with the twistor concept introduced in §B2, we must think of ω^A (and $\pi_{A'}$) as *not* being defined with respect to a fixed origin point O, as in §B2, but now taken with respect to a *variable* point Q in $\mathcal{M}$. Recall from §B5 that in $\mathbb{M}$, when the origin O is displaced to a *general* point Q, with position vector q^a with respect to O in $\mathbb{M}$, the twistor $(\omega^A, \pi_{A'})$ defined with respect to O becomes $(\omega^A - \mathrm{i}q^{AA'}\pi_{A'}, \pi_{A'})$ with respect to Q. The local twistor perspective on this is that $(\omega^A, \pi_{A'})$, defined at O, when carried to Q by *local twistor transport*, becomes $(\omega^A - \mathrm{i}q^{AA'}\pi_{A'}, \pi_{A'})$ at Q, and this will enable us to extend this local twistor concept, in a conformally invariant way, to any entire *ray* in a general $\mathcal{M}$.

This is achieved via the definition of *local twistor transport* along a smooth curve γ in $\mathcal{M}$ with tangent vector t^a, which is

$$t^a\nabla_a\omega^B = -\mathrm{i}t^{BB'}\pi_{B'}, \qquad t^a\nabla_a\pi_{B'} = -\mathrm{i}t^{AA'}\mathrm{P}_{AA'BB'}\omega^B,$$

where

$$P_{ab} = \frac{1}{12}Rg_{ab} - \frac{1}{2}R_{ab}, \text{ with } R_{ac} = R^{b}_{abc}$$

(sign conventions as in [31, 41]). Taking γ to be a *ray*—which is simply-connected (by $\mathcal{M}$'s global hyperbolicity)—we use local twistor transport to propagate $(\omega^A, \pi_{A'})$ uniquely all along γ, thereby providing us with our canonical *twistor space* $\mathbb{T}_\gamma$, assigned to γ. Correspondingly, we shall have spaces $\mathbb{PT}_\gamma$, $\mathbb{N}_\gamma$, and $\mathbb{PN}_\gamma$, just as in §B3. When $\mathcal{M}$ is *conformally flat* (and simply-connected), these spaces are all independent of the choice of any curve γ connecting any pair of points in $\mathcal{M}$, so the local twistor spaces are all canonically identifiable, and may be referred to simply as spaces $\mathbb{T}$, $\mathbb{PT}$, $\mathbb{N}$, and $\mathbb{PN}$, respectively, and we have a global twistor theory just as for $\mathbb{M}$, but this does *not* hold if $\mathcal{M}$ is conformally curved. However, for a general $\mathcal{M}$, we can regard the local twistor space $\mathbb{T}_\gamma$ as a kind of *pseudo-tangent space*, at the point Γ of $\mathbb{P}\mathcal{N}$ which represents γ. The twistor space $\mathbb{T}_\gamma$ would have its algebra $\mathbb{A}_\Gamma$ and the idea would be that these algebras would form a bundle over the various points Γ of $\mathcal{N}_j$. If it could be assured that these bundles always have (holomorphic) cross sections, for simple enough, e.g. having Euclidean topology and being appropriately holomorphically (pseudo-)convex (see [51]), then this might provide an appropriate candidate for the $\mathbb{A}_{\mathcal{N}_i}$ required for §C7.

Not surprisingly, there is a lot of arbitrariness in this proposed procedure, but this may be argued to be similar to the arbitrariness in a choice of coordinate system when coordinate patches are used in a normal ("Čech") procedure for constructing a (complex) manifold, as with the non-linear graviton construction. Each such flat $\mathbb{A}_{\mathcal{N}_i}$ is to be thought of in the spirit of a "coordinate patch". As already remarked in §C7, as the patching goes, it is perfectly legitimate for the "intersections" of the algebras $\mathbb{A}_{\mathcal{N}_i}$ and $\mathbb{A}_{\mathcal{N}_j}$ to relate to the actual pointwise intersections $\mathcal{N}_i \cap \mathcal{N}_j$ of the regions $\mathcal{N}_i$ and $\mathcal{N}_j$ (since there is no issue of "α-surfaces" arising here). With regard to the *intersection* $\mathcal{N}_i \cap \mathcal{N}_j$ of two simple regions $\mathcal{N}_i$ and $\mathcal{N}_j$ we require consistency of the algebras $\mathbb{A}_{\mathcal{N}_i}$ and $\mathbb{A}_{\mathcal{N}_j}$, retaining a consistent ket space on the intersection, but we do not require a common ket-space to be present for the whole of their *union* $\mathcal{N}_i \cup \mathcal{N}_j$. Such consistency would not generally be possible globally. Instead, our fully "patched together" algebra $\mathcal{A}$ would *not* have a consistent ket space (unless $\mathcal{M}$ is conformally flat). The idea would be that a measure of the *departure* from global consistency of a ket space, over the whole of $\mathcal{N}$, would be something of the nature of a (non-linear) 1-function—as with the inconsistency expressed in Fig. 9 and which, in space-time terms, would express the presence of a non-zero *Weyl conformal tensor*, i.e. conformal curvature for $\mathcal{M}$.

We need to be able to identify the *points* of $\mathcal{M}$ in terms of the algebra $\mathcal{A}$. These have to arise by *non-local* considerations (as was the case for the non-linear graviton construction). Corresponding to any particular point P of $\mathcal{M}$ there would be a locus $\mathcal{P}$ in $\mathbb{P}\mathcal{N}$ representing P, namely the family of all rays through P, which is topologically S^2, to be thought of as a projective 2-spinor space. We need something of the nature of a completely commuting self-conjugate sub-algebra of $\mathcal{A}$ as exemplified by the algebra generated by the four elements Z^2, Z^3, $\partial/\partial Z^0$, $\partial/\partial Z^1$, to define the origin point O for $\mathbb{M}$. For this to work in general, we would certainly need some suitable generalization of the Kodaira theorem [53] that was central to the non-linear graviton construction.

None of this yet encodes the formulation of Einstein's equations. It is perhaps remarkable, therefore, to find that Einstein's Λ-vacuum equations are themselves very simply incorporated into this kind of structure. For these equations provide *precisely* the necessary and sufficient condition that local twistor transport globally preserves the infinity twistor $I_{\alpha\beta}$ (and its dual $I^{\alpha\beta}$; see [41]. Accordingly, all we require for the satisfaction of the Einstein Λ-vacuum equations is that the "matching" of the algebras $\mathbb{A}_{\mathcal{N}_i}$ from one patch to another preserves the infinity twistors. If all these procedures (or something like them) indeed work roughly as intended (with generalizations to the Yang–Mills equations and other aspects of physics, such as the incorporation of particles' rest-masses [45, 61–63]), then there could appear to be possible openings for twistor theory applicable to basic physics generally. Among many other things, this might perhaps directly incorporate the non-local features of quantum theory in a natural way, and other matters not envisaged before. Yet, much work needs to be done to decide whether or not the ideas outlined here can really be made to hang together, and if they do not, then we need to know what might replace them.

Acknowledgements I'm grateful for Joseph Kouneiher, for his comments, suggestions throughout the process of writing and the preparation of this paper.

References

1. R. Penrose, On the origins of twistor theory, in *Gravitation and Geometry: A Volume in Honour of I. Robinson*. ed. by W. Rindler, A. Trautman (Bibliopolis, Naples, 1987)
2. L.J. Mason, N.M.J. Woodhouse, *Integrability, Self-Duality, and Twistor Theory* (Oxford University Press, Oxford, 1996)
3. S.A. Huggett, L.J. Mason, K.P. Tod, S.T. Tsou, N.M.J. Woodhouse, *The Geometric Universe: Science, Geometry, and the Work of Roger Penrose* (Oxford University Press, Oxford, 1998)
4. R. Penrose, Twistor algebra. J. Math. Phys. **8**(2), 345–66 (1967)
5. R. Penrose, M.A.H. MacCallum, Twistor theory: an approach to the quantization of fields and space-time. Phys. Repts. **6C**, 241–315 (1972)
6. A.P. Hodges, S. Huggett, Twistor diagrams. Surv. High Energy Phys. **1**, 333–53 (1980)
7. A.P. Hodges, Twistor diagrams. Physica **114A**, 157–75 (1982)
8. A.P. Hodges, A twistor approach to the regularization of divergences. Proc. Roy. Soc. Lond. **A397**, 341–74 (1985)
9. S. Parke, T. Taylor, Amplitude for n−gluon scatterings. Phys. Rev. Lett. **56**, 2459 (1986)
10. V. Nair, A current algebra for some gauge theory amplitudes. Phys. Lett. B **214**, 215 (1988)
11. W.T. Shaw, L.P. Hughston, Twistors and strings, in *Twistors in Mathematics and Physics*, LMS Lect. Note Ser. 156, ed. by T.N. Bailey, R.J. Baston (Cambridge University Press, Cambridge, 1990)
12. E. Witten, Perturbative gauge theory as a string theory in twistor space. Commun. Math. Phys. **252**, 189–258 (2004). arXiv:hep-th/0312171v2
13. A.P. Hodges, *Twistor Diagrams for All Tree Amplitudes in Gauge Theory: A Helicity-independent Formalism*, 2006. arXiv:hep-th/0512336v2
14. A.P. Hodges, *Scattering Amplitudes for Eight Gauge Fields* (2006). arXiv:hep-th/0603101v1
15. A. Hodges, in *Eliminating Spurious Poles from Gauge-theoretic Amplitudes* (2009). arXiv:0905.1473v1
16. A.P. Hodges, Nat. Phys. **9**, 205–206 (2013). https://doi.org/10.1038/nphys2597
17. L.J. Mason, D. Skinner, *Scattering Amplitudes and BCFW Recursion in Twistor Space* (2009). arXiv:0903.2083v3
18. L. Mason, D. Skinner, *Dual superconformal invariance, momentum twistors and grassmannians* (2009). arXiv:0909.0250v2 [hep-th]

19. N. Arkani-Hamed, F. Cachazo, C. Cheung, J. Kaplan, The S-matrix in twistor space. J. High Energy Phys. **1003**, 020 (2010)
20. N. Arkani-Hamed, A. Hodges, J. Trnka, *Positive Amplitudes in the Amplituhedron* (2014). qrXiv:1412.8478v1 [hep-th]
21. R. Penrose, Non-linear gravitons and curved twistor theory. Gen. Rel. Grav. **7**, 31–52 (1976)
22. R.S. Ward, Self-dual space-times with cosmological constant. Comm. Math. Phys. **78**, 1–17 (1980)
23. R. Penrose, Palatial twistor theory and the twistor googly problem; in theme issue 'New geometric concepts in the foundations of physics'. Phil. Trans. R. Soc. (Lond.) **A 373**, 20140237 (2015)
24. R. Penrose, Towards an objective physics of bell non-locality: palatial twistor theory, in *Quantum Nonlocality and Reality* ed. by M. Bell, S. Gao, Core - History, Philosophy and Foundations of Physics, Chinese Academy of Sciences, Beijing. (Cambridge University Press, Cambridge, 2016)
25. B. Greene, *The Elegant Universe; Superstrings, Hidden Dimensions, and the Quest for the Ultimate Theory* (Jonathan Cape, Random House, London, 1999)
26. R. Penrose, The apparent shape of a relativistically moving sphere. Proc. Camb. Phil. Soc. **55**, 137–9 (1959)
27. É. Cartan, *The Theory of Spinors* (Hermann, Paris, 1966)
28. B.L. van der Waerden, Spinoranalyse. Nachr. Akad. Wiss. Götting. Math.-Physik Kl, 100–109 (1929)
29. R. Penrose, A spinor approach to general relativity. Ann. Phys. (New York) **10**, 171–201 (1960)
30. W.T. Payne, Elementary spinor theory. Am. J. Phys. **20**, 253–62 (1952)
31. R. Penrose, W. Rindler, *Spinors and Space-Time, vol. 1: Two-Spinor Calculus and Relativistic Fields* (Cambridge University Press, Cambridge, 1984)
32. L. Witten, Invariants of general relativity and the classification of spaces. Phys. Rev. **113**, 357–62 (1959)
33. P.A.M. Dirac, Relativistic wave equations. Proc. Roy. Soc. (Lond.) **A155**, 447–59 (1936)
34. E.M. Corson, *Introduction to Tensors, Spinors and Relativistic Wave Equations* (Blackie, Glasgow, 1953)
35. R. Penrose, Zero rest-mass fields including gravitation: asymptotic behaviour. Proc. Roy. Soc. Lond. **A284**, 159–203 (1965)
36. R. Penrose, Null hypersurface initial data for classical fields of arbitrary spin and for general relativity. Gen. Rel. Grav. **12**, 225–264 (1980)
37. R.P. Kerr, Gravitational field of a spinning mass as an example of algebraically (1963)
38. R.H. Boyer, R.W. Lindquist, Maximal analytic extension of the Kerr metric. J. Math. Phys. **8**, 265–281 (1967)
39. E.T. Newman, E. Couch, K. Chinnapared, A. Exton, A. Prakash, R. Torrence, Metric of a rotating charged mass. J. Math. Phys. **6**, 918–9 (1965)
40. I. Robinson, A. Trautman, Some spherical gravitational waves in general relativity. Proc. Roy. Soc. (Lond.) **A265**, 463–73 (1962)
41. R. Penrose, W. Rindler, *Spinors and Space-Time, vol. 2: Spinor and Twistor Methods in Space-Time Geometry* (Cambridge University Press, Cambridge, 1986)
42. R. Penrose, Twistor quantization and curved space-time. Int. J. Theor. Phys. **1**, 61–99 (1968)
43. P.A.M. Dirac, *The Principles of Quantum Mechanics*, 3rd edn. (Clarendon Press, Oxford, 1947)
44. R. Penrose, Solutions of the zero rest-mass equations. J. Math. Phys. **10**, 38–9 (1969)
45. L.P. Hughston, *Twistors and Particles,* Lecture Notes in Physics No. 97 (Springer, Berlin, 1979)
46. E.T. Whittaker, On the partial differential equations of mathematical physics. Math. Ann. **57**, 333–55 (1903)
47. H. Bateman, The solution of partial differential equations by means of definite integrals. Proc. Lond. Math. Soc. **2**(1), 451–8 (1904)
48. H. Bateman, *Partial Differential Equations of Mathematical Physics* (Dover, New York, 1944)
49. R.F. Streater, A.S. Wightman, *PCT, Spin and Statistics, and All That*, 5th edn. (Princeton University Press, Princeton, 2000)

50. R. Penrose, On the cohomology of impossible figures [La cohomologie des figures impossibles]. Struct. Topol. [Topologie structurale] **17**, 11–16 (1991)
51. R.C. Gunning, R. Rossi, *Analytic Functons of Several Complex Variables* (Prentice-Hall, Englewood Cliffs, 1965)
52. R.S. Ward, On self-dual gauge fields. Phys. Lett. **61A**, 81–2 (1977)
53. K. Kodaira, On stability of compact submanifolds of complex manifolds. Am. J. Math. **85**, 79–94 (1963)
54. K. Kodaira, D.C. Spencer, On deformations of complex analytic structures I, II. Ann. Math. **67**(328–401), 403–466 (1958)
55. E.T. Newman, Heaven and its properties. Gen. Rel. Grav. **7**, 107–11 (1976)
56. E.T. Newman, Deformed twistor space and H-space, in *Complex Manifold Techniques in Theoretical Physics*, ed. by D.E. Lerner, P.D. Sommers (Pitman, San Francisco, 1979) pp. 154–165
57. R. Penrose, A new angle on the googly graviton, in *Further Advances in Twistor Theory, vol. III: Curved Twistor Spaces* Chapman & Hall/CRC Research Notes in Mathematics 424, ed. by L.J. Mason, L.P. Hughston, P.Z. Kobak, K. Pulverer (Chapman & Hall/CRC, London, 2001). pp. 264–269. ISBN 1-58488-047-3
58. C.R. LeBrun, Ambi-twistors and Einstein's equations. Class. Quantum Grav. **2**, 555–63 (1985)
59. C.R. LeBrun, Twistors, ambitwistors, and conformal gravity, in *Twistors in Mathematical Physics*, LMS Lec. note ser. 156, ed. by T.N. Bailey, R.J. Baston (Cambridge University Press, Cambridge, 1990)
60. R. Geroch, Domain of dependence. J. Math. Phys. **11**, 437 (1970)
61. R. Penrose, Twistors and particles: an outline, in *Quantum Theory and the Structures of Time and Space*, ed. by L. Castell, M. Drieschner, C.F. von Weizsäcker (Carl Hanser Verlag, Munich, 1975)
62. Z. Perjés, Introduction to twistor particle theory, in *Twistor Geometry and Non-Linear Systems*, ed. by H.D. Doebner, T.D. Palev (Springer, Berlin, 1982) pp. 53–72
63. Z. Perjés, G.A.J. Sparling, The twistor structure of hadrons, in *Advances in Twistor Theory*, ed. by L.P. Hughston, R.S. Ward (Pitman, San Francisco, 1979)

What Are We Missing in Our Search for Quantum Gravity?

Lee Smolin

Abstract Some reflections are presented on the state of the search for a quantum theory of gravity. I discuss diverse regimes of possible quantum gravitational phenomenon, some well explored, some novel.

1 Introduction

Despite enormous effort from thousands of dedicated researchers over a century,[1] the search for the quantum theory of gravity has not yet arrived at a satisfactory conclusion. We have indeed several impressive proposals, each of which partly succeeds in describing plausible quantum gravitational physics. Each tells a compelling story that has, for good reason, won it advocates. Each has also run into persistent roadblocks, which are pointed to by their skeptics. Looking back, before strings and loops, before causal sets, causal dynamical triangulations, asymptotic safety, amplitudes, twisters, shape dynamics, etc, to the early days of Bergman, Deser, DeWitt, Wheeler and their friends, who would have thought that there would turn out to be at least half dozen ways to get part way to quantum gravity?

Perhaps we might, for a moment, consider that the approaches so far pursued are not really theories, in the sense quantum mechanics, general relativity and Newtonian mechanics are theories. For those are based on principles and perhaps we can agree that we don't yet know the principles of quantum gravity. There are of course proposals for quantum gravity principles, and part of the reason for this paper is to prepare the ground for proposals of new principles (or, in some cases, such as the

[1] The idea that there might be quanta of gravitational waves was first mentioned by Einstein in a paper in 1917 [1].

L. Smolin (✉)
Perimeter Institute for Theoretical Physics, 31 Caroline Street North,
Waterloo, ON N2J 2Y5, Canada
e-mail: lsmolin@perimeterinstitute.ca

J. Kouneiher (ed.), *Foundations of Mathematics and Physics One Century After Hilbert*,
https://doi.org/10.1007/978-3-319-64813-2_11

holographic principle, sharpening up their formulation).[2] Instead, let us, just for a moment, think of the current approaches as models, which each describe some plausible quantum gravitational phenomena. The different approaches can then be thought of as complementary, rather than in conflict, as they investigate diverse regimes of possible new physics. Could we hope that taking this view may open up discussions between people working on different approaches, to the benefit of all of us?

This frees us up to consider that, despite genuine achievements on several sides, we have yet to see a real theory of quantum gravity. Can we then begin to look for one? If we adopt this view, we can learn from all that has been done, while taking a clean slate. How then do we proceed to look for a theory?

Of course, we work under an obvious handicap, which is that there are few experiments whose results can guide us by winnowing down the possibilities. But there are a few real Planck scale experiments, which have yielded clues, and which are on the threshold of constraining possible quantum gravity effects at order of $\frac{\text{Energies}}{E_{Pl}}$. Even now we can be fairly sure that Lorentz invariance is not simply broken at that order [3].[3] There is also a window into possible quantum gravitational effects in cosmology, such as low l anomalies [5] or parity breaking in B modes [6]. These represent opportunities that must be explored.

In situations like this, it can be good to pause and take stock of where we are [7]. The following are some reflections on what we may be missing in our search for quantum gravity.

2 What Is Missing from Attempts to Discover Quantum Gravity?

Approaches are great, and we have good reasons for affection for particular ones. But can we put aside the different approaches and, especially, their unfortunate sociologies, and just talk physics?

We can start with a simple question:

2.1 Where Are the Zeroth Order Quantum Gravity Phenomena?

In many prior revolutionary transitions there was a key first step where some well studied phenomena, which are already understood in the then current theoretical framework, were reinterpreted in terms of new concepts and principles. This often leads to surprising new insights, by giving us a Rosetta stone for translating between

---

[2] Hence, this is a companion paper to [2].

[3] But see [4].

the old and new theoretical languages. That is, at first the correspondence is a mere reinterpretation of phenomena, already explained by the old theory, in a surprising new language.

Associated with this dual description is a new parameter, which controls novel phenomena. The correspondence establishes at order zero in the new parameter a translation between the languages of the new and old theory. But as soon as this is established we notice that the correspondence holds in a limited domain. There is then a space to move beyond the zone of correspondence to novel phenomena whose scale is set by the new parameter. By doing so we adventure into a new regime of phenomena, but one with clear connections to established knowledge.

Here are some examples of how a transition to a new theory was initiated by reinterpreting a familiar, well understood phenomena in new terms.

- Galileo's reinterpretion of the tower experiment [8]. Consider the fact that a ball dropped from the top of a tower falls to the bottom. This simple fact has two interpretations, which lead to opposite conclusions. If you are an Aristotelean, you interpret this result as evidence that the earth doesn't move. But if you are Galileo, and believe in the principle of inertia, you interpret the same result as a confirmation that the Earth could be moving without our experiencing any effects. After all, he argued, a ball dropped from the top of a ship's mast, while sailing smoothly in the harbour of Venice, falls to the bottom of the mast.
- In special relativity; mass is reinterpreted as energy. One new phenomena this allows is pair creation i.e. the transformation between matter-energy and other forms of energy. The new parameter is $\frac{v^2}{c^2}$.
- In general relativity, the equivalence principle explains in a radically new way the old fact that all massive objects fall with the same acceleration. Newton understood this as a consequence of the equality of inertial and gravitational mass, which seemed to be a coincidence. Einstein explained this as a necessary consequence of a new principle. This allows gravity to be reinterpreted as the curvature of spacetime. The new parameter is $\frac{GM/c^2}{r}$.
- Matter was initially modelled as a continuum, i.e. fluids, gases and solids. Boltzmann, Kelvin and others reinterpreted continuous, thermodynamic phenomena in terms of the atomic hypothesis. At first they were able to work out many correspondences, such as the ideal gas law and the laws of thermodynamics. These correspondences were exact in the thermodynamic limit, in which Avogadro's number would go to infinity. Then, Einstein and others noted that if Avogadro's number were finite there would be novel phenomena such as Brownian motion.

Let us then ask, *Where are the zeroth order quantum gravity phenomena?* Can we find zeroth order correspondences between some well known phenomena and quantum gravity? We don't have a real start on quantum gravity unless we can provide an answer to this. Here are some proposals for zeroth order quantum gravity phenomena.

1. *The zeroth order phenomena is locality itself.* This must be the case if as is sometimes hypothesized, locality is emergent in the classical or continuum limit of a

fundamental quantum theory of gravity, whose states are networks living in no space, perhaps spin networks or records of entanglement. The first order departures from locality are quantum phenomena, especially entanglement. Indeed one version of this proposal is that spatial relations are emergent from entanglement [9–12].
The second order departures from locality are then disordered locality [13] and relative locality [14, 15].

2. *More precisely, the zeroth order quantum gravity phenomena is space itself,* specifically its low dimensionality and fantastically low curvature, compared to the Planck scale. Julian Barbour used to emphasize that space itself, and especially the low dimensionality is a highly nonlocal phenomena [16]. This is seen if you try to express the physics of N particles in terms of their relative distances, r_{jk}, alone. For these $\frac{N(N-1)}{2}$ quantities are determined in terms of $3N - 6$ coordinates, which is many fewer. This means the r_{jk} are subject to a large number, $C = \frac{N(N-7)}{2} + 6$ constraints. These can be understood as the vanishing of the volumes of all independent n-simplices, with $n > 3$, made from the r_{jk}.
Indeed, the AdS/CFT correspondence succeeds in generating one dimension of space from d others, in the special case that the cosmological constant is negative [9–11, 17]. This provides many interesting examples to study, and provides a partial Rosetta stone for translating between conformal field theory phenomena and gravitational phenomena. It however remains to be seen whether the construction helps us do what we would really like to do which is to understand how all he dimensions of space emerge from something more fundamental.
3. *The zeroth order phenomena is gravity.* This is suggested by the thermodynamic derivations of general relativity by Jacobson [18, 19] and elaborations on it [2, 20, 21]. In the course of the derivation one refers to quantum phenomena such as the Unruh temperature [22], so $\hbar$ appears. Another $\hbar$ appears explicitly in expressing an entropy proportional to area. These $\hbar$'s cancel in the resulting Einstein's equation. This applies even more to Verlinde's entropic derivation of Newtonian gravity, in which $\hbar$'s and c's are both present, but cancel [23].
4. *Could another zeroth order phenomena be MOND?* Perhaps MOND [24] is a quantum gravity effect, for positive cosmological constant, Λ, which arises in a regime, or phase, of accelerations small compared to

$$a_\Lambda = c^2\sqrt{\Lambda} \tag{1}$$

which also involves the cancellation of $\hbar$'s and c's. This idea is explored from diverse perspectives in [2] and in [25–34].
5. *The zeroth order phenomena is the universe itself, its vast scale and stability as well as the relative stability of the laws.*

The hypotheses just mentioned may serve as bridges to the true theory of quantum gravity. While we are looking for such bridges, let's keep in mind theories and hypotheses that are clearly transitional and incomplete, but nonetheless may capture some of the truth.

1. The proposal that space is emergent from entanglement [9–12].
2. The causal set hypothesis that spacetime is really a discrete causal set made up of discrete events and their causal relations [35–40]. This is very like the hypothesis that matter is made of large but finite collections of atoms. The first order phenomena would be analogous to Brownian motion. Two have been proposed: (1) the hypothesis that the cosmological constant is the result of a fluctuation [41] and (2) covariant dispersion [42].
3. The shape dynamics hypothesis that the universe is not a four dimensional spacetime mod spacetime diffeomorphisms [43, 44]. It is instead reinterpreted as an evolving three dimensional geometry mod spatial diffeorphisms and Weyl transformations (i.e. local conformal rescaling). This is analogous to Galileo's reinterpretation of the tower experiment from an Aristotelean demonstration of the Earth's stationarity to a demonstration of the principles of relativity and inertia. A proposed first order phenomena where the two are no longer different interpretations, but differ substantially is in black hole singularities, which are eliminated in favour of bounces to baby universes in shape dynamics [45].
4. In relative locality [14, 15] and its discrete version, energetic causal sets [37–39], a picture of particle dynamics in which relativistic particles propagate on a fixed background spacetime is replaced by an apparently equivalent picture in which particles propagate on a fixed momentum space. Interactions which happen locally at spacetime events in the old picture become events, elements of a causal set, at which energy-momentum conservation is imposed. In this new picture spacetime emerges as an auxiliary description. The first order phenomena where they diverge is gotten by curving or adding torsion or non-metricity to momentum space, which leads to the novel phenomena of relative locality.
5. The AdS/CFT hypothesis in the planar limit in which $N \rightarrow \infty$, is a precise dictionary for translating some classical gravitational phenomena into an equivalent, non-gravitational language [17]. There are many interesting correspondences. And there are clear paths for going beyond zeroth order to give novel quantum gravitational phenomena.

   It is then urgent to understand this correspondence in terms that both don't rely on a negative cosmological constant and apply very generally, without the need for supersymmetry or special dimensions. Some suggestions to explore are in [46] and also in the companion paper [2], where I suggest that the AdS/CFT correspondence is an expression of a more general quantum equivalence principle.

2.2 Phenomenological Limits and Regimes of Quantum Gravity

Whatever the quantum theory of gravity is, it will depend on four dimensional constants, $\hbar$, G, c and Λ. We are familiar with the commonsense idea that the limit of $\hbar \rightarrow 0$ with the others fixed should define general relativity, while the limit in which G and Λ are taken to vanish should give quantum field theory. But, there are several other interesting limits of the three parameters $\hbar$, G and c, which each define a regime of quantum gravity phenomenology.

Two interesting regimes come from taking $\hbar \to 0$; while c is held fixed; these may be called the non-quantum regimes of quantum gravity.

2.2.1 The Relative Locality Regime

We can recall first the relativity locality regime in which G and $\hbar$ are both taken to zero, with c held fixed, but with the Planck mass also held fixed, giving us [14, 15]

$$m_p^2 = \frac{\hbar c}{G} \tag{2}$$

This defines a regime of phenomena in which m_p and c are fixed while both quantum and gravitational effects are negligible, because $G = \hbar = 0$. Since $l_p = \sqrt{\frac{\hbar G}{c^3}} \to 0$ there is no quantum geometry. In this regime the propagation and scattering of particles may be deformed due to curving momentum space [14, 15]. But there is no corresponding deformation of wave propagation. Indeed, as $\hbar = 0$ the correspondence between waves and particles is lost.

Notice that in this regime the entropy of black holes goes to infinity, while their temperature remains finite [47].

2.2.2 The Strong Gravity Regime

Alternatively we can explore phenomenology where $G \to \infty$ as $\hbar \to 0$, such that

$$\frac{\hbar G}{c^3} \to l_p^2, \quad \text{fixed} \tag{3}$$

again with c held fixed. It follows that $m_p \to 0$. Now there is no deformation of particle dynamics, while wave propagation can be modified, for example as,

$$\left(\frac{2}{c^2}\frac{\partial^2}{\partial t^2} - \nabla^2 - l_p^2 \nabla^4 + \cdots \right)\phi = 0 \tag{4}$$

Now the black hole entropy stays finite while the temperature of black holes goes to zero.

An unusual feature of this limit is that it is the opposite of the semiclassical limit. In this limit the commutation relations of quantum gravity are unchanged, because they involve l_p^2.

$$[A_b^j(x), \tilde{E}_i^a(y)] = \hbar G \delta_b^a \delta_i^j \delta^3(x, y) \tag{5}$$

Meanwhile, the commutators of matter degrees of freedom go to zero

$$[\phi(x), \pi(y)] = \imath\hbar\delta^3(x, y) \rightarrow 0 \tag{6}$$

2.3 The Holographic Regime

A very interesting regime about which a lot is known is the one studied in most calculations in the AdS/CFT correspondence [17]. This is based on a limit in which one takes N large, where N measures the ratio of the cosmological constant scale to the Planck scale.

$$N = \frac{R^2}{l_{pl}^2} = \frac{1}{\hbar G \Lambda} \tag{7}$$

Here $\Lambda = -\frac{1}{R^2}$. N can be seen as counting the number of degrees of freedom defined on a boundary in an asymptotically anti-deSitter spacetime.

2.4 The Loop Quantum Cosmology Regime

A second cosmological regime is related to quantum cosmology [48, 49], and is defined by the limit in which the Planck energy E_p is taken to infinity, while the Planck energy density

$$\rho_p = \frac{E_p}{l_p^3} = \frac{c^7}{\hbar G^2} \tag{8}$$

is held fixed.The speed of light, c is also kept fixed. In terms of $\hbar$ and G this limit is defined by scaling by a dimensionless t

$$\hbar = \hbar_0 t^4, \qquad G = G_0 \frac{1}{t^2} \tag{9}$$

where t is then taken to infinity, so that E_p and l_p both diverge.

This is justified by the FRW equation or Hamiltonian constraint, which can be read as

$$(a')^2 = \frac{\rho^{matter}}{\rho_p} + \frac{1}{E_p} \int d^3x (\partial h_{ij})^2. \tag{10}$$

Here the spatial metric is expanded as

$$g_{ij} = a^2(\delta_{ij} + \sqrt{\frac{l_P}{m_p}} h_{ij}) \tag{11}$$

and $'$ denotes differentiation with respect to dimensionless conformal time. In the regime defined by holding ρ_p fixed while E_p is taken to infinity the spatial inhomogeneities decouple, $g_{ij} \rightarrow a^2\delta_{ij}$, and we are left with the homogeneous equation studied to good effect in papers on loop quantum cosmology [48, 49].

2.5 Are There Any Newtonian Quantum Gravity Phenomena?

The regimes we mentioned previous are relativistic in that the speed of light, c, is held fixed. But, are there phenomena which are measured in units of hG or h/G with no c's? i.e. is there a Newtonian regime of quantum gravity?

Here is a curious fact: combinations of just $\hbar$ and G without c are not simple. To get simple quantities like a mass or a length we need to combine them with c. Indeed the Planck mass and Planck length involve all three constants, $\hbar$, G and c. This is a simple truism but it means that any of the characteristic phenomena we associate with l_p and m_p, such as the discreteness of quantum geometry or the unification of the forces, will go away in the limit $c \rightarrow \infty$ and so they will not have Newtonian analogues.

So it is worth asking whether there are any characteristic Newtonian quantum gravity phenomena, which occur at scales parametrized by combinations of $\hbar$ and G alone, without c.

Indeed, there are such characteristic phenomena associated with combinations of the other two pairs. $\hbar$ and c go together well to convert length to momentum or time to energy. Together with e^2 they give us the dimensionless fine structure constant, which organizes the scales of phenomena in quantum electrodynamics. G and c together convert a mass to a length $R_{Schw} = \frac{2G}{c^2}M$. But simple combinations of $\hbar$ and G don't seem to make anything that parameterizes a new, unexpected phenomena.

By dimensional analysis, the phenomena exhibited by a Newtonian quantum gravity regime would involve some peculiar physical dimensions. Without c, there would be no Planck mass, nor is there a Planck length. There is a unit of *mass per square root of velocity.*

$$\mathcal{A}^2 = \frac{\hbar}{G} = \frac{mass^2}{speed} \tag{12}$$

There is also a unit of *length to the fifth per time cubed.*

$$\mathcal{B} = \hbar G = \frac{length^5}{time^3} \tag{13}$$

This suggests Lifshitz scaling at small velocities. Perhaps a connection to MOND [24]?

It would be very interesting to discover a regime of phenomena where c has been taken to infinity, but where the new quantities $\mathcal{A}$ and $\mathcal{B}$ are measurable. Presumably this involves heavy, slow quantum gravity objects.

Note that it can't involve an analogue black holes as c has gone to infinity.

If we look at more complex combinations of $\hbar$ and G, we find a conversion from $mass^{-3}$ to length, given by

$$r_g = \frac{\hbar^2}{G}\frac{1}{m^3} \tag{14}$$

This is the "gravitational Bohr radius", i.e. from the Schroedinger equation, the ground state of an atom held together by a Newtonian gravitational potential has a wave function

$$\psi(r) = e^{-\frac{r}{r_g}} \tag{15}$$

A Newtonian gravitational atom would be a good trick to play with, but it wouldn't teach us anything about quantum gravity. In any case they will be prohibitively big for atoms and small for planets, as

$$r_g = l_p \frac{m_p^3}{m^3} \tag{16}$$

with a correspondingly tiny binding energy:

$$E_g = -\frac{1}{2}\frac{G^2 m^5}{\hbar^2} = -\frac{mc^2}{2}\left(\frac{m}{m_p}\right)^4 \tag{17}$$

2.6 *Newtonian Quantum Cosmology*

Of course there is another dimensional constant in quantum gravity, the cosmological constant, Λ and, with it together with $\hbar$ and G one can form a complete set of units without c. These depend on whether you take the fixed constant to be the inverse length-squared $\Lambda = \frac{3}{R^2}$ or the Hubble time $T = H^{-1}$.

There is a third possibility, which is to hold the cosmological acceleration $a_\Lambda = \frac{c^2}{R}$ fixed. These three regimes are inequivalent as c has been taken to ∞.

Note that we take care to distinguish the empirically measured $MOND$ acceleration, a_0, which is determined to be roughly $1.2 \times 10^{-8}\ cm/s^2$ from the cosmological acceleration a_Λ, which is related to the cosmological constant.

- **Newtonian quantum cosmology**: $\Lambda = \frac{1}{R^2}$ fixed as $c \to \infty$.

$$m_g = \left(\frac{\hbar^2}{GR}\right)^{\frac{1}{3}} = m_p\left(\frac{l_p}{R}\right)^{\frac{1}{3}} = 10^{-20} m_p \tag{18}$$

$$t_g = \left(\frac{R^5}{\hbar G}\right)^{\frac{1}{3}} = \frac{R}{c}\left(\frac{R}{l_p}\right)^{\frac{2}{3}} \tag{19}$$

where in the right hand expressions the c's cancel.
Note that m_p is roughly the proton mass!. This is a $c \to \infty$ residue of Dirac's large number phenomena.

- **Hubble-time Newtonian quantum cosmology** $T = H^{-1}$ fixed as $c \to \infty$.

$$r_H = \left(\hbar G T^3\right)^{\frac{1}{5}} = R\left(\frac{l_p}{R}\right)^{\frac{2}{5}} = 10^{-24} R \tag{20}$$

$$m_H = \left(\frac{\hbar r_H}{GT}\right)^{\frac{1}{2}} = 10^{-12} m_p = 10^{-20} m_p \tag{21}$$

- **MOND quantum cosmology** $a_\Lambda = \frac{c^2}{R}$ fixed as $c \to \infty$ and $R \to \infty$.
Here there is a unit of velocity

$$v_0 = \left(\hbar G a_\Lambda^2\right)^{\frac{1}{7}} = c\left(\frac{l_p}{R}\right)^{\frac{2}{7}} \approx 10^{-7} cm/s \tag{22}$$

We can study a MOND bound atom, which has a potential

$$U_{MOND} = -\frac{GmM}{r} + m v_{TF}^2 \ln r/\rho_0 \tag{23}$$

where $v_{TF}^2 = \sqrt{GMa_0}$, expressing the Tully-Fischer relation [50]. This gives a MOND-Bohr radius of

$$r_{MOND} = \frac{\hbar}{m v_{TF}} \tag{24}$$

The binding energy is order

$$E_0 \approx -m v_{TF}^2 \tag{25}$$

which we note has no $\hbar$ in it

- **The classical MOND limit** $a_0 \approx a_\Lambda = \frac{c^2}{R}$ fixed as $c \to \infty$, $R \to \infty$ and $\hbar \to 0$, with G fixed.
The key constant here is $\mathcal{A}_0 = G a_0$, which is a conversion between mass and velocity to the fourth power. This constant is actually measured, by observations of the Tully-Fischer relation [50]

$$v^4 = G a_0 M_b \tag{26}$$

where M_b is the baryonic mass of a galaxy, and v is the stellar rotational velocity in the outer disk where the rotation curve flattens out. Fits to data find

$$a_0 = 1.2 \times 10^{-10} ms^{-2} \tag{27}$$

which is not far from $a_\Lambda = c^2\sqrt{\frac{\Lambda}{3}}$.
Note that this is a limit in which $\Lambda \rightarrow 0$ from above.
A reason we might expect to see novel phenomenon in this limit is that for small accelerations,

$$a < a_\Lambda \tag{28}$$

the equivalence principle need not apply. One way to say this is that an observer's acceleration horizon-the horizon created by their own acceleration, falls near their cosmological horizon when $a < a_\Lambda$. In [2, 34] I argue that this could be the origin of MOND.

2.7 *Energy and Its Positivity*

A key issue for any non-perturbative approach to quantum gravity is the role of energy. The puzzle originates from the equivalence principle, which forbids there from being a local measure of the gravitational field. This is so that a freely falling observer has no way to distinguish their situation from an inertial observer in Minkowski spacetime, to leading order in separations over curvatures.

As a consequence, the energy of the gravitational field can only be measured quasi-locally, or at infinity. So there is no expression of the energy of a spacetime, or region thereof, expressed as a volume integral over a positive definite expression.

It turns out that the energy of the gravitational field is still positive [51]. This is very fortunate, otherwise flat empty spacetime would be unstable. But this positivity is an on-shell property. It only holds in the presence of the field equations or, in the Hamiltonian formulation-of the constraints.

We the must ask whether there is in quantum gravity an operator on the space of physical states that represents the energy which is both positive definite and Hermitian, in the physical inner product. Such an operator cannot be just the sum of squares of local operators.

In [52] I have explored conditions on the physical inner product that must be satisfied if there is to be a positive definite and Hermitian operator representing the ADM mass.

2.8 *Where Does the Planck Mass Come From?*

There is another issue regarding energy that challenges quantum gravity theories. This is that the Planck area, $l_p^2 = \frac{\hbar G}{c^3}$ turns up easily and naturally, while the Planck mass

$$m_p = \frac{\hbar}{l_p}. \tag{29}$$

does not easily turn up. The reason is the following.

The action principle for the gravitational field in general relativity, including the boundary term, is proportional to $\frac{c^3}{G}$.

$$S^{gr} = -\frac{c^3}{8\pi G}\left[\int_{\mathcal{M}} d^4x(R-2\Lambda) - \int_{\partial\mathcal{M}} d^3\sigma\kappa\right] - \int_{\mathcal{M}} d^4x\mathcal{L}^{matter} \tag{30}$$

This is indeed the only place that G appears in the action. A little dimensional analysis tells us why. The Riemann curvature scalar, R, and the intrinsic curvature κ are both purely geometrical. R has units of inverse length-squared while the intrinsic curvature κ has units of inverse length. If the action is to be, well, an action, this has to be turned into a mass. The conversion factor $\frac{1}{G}$ is needed to convert a length into a mass. The same is true for the boundary term, it has a $\frac{1}{G}$ in front of it.

The phase factor of the path integral then is of the form

$$e^{\frac{\imath S}{\hbar}} = e^{-\imath\frac{1}{8\pi\hbar G}[\int_{\mathcal{M}} d^4x(R-2\Lambda)-\int_{\partial\mathcal{M}} d^3\sigma\kappa]-\frac{1}{\hbar}\int_{\mathcal{M}} d^4x\mathcal{L}^{matter}} \tag{31}$$

so we see that in the absence of matter, G only appears in the combination $l_p^2 = \hbar G$. Without matter, the gravitational action is invariant under a scaling

$$\hbar \rightarrow \lambda\hbar, \quad G \rightarrow \frac{G}{\lambda} \tag{32}$$

The same is true of the commutation relations between the Ashtekar connection, A_a^i and the inverse sensitized from field, $\tilde{E}_j^a$ which represents information about the three geometry

$$[A_a^i(x), \tilde{E}_j^b(y)] = -\hbar G\delta^3(x, y)\delta_a^b\delta_j^i \tag{33}$$

It then seems impossible without matter to produce the expression (29) for m_p. It is the same for the spin foam action, which differs from a topological field theory by the imposition of the simplicity constraint. The latter is a constraint on representations and is dimensionless.

We see the same story when the boundary term comes from the boost Hamiltonian of FGP [21, 53, 54], which has dimensions of action (since it is conjugate to a dimensionless boost), and is equal to,[4]

$$H_B = \frac{c^3}{8\pi G}A(W) \tag{34}$$

[4]Note that this is the contribution to the action from a corner of a causal diamond, and hence comes into an action directly, without being integrated over time.

where $A(W)$ is the area of the horizon as seen by the boosted observer.

Indeed, in LQG the Planck area and Planck volume appear easily and give the scale of the spectrum of quantum geometry. But it appears that to talk about the Planck energy, we need to couple the quantum geometry to matter. Matter terms in the action will bring in independent factors of $\hbar$.

One place the Planck mass appears is the energy for the Schwarzchild black hole in the isolated horizon approach [55], which is taken to be

$$H = M = \frac{c^2}{4\pi G}\sqrt{A} \approx \sqrt{\frac{\hbar}{G}} n \frac{c^2}{4\pi} \tag{35}$$

Here n stands for the area quantum numbers. But this is written down by definition, it is not derived from a Hamiltonian operator defined on the whole Hilbert space.

If we want LQG to make predictions for quantum gravity phenomenology, it has to be able to speak to us about corrections of the form of energies in units of $\frac{1}{m_p} \approx \sqrt{\frac{G}{\hbar}}$.

This is connected to the assumption that the quantum theory of gravity predicts phenomena associated with gravitons such as the gravitational analogue of the photo-electric effect, at least for wavelengths long compared to l_p. This must ultimately be due to a normalization of the linearized hamiltonian which gives to each graviton an energy $\hbar\omega$, which is independent of G. Linearized and perturbative quantum gravity introduces the Planck mass, when it gives the perturbed metric, h_{ab}, defined by

$$g_{ab} = \eta_{ab} + \sqrt{\frac{G}{c^2}} h_{ab} \tag{36}$$

canonical dimensions of square root of energy per length. This and the assignment of $\hbar$ to loops separates the dependence on G and $\hbar$. This separation occurs also in the semiclassical approximation [56]. But I know of no mechanism for producing those factors from the non-perturbative quantum theory without invoking matter.

But if m_p is missing in the bulk dynamics of LQG and spin foam models then it may be because these describe, not full quantum gravity, but a strong coupling limit of the theory in which $\hbar \to 0$ and $G \to \infty$ with l_p held fixed and m_p taken to zero. Indeed we can see this, from the form of the Ashtekar connection [57],

$$A = \Gamma(e) + \imath G \Pi \tag{37}$$

so in the limit $G \to \infty$ we pick up the ultra local limit in which all spatial derivatives go away.

In the light of these comments we can consider the derivations of black hole thermodynamics by Perez et al. [53] and Bianchi [54]. At two crucial steps in their derivations they introduce $\hbar$ independently when they relate the simplicity constraint to the first law of thermodynamics.[5] This requires identifying $B^a = \hbar K^a$ as the boost

[5] See [21] for discussion of how the first law of thermodynamics plays a role in these arguments.

Hamiltonian in the Hilbert space of a triangle. Similarly they identify $T_U = \frac{\hbar}{2\pi c}$ as the boost temperature. These independent introductions of $\hbar$ make it possible to extract Newton's constant from the ratio $\frac{l_p^2}{\hbar} = G$. Without this they couldn't derive the classical Einstein equations (with matter) from the quantum statistical physics of the horizon.

2.9 Gravity Is Missing

If LQG and other approaches fail to talk about energy, they fail too when it comes to gravity. That is, in classical general relativity there is a straightforward way to derive Newton's gravitational theory as the non-relativistic approximation to general relativity. If one pulls a scalar field, ϕ out of the metric by a rescaling $g_{ab} \rightarrow g'_{ab} = e^{\phi} g_{ab}$ it follows right away that

$$\nabla^2 \phi = 4\pi G \rho \tag{38}$$

I know of only one way to get this out of the Hamiltonian approach to LQG, which is by an indirect entropic argument [58].

This is of course consistent with the hypothesis that the Hamiltonian approach to LQG describes quantum general relativity only in the strong coupling limit in which $G \rightarrow \infty$ while $\hbar \rightarrow 0$.

The situation is better in spin foam models where one can recover the $\frac{1}{q^2}$ behaviour of the graviton propagator from correlation functions of boundary excitations [59].

2.10 Maximal CPT Violation

There is another line of thought that should be connected to this one. This is the set of arguments that lead to the conclusion that irreversibility and time reversal invariance breaking are fundamental. These are discussed in [37–39, 60–63]. These lead to the conclusion that the familiar time symmetric laws hold in a limited regime, beyond which we should see the effects of a preferred fundamental arrow time. Models for how time reversal physics might emerge in a limited regime from a more fundamental time irreversible physics are described in [37, 64].

One consideration along these lines begins by noting that according to the CPT theorem, CPT must be a symmetry of any Lorentz invariant relativistic QFT. But global Lorentz invariance is an accidental or emergent symmetry of the ground state of the gravitational field-Minkowski spacetime. Thus we may hypothesize that CPT is enforced only to the extent that the assumptions of the CPT theorem hold. This CPT regime should then be delimited by, R the radius of curvature of spacetime. Thus we should expect to find CPT violation on the order of

$$\Delta^{XCPT} = \frac{\lambda}{R} \tag{39}$$

where λ is a wavelength.

Here is one idea: assume the fundamental theory is irreversible, but there is an emergent theory which is a local, lorentz invariant QFT. Then by the CPT theorem the emergent theory has CPT symmetry. This suggests that CPT is maximally broken, given that CPT is enforced by the lorentz invariance of the emergent theory. Now lorentz invariance is broken if the metric is curved, so the CPT breaking should be proportional to the curvature tensor. So they could be given by effective actions like:

$$\Delta S \approx \frac{1}{m_p} R_{ab} \bar{\psi} \gamma^a \gamma^b \psi \tag{40}$$

The second idea is that my precedence theory of quantum dynamics, introduced in [65], has no need to be time reversal invariant, so if its true we should see rare processes which break time reversal.

3 A New Strategy: Quantum Gravity as a Principles Theory

In light of these reflections we might consider novel strategies for searching for quantum gravity. One is to stop asking for a specific model of quantum spacetime but, instead, to search for general principles which might constrain the choice of models to investigate. That is, following Einstein, we seek a *principle theory*, rather then a *constitutive theory.* This strategy is explored in a companion paper [2].

Acknowledgements I am grateful to Joseph Kouneiher for including me in this volume. I would like to thank Andrzej Banburski, Linqing Chen, Bianca Dittrich, Laurent Freidel, Henriques Gomes, Jerzy Kowalski-Glikman, Joao Magueijo and Yigit Yargic for very helpful discussions and encouragement.

I am also indebted to Stacy McGaugh, Mordehai Milgrom and Maurice van Putten for very helpful correspondence.

This research was supported in part by Perimeter Institute for Theoretical Physics. Research at Perimeter Institute is supported by the Government of Canada through Industry Canada and by the Province of Ontario through the Ministry of Research and Innovation. This research was also partly supported by grants from NSERC and FQXi. I am especially thankful to the John Templeton Foundation for their generous support of this project.

References

1. A. Einstein, Preussische Akademie der Wissenschaften, Sitzungsberichte, 1916 (part 1), 688–696
2. L. Smolin, *Four Principles for Quantum Gravity*. arXiv:1610.01968, contribution to Paddy@60, a book in honour of Thanu Padmanabhan

3. G. Amelino-Camelia, L. Smolin, Prospects for constraining quantum gravity dispersion with near term observations, Phys. Rev. D 80, 084017 (2009). arXiv:0906.3731; S. Hossenfelder, L. Smolin, Phenomenological quantum gravity, Nov. 2009. Phys. Canada **66**(2), 99-102 (2010) (Apr-June). ArXiv:0911.2761
4. G. Amelino-Camelia, G. D'Amico, G. Rosati, N. Loret, *In-vacuo-dispersion Features for GRB Neutrinos and Photons*. arXiv:1612.02765
5. C.J. Copi, D. Huterer, D.J. Schwarz, G.D. Starkman, Large-angle anomalies in the CMB. Adv. Astron. **2010**, Article ID 847541 (2010). arXiv:1004.5602
6. C.R. Contaldi, J. Magueijo, L. Smolin, Anomalous CMB polarization and gravitational chirality. Phys. Rev. Lett. **101**, 141101 (2008). arXiv:0806.3082
7. L. Smolin, *What is the Problem of Quantum Gravity?*. Introduction to Ph.D. thesis, also IAS preprint, September 1979
8. P. Feyerabend, **Against Method**
9. J.-W. Lee, H.-C. Kim, J. Lee, Gravity from quantum information. J. Korean Phys. Soc. **63**, 1094 (2013). https://doi.org/10.3938/jkps.63.1094. arXiv:1001 5445
10. H. Casini, M. Huerta, R.C. Myers, Towards a derivation of holographic entanglement entropy. JHEP **1105**, 036 (2011). https://doi.org/10.1007/JHEP05(2011)036. arXiv:1102.0440
11. V. Balasubramanian, B.D. Chowdhury, B. Czech, J. de Boer, M.P. Heller, A hole-ographic spacetime, J. Phys. Rev. D 89, 086004 (2014). arXiv:1305.085; V. Balasubramanian, B. Czech, B.D. Chowdhury, J. de Boer, The entropy of a hole in spacetime. JHEP **1310**, 220 (2013)
12. F. Markopoulou, L. Smolin, Quantum theory from quantum gravity. Phys. Rev. D **70**, 124029 (2004). arXiv:gr-qc/0311059
13. F. Markopoulou, L. Smolin, Disordered locality in loop quantum gravity states. Class. Quantum Gravity **24**, 3813–3824 (2007). arXiv:gr-qc/0702044; C. Prescod-Weinstein, L. Smolin, Disordered locality as an explanation for the dark energy. Phys. Rev. D **80**, 063505 (2009). arXiv:0903.5303 [hep-th]
14. G. Amelino-Camelia, L. Freidel, J. Kowalski-Glikman, L. Smolin, The principle of relative locality. Phys. Rev. D **84**, 084010 (2011). arXiv:1101.0931 [hep-th]
15. L. Freidel, L. Smolin, Gamma ray burst delay times probe the geometry of momentum space. arxiv:hep-th/arXiv:1103.5626
16. J. Barbour, personal communication (1980s)
17. J.M. Maldacena, hep-th/9711200, hep-th/9803002; E. Witten, hep-th/9802150, hep-th/9803131; S.S. Gubser, I.R. Klebanov, A.M. Polyakov, Phys. Lett. B **428**, 105-114 (1998), hep-th/9802109. For a review, see, O. Aharony, S.S. Gubser, J. Maldacena, H. Ooguri, Y. Oz, Large N field theories, string theory and gravity hep-th/9905111
18. T. Jacobson, Thermodynamics of spacetime: the Einstein equation of state. Phys. Rev. Lett. **75**, 1260–1263 (1995). gr-qc/9504004
19. T. Jacobson, Entanglement equilibrium and the Einstein equation. arXiv:1505.04753
20. T. Padmanabhan, Thermodynamical Aspects of gravity: new insights. Rep. Prog. Phys. **73**, 046901 (2010). arXiv:0911.5004; Entropy of static spacetimes and microscopic density of states. Class. Quantum Gravity **21**, 4485–4494 (2004). arXiv:gr-qc/0308070; Equipartition of energy in the horizon degrees of freedom and the emergence of gravity. Mod. Phys. Lett. **A25**, 1129–1136 (2010). arXiv:0912.3165
21. L. Smolin, The thermodynamics of quantum spacetime histories. arXiv:1510.03858; General relativity as the equation of state of spin foam. arXiv:1205.5529
22. W.G. Unruh, Notes on black hole evaporation. Phys. Rev. D **14**, 870 (1976)
23. E.P. Verlinde, On the origin of gravity and the laws of Newton. JHEP **1104**, 029 (2011). arXiv:1001.0785
24. M. Milgrom, ApJ **270**, 365 (1983). For reviews, see B. Famaey. S. McGaugh, Liv. Rev. Rel. **15**, 10 (2012); M. Milgrom, Scale invariance at low accelerations (aka MOND) and the dynamical anomalies in the Universe, arXiv:1605.07458v2; M. Milgrom, Scholarpedia **9**(6), 31410 (2014); S. McGaugh, F. Lelli, J. Schombert, The radial acceleration relation in rotationally supported galaxies. arXiv:1609.05917v1
25. E.P. Verlinde, Emergent gravity and the dark universe. arXiv:1611.02269

26. M. Milgrom, The modified dynamics as a vacuum effect. Phys. Lett. A **253**, 273–279 (1999). https://doi.org/10.1016/S0375-9601(99)00077-8. arXiv:astro-ph/9805346; Dynamics with a nonstandard inertia-acceleration relation: an alternative to dark matter in galactic systems. Ann. Phys. **229**(2), 384–415. arXiv:astro-ph/9303012
27. M.H.P.M. van Putten, Galaxy rotation curves in de Sitter space. arXiv:1411.2665
28. S. Hossenfelder, A covariant version of Verlinde's emergent gravity. arXiv:1703.01415
29. R.P. Woodard, Nonlocal metric realizations of MOND. Can. J. Phys. **93**(2), 242–249, arXiv:1403.6763
30. L. Modesto, A. Randono, Entropic corrections to Newton's law. arXiv:1003.1998v1 [hep-th]
31. C.M. Ho, D. Minic, Y.J. Ng, Quantum gravity and dark matter, 35. Gen. Relativ. Gravit. 43, 2567–2573 (2011). arXiv:1105.2916
32. S.H. Hendi, A. Sheykhi, Entropic corrections to Einstein equations. Phys. Rev. D **83**, 084012 (2011).arXiv:1012.0381
33. M.E. McCulloch, Quantised inertia from relativity and the uncertainty principle. arXiv:1610.06787; A toy cosmology using a Hubble-scale Casimir effect. Galaxies **2**(1), 81–88 (2014).https://doi.org/10.3390/galaxies2010081
34. L. Smolin, MOND as a regime of quantum gravity. arXiv:1704.00780
35. Luca Bombelli, Joohan Lee, David Meyer, Rafael D. Sorkin, Spacetime as a causal set. Phys. Rev. Lett. **59**, 521–524 (1987)
36. A. Criscuolo, H. Waelbroeck, Causal set dynamics: a toy model. Class. Quantum Gravity **16**(1999), 1817–1832 (1998). arXiv:gr-qc/9811088
37. M. Cortês, L. Smolin, The universe as a process of unique events. Phys. Rev. D **90**, 084007 (2014). arXiv:1307.6167 [gr-qc]
38. M. Cortês, L. Smolin, Energetic causal sets. Phys. Rev. D **90**, 044035. arXiv:1308.2206 [gr-qc]
39. M. Cortês, L. Smolin, Spin foam models as energetic causal sets. Phys. Rev. D **93**, 084039 (2016). https://doi.org/10.1103/PhysRevD.93.084039.. arXiv:1407.0032
40. C. Furey, Handwritten note, Sept 2 2011, personal communication to Lee Smolin, available at: http://www.perimeterinstitute.ca/personal/cfurey/2011notes__.pdf. See also www.perimeterinstitute.ca/personal/cfurey/essay_excerpts_2006.pdf
41. R.D. Sorkin, Is the cosmological "constant" a nonlocal quantum residue of discreteness of the causal set type? in *AIP Conference Proceedings*, vol. 957 (2007), pp. 142–153. https://doi.org/10.1063/1.2823750.
42. F. Dowker, J. Henson, R. Sorkin, Discreteness and the transmission of light from distant sources. Phys. Rev. D **82**, 104048 (2010). arXiv:1009.3058
43. J. Barbour, *Shape dynamics. An introduction*, arXiv:1105.0183
44. H. Gomes, S. Gryb, T. Koslowski, Einstein gravity as a 3D conformally invariant theory. Class. Quantum Gravity **28**, 045005 (2011). arXiv:1010.2481
45. F. Mercati, H. Gomes, T. Koslowski, A. Napoletano, Gravitational collapse of thin shells of dust in asymptotically flat Shape Dynamics. arXiv:1509.00833; H. Gomes, A Birkhoff theorem for Shape Dynamics. Class. Quantum Gravity **31**, 085008 (2014). arXiv:1305.0310
46. H. Gomes, S. Gryb, T. Koslowski, F. Mercati, The gravity/CFT correspondence. Eur. Phys. J. C 75, 2275 (2013). arXiv:1105.0938
47. L. Smolin, The black hole information paradox and relative locality, gr-qc/ArXiv:1108.0910, to appear in Physical Review D
48. M. Bojowald, Living Rev. Relativ. **11**: 4 (2008). https://doi.org/10.12942/lrr-2008-4
49. A. Ashtekar, P. Singh, Loop quantum cosmology: a status report. Class. Quantum Gravity **28**, 213001 (2011). https://doi.org/10.1088/0264-9381/28/21/213001. arXiv:1108.0893
50. R.B. Tully, J.R. Fisher, Astron. Astrophys. **54**, 661 (1977); S. McGaugh, F. Lelli, J. Schombert, *The small scatter of the baryonic Tully-Fisher relation*. Astrophys. J. Lett. 816(1), article id. L14, 6 pp. (2016), arXiv:1512.04543
51. P. Schoen, S.T. Yau, Phys. Rev. Lett. **43**, 1457 (1979); R. Schoen, S.-T. Yau, Commun. Math. Phys. **79**, 231 (1981); E. Witten, Commun. Math. Phys. **80**(3), 381–402 (1981)
52. L. Smolin, Positive energy in quantum gravity. Phys. Rev. D 90, 044034. arXiv:1406.2611; L. Smolin, A. Starodubtsev, Positive energy theorem and the factor ordering problem in quantum gravity, unpublished manuscript (2004)

53. E. Frodden, A. Ghosh, A. Perez, A local first law for black hole thermodynamics. arXiv:1110.4055; A. Ghosh, K. Noui, A. Perez, Statistics, holography, and black hole entropy in loop quantum gravity. arXiv:1309.4563; A. Ghosh, A. Perez, Black hole entropy and isolated horizons thermodynamics. arXiv:1107.1320
54. E. Bianchi, Entropy of non-extrenal black holes from loop gravity. arXiv:1204.5122
55. A. Ashtekar, J. Baez, A. Corichi, K. Krasnov, Quantum geometry and black hole entropy. Phys. Rev. Lett. 80(5), 904–907 (1998). arXiv:gr-qc/9710007
56. L. Smolin, Falsifiable predictions from semiclassical quantum gravity. Nucl. Phys. B **742**, 142–157 (2006). hep-th/0501091
57. A. Ashtekar, New variables for classical and quantum gravity? Phys. Rev. Lett. **57**(18), 2244–2247 (1986)
58. L. Smolin, Newtonian gravity in loop quantum gravity. arXiv:1001.3668
59. E. Bianchi, L. Modesto, C. Rovelli, S. Speziale, Graviton propagator in loop quantum gravity. Class. Quantum Gravity **23**, 6989–7028 (2006). https://doi.org/10.1088/0264-9381/23/23/024, arXiv:gr-qc/0604044
60. R.M. Unger, L. Smolin, *The Singular Universe and the Reality of Time* (Cambridge University Press, Cambridge, 2015)
61. L. Smolin, *Time Reborn* (Houghton Mifflin Harcourt, Penguin and Random House Canada, 2013)
62. M. Cortes, H. Gomes, L. Smolin Time asymmetric extensions of general relativity. Phys. Rev. D **92**, 043502 (2015). arXiv:1503.06085; M. Corts, A.R. Liddle, L. Smolin, Cosmological signatures of time-asymmetric gravity. arXiv:1606.01256
63. L. Smolin, Dynamics of the cosmological and Newton's constant. Class. Quantum Gravity **33**(2) (1977). https://doi.org/10.1088/0264-9381/33/2/025011, arXiv:1507.01229
64. M. Cortês, L. Smolin, in draft
65. L. Smolin, Precedence and freedom in quantum physics. arXiv:1205.3707

A Schema for Duality, Illustrated by Bosonization

Sebastian De Haro and Jeremy Butterfield

Abstract In this paper we present a schema for describing dualities between physical theories (Sects. 2 and 3), and illustrate it in detail with the example of bosonization: a boson-fermion duality in two-dimensional quantum field theory (Sects. 4 and 5). The schema develops proposals in De Haro (Space and Time after Quantum Gravity, 2016 [15]; Duality and Physical Equivalence, 2016a [16]): these proposals include construals of notions related to duality, like representation, model, symmetry and interpretation. The aim of the schema is to give a more precise criterion for duality than has so far been considered. The bosonization example, or boson-fermion duality, has the feature of being *simple* yet *rich enough* to illustrate the most relevant aspects of our schema, which also apply to more sophisticated dualities. The richness of the example consists, mainly, in its concern with two non-trivial *quantum field theories*: including massive Thirring-sine-Gordon duality, and non-abelian bosonization. This prompts two comparisons with the recent philosophical literature on dualities. (a) Unlike the standard cases of duality in quantum field theory and string theory, where only specific simplifying limits of the theories are explicitly known, the boson-fermion duality is known to hold *exactly*. This exactness can be exhibited explicitly. (b) The bosonization example illustrates both the cases of isomorphic and *non-isomorphic* models: which we believe the literature on dualities has not so far discussed.

S. De Haro (✉) · J. Butterfield
Trinity College, Cambridge CB2 1TQ, United Kingdom
e-mail: sd696@cam.ac.uk

J. Butterfield
e-mail: jb56@cam.ac.uk

S. De Haro
Department of History and Philosophy of Science, University of Cambridge,
Free School Lane, Cambridge CB2 3RH, United Kingdom

J. Kouneiher (ed.), *Foundations of Mathematics and Physics One Century After Hilbert*,
https://doi.org/10.1007/978-3-319-64813-2_12

1 Introduction

In this paper we present a schema for describing dualities between physical theories (Sects. 2 and 3). Then we illustrate it in detail with the example of bosonization: a boson-fermion duality in two-dimensional conformal field theories (Sects. 4 and 5).

Before we introduce these two parts in turn (Sects. 1.1 and 1.2), we briefly set our project in the context of the legacy of Hilbert's work, a hundred and more years ago, on the foundations of physics and its axiomatisation—work which it is an honour to commemorate. This legacy is of course so broad and deep that we can only touch on it. We will confine ourselves to recalling two Hilbertian ideas about the role of axiomatizing a theory (either mathematical or physical): ideas which obviously relate to our project, and which we will return to in Sects. 2.4 and 5.4.

The background for both ideas, indeed for all Hilbert's work in axiomatisation (such as his axiomatisation in 1899 of Euclidean geometry, and his choosing as the sixth Problem in his famous 1900 'To Do' list, the axiomatisation of mechanics and geometry) was, of course, the development of formal methods, in particular axiomatic studies, in all of mathematics from about 1850.[1]

First, there is the idea that an axiom system can be realized, i.e. made true, by very different models. (Recall Hilbert's famous remark that 'one must be able to say at all times—instead of points, straight lines and planes—tables, chairs and beer-mugs'.) We shall see that duality, in the sense nowadays used in physics, gives illustrations of this idea. Indeed, very *vivid* illustrations. For duality, in physicists' current jargon, involves there being two theories that look very different (not just in their formulation and concepts, but also apparently in the topics they are about) that are in some sense equivalent. In particular, there is a 'dictionary' that pairs off the concepts in one theory with those in the other. Thus in our example of bosonization, one theory will describe fermions, while the other describes bosons: very different field-contents. So duality illustrates this Hilbertian idea: but on a grand scale! For now, it is entire physical theories that are the very different realizations of some *common core* axiom system. Indeed, as explained in Sect. 1.1: we shall call the two sides, i.e. the items that are dual to each other, *models* (viz. models of the single common core), rather than *theories*. So this usage echoes the Hilbertian idea.[2]

Second, Hilbert sees the activity of axiomatisation, *not* as giving a theory its final form and so best undertaken when (one hopes!) it is fully understood, but as worthwhile even when we recognize that the theory is far from its final form. For it is worthwhile precisely in order to deepen our understanding of the theory. Again this

---

[1]Again, we can only touch on the vast literature. For Hilbert's Problems of 1900, cf. e.g. [32, 33]. For the sixth Problem, cf. [11–14, 51]. For some context for Hilbert's famous 'beer-mug' remark, cf. [38]. Finally, we note that Gray [34] makes an interesting case that this broad development represented a rise of 'modernism', in a sense analogous to that in art and literature: cf. also [35].

[2]The Hilbertian idea has, of course, other important facets: for example, in fostering the idea that an axiom system—or more generally, a doctrine expressed in language—'implicitly defines' its terms. This has been very influential in the foundations of logic and mathematics, beginning with Hilbert's debate with Frege. It has also of course been contested: in the face of non-categoricity, the claim to 'define' terms by a body of doctrine containing them is questionable.

idea has been very influential. In physics, the best known example of its influence is no doubt von Neumann's monumental treatise on quantum mechanics [56], which over the decades has spawned so many axiomatic studies of quantum mechanics, most directly the quantum logic approach. But also in philosophy, the idea was very influential. Reichenbach and other logical empiricists saw axiomatisation as the way by which philosophers could clarify scientific theories (and in particular, distinguish their factual and conventional contents—a project that, for the logical empiricists, was the distinctive task of philosophy). Thus we think of our own project—to formulate in general, almost formal, terms, the notion of duality (Sects. 2 and 3), and to illustrate this in bosonization (Sects. 4 and 5)—as an exercise in the tradition of this idea.[3]

In Sect. 1.1, we briefly introduce our notions of theory and model, and of duality as an isomorphism between models. We motivate our usage and compare the notion of duality to the analogous notion of symmetry. In Sect. 1.2, we introduce our main example, of bosonization, and compare this example to other examples used in the literature on dualities.

1.1 The Schema

The schema develops proposals in De Haro [15, 16]. Like other authors, we take duality to be a suitable relation of equivalence between physical theories. The main features of our schema are that:

(1): we distinguish uninterpreted theories, which we call *bare theories*, from interpreted theories;

(2): we emphasize that, wholly independently of issues of interpretation, a bare theory can have many realizations, which we call *models*;

(3): we take duality to be an isomorphism between two models of a single bare theory.

Of these three features, it is (2) and (3) that are the distinctive ones. For several authors also define duality in terms of uninterpreted theories. This has the advantage of making verdicts of duality not beholden to semantic issues, and so less vague or even controversial. And it allows cases of duality without any sort of physical or semantic equivalence—which certainly occur, e.g. Kramers-Wannier duality between the high and low temperature regimes of the statistical mechanics of a lattice. But features (2) and (3) make duality an equivalence (formally: an isomorphism) between items that are not only uninterpreted, but also *more specific* than an uninterpreted theory: viz. realizations—which we will call 'models'—of a (single) bare theory. A prototypical example is: taking a bare theory to be an abstract algebra of quantities (maybe also equipped with a dynamics, viz. as a 1-parameter group of

[3]For some 'post-Hilbert' history of axiomatisation as 'deepening the foundations', cf. [52, 53]. But we should add that we do not endorse the logical empiricist project of distinguishing, once and for all, the factual and conventional parts of a theory: our misgivings are essentially those of Putnam [47].

automorphisms, and a set of abstract states, i.e. rules for evaluating (i.e. assigning values to) the quantities): a model or realization is given by a representation (in the mathematical sense) of the algebra, together with a realization of the rules for evaluating the quantities, for the representation in question, i.e. a set of maps to the relevant field, of complex or real numbers.

We shall say 'model' rather than 'realization', not least for brevity. But we should disavow, here at the outset, some misleading connotations of the word 'model'. Indeed, there are three misleading connotations. The word 'model', as contrasted with 'theory', often connotes:

(i): a specific solution (at a single time: or for all times, i.e. a possible history) for the physical system concerned, whereas the 'theory' encompasses all solutions—and in many cases, for a whole class of systems;

(ii): an approximation, in particular an approximate solution, whereas the 'theory' deals with exact solutions;

(iii): being part of the physical world (in particular, being empirical, and-or observable) that gives the interpretation, whereas the 'theory' is of course not part of the physical world, and so stands in need of interpretation.

So we stress that our use of 'model' rejects all three connotations. As we just said: for us, a model is a specific realization—one might say 'version'—of a theory. But it is a version of a bare, i.e. uninterpreted, theory, and the version is itself bare, i.e. uninterpreted. So a model adds details—we shall say: 'specific structure'—to its bare theory. But these details are *not* a matter of specifying: (i) a solution or history of the system; or (ii) approximation(s); or (iii) interpretation(s). Rather, the extra details are extra mathematical structure: just like a representation of a group or an algebra has extra details or structure, beyond that of the group or algebra of which it is a representation.[4]

Indeed, for clarity later on, we should distinguish two broad kinds of extra detail or structure that a model adds. Again, group representations provide obvious—and countless—examples.

(A): The 'concreteness' of a specific mathematical object: such as GL$(n,\mathbb{C})$, the general linear group over $\mathbb{C}^n$, or any subgroup of it—any of which is a 'concrete', not abstract, group. (Agreed, the concrete vs. abstract contrast is flexible; but this will not matter for anything that follows.)

(B): The fact that the mathematical notion of representation requires homomorphism, not isomorphism: i.e. it allows non-injectivity and non-surjectivity. Thus two

[4]We agreed that for our notion, the word 'model' has disadvantages. But note that other words also have disadvantages. For example: 'formulation' connotes that any two formulations of a theory are 'notational variants', i.e. fully equivalent: they say exactly the same thing about the world. But that is far from true for our notion (and this matches the connotations of 'model'): for us, two models of a bare theory are in general not isomorphic, and not in any sense equivalent; and so typically, it is surprising to find two isomorphic models, i.e. to find a duality. Other examples: 'realization', 'instance' and 'instantiation' connote being part of the physical world, as in 'the mechanism/hardware which realizes some specific function/software', or 'the object is an instance/instantiation of the predicate'—which is the misleading connotation (iii) above.

Notice that in theoretical physics, the use of model is, roughly, between: (a) our use, and (b) (ii) and (iii) above: e.g. the 'massive Thirring model' or the 'sine-Gordon model'.

representations of an abstract group G can be non-isomorphic as groups—i.e. different, even as described in only abstract group-theoretic terms—to G; and of course also, non-isomorphic to each other.
Of course, these kinds (A) and (B) of 'extra detail' usually occur together: just think of how every abstract group can be represented by the trivial one element subgroup of GL(n,$\mathbb{C}$), the $n \times n$ identity matrix.

Our picture is therefore of a bare theory, that can be realized (we will say: modelled) in various ways: like the different representations of an abstract group or algebra. And these models are in general *not* isomorphic, since they differ from one another in their specific structure: like inequivalent representations of a group. But we say: *when the models are isomorphic, we have a duality.*

In Sects. 2 and 3, we will develop this view of duality (with Sect. 2 dealing with theories, models and interpretations, and Sect. 3 with symmetries). We end this Subsection with two further remarks about our schema. The first motivates our usage of 'theory' and 'model'; the second compares duality with that more familiar topic, symmetry.

(1): *Motivating our usage*:— Dualities in physics give a rationale for our usage of 'theory' and 'model', as introduced above. (This rationale does not depend on the contrast between interpreted and uninterpreted (bare) theories; and so we temporarily set that aside.) Recall that in both physics and philosophy of physics, 'theory' is usually taken as something like a set of differential equations, and 'model' is usually taken as something like a solution to such a set. But a duality often shows us that what we first considered as distinct theories can, or should, be seen as the same theory, in two guises. Agreed, that is very rough speaking: which will of course be clarified in what follows. But for now, we only need the point that this kind of surprising discovery prompts us to move our usage of 'theory' "one level up". After all: if two sets of differential equations somehow express the same theory, then a theory cannot be identified with such a set. Besides: if we thus move our usage of 'theory' one level up, we can still keep the usual intuitive idea of how 'theory' and 'model' are related—viz. that a model is a realization, or instance, of a theory—by correspondingly moving our usage of 'model' one level up. And this is what we have proposed.

To sum up: the broad and widely-agreed idea, that in physics a duality often suggests that the two theories concerned, though they look different, are in fact 'the same', motivates our proposed usage of 'theory' and 'model'.

(2): *Analogy with symmetry*:— The analogy is (as is often remarked) that 'a duality is like a symmetry, but at the level of a theory'. Here, and for the rest of this Subsection, we will *temporarily* set aside our jargon just announced, of 'theory' versus 'model'. We will temporarily join the literature's usual jargon of taking a theory to be interpreted, and a model to be—not a 'version' of the theory with some specific structure of its own—but a solution (or representative of a solution) of the theory.

That is, the analogy is: while a symmetry carries a state to another state that is 'the same' or 'matches it', a duality carries a theory to another theory that is 'the same' or 'matches it'. We will endorse this analogy. So the interesting questions, for both sides of the analogy, will concern the different ways to make precise 'the same' or 'matches'. We give details (respecting our proposed 'theory' vs. 'model' usage!) in Sects. 3.1 and 3.2. But the questions about making precise 'the same'/'matches' can be introduced as follows.

A symmetry a (we write a for 'automorphism') carries a state s in a state space $\mathcal{S}$ to another state $a(s)$: thanks to a being a symmetry, the two states s and $a(s)$ assign the same values to all the quantities (i.e. magnitudes) in some salient, usually large, set of quantities. The question then arises: do s and $a(s)$ represent the very same physical state of affairs, or scenario—or in philosophers' jargon: the same possible world?

The answer, in full generality, is of course: 'No'. That is: not always. But for a large enough set of quantities being preserved; and in particular for a theory that is a 'toy cosmology' (i.e. a theory whose system of interest is a cosmos, with no external environment, so that there are no relational quantities whose values are *not* preserved by a): there is a tradition of answering 'Yes'.

Debate then ensues about:

(i) what are the general conditions for the 'Yes' answer being correct? and

(ii) what does the 'Yes' answer imply about the propriety of—perhaps even the requirement of—moving to a reduced formalism, i.e. one in which states are taken as the orbits, in the given formalism, of the action of the symmetry a?[5]

So, turning to our topic of dualities: we endorse the analogy. We will say, roughly speaking, that: a theory T is mapped by duality d to a theory $d(T)$ which is 'the same' as T. This will be made precise in various ways. But it is worth stressing now, in line with the three features (1), (2) and (3) we listed at the start of this Subsection, that:

(a): We take theories to be initially uninterpreted: so it will not follow from the existence of a duality map d that T and $d(T)$ are wholly equivalent ('state the very same propositions').

(b): We make explicit the interpretation of a theory's formalism: so there will be interpretation maps I acting on both the theory T and its dual $d(T)$.

(c): For a given theory, we distinguish different realizations of it, which we call 'models'. Duality is an isomorphism between such models: an isomorphism that is often surprising since the models, despite their common core, "look different".

[5] A bit more precisely: states would be taken as the union of the orbits for all the symmetries for which the 'Yes' answer is true. For recent work on the debate about (i) and (ii), cf. [8, 22, 57].

1.2 Bosonization and Other Dualities

In this Subsection, we motivate our choice of bosonization as the illustration of our schema. We first sketch a spectrum of examples of dualities (Sect. 1.2.1). Then we describe how bosonization strikes a balance between mathematical rigour and physical interest, and introduce its main features (Sect. 1.2.2).

1.2.1 Examples of Dualities

Recent philosophical literature on duality and theoretical equivalence has dealt with three main kinds of examples:—

(a): equivalence between models (in our sense!) formulated in *first-order* (maybe *many-sorted*) *logic*: e.g. definitional equivalence, Morita equivalence, and-or categorical equivalence (e.g. [4, 5]);

(b): categorical equivalence of models (in our sense) of *classical theories* (e.g. [55, 57]);

(c): dualities between models (in our sense) of *quantum theories* whose classical descriptions are very different (e.g. [17, 20, 23, 25, 36, 46, 49]).

The classification (a)–(c) is arranged in increasing order of physical (not mathematical!) sophistication. Consequently, there is also decreasing mathematical rigour, as one moves from kind (a) to kind (c):

Kind (a): These examples have the advantage of being very simple, in their reliance on first-order logic only: and so, the notions of equivalence in question can be defined rigorously. But in their simplicity, the notions developed, and the examples given, generally do not seem to have sufficient structure that they could describe in detail the sorts of examples that physicists would be interested in. (At any rate, the authors cited do not describe how such logical models can illustrate even the simplest physical models of, say, classical, source-free Maxwell theory: which is, of course, not to claim that this is impossible!).

Kind (b): These examples include some important models of *classical theories*, such as Newtonian gravitation, general relativity, and Yang-Mills theory models. But these examples also have some limitations. (1): To physicists, the example is, typically, not surprising (e.g. Newtonian gravitation being equivalent to *geometrized* Newtonian gravitation). (2): When it *is* surprising (e.g. [55]), it is not a case of equivalence, but rather of analogy. Furthermore, (3): categorical equivalence has been criticised by Barrett and Halvorson [5] for being 'too liberal'. In our view, the element of 'surprise' (see Sect. 2.1) seems to come with models of *quantum theories*, i.e. examples of (c):

Kind (c): Typical examples of this kind are dualities between very different-looking models of *quantum field theories*, or of *string theories* (cf. [59]).

This explains the recent interest, shown by both physicists and philosophers of physics, in such dualities. Physicists tend to view dualities as powerful epistemic statements: the epistemic gain being both mathematical and physical. As the mathematical aspect: mirror symmetry is the prime example.[6] Michael Atiyah has characterised the discovery of mirror symmetry as 'spectacular': since it established a new link between complex geometry and symplectic geometry, later proven (in one of its simplified versions) by mathematicians [2, p. 83]. As to the physical aspect: gauge-gravity duality is an important example, which has led to both new theoretical developments in quantum gravity, and to new experimental results and ideas (like the explanation of the shear viscosity-to-entropy ratio in a quark-gluon plasma, and recent applications to cosmology: cf. e.g. [1, 15]). Other examples are T-duality (related to mirror symmetry) and S-duality: which falls under the same class of dualities as our bosonization example, viz. exchanging Noether charges and topological charges [7]. An important idea of these dualities is that it is the models of the *quantum theories* which are equivalent, while their classical limits are very disparate (differing in the number or the size of the dimensions, the matter content, etc.). These two aspects—physical and mathematical—will be developed in Sect. 2.1's discussion of the *scientific importance* of dualities.

But there is a second reason these dualities interest philosophers of physics: which the recent literature has emphasised. Namely, these dualities obviously bear on philosophical questions such as the distinction between theoretical and physical equivalence, emergence (of spacetime, and-or other entities), and realism versus structuralism. We will return to these questions in another paper.

Agreed: examples of kind (c) also have limitations, as follows. (1): The models (in our sense) are mathematically very difficult; and typically, no exact formulation of the models that are dual is yet known. So the duality, e.g. in the case of gauge-gravity duality, is still—despite all the favourable evidence, in various limits etc.—a conjecture. (2): The physics involved is not yet established, since the models involved either deal with quantum gravity situations (a regime of energies about which experiment has so far given no direct clues: cf. [50]), or involve simplifying assumptions about the world, typically a high degree of symmetry (e.g. supersymmetric quantum field theory models).

1.2.2 Bosonization Introduced

It is clear, in the light of Sect. 1.2.1, that to illustrate our schema, we should choose an example that judiciously balances the desiderata: on the one hand, (i): mathematical precision and established physics, as in kinds (a) and (b); on the other, (ii): scientific importance, as in kind (c). As we will see in detail in Sects. 4 and 5: bosonization or, more precisely, boson-fermion duality in two dimensions, is just such an example.

[6]Despite the name of 'symmetry', mirror symmetry falls under what we here call a duality. That mirror symmetry is a case of duality—of two different models, rather than a single model, being related—is uncontroversial, and reflected in the literature.

As to (i): Boson-fermion duality allows a treatment of the quantum theory model that does not need to rely on techniques of approximation such as perturbation theory. For this reason, boson-fermion duality explicitly illustrates our schema: the common core *theory* can be formulated according to our construal in Sect. 2.2.1, and the two sides of the duality are *models* in the sense of Sect. 2.2.2. The physics of these models is not speculative, and these 1+1-dimensional models describe systems that can be realised in the lab, e.g. as one-dimensional spin chains [29, Chap. 2], [3, Sects. 4.3 and 9.4.4].

As to (ii): The scientific importance of the duality is witnessed by three facts. (1): Bosonization involves rich models of *quantum field theories*, and not just classical theories; (2) it is an active area of physics (see e.g. [31, 39]); and (3) it illustrates the surprise that we discuss in Sect. 2.1, viz. by relating a model of bosons and a model of fermions.

Bosonization was discovered in two papers by Coleman [9] and Mandelstam [45], which built on previous work on the sine-Gordon and Thirring models [21]. Coleman discovered that the sine-Gordon model (a scalar field whose interaction potential is the cosine of the field) in 1+1 dimensions was equivalent to the *charge-zero sector* of the massive Thirring model (a massive Dirac fermion field with quartic interaction) in 1+1 dimensions.[7] 'Charge-zero sector' here refers to the restriction of the physical quantities of the model to pairs of fermionic fields. Thus Coleman wrote:

> ... under the assumption that one can only make particle-antiparticle pairs out of the vacuum, not single particles ... For massless particles in two dimensions, it is quite possible to make a pair that never separates. Such a pair consists of two particles moving in the same direction. The wave functions do not spread; they just move on steadily at the speed of light, and the particles never get away from each other [since there is no other direction in which they could turn]. If the particles had a mass, or if the world were of greater than two dimensions, this would not be possible. (p. 2094).

While Coleman's analysis was perturbative, Mandelstam constructed a map which was exact, and went both ways. Not only could a boson be mapped to a pair of fermions; but also the map could be inverted, so as to map a single fermion to a coherent state of bosons. The construction was non-perturbative, i.e. it did not use perturbation theory. This is related to the fact, already recognised by Coleman [9, p. 2088], that all divergences that occur in perturbation theory can be removed by normal-ordering the Hamiltonian. Mandelstam also did a canonical treatment of the model, working out canonical commutation relations between the fields and the currents constructed from them, using regularisation and renormalisation. Therefore, boson-fermion duality was proven to be *exact*.

There are three significant features of this duality: features that both (a) justify our claim, two paragraphs above, that boson-fermion duality balances the desiderata (i) and (ii), and (b) bear out the conceptual relevance of the example.

---

[7] See Sect. 5.5.1; for the simple, free case, see Sect. 4. Here of course we adopt the usual theoretical physics usage of 'model': cf. the end of footnote 4.

(A): *The duality is exact*. That is: it is valid for all physically interesting values of the parameters, and it does not require the use of perturbation theory. In this respect, boson-fermion duality is closer to kinds (a) and (b) than kind (c). Yet the models related as duals are non-trivial (because massive, or massless, and interacting!: see Sect. 5.5) quantum field theory models—and in *that* sense they are in kind (c)!

(B) *The duality goes both ways*. It relates boson operators to fermion operators, and vice versa. This is, of course, surprising, since these two kinds of operators have very different properties: both mathematically (e.g. different statistics) and physically (they describe particles with distinct properties). We will explain, in Sect. 4, how two models with such disparate formulations can nevertheless be isomorphic to each other.

(C) *The duality maps the weak-coupling regime of one model to the strong-coupling regime of the other,* and vice versa [9, p. 3027]:

$$\frac{g}{\pi} = \frac{4\pi}{\beta^2} - 1 \,. \tag{1}$$

Here, β is the coupling constant of the bosonic model (the sine-Gordon model) and g is the coupling constant of the fermionic model (the Thirring model: for more details, see Sect. 5.5.1). Clearly, when $\beta \to 0$, $g \to \infty$. This attests to the physical richness and, indeed, the non-trivial character of the duality. This weak coupling/strong coupling correspondence has later been found to be a feature of most dualities of kind (c), i.e. dualities in models of quantum field theory and string theory: especially S-duality and gauge-gravity duality.[8] This richness is the main reason why physicists are interested in dualities: since they can learn about the strong-coupling regime of one model (where perturbation theory cannot be used effectively nor reliably, so that the model is in general much harder to deal with) from the weak-coupling regime of the other model (where perturbation theory is usually a good guide). For more discussion of how Eq. (1) contributes to scientific importance, see Sect. 2.1-(2).

Features (A) and (B) are needed in order that the example illustrate our schema with mathematical precision. We will spell this out in Sect. 5. Indeed, we believe this is the first conceptual and technical exposition in the philosophical literature of a duality combining the physical interest of kind (c), with features (A)–(B).

Feature (C) relates to another important topic relating to dualities, viz. that of *emergence*. Indeed, a recent theme in the philosophy of physics literature has been the close connection between duality and emergence.[9] A framework for understanding the connection between dualities and emergence was developed in De Haro [15]: it was argued that the two notions (duality as isomorphism, and emergence as novel and robust behaviour relative to a comparison class), while close to each other, also exclude one another. But we must leave the topic of emergence for another occasion.

---

[8]In T-duality and mirror symmetry, it is not the size of the *couplings* that is inverted by the duality map but, roughly speaking, the sizes of the *spaces*.

[9]On the connection between duality and emergence, see: [23, 48, 54].

Feature (B) also prompts the question of *fundamentality*. Coleman himself wrote:

> I am led to conjecture a form of duality, or nuclear democracy in the sense of Chew, for this two-dimensional theory. A single theory has two equally valid descriptions in terms of Lagrangian field theory: the massive Thirring model and the quantum sine-Gordon equation. The particles which are fundamental in one description are composite in the other: In the Thirring model, the fermion is fundamental and the boson a fermion-antifermion bound state; in the sine-Gordon equation, the boson is fundamental and the fermion a coherent bound state (1975: p. 2096).

The issue of fundamentality in boson-fermion duality, and in electric-magnetic duality, has been addressed by Castellani [7]. Her account is, in philosophers' jargon, deflationary. That is: she argues that all the manifestations of the fields (as bosonic or as fermionic; as electric or magnetic) are ontologically equally fundamental: 'What the duality specifically implies here, concerns, not mutual composition of the particles, but rather their different modes of appearance when considering the different classical limits of the quantum theory, i.e. the dual perspectives' (2017: Sect. 3.3).

Our construal of duality as an isomorphism, in Sects. 2 and 3, is in agreement with such a deflationary account. For the content of the theory will be taken to be based on the *common core* of the models: and this common core includes *both* bosons and fermions, on an equal footing. We will discuss some of these issues in Sect. 3.2.3, but we will not emphasise this point: (for it was worked out in detail for a general duality, and illustrated for gauge-gravity duality, in De Haro [16, Sect. 1], under the heading of 'physical equivalence').

2 The Schema Introduced: Theories and Models

In this Section and the next, we develop the treatments of theory, model, interpretation, symmetry and duality, given in De Haro [15, Sect. 1], [16, Sect. 1] (and foreshadowed in De Haro, Teh, and Butterfield [19]). This Section deals with theory, model and interpretation; Section 3 will deal with symmetry and duality itself.

We begin with the scientific importance of dualities, and the comparison of duality with gauge (Sect. 2.1). Then we further specify our notions of theory and model (Sect. 2.2). Then we discuss: interpretations (Sect. 2.3), representations and isomorphisms (Sect. 2.4).

2.1 Duality's Scientific Importance

Recall from Sect. 1, our overall proposal. A bare theory can be realized (we will say: modelled) in various ways, like the different representations of an abstract algebra. These models are in general *not* isomorphic, since they differ from one another in their specific structure. But when they are isomorphic, we have a duality.

To develop this proposal, we begin with four clarifying remarks. Each remark leads in to the next. The first three defend our taking duality as a notion that is both logically weak and independent of a theory's interpretation. The first is, in effect, just the point that 'duality' is a term of art; so one can choose how to use it: and our choice of a logically weak definition makes for a *strong* physical notion! But the second and third are substantive—about the scientific importance of dualities. The fourth remark is a contrast with the notion of *gauge*.

(1): *A logically weak but physically strong definition*:— We agree that at first sight, it looks profligate to say that there is duality whenever two models are isomorphic. For it means there are countless dualities. For example: if a group or an algebra, endowed with a set of rules for evaluating quantities, can be a bare theory, any two isomorphic representations will yield a duality, as long as the isomorphism preserves the values of the quantities. Accordingly, the notion of duality is sometimes narrowed by adding physical conditions, not just on 'bare theory', but also on the isomorphism between models, e.g. by requiring the isomorphism to link the weak and strong coupling regimes of the two models [see Eq. (1)]. But we will maintain in (2) and (3) below that it is best to leave 'duality' broadly defined, as we have done: with such extra conditions being articulated in individual cases as the need arises. As we will see in Sect. 3.2.2, the strengthening will be given by the kind of physical degrees of freedom that one wishes to describe. And so, our notion of duality will be physically strong. In particular, it cannot be argued that two given models which share some structure are dual, unless the common structure is exactly equal to what the models regard as physical. In short: this apparently profligate verdict can be accepted.

(2): *Duality as surprising*:— So far we have spoken mainly of logico-semantic issues, and ignored epistemological ones: we have said what a duality is, but not how surprising and fruitful it can be. Our case-study, in Sect. 4 et seq., will of course bring out these issues. It is surprising indeed to learn that a theory we thought of as having as its quantum particles fermions also contains bosons—and even more surprising to learn that conversely the theory can be presented in the first place as having bosons, and then shown to contain the fermions with which we first began. For the moment, we note three clarifying comments—which are suggested by phrases like 'a theory we thought of', and 'the theory can be presented'. Each comment leads in to the next.

(i): We usually discover a duality in the context, not of a bare theory, but of an interpreted theory; for of course we work with interpreted theories.[10]

(ii): Indeed, we usually work with what we have called 'a model of the theory', indeed an interpreted model. That is: usually, before the duality is discovered, we have two interpreted models (usually called 'physical theories'!) which we do not believe to be isomorphic in any relevant sense.

[10]Agreed, pure mathematicians sometimes work with uninterpreted theories; and duality is a grand theme in mathematics, just as it is in physics. But although comparing duality in mathematics and in physics would be a very worthwhile project, we set it aside. Cf. [10].

(iii): Usually, we do not initially believe the two models are models of any single relevant theory (even of a bare one: i.e. even if we let ourselves completely suspend our antecedent interpretation of the models). The surprise is to discover that they are such models—indeed are isomorphic ones.

The word 'relevant' in (ii) and (iii) signals the fact that of course 'isomorphism', 'model' and 'theory' are very flexible words. For example: almost any two items can be considered isomorphic, i.e. as having a common structure, under a weak enough construal of 'structure'. Thus physicists might well in some specific context notice that the two models in question are both groups, or both algebras. But they rightly do not announce this as discovering a duality: not even if they also notice that the two groups (or algebras) are isomorphic. They set it aside as irrelevant, since the abstract notion of group or of algebra is so general that having it identified as a bare theory in common between the models is scientifically useless.

On the contrary, what is surprising, and scientifically valuable, is to find very specific, *not* general, structures in common between different models: especially when

(a) the models as presented (so: as interpreted) are very disparate, and-or

(b) the common structure is not only detailed (like '10-dimensional semisimple Lie group', as against 'group') but amounts to an isomorphism of that detailed structure (like 'isomorphic as 10-dimensional semisimple Lie groups').

As noted above, what will give physical theories their specificity, thus making duality a more powerful tool than its logically weak definition might make it seem, is the fact that physical theories, even bare ones, come with sets of maps from groups and algebras to appropriate fields (in the mathematical, not physical, sense!), i.e. maps that assign values to the physical quantities. These maps are defined at the level of the abstract structure, but must also be instantiated in each of the models (according to the relevant sense of instantiation, as either 'representation' or 'realization': cf. Sect. 2.2.2). And this set of maps is usually so rich, that it often suffices to reconstruct a model. And so, the fact that duality preserves these maps can be very non-trivial, and surprising, especially when combined with (a)–(b) above.

This discussion of (a)–(b) returns us to (1) above. We doubt that there can be a general characterization of when the models as presented are disparate enough, and-or the discovered isomorphism is detailed enough, for scientific importance. Instead, one can only articulate in any specific case how the disparity and-or the details are enough: e.g. because the isomorphism links the weak and strong coupling regimes of the two models. So it is not worth trying to tighten the *definition* of 'duality' with conditions beyond the logically weak ones we advocate. One just needs to use one's judgment about which cases count as scientifically important enough to analyse.

(3): *Examples*:— The conclusion of (2) is supported by some famous examples of duality in physics. Apart from boson-fermion duality, which we already introduced in Sect. 1.2, it is worth illustrating this with two other examples.

(A): Gauge-gravity duality. In this case, the models differ in the dimensions they assign to spacetime (d in the gravity model, $d-1$ in the gauge model), in their field content and classical equations of motion (Einstein's equations coupled to matter in the gravity model, the Yang-Mills equations in the gauge model), and in much more. In this case, the common core consists only in a class of asymptotic operators and a conformal class of $(d-1)$-dimensional metrics. Of course, it is very surprising to learn that a gauge theory model in $d-1$ dimensions, and a model of quantum gravity in d dimensions, despite their very disparate guises, nevertheless have the same common core, and represent the same theory. See De Haro [15, 16] for a discussion in the context of our schema.

(B): Electric-magnetic, or S-duality. This relates two models by mapping the electric charges of one model to the magnetic charges of the other. Furthermore, it does so by mapping a small electric charge to a large magnetic charge, analogously to Eq. (1) (since the charges play the role of couplings, in gauge models). Nevertheless, the common structure is the same in the two models, i.e. the quantum theory is invariant under the replacement of one gauge group by its dual.

(4): *A contrast with 'gauge'*:— This discussion of dualities' scientific importance brings out a contrast between our treatment of duality, and the notion of gauge. Physicists sometimes make remarks like: 'two dual theories are like different gauge formulations of a single theory'. We agree that this remark is *analogous* to our view: indeed, in two ways.

(i): A gauge formulation of a theory has specific structure (viz. the gauge variables) going beyond that mandated by the ideas (gauge-invariant ideas!) of the theory; just like for us, a model has specific structure going beyond that mandated by the bare theory.

(ii): The idea of gauge as 'descriptive redundancy' means that two gauge formulations of a single theory must 'say the same thing'; just like we say that in a duality, two models are isomorphic, and so (if interpreted: could) 'say the same thing'.

But we submit that this is *only* an analogy. There are two differences. First, we want to allow for cases where the two duals are not physically equivalent (as in Kramers-Wannier duality, mentioned above): *pace* the suggestion in (ii). Second (and more importantly), the extra structure in a model is usually *not* gauge, i.e. descriptively redundant: think of how the extra structure in a representation of a group usually carries physical information (e.g. a representation of the Poincare group carrying mass and spin information). Again, as stressed in (2) above: the surprising and scientifically important discovery is that in two models, with apparently very disparate structures, there is in fact an exact correspondence of structures. We shall return to these two differences in Sect. 3.2.4, comment (3).

2.2 Theories and Models

In this subsection, we add details about the ideas of a *bare theory* and its *models*, already introduced in Sect. 1.

2.2.1 Bare Theories

Following De Haro [15, 16], we take a *bare theory* to be a triple $T := \langle \mathcal{S}, \mathcal{Q}, \mathcal{D} \rangle$ comprising a structured set of states, a structured set of quantities, and a dynamics: together of course with the rules for evaluating the physical quantities on the states. (We will later discuss symmetries, which we will take as automorphisms $a : \mathcal{S} \to \mathcal{S}$ of the set of states; or as the dual maps on the set $\mathcal{Q}$ of quantities.)

Two immediate points of clarification:—

(1): We stress that, despite the physical connotations of the words 'states', 'quantities', and 'dynamics', a bare theory is *not* interpreted physically. Interpretation comes later (cf. Sect. 2.3). Thus it will help to think of a bare theory as given by an entirely abstract algebra of quantities, together with a similarly abstract state-space and dynamics. For example, the quantities might be (the self-adjoint elements of) an abstract C*-algebra A, the state-space might be determined by A, viz. as the positive linear functionals on A, and the dynamics might be an arbitrary automorphism of A.

(2): Indeed, it will help to think of a bare theory yet more generally. The reason is that most of what we need to say throughout Sect. 2, about theories and accordingly about their models, is independent of taking a bare theory as a triple—even an uninterpreted one *a la* (1). It depends only on a bare theory having two features:

(a) being uninterpreted, yet ready to be interpreted as a physical theory (hence the idea of the abstract set of states, quantities etc. 'standing ready' for interpretation);

(b) being augmentable, i.e. able to be supplemented with extra (again: uninterpreted) structure, in various ways, yielding different realizations, which we will call 'models': (to which we will turn directly, in Sect. 2.2.2).

Clearly, a theory does not need to be a triple $\langle \mathcal{S}, \mathcal{Q}, \mathcal{D} \rangle$ in order to have features (a) and (b). It could, for example, be a theory in logicians' traditional sense of a deductively closed set of formulas in a formal language: such a theory is uninterpreted, and can be augmented in many ways, for example just by adding an extra set of formulas and then closing under deduction. (Here we again connect with the Hilbertian and logical empiricist tradition of formulating physical theories axiomatically, mentioned at the start of Sect. 1. But we must postpone discussion till Sect. 2.4.)

2.2.2 Models

So we define a *model* M of a bare theory T to be a realization of T. The word 'realization' can be taken in two senses:—

(i): In the broad sense of a mathematical instantiation: i.e. a mathematical entity having the structure of the theory, and usually of course some specific structure of its own (cf. (2)(b) in Sect. 2.2.1). Thus if T is a theory in the logicians' sense of a deductively closed set of formulas, a realization is an entity that in some sense 'satisfies' all the formulas of T, and usually of course some formulas of its own. So any deductively closed superset of T would count as a realization of T; but so of course would any model of T in logicians' usual sense of 'model'.

(ii): In the mathematical sense of 'representation', as in representation theory. This requires T to have some structure, so that a representation is a homomorphism (of that structure) from T to some given, structured object. Since homomorphism need not be isomorphism, the homomorphism's range—the structured object that represents—may have only a 'coarse-grained' version of T's structure. But like in (i): since the representing object is 'given', it will also have specific structure of its own. (Recall the two kinds, (A) and (B), of 'extra detail' in Sect. 1.1.) Again, the obvious examples are when T is an abstract group or algebra, endowed with a set of rules for evaluating quantities, and a model is a group/algebra representation: for example a subgroup of the general linear group on a complex vector space, $\mathrm{GL}(n,\mathbb{C})$, endowed with a set of maps to the complex numbers, invariant under similarity transformations, e.g. the group characters.

In our specific example of duality, bosonization (Sect. 4 et seq.), we will use the second more specific notion, i.e. representation, (ii). But again: much of what we say in this Section needs only (i): the first, more general, sense of realization. And we believe that this notion applies more generally, to the dualities in quantum field theory and string theory: gauge-gravity duality [16], mirror symmetry, T-duality, and S-duality; cf. Sects. 2.4 and 5.4.

2.2.3 Notations for Models; Model Roots and Model Triples

It is helpful to have a schematic notation for models that exhibits how they augment the structure of a theory with specific structure of their own. This will also introduce some jargon which will be important for us.

One's first thought is to write the model as the ordered pair of the theory and its specific structure, $\bar{M}$ say: $M = \langle T, \bar{M} \rangle$. But we need to reflect the fact that (in almost all cases) the way that a model incorporates the theory's structure is not by simply containing a 'copy' of the theory 'beside' its specific structure $\bar{M}$: but instead, by using $\bar{M}$ to build a realization—in one or other of Sect. 2.2.2's two senses—of the theory's structure. Again, the obvious example of group representations illustrates. We should not think of a matrix representation of, say, the symmetric group S_N as containing a copy of S_N 'beside' its specific structure of a vector space V and $N!$ linear maps on V; (or maybe less than $N!$ maps—recall that a representation need only be a homomorphism). Rather, V and the (upto!) $N!$ linear maps realize, give a 'concrete copy' of, S_N.

Similarly for examples of dualities in physics, including our example of bosonization. In a typical physics example, the specific structure $\bar{M}$ consists of a set of fields, endowed with a set of symmetries, a dynamics for the fields, and a set of states of the fields. (So here, fields play the role of quantities in our conception of theories as triples, though of course not all fields are observable: in quantum theories, self-adjoint.) These fields etc. are used to build a 'concrete copy' of the bare theory's structure (or maybe a 'coarse-grained' copy). In our own example: a concrete copy of the enveloping affine Lie algebra or Kac-Moody algebra (see Eq. (41) in the Appendix), which is the algebra of which both the bosons and the fermions are representations.

So a better notation that reflects how $\bar{M}$ is used to build a realization of T is to write: $M = \langle T_M, \bar{M} \rangle$. The occurrence of T in the notation encodes that the model M is indeed a model of T. But the subscript M on T reflects that the specific structure $\bar{M}$ is used to realize T. In short, T_M is not 'given before' M itself: rather, T_M realizes T—in one or other of Sect. 2.2.2's two senses—by making use of $\bar{M}$.

So one should not think of the model as an ordered pair made from two already-given items T_M and $\bar{M}$. Rather, the decomposition $M = \langle T_M, \bar{M} \rangle$ is conceptual.[11]

It will be convenient to have a word for T_M, the 'part' of M that realizes T. We call it the **model root**. It will also be convenient to have a notation for the model root that does not mention T, just for simplicity in discussions where it is obvious that one theory T is in question. We use m. Thus a theory T can have various models and model roots, M_i and m_i, where i is in some index set I. This notation m, i.e. without mention of T, will be useful in Sect. 2.4.

This discussion carries over intact to the more detailed conception of a theory as a triple, $T = \langle \mathcal{S}, \mathcal{Q}, \mathcal{D} \rangle$. We write a model, not as a pair, but as a quadruple:

$$M = \langle \mathcal{S}_M, \mathcal{Q}_M, \mathcal{D}_M, \bar{M} \rangle =: \langle m, \bar{M} \rangle \,, \tag{2}$$

where $m := T_M := \langle \mathcal{S}_M, \mathcal{Q}_M, \mathcal{D}_M \rangle$ will be dubbed the **model triple**, as well as *model root*. As before, $\bar{M}$ is the specific structure that distinguishes one model, and so model root/triple, from another; and it is $\bar{M}$ that is used to build the model triple.

This jargon of model root, and model triple, will be important in relation to our proposed schema for duality, that a duality is an isomorphism of models. For this isomorphism is of course isomorphism as regards the bare theory's structure, rather than any other structure: in particular, this isomorphism sets aside the specific structure (even though the model's realizing the bare theory is built from its specific structure). Accordingly, we will often talk of duality as an isomorphism of model roots; and also, when a theory is conceived as a triple, as an isomorphism of model triples.

Finally, it will also be convenient (especially in Sect. 3.1) to have notation for a model considered in itself, not by comparison with the bare theory of which it is a model. A model is of course *itself* also a triple of a set of states, quantities and a dynamics: i.e. its own states etc., not that of the bare theory. And we will again use the overbar to indicate what is specific to the model. So we write: $M = \langle \bar{\mathcal{S}}, \bar{\mathcal{Q}}, \bar{\mathcal{D}} \rangle$.

This prompts the question: what is the relation between the unbarred items $\mathcal{S}$, $\mathcal{Q}$, $\mathcal{D}$ that make up a bare theory T, and the barred items $\bar{\mathcal{S}}$, $\bar{\mathcal{Q}}$, $\bar{\mathcal{D}}$? We will not attempt a general answer to this: we will not need one, and indeed we doubt that there is one. But bearing in mind the prototypical cases of representations of groups or algebras

[11] Besides, in physics examples the actual mathematical structure of a model is often very rich, e.g. in gauge-gravity dualities the metric of a $(d+1)$-dimensional spacetime will belong to the specific structure $\bar{M}$, but this metric will be a *fibration* over the metric of a d-dimensional spacetime, which belongs to T_M: see [15, Sect. 2.1], [16, Sect. 2.2].

(e.g. representations of the Poincare or conformal algebras which add a specific field content), the tempting, broad answer is that the barred items "are bigger"/"have more structure". (This is another aspect of the analogy between duality and gauge, despite the contrasts we stressed in remark (4) at the end of Sect. 2.1: both for two dual models and for a gauge formulation of a theory, there is the intuition of "being bigger"/"having more structure".)

In the example of boson-fermion duality, as we will see in Sects. 4.1 and 4.2, the theory consists of a specific enveloping algebra of the affine algebra (see Eq. (41) in the Appendix), together with a representation space for this algebra and appropriate functionals to the real numbers, which represent the quantities that (in the interpreted theory) become the physical quantities. But the models contain a complex variable z, which represents two-dimensional spacetime. Using this variable, one can construct, from the operators of the algebra, the bosonic (Klein-Gordon) or fermionic (Dirac fermion) field of the bosonic and the fermionic model, respectively. We shall argue in Sect. 5.4:(c) that, at the level of the bare theory, the specific dependence of the fields on the spacetime variable z has no particular significance, and is not needed: so that this suggests the formulation of a more general theory, with less structure.

2.3 *Interpretations of Theories and Models*

So far, so abstract! Both theories and models have so far been bare, i.e. uninterpreted. We now sketch how we envisage their physical interpretations. We will adapt the elementary ideas of the Frege-Carnap-Lewis framework for semantics [6, 27, 41]. This will be easy work: two interpretation maps, I_{Int} and I_{Ext}, will map from our theories and models, to 'meanings' and to 'the world' respectively. These maps will later be useful for discussing symmetries (in Sect. 3.1.2).

We recall that according to the Frege-Carnap-Lewis framework:—

(i) A word gets assigned: first, an intension (Carnap's word: Frege's is 'Sinn', or in English, 'sense': roughly, 'linguistic meaning'); and second, an extension (Carnap's word: Frege's is 'Bedeutung', or in English, 'reference/referent': roughly, 'the object or worldly item mentioned'). Roughly speaking: our interpretation maps, I_{Int} and I_{Ext}, will assign intensions and extensions, respectively.

(ii) A word's intension is assigned to it, once and for all (making the simplifying idealization that all words are univocal, and their linguistic meanings do not change). But a word's extension is assigned to it, relative to a possible world and to other features of the context of use that together determine the reference. For example, the reference of 'the tallest Swede alive today' depends not just on the possible world, but also on the day of use. And in general, the set of features that together determine the reference is large and open-ended (cf. [43]).

(iii) A singular term (such as a proper name 'Aristotle' or a definite description 'the capital of Denmark') has as its extension, its bearer (in these examples: the man, the city Copenhagen); while a one-place predicate (such as '...walks') has as its extension its set of instances (the set of walkers at the possible world and time in question); while similarly, a two-place predicate (such as '... loves ...') has as its extension *its* set of instances, i.e. the set of ordered pairs where the first loves the second (at the possible world and time in question), e.g. ⟨Romeo, Juliet⟩; and so on for predicates with three or more places.

(iv) Compositional rules describe how to assign intensions and extensions to grammatical phrases and thus to complete sentences, in terms of the intensions and extensions of the component words. For example, 'John walks' is assigned, at a world W and time t, the extension True (as against False) iff the reference of 'John' for (W, t) is in the set of walkers for (W, t).

To adapt these ideas, (i) to (iv), to our theories and models, there are two points to bear in mind. They seem to be stumbling blocks, or sources of confusion. But they are easily surmounted and dispelled.

(a): Notice how the ideas above express *contingency* and *transience*: by postulating a background set of possible worlds and times, they secure that a sentence's truth-value can be contingent (vary across the worlds) and transient (vary across the times). In philosophical discussion of physics, contingency and transience are often expressed in a corresponding way: the theory has many solutions, and typically a solution changes with time. This is often expressed using the word *model*: a state at a time, or a temporal sequence of states (a trajectory through the state-space) is called 'a model of the theory'. Thus recall that this was connotation (i) at the start of Sect. 1.1. The stumbling block is of course that since Sect. 1, we have reserved 'model' for a very different use: for what many would call 'specific theory', i.e. for a notion that encompasses many solutions throughout time. But we take it that one can surmount this stumbling block, and avoid confusion, just by recognising our stipulated usage.

(b): The ideas, (i) to (iv), were of course developed to give semantics for language about ordinary objects, such as people and towns, like Aristotle, Romeo and Copenhagen. But when one considers one of Sect. 2.2's theories or models, one is hard pressed to find mention of objects: at least, of objects in the plural. For undoubtedly, 'most of the talk' in the theory or model is about the various states and quantities, about which so many details are given. But these are surely not objects, but rather properties. Namely, properties of the one object—the physical system itself—being theorized about. Agreed: if the system is composite, one naturally regards its component systems as objects in their own right.[12] But the main point remains: most of the talk in a theory or model is about (numerically quantifiable) properties, and their

[12] Also agreed: it is common, and mathematically natural, to consider the set $\mathcal{S}$ of all states, with quantities as extra structure on $\mathcal{S}$: e.g. in classical mechanics, as real-valued functions on the phase space $\mathcal{S}$, and in quantum mechanics, as linear operators on the Hilbert space $\mathcal{S}$. But this does *not* make it compulsory to treat states as the basic objects in a semantics of a physical theory. For it is equally legitimate, though less common in textbooks, to start with the set $\mathcal{Q}$ of quantities, and take states as extra structure on $\mathcal{Q}$. And the legitimacy of both these approaches shows that *au fond*, a state is an assignment of numerical values to all quantities; and *mutatis mutandis* a quantity is an assignment of numerical values to all states. For now, the point is just that if one is asked to

intricate (quantitative) relationships, not about objects. But this second stumbling block is, like the first, minor. For nothing prohibits interpretations using just one object, viz. the system; or just the system and its component subsystems. (Besides, though we will not go into details: nothing prohibits interpretations treating as objects what are in fact properties; cf. e.g. [42, p. 429].)

Bearing in mind (a) and (b), we can now spell out how interpretation maps, I_{Int} and I_{Ext}, assign intensions and extensions, respectively—once we are given a bare theory T, or a bare model M. We will discuss the assignment to an element of the set of quantities: i.e. for T, an element of $\mathcal{Q}$ in the triple $T = \langle \mathcal{S}, \mathcal{Q}, \mathcal{D} \rangle$; and for M, an element of $\bar{\mathcal{Q}}$ in the triple $M = \langle \bar{\mathcal{S}}, \bar{\mathcal{Q}}, \bar{\mathcal{D}} \rangle$. (Recall from the end of Sect. 2.2.3 that $\bar{\mathcal{Q}}$ is all the quantities in the model, and is intuitively 'larger' than $\mathcal{Q}_M$, which is the realization of T's $\mathcal{Q}$ in the quadruple $M = \langle \mathcal{S}_M, \mathcal{Q}_M, \mathcal{D}_M, \bar{M} \rangle$.) But to stress that this element is uninterpreted/abstract, we will call it a, not Q. It will be obvious how the corresponding assignments get made for an element interpreted as a state, or as a dynamics.

For I_{Int}, the idea is: I_{Int} assigns to an element a of T or M, a physical quantity understood in general terms. For example, a could be an element of an abstract C*-algebra, abelian for classical mechanics and non-abelian for quantum mechanics. And $I_{\text{Int}}(a)$ could be the quantity, position: which in philosophical terms, is a property with numerically measurable degrees. Or perhaps $I_{\text{Int}}(a)$ is, more specifically, position in the x direction, using such-and-such point as spatial origin, and using axes and length-unit thus and so.

Two comments are in order, here. First: We thus envisage a 'Platonic realm' of numerically measurable properties as the codomain of the function I_{Int}. But our 'realism' about quantities is milder than it might appear; and anyway, nothing in what follows will depend on it. In particular, (i): nothing will depend on our taking intensions to encode conventional choices such as spatial origin, axis-direction and length-unit. Besides, (ii): we do not need 'trans-theoretic identity' for quantities. That is, a quantity like position need not be 'the very same quantity' in different theories: especially if they are radically different, e.g. classical mechanics and quantum mechanics. Despite the single word 'position', $I_{\text{Int}}(a)$ can be different quantities, according as a is in an abelian, or non-abelian, C*-algebra.

Second: Note that we do not need to be precise about the exact domain of definition of I_{Int}. For we are only sketching how we envisage interpretation proceeding; and there is of course a great deal of convention about how to formalize both T and M, and thus about what is the exact domain of definition of I_{Int}.

For I_{Ext}, the idea is similar, except that we need to allow for the fact that extensions are assigned relative to a possible world and to all the other features of the context of use that together determine reference. Interpreting any theory or model T or M means taking it to be used to describe some empirical phenomena: i.e. taking there

---

classify states and quantities in either of the philosophical categories of 'object' and ' property', undoubtedly one should classify both states and quantities as properties.

to be, in an appropriate possible world, a context of use with a rich enough set of features to determine reference for the elements of T or M.[13]

For example, let a be again an element of an abstract C*-algebra: say, an abelian one, because T is a bare theory that is to be interpreted as classical mechanics. Suppose that the possible world W contains two classical point particles; and we take T to be used in a context sufficiently rich that a successfully refers to the position of the more massive particle; or perhaps more specifically, its position in the x direction, using such-and-such place in W as spatial origin, and using axes and length-unit thus and so. Then relative to W and this assumed context of use, $I_{\text{Ext}}(a)$ is defined to be: the heavier particle's position.

2.4 Isomorphisms: Defining Theories by Abstraction from Models

So far, we have taken a theory T as given, and then considered its models M. In this Subsection, we note that one can argue in the opposite direction: i.e. one can approach defining a theory starting from a class of models. The idea is a widespread one: to define a notion (here, a theory) as those features in common between a suitably varied class of examples (here, models).[14] This is what earlier we called the 'common core' of the models (in Sects. 1.1-(c), and 1.2.2, 2.1-(3)). We will begin within our previous perspective, i.e. with a theory as given, to introduce notation; then we will sketch how to define a theory, 'arguing in the opposite direction'.

So recall from Sects. 2.2.2 and 2.2.3 that a bare theory T can have various models and model roots, M_i and m_i, where i is in some index set I. These 'realize' the theory, in the senses ('instantiate' or 'represent') discussed in Sect. 2.2.2. But they are in general not isomorphic to each other, nor to the theory: we in general do not have $m_i \cong m_j \cong T$. And even if there *is* an isomorphism: it is not an *identity*, because each m_i is a realization of T built using the model's specific structure $\bar{M}_i$: it is not a 'pure copy' of T.

13We say 'appropriate' so as to signal that of course, for any T or M, not every possible world has a context rich enough to determine reference for all T's or M's elements. Indeed: for many worlds, all their contexts will determine reference for none of T's or M's elements. For example, take T or M to be supersymmetric theories, and a world with no supersymmetric physics. So whatever our precise definition of the domain (i.e. set of arguments) of the map I_{Ext}, the map will surely be partial, i.e. undefined on some, maybe the majority, of its arguments. But that is no problem. Formal semantics and philosophy of language in the Frege-Carnap-Lewis framework have long had various proposals for how to treat words and phrases that lack extensions (called 'bearerless terms'); and these proposals can be adapted to our T or M.

14This line of thought is not only widespread, but has a long tradition: for thousands of years in philosophical accounts of abstraction; and for a hundred and fifty years in mathematics, with e.g. Frege's proposal to define notions as equivalence classes (e.g. a direction as an equivalence class of parallel lines).

Supposing there is an isomorphism between a model root m_i and the theory T, we denote it by $f_i : m_i \to T$, $i \in I$. Of course, m_i is in general not a single set, but an n-tuple or family of sets: often endowing some base set in the family with structure, e.g. the structure of an algebra. In particular, with a theory taken as a triple, $T = \langle \mathcal{S}, \mathcal{Q}, \mathcal{D} \rangle$, each m_i is a triple $m_i = \langle \mathcal{S}_{M_i}, \mathcal{Q}_{M_i}, \mathcal{D}_{M_i} \rangle$. So the domain and codomain of f_i are triples, so that f_i is actually a triple of maps that map respectively: states to states, quantities to quantities, and dynamics to dynamics.[15] We will, however, not need to indicate this in our notation.

Likewise, let the maps $f_{ij} : m_i \to m_j$ $(i, j \in I)$ denote isomorphisms—when they exist—between two model roots. But again, our notation will not need to indicate the precise domains and codomains of these maps.

Now in the opposite direction. Suppose we are given, not a theory T, but a set of models indexed by I: we write this set as $\{M_i : i \in I\}$. Similarly, we write the set of model roots indexed by I as $\{m_i : i \in I\}$. Thus we are not assuming that these models, or these model roots, are given as realizing a theory. But we do assume that the model roots (and the models) have some kind of structure, so that it makes sense to say that some pair of model roots is isomorphic with respect to that kind of structure. The idea now is to define a theory as 'what is in common' among a suitable set of model roots; where 'what is in common' will be expressed as an equivalence class of an appropriate equivalence relation.

The most obvious implementation of this idea is to define equivalence as just the given notion of isomorphism that we assumed applies to the model roots. Thus we might define two model roots m_i, m_j $(i, j \in I)$ to be equivalent, $m_i \sim m_j$, just in case there is an isomorphism $f_{ij} : m_i \to m_j$ between them. Then the proposal would be: a theory is an equivalence class under this equivalence relation. Recall for example the way in which Frege [26, Sects. 64–67], defined a *direction* as an equivalence class of straight lines under the relation of being parallel.

But if we apply this proposal to model roots, it has the trivial consequence that all model roots of a theory thus defined *are* isomorphic. And as we have discussed above, we must allow a theory to have non-isomorphic model roots.[16]

So the obvious implementation of the initial idea stumbles. If we want to define a theory as 'what is in common' among model roots, and express 'what is in common' as an equivalence class, then we need a more judicious—no doubt, a weaker—choice

[15] The first two maps will be isomorphisms, the last an equivariance condition: we will say more about this in Sect. 3.2.1.

[16] Similarly, if we apply this proposal to models: all models of a theory thus defined *are* isomorphic. Of course: we expect that since a model M has specific structure $\bar{M}$ going beyond its model root m, isomorphism for models will in general be stronger—i.e. lead to smaller equivalence classes—than does isomorphism of model roots. But in this Section we will not need to linger on this model versus model root contrast. For our main concern is defining a theory using isomorphism of model roots. As we will argue below (at the end of this Subsection), there is a natural constraint that model roots must be *sufficiently varied*. For mistaking the presence of accidental similarities between the model roots one happens to have at hand for necessary similarities between *all* the model roots of the theory one is trying to define, leads to unnecessarily, or perhaps undesirably, restrictive theories. This is *a fortiori* true of the models M_i of the theory: since the specific structure $\bar{M}_i$ is *specific* to M_i, and so in general not shared with the another model.

of equivalence relation than isomorphism. We will not go into details about how to make this choice. We doubt that there are general rules. But we note that it will be guided by two desiderata:

(i): our previous understanding (maybe partial understanding) of the theory we are trying to define: the choice is meant to pick out just the structure of the theory we intuitively intend;

(ii): our expectation that the model roots will end up being representations, in the mathematical sense, of the theory thus defined (and so, in general, non-isomorphic to each other).

We will close by mentioning another constraint. It is partly independent of the question of choosing an equivalence relation. For it is about the membership of the class of models or model roots one begins with, rather than the judicious choice one must make of an equivalence relation over it.

Namely: if this whole approach to defining theories is to work—i.e. is to define theories of the kind we intended in our original conception—then the model roots must be sufficiently varied that there are no 'accidental commonalities' between them, which would then be inadvertently encoded in the theory defined as an equivalence class—thereby limiting the theory's possibilities of representation. The point is familiar, e.g. from Frege's example. If one imagines the lines can be coloured, then Frege's approach to defining direction needs the lines to vary in colour sufficiently. For if all the lines in a given parallelism equivalence class were the same colour, Frege's definition of the corresponding direction, viz. as that class, would inadvertently be ambiguous between (i) the direction, as we originally intended it, and (ii) the unintended common colour. Hence our mentioning, in the opening paragraph of this Subsection, 'a suitably varied class of examples'.

We will not go into details about this constraint. But we stress that obviously, it is substantive. For recall how on the original 'theory-first' approach in Sect. 2.2, we stressed that a model root $m = T_M$ is not a 'pure copy' of the theory T, but is built from the model M's specific structure $\bar{M}$. So on the present 'reverse', or 'models-first', approach: the danger is that if a class of model roots is not sufficiently varied, they may have considerable specific structure in common (like colour for Frege's lines)—which will therefore be inadvertently encoded in theories defined as equivalence classes of model roots. We will come back to this point in Sect. 5.4.[17]

[17]Agreed: one does not always get to 'choose' one's model roots (or models), and so this constraint cannot always be implemented. Thus there is judgment involved in this process of abstraction, viz. of (i) how many, and how varied, the model roots should be, to provide representations of one's theory, and (ii) how to make the distinction, for a given model, between model root and specific structure (since part of the specific structure of a model could be mistaken for e.g. additional information about the theory). We therefore maintain that this reverse approach, from model roots to theory, is not deductive but inductive—which brings us back to our Hilbertian theme from Sect. 1. It only stops when one is happy with the theory—based on whatever independent criteria one uses to judge one's theory and models.

3 Duality and Symmetry

In this Section, we first develop our treatment of symmetry (Sect. 3.1). This, together with our discussion in Sect. 2, sets us up to present (at last!) our schema for duality as isomorphism of models (Sect. 3.2).

3.1 *Symmetries of Theories and Their Models*

We mentioned symmetry in Sect. 1.1's closing analogy between a symmetry mapping a state to a state that is 'the same', and a duality mapping a theory to a theory that is 'the same'. But we now can say more about symmetries, using: (A) our distinction between theories and models (from Sect. 2.2); and (B) our interpretation maps (from Sect. 2.3). We take up these topics in Sects. 3.1.1 and 3.1.2, respectively.

About (A), our main point will be that the symmetries of a bare (or indeed, an interpreted) theory, and the symmetries of a model of it, are in general overlapping, but distinct, sets. In particular, the symmetries of a theory can be a proper subset of the symmetries of its model: and beware—this inclusion is in the opposite direction from that for the other, more common, use of 'model', viz. as a solution, or representative of a solution, of a theory.

About (B), our main point will be that a symmetry must 'commute' in an appropriate sense with interpretation. Both these points, and our other ones, will be uncontroversial.

3.1.1 Symmetries of Uninterpreted Theories and Models

Recall the usual conception of symmetry as a map a on states that preserves the values of a salient, usually large, set of quantities: the state s and the image-state $a(s)$ have the same values for quantities. This prompts three immediate comments.

(i): Agreed, it is also usual to think of a symmetry as a map on quantities that preserves values on a salient, usually large, set of states: i.e. for a given state, the value of the argument-quantity equals the value of the image-quantity. But there is no conflict here: the two conceptions are related by duality—in the mathematical, not physical, sense! That is: one map is the (mathematical) dual of the other.

In more detail: given any map $a : \mathcal{S} \to \mathcal{S}$, we can define its *dual map* (not to be confused with a 'duality map!') on quantities, $a^* : \mathcal{Q} \to \mathcal{Q}$, by requiring that for any $s \in \mathcal{S}$ and $Q \in \mathcal{Q}$: $\langle a^*(Q), s\rangle := \langle Q, a(s)\rangle$. And similarly, starting with quantities: given any map $a : \mathcal{Q} \to \mathcal{Q}$, we say that its dual map on states, $a^* : \mathcal{S} \to \mathcal{S}$, is defined by requiring for all arguments: $\langle Q, a^*(s)\rangle := \langle a(Q), s\rangle$.

(ii): Recall the question at the end of Sect. 1.1: do a state s, and its image $a(s)$ under a symmetry, represent the same physical state of affairs ('possible world')?

Our answer there was, roughly: 'In general, No: but the 'Yes' cases are a natural focus of interest'. We will return to this when discussing interpretation, in Sect. 3.1.2.[18]

(iii): We have discussed symmetries as preserving values. But it is common to also require that a symmetry 'preserves the dynamics'. Taking a symmetry as a map on states, this means, roughly: if a sequence of states is possible under the dynamics, so is the sequence of image-states. That is: if a possible time-evolution is represented by a temporal sequence of states, with the state at time t being $s(t)$ (a 'Schrödinger picture' of time-evolution), then the sequence $a(s(t))$ of states is also possible.[19] A corresponding definition can be given for when we take a symmetry as a map on quantities, and use a 'Heisenberg picture' of time-evolution as given by a sequence of quantities, i.e. by the sequence of their values on a single 'fixed' state.

These comments (i)–(iii) bring out two points. The second is longer: it addresses symmetries of *models*, pointing out that these are in general overlapping but distinct (as abstract groups) from symmetries of theories; and that in the special case where a model triple is isomorphic to the bare theory, the symmetries of the theory can be a proper subset of the symmetries of its model.

First: it is clear that discussing symmetries returns us to Sect. 2.2's more detailed conception of a theory, even a bare one, as a triple comprising a set of states, a set of quantities and a dynamics: $T := \langle \mathcal{S}, \mathcal{Q}, \mathcal{D} \rangle$, together with a set of rules for evaluating quantities.

Second: it is clear that comments (i)–(iii) carry over exactly to *models* in our sense, viz. a realization ('a more detailed version') of a theory T, with specific structure of its own. Recall that Sect. 2.2.3 introduced two notations for models in this sense.

---

[18]For the moment, we just note that it is also common to think that a symmetry as a map on states is 'active', i.e. the image-state must be a different physical state of affairs (so the question's answer is 'No'), while a symmetry as a map on quantities is 'passive', i.e. the image-quantity and the argument-quantity (each with their common value) describe the single given physical state (so that now the question's answer is 'Yes').

We will *deny* this. There is no universal association of symmetry as a map on states as 'active', and symmetry as a map on quantities as 'passive'. The reason lies, essentially, in the distinction between a mathematical state and a physical state: (in the jargon of 'gauge', the latter is a gauge-equivalence class of the former). That is: we of course concede that a symmetry as a map on states is 'active', in the sense that it changes the states. That is a tautology: (except for the degenerate case where the symmetry is given as being the identity map!). But this concession does not imply that a symmetry as a map on states must change the physical state of affairs represented: for the states in question could yet be 'merely' mathematical. That is: one still needs a further argument why a difference of these states must imply a difference of physical state (and thus why the question's answer is 'No'). This distinction, between a mathematical and a physical symmetry, was labelled, in De Haro et al. [20, Sect. 2], with the label (Redundant); and in De Haro [16, Sect. 1.1.2.b], as (Physical)-(Redundant). It also roughly corresponds to the distinction, in Caulton [8], between an 'analytic' and a 'synthetic' symmetry.

[19]If the dynamics is deterministic, we can write $s(t) = D_{t,t_0}(s(t_0))$ where D_{t,t_0} represents the deterministic dynamics; and then 'preserving the dynamics' is equivalent to the commutation i.e. equivariance condition, $a(s(t)) \equiv a(D_{t,t_0}(s(t_0))) = D_{t,t_0}(a(s(t_0)))$.

$$\begin{array}{ccc} \mathcal{S} & \xrightarrow{a} & \mathcal{S} \\ \downarrow{\scriptstyle \theta} & & \downarrow{\scriptstyle \theta} \\ \mathcal{S}_M & \xrightarrow{a_M} & \mathcal{S}_M \end{array}$$

Fig. 1 Commutativity diagram of the symmetry a with the representation map θ

$$\begin{array}{ccc} G_1 & \xrightarrow{a_1} & G_1 \\ \downarrow{\scriptstyle \theta} & & \downarrow{\scriptstyle \theta} \\ G_2 & \xrightarrow{h} & G_2 \end{array}$$

Fig. 2 Commutativity of group automorphism a_1 with group homomorphism θ

Both notations will be useful in what follows: the first notation immediately, the second notation in the next Subsection. (The second notation was simpler: since a model is itself a triple of its own sets of states, quantities and a dynamics, we wrote: $M = \langle \bar{\mathcal{S}}, \bar{\mathcal{Q}}, \bar{\mathcal{D}} \rangle$. It means we can define and discuss 'symmetry of a model' just as we did symmetries of theories, e.g. as an automorphism of the state-space $\bar{\mathcal{S}}$ preserving values of a large salient subset of $\bar{\mathcal{Q}}$.)

Section 2.2.3's first notation distinguishes the realization of the theory's triple from the specific structure $\bar{M}$, and gives a subscript M to the former to signal that it is built out of the latter: $M = \langle \mathcal{S}_M, \mathcal{Q}_M, \mathcal{D}_M, \bar{M} \rangle$ [Eq. (2)]. We also wrote this as $M =: \langle m, \bar{M} \rangle$, where $m := T_M := \langle \mathcal{S}_M, \mathcal{Q}_M, \mathcal{D}_M \rangle$ is the *model triple*. This notation brings out that for any theory T and any of its models M, there is a natural condition for a symmetry a of T to be itself realized in M: for it to have, so to speak, a 'shadow' in the model M, i.e. in the model triple. This condition is that a diagram should commute, and is not automatic.

To state this condition, however, we need a bit more notation about realization. We will write it as a map θ. And we will suppose that in T we treat symmetries as maps on states, so that $a : \mathcal{S} \to \mathcal{S}$ preserves the value of all quantities in a salient subset, say $\mathcal{Q}^0$, of the set of all quantities $\mathcal{Q}$. Then in the usual case where 'realization' means 'representation', we can take θ as an appropriate structure-preserving map: from $\mathcal{S}$ in the theory T itself, to $\mathcal{S}_M$ in the representing model triple $m = T_M = \langle \mathcal{S}_M, \mathcal{Q}_M, \mathcal{D}_M \rangle$. Then the condition in question—that the symmetry a is itself realized in M—is that there should be a map $a_M : \mathcal{S}_M \to \mathcal{S}_M$, such that the diagram in Fig. 1 commutes.

To convey this idea less abstractly, think of the simplest case. Let the bare theory be just a group G_1,[20] with an automorphism $a_1 : G_1 \to G_1$; and suppose a group G_2 represents G_1 thanks to the existence of a homomorphism $\theta : G_1 \to G_2$. So $G_2 \cong G_1/\ker\theta$. For there to be a homomorphism of G_2, $h : G_2 \to G_2$ (even homomorphism: let alone automorphism), that realizes a_1 (counts as a_1's 'shadow' in G_2) requires commutation: i.e. for all $g_1 \in G_1$, $\theta(a_1(g_1)) = h(\theta(g_1))$. Or as a diagram, see: Fig. 2.

[20]Together with a set of maps to the real numbers, to express evaluation of the quantities. But for simplicity we ignore these maps for the moment.

In the special case where the model triple is isomorphic to the bare theory, this discussion of course simplifies. Then the map θ will be an isomorphism, and the map a_M (or h in the toy example of groups) will trivially exist, and both the above diagrams will trivially commute. In this case, a symmetry of the theory has a 'duplicate', or 'replica', in the symmetries of the model—so that in effect, the symmetries of the theory are a *subset* of the symmetries of the model. We say 'in effect' just because of the different domains of definition: $\mathcal{S}$ versus $\mathcal{S}_M$. Apart from this 'in effect', there are two comments to make about this special, simplified, case.

(1): First, we note that on the other use of 'model' as an individual solution of a theory, a model is in general *less* symmetric than its theory—as is often remarked, with the buzz-word 'symmetry-breaking'. A solution of a dynamics with a spherically symmetric Hamiltonian need not be spherically symmetric; a cubical crystal lattice with one particular placing of its lattice points, and one particular orientation of its edges, can be a solution of a dynamics that is translation-invariant and isotropic; and so on. So: the subset-inclusion in our case above is in the opposite direction from that holding for the other use of 'model'.

(2): Besides, 'subset' here will usually mean *proper subset*. That is: a model's specific structure $\bar{M}$—its 'content' that goes beyond its being a model/realization of T—will mean the model has symmetries additional to those that are 'duplicates' of the symmetries of T. And we expect that if these additional symmetries are well-defined on the model triple, or if they naturally induce a symmetry there, that symmetry is trivial, i.e. just the identity map on the model triple. Our prototypical cases of representations of a group or algebra give examples. Perhaps the simplest is as follows. Let T be the real numbers $\mathbb{R}$; and let M be the complex numbers $\mathbb{C}$ which of course represents $\mathbb{R}$ as the real axis, i.e. the complex numbers with zero imaginary part, $\{z \in \mathbb{C} \mid z = x + i0,\ x \in \mathbb{R}\}$. So this latter set, the real axis, is like the model triple. Then M has the symmetry of complex conjugation $z \mapsto \bar{z}$ which is indeed well-defined on the real axis: but there, it is just the identity map.

And there are examples in interesting cases of dualities. In gauge-gravity dualities, De Haro [18] showed that a certain subgroup of the diffeomorphism group of the gravity model of the theory (roughly, the diffeomorphisms which preserve the asymptotic boundary conditions) was 'invisible' to the gauge model of the theory, in the sense of not representing any difference on that model: and so these diffeomorphisms are not in the common core between the two models, and they are trivially represented on the theory. The same verdict was made in De Haro [16, Sect. 2.2.3] for the 'gauge symmetries' of the gauge side of the duality. These are not visible on the gravity side: they are symmetries of the formulation of the gauge model of the theory, and are trivially represented on the theory.

To sum up this discussion of the symmetries of a bare theory, and those of its models, and of its model-triples: there are really three points here:

(i): A bare theory T is realized—typically: represented in the mathematical sense— by one of its model triples, m. The model M then consists of m and some specific structure $\bar{M}$; (cf. Sect. 2.2.3). And representation requires only a homomorphism, not an isomorphism. Hence our articulating in this Section the condition—in terms of a commuting diagram—for a symmetry of T to be itself realized in m.

(ii): And even if in some given case, the representation is an isomorphism, i.e. the representing model triple is isomorphic to the theory T, so that any symmetry a of T will indeed have a 'duplicate' or 'replica' symmetry in the model triple: still, we must expect that the *model* (as against the model triple) has its own specific structure $\bar{M}$. And this specific structure may have symmetries that m, and the theory T, 'knows nothing of': (cf. comment (2) just above).

(iii): Furthermore, different models, and therefore model triples, of a bare theory are in general *not isomorphic* (as we also discussed in Sect. 2.4). However, our example of boson-fermion duality, in Sect. 4, will not illustrate this in detail: i.e. all the model triples *will* be isomorphic. (But see the comments in Sect. 5.4.)

3.1.2 Interpretations Respect Symmetries

Recall from Sect. 2.3 that once we are given a bare theory T, or a bare model M, the interpretation maps I_{Int} and I_{Ext} assign intensions and extensions (respectively) to, for example, an element a of the set of quantities: i.e. for T, an element of $\mathcal{Q}$ in the triple $T = \langle \mathcal{S}, \mathcal{Q}, \mathcal{D} \rangle$; and for M, an element of $\bar{\mathcal{Q}}$ in the triple $M = \langle \bar{\mathcal{S}}, \bar{\mathcal{Q}}, \bar{\mathcal{D}} \rangle$. (And similarly for assigning intensions and extensions to states and dynamics.) Thus I_{Int} assigns to an element a of T or M, a physical quantity, e.g. position (or more specifically, position in the x direction, using such-and-such spatial origin), understood in general terms. Similarly, I_{Ext} assigns a an extension relative to a possible world W and the other features of the context of use that together determine reference. For example, W and the context of use may determine that a is assigned the position of the more massive of two classical point particles that are in W.

We now propose that these interpretation maps should satisfy appropriate meshing conditions whereby they form commuting diagrams with symmetry maps. The reason is simply that this reflects what one usually means by 'interpretation' of the formalism of a physical theory. And this is so whether 'formalism of a physical theory' is (in our senses) a bare theory, or a bare model; and whether 'interpretation' refers to intension or extension.

Thus to take a very simple example: suppose that a state describes three classical point particles forming a scalene triangle, stationary in an absolute Newtonian space, and (for more simplicity) that the particles do not interact; and suppose the bare theory, or model, at issue has spatial translation and rotation as symmetries. Taking symmetries as maps on states (as usual): these suppositions mean that a spatially translated and-or rotated state has the same values as the given state, for a large and salient set of quantities. So far, these suppositions are, officially, at the level of a bare theory or model—though of course words like 'point particles', 'scalene triangle' and 'Newtonian space' suggest interpretation. And indeed: reading these suppositions, one tends to read them as interpreted, i.e. to unconsciously apply the interpretation maps I—where I is short for both I_{Int} and I_{Ext}. In any case: applying these maps to the two bare states, say s and $T(s)$ ('T' for 'transform' or 'translate and-or rotate'), one concludes that for the act of interpretation to respect the given symmetries, the interpretations $I(s)$ and $I(T(s))$ (where again: I is short for both $I_{\rm Int}$ and $I_{\rm Ext}$) must have the same values for a large and salient set of quantities—namely, of course, for the interpretations of the bare quantities. This is exactly the condition that symmetry and interpretation form a commuting diagram. But to write such diagrams down, we will need a bit more notation.[21]

The reasons we need more notation are as follows. So far:

(1): We did not spell out that the domain of each of $I_{\rm Int}$ and $I_{\rm Ext}$ would include states and quantities and dynamics; so that each of $I_{\rm Int}$ and $I_{\rm Ext}$ is really a triple of three maps from states to states, from quantities to quantities, and from dynamics to dynamics. (This is just like an isomorphism f_i in Sect. 2.4 being really a triple of such maps; cf. footnote 15.)

(2): We have not introduced notation for the codomains of the interpretation maps: for what one might call the 'realm of intension', or 'meanings', for $I_{\rm Int}$, and for what one might call the 'realm of extension', or the 'world', for $I_{\rm Ext}$.

(3): Nor did we introduce a notation for symmetry maps defined on the 'realm of intension' or on the 'realm of extension'. Indeed, we did not yet do this: neither (a) on the states therein (generally understood, in the realm of intension, and specific to a particular physical system, in the realm of extension), analogous to the symmetry maps $a : \mathcal{S} \to \mathcal{S}$ on the states of a bare theory; nor (b) on the quantities therein (generally understood, in the realm of intension, and specific to a particular physical system, in the realm of extension), analogous to the symmetry maps $a^* : \mathcal{Q} \to \mathcal{Q}$ on the quantities of a bare theory (which are mathematical duals of the maps $a : \mathcal{S} \to \mathcal{S}$).

To avoid a lot of extra notation, we shall only act on (2) and (3) above. That is:

(2'): We now denote the codomains of the interpretation maps $I_{\rm Int}$ and $I_{\rm Ext}$ by, respectively: 'Sinn' (in honour of Frege's German word for the realm of intension) and 'Bed' (short for 'Bedeutung', which was Frege's word for referent, such as the bearer of a name).

(3'): We confine ourselves to treating symmetries as maps on states (treating them as maps on quantities would be parallel). So we now denote a symmetry on the states in 'Sinn' as $a^{\rm Int}$, where the superscript $_{\rm Int}$ corresponds to the subscript in $I_{\rm Int}$. And we now denote a symmetry on the states in 'Bed' as $a^{\rm Ext}$, where the superscript $_{\rm Ext}$ corresponds to the subscript in $I_{\rm Ext}$.

Putting (2') and (3') together, we write a state-space in the realm of intension as $\mathcal{S}_{\rm Sinn}$, and a state-space in the realm of extension as $\mathcal{S}_{\rm Bed}$. So we write: $a^{\rm Int} : \mathcal{S}_{\rm Sinn} \to \mathcal{S}_{\rm Sinn}$; and we write $a^{\rm Ext} : \mathcal{S}_{\rm Bed} \to \mathcal{S}_{\rm Bed}$.

But we shall not act on (1) above: the notation would be cumbersome, and without compensating advantages. In short: acting only on (2) and (3) above—i.e. introducing 'Sinn' and 'Bed', with symmetry maps $a^{\rm Int}$ and $a^{\rm Ext}$ respectively—is enough to enable us to draw the required commuting diagrams.

---

[21] Agreed, to impose this commutation condition for every symmetry and every interpretation is contentious. It seems best justified when we envisage that the bare theory or model describes the whole universe; so that for the example of three point particles, there are no other material bodies in the universe. But in this paper, we do not need to assess exactly when the commutation condition is justified. See De Haro [16, Sect. 1.3–1.4], especially the condition called 'unextendability'.

$$\begin{array}{ccc} \mathcal{S} & \xrightarrow{a} & \mathcal{S} \\ \big\downarrow I_{\text{Int}} & & \big\downarrow I_{\text{Int}} \\ \mathcal{S}_{\text{Sinn}} & \xrightarrow{a^{\text{Int}}} & \mathcal{S}_{\text{Sinn}} \end{array}$$

Fig. 3 Commutativity of the symmetry a with the interpretation map I_{Int} for T

$$\begin{array}{ccc} \bar{\mathcal{S}} & \xrightarrow{a} & \bar{\mathcal{S}} \\ \big\downarrow I_{\text{Int}} & & \big\downarrow I_{\text{Int}} \\ \bar{\mathcal{S}}_{\text{Sinn}} & \xrightarrow{a^{\text{Int}}} & \bar{\mathcal{S}}_{\text{Sinn}} \end{array}$$

Fig. 4 Commutativity of the symmetry a with the interpretation map I_{Int} for M

$$\begin{array}{ccc} \mathcal{S} & \xrightarrow{a} & \mathcal{S} \\ \big\downarrow I_{\text{Ext}} & & \big\downarrow I_{\text{Ext}} \\ \mathcal{S}_{\text{Bed}} & \xrightarrow{a^{\text{Ext}}} & \mathcal{S}_{\text{Bed}} \end{array}$$

Fig. 5 Commutativity of the symmetry a with the interpretation map I_{Ext} for T

We spell these out: first for the realm of intension, then for the realm of extension. In each case, we first draw the diagram for a bare theory T taken as a triple, $T = \langle \mathcal{S}, \mathcal{Q}, \mathcal{D} \rangle$; and then draw the diagram for a bare model M taken as a triple $M = \langle \bar{\mathcal{S}}, \bar{\mathcal{Q}}, \bar{\mathcal{D}} \rangle$. (Note that these diagrams do not reflect Sect. 3.1.1's discussion about the overlap but distinctness (in general) of symmetries of a theory T, and symmetries of its model M.)

Thus for the realm of intension, we have: for a bare theory T with state-space $\mathcal{S}$, the diagram in Fig. 3. For a bare model M with state-space $\bar{\mathcal{S}}$, we have the diagram in Fig. 4.

Strictly speaking, we should in diagrams in Figs. 4 and 3 distinguish two interpretation maps—both of which we have in fact written as I_{Int}—according as the domain of definition is the state-space of a bare theory or of a bare model. But like in comment (1) above: the precision is not worth the burden of extra notation. And similarly of course, for the next two diagrams about the realm of extension.

For the realm of extension, we have, at some given possible world W and context of use sufficiently rich to determine references (i.e. to avoid the interpretation maps being undefined on the given arguments): for a bare theory T with state-space $\mathcal{S}$, the diagram in Fig. 5.

For a bare model M with state-space $\bar{\mathcal{S}}$, we have the diagram in Fig. 6.

$$\begin{array}{ccc} \bar{\mathcal{S}} & \xrightarrow{a} & \bar{\mathcal{S}} \\ \Big\downarrow I_{\rm Ext} & & \Big\downarrow I_{\rm Ext} \\ \bar{\mathcal{S}}_{\rm Bed} & \xrightarrow{a^{\rm Ext}} & \bar{\mathcal{S}}_{\rm Bed} \end{array}$$

Fig. 6 Commutativity of the symmetry a with the interpretation map I_{Ext} for M

3.2 *Duality as Isomorphism of Models*

We turn in this last Subsection to our proposal that a duality is an isomorphism of models of a bare theory. To be precise: it is *an isomorphism of model triples* of a bare theory. Indeed, after all the stage-setting of the previous Subsections (!), the proposal is straightforward. We first give its details, using the notations we have established (Sect. 3.2.1). In Sect. 3.2.2, we argue that our notion of duality is logically weak but physically strong. Then in Sects. 3.2.3 and 3.2.4, we turn to how duality relates to the topics of Sects. 2.3 and 3.1: interpretations and symmetries.

3.2.1 Duality as Isomorphism

Our basic idea is that a duality is an isomorphism between two triples, each comprising a state-space endowed with appropriate structure, a set of quantities endowed with appropriate structure, and a dynamics, consistent with that structure.[22] 'Isomorphism between triples' is of course short for a triple of maps: an isomorphism between the two state-spaces, and isomorphism between the sets (almost always: algebras, cf. footnote 22) of quantities, and an equivariance condition on the dynamics.[23] In addition, the isomorphism must commute with the symmetries of the theory, as sketched in Sect. 3.1.

[22]As we mentioned in Sect. 2.2.1, 'appropriate structure' here refers to: (i) the structure of the sets of spaces and quantities, (ii) the rules for evaluating quantities, (iii) the structure which the dynamics satisfies, (iv) the set of symmetries of the theory. We can now be more specific about these, for the examples of quantum theories, which will illustrate our schema: (ia) the set of states will be a separable Hilbert space; (ib) the quantities will be elements (normally the self-adjoint, renormalisable elements) of an algebra; (ii) the rules for evaluating quantities are maps to the appropriate field: for most quantum theories, the inner product on the Hilbert space, and the usual rules for evaluating matrix elements; (iii) dynamical evolution will usually be a (unitary) map, satisfying appropriate commuting diagrams with the other maps in the theory; (iv) the group of symmetries will comprise the automorphisms of the algebra: and possibly additional symmetries, on the states and on the quantities. For classical theories, these comments get modified in familiar ways: e.g. (ia) would say that the set of states is a manifold, with structure appropriate to e.g. Lagrangian or Hamiltonian mechanics.

[23]Our proposal does not depend on the formulation of models as triples. A model root can be presented in many different forms, and the isomorphism should then preserve the corresponding structure. Even for triples, one can envisage isomorphisms which do not respect the triple structure, though they map the model roots isomorphically. Compare Sect. 3.2.2. But it will suffice for our purposes to restrict to model roots defined as triples, whose structure is preserved by the duality.

More important is the question of *which* kinds of triples are related by duality. Recalling our distinction between bare theories and their more specific models, the answer is clear: *a duality relates two model triples of a single bare theory*.

The crucial point here is that the model triple is separated from the model's own specific structure, and expresses only the model's realizing (typically: representing in the mathematical sense) the bare theory. Recall the notation from Eq. (2) in Sect. 2.2.3: $M = \langle \mathcal{S}_M, \mathcal{Q}_M, \mathcal{D}_M, \bar{M}\rangle =: \langle m, \bar{M}\rangle$, where $m := T_M := \langle \mathcal{S}_M, \mathcal{Q}_M, \mathcal{D}_M\rangle$ is the model triple. We emphasised already in Sects. 2.2.2 and 2.2.3 that two model triples are in general *not* isomorphic to each other, nor to the bare theory. So the assertion of duality is substantive: it asserts that two model triples are in fact isomorphic.

But this is *not* to say that the two *models*, each 'considered in their entirety', are isomorphic. They each have their own specific structure, and are (in almost all cases) *not* isomorphic. Recall our other notation from Sect. 2.2.3 for models 'considered in their entirety': $M = \langle \bar{\mathcal{S}}, \bar{\mathcal{Q}}, \bar{\mathcal{D}}\rangle$. Indeed, their being non-isomorphic is usually part of what makes the duality surprising and (if Nature is kind to us!) empirically fruitful, i.e. of scientific importance. And 'the more non-isomorphic'—i.e. the more disparate the two models, considered in their entirety, are—the more surprising, and (one hopes) empirically fruitful, is the duality (cf. (2) and (3) in Sect. 2.1).[24]

We now introduce some notation for dualities as isomorphisms between model triples. This will require first giving:

(1) some new notation for the value of a quantity on a state, and

(2) a more detailed discussion of dynamics (in both the 'Schrödinger' and 'Heisenberg' pictures).

Both (1) and (2) can be given wholly independently of our distinctions (i) between theories and their models, and (ii) between interpreted and uninterpreted theories. So for the moment, please consider a generic triple of a state-space, a set of quantities, and a dynamics: $\langle \mathcal{S}, \mathcal{Q}, \mathcal{D}\rangle$.[25]

(1): Suppose we are given a set of states $\mathcal{S}$, a set of quantities $\mathcal{Q}$ and a dynamics $\mathcal{D}$: $\langle \mathcal{S}, \mathcal{Q}, \mathcal{D}\rangle$. We will write $\langle Q, s\rangle$ for the value of quantity Q in state s. This prompts two further general points.

---

[24]We should put this last point more precisely, since our notion of bare theory is logically weak, with even a group or an algebra, together with a set of maps to the real numbers, counting as a legitimate bare theory: (cf. (1) in Sect. 2.1). And it is in general not surprising, nor likely to be empirically fruitful, to learn that two very disparate models are both groups, or both algebras: (unless the maps to the real numbers are so disparate that the existence of an isomorphism is not easy to guess). Thus the point here, more precisely, is that, for a given degree of detail or logical strength in the bare theory (and the more, the better!): the more disparate its models (considered in their entirety), the more surprising, and one hopes fruitful, is their both realizing the bare theory. That is: the more surprising is the duality.

[25]This simpler idea of a triple was used in our earlier—cruder!—discussion of duality: cf. [20, Sect. 3.2]. There, the simplicity engendered no errors, since our general description of duality was but a preamble to a specialist topic: an assessment of gauge symmetries in gauge-gravity duality.

First: it is common to think of a state $s \in \mathcal{S}$ as a maximal specification of the instantaneous properties of the system in question; and a quantity $Q \in \mathcal{Q}$ as a numerically measurable property of it. In effect, this makes states and quantities nothing but assignments of values to each other. Second: for classical physics, one naturally takes quantities as real-valued functions on states, so that $\langle Q, s\rangle := Q(s) \in \mathbb{R}$ is the system's possessed or intrinsic value of the quantity; and for quantum physics, one naturally takes quantities as linear operators on a Hilbert space of states, so that $\langle Q, s\rangle := \langle s|\hat{Q}|s\rangle \in \mathbb{R}$ is the system's Born-rule expectation value of the quantity. But for quantum physics it is often important to consider the non-diagonal matrix elements of a given quantity/operator $\hat{Q}$, without requiring this to be adequately encoded in the Born-rule expectation values of various other quantities. So for a quantum theory—as in the bosonization example of Sect. 4 et seq.!—we should understand a value written schematically as $\langle Q, s\rangle$ to also represent all the matrix elements $\langle s_1|\hat{Q}|s_2\rangle$. Thus $\langle Q, s\rangle$ is a short-hand for an expression like $\langle Q; s_1, s_2\rangle := \langle s_1|\hat{Q}|s_2\rangle$,[26] i.e. Q is regarded as a map: $\mathcal{S} \times \mathcal{S} \to \mathbb{C}$.

(2) We turn to the dynamics $\mathcal{D}$, i.e. a specification of how the values of quantities change over time. We will keep the discussion very simple. First, we assume the dynamics is deterministic: also in quantum theories, despite the threat of Schrödinger's cat. Then it can be presented in two ways, for which we adopt the quantum terminology, viz. the 'Schrödinger' and 'Heisenberg' pictures; (though the ideas occur equally in classical mechanics: for example the remark, frequent in the textbooks, that in Hamiltonian mechanics time-evolution can be regarded as a sequence of canonical transformations, is in effect a statement of the Heisenberg picture). But we shall not need to distinguish otherwise between the different detailed formalisms for dynamics, such as Hamiltonian versus Lagrangian, and the path-integral. Besides, we will adopt for simplicity the Schrödinger picture.

So we say: D_S is an action of the real line $\mathbb{R}$ representing time on $\mathcal{S}$. There is an equivalent Heisenberg picture of dynamics with D_H, an action of $\mathbb{R}$ representing time on $\mathcal{Q}$. The pictures are related by, in an obvious notation:

$$\begin{aligned} D_S : \mathbb{R} \times \mathcal{S} \ni (t, s) \mapsto D_S(t, s) =: s(t) \in \mathcal{S} \text{ iff} \\ D_H : \mathbb{R} \times \mathcal{Q} \ni (t, Q) \mapsto D_H(t, Q) =: Q(t) \in \mathcal{S} \end{aligned} \tag{3}$$

where for all $s \in \mathcal{S}$ considered as the initial state, and all quantities $Q \in \mathcal{Q}$, the values of physical quantities at the later time t agree in the two pictures:

$$\langle Q, s(t)\rangle = \langle Q(t), s\rangle \; . \tag{4}$$

With the notations and notions of remarks (1) and (2) in hand, we can now present the notation for dualities as isomorphisms between model triples. Let M_1, M_2 be two models, with model triples $m_1 = \langle \mathcal{S}_{M_1}, \mathcal{Q}_{M_1}, \mathcal{D}_{M_1}\rangle$ and $m_2 = \langle \mathcal{S}_{M_2}, \mathcal{Q}_{M_2}, \mathcal{D}_{M_2}\rangle$. We

[26]Therefore duality will imply unitary equivalence.

$$\begin{array}{ccccccc} \mathcal{S}_{M_1} & \xrightarrow{d_s} & \mathcal{S}_{M_2} & \quad & \mathcal{Q}_{M_1} & \xrightarrow{d_q} & \mathcal{Q}_{M_2} \\ \downarrow D_{S:1} & & \downarrow D_{S:2} & & \downarrow D_{H:1} & & \downarrow D_{H:2} \\ \mathcal{S}_{M_1} & \xrightarrow{d_s} & \mathcal{S}_{M_2} & & \mathcal{Q}_{M_1} & \xrightarrow{d_q} & \mathcal{Q}_{M_2} \end{array}$$

Fig. 7 Equivariance of duality and dynamics, for states and quantities

can suppose that M_1, M_2 are both models of a bare theory T. Or we can proceed in the 'opposite direction' discussed in Sect. 2.4: that is, we can suppose that M_1, M_2 are given independently of a bare theory T, but their model triples (model roots in the more general language of Sect. 2.4) are isomorphic. Either way, the notation for dualities is as follows.

To say that the model triples m_1, m_2 are isomorphic is to say, in short, that: there are isomorphisms between their respective state-spaces and sets of quantities, that (i) make values match, and (ii) are equivariant for the two triples' dynamics (in the Schrödinger and Heisenberg pictures, respectively). We now spell this out. Though retaining the Ms in the subscripts is cumbersome, we will do so, in order to emphasise our main conceptual point: that duality is a relation between model triples in our sense—it is *not* between theories, or between generic triples $\langle \mathcal{S}, \mathcal{Q}, \mathcal{D} \rangle$ as in remarks (1) and (2).

Thus we say:— A duality between $m_1 = \langle \mathcal{S}_{M_1}, \mathcal{Q}_{M_1}, \mathcal{D}_{M_1} \rangle$ and $m_2 = \langle \mathcal{S}_{M_2}, \mathcal{Q}_{M_2}, \mathcal{D}_{M_2} \rangle$ requires[27]:

an isomorphism between the state-spaces (almost always: Hilbert spaces, or for classical theories, manifolds):

$$d_s : \mathcal{S}_{M_1} \rightarrow \mathcal{S}_{M_2} \quad \text{using } d \text{ for 'duality'} \; ; \tag{5}$$

and an isomorphism between the sets (almost always: algebras) of quantities

$$d_q : \mathcal{Q}_{M_1} \rightarrow \mathcal{Q}_{M_2} \quad \text{using } d \text{ for 'duality'} \; ; \tag{6}$$

such that: (i) the values of quantities match:

$$\langle Q_1, s_1 \rangle_1 = \langle d_q(Q_1), d_s(s_1) \rangle_2 \, , \quad \forall Q_1 \in \mathcal{Q}_{M_1}, s_1 \in \mathcal{S}_{M_1} . \tag{7}$$

and: (ii) d_s is equivariant for the two triples' dynamics, $D_{S:1}$, $D_{S:2}$, in the Schrödinger picture; and d_q is equivariant for the two triples' dynamics, $D_{H:1}$, $D_{H:2}$, in the Heisenberg picture: see Fig. 7.

Equation (7) appears to favour m_1 over m_2; but in fact does not, thanks to the maps d being bijections.

It is already clear that a duality reduces to a symmetry, in the case where there is just one model, and one model triple, at issue, i.e. $M_1 = M_2$ and $m_1 = m_2$. We

[27] See footnote 23 and Sect. 3.2.2 for a brief discussion of more general cases.

shall return to this topic in Sect. 3.2.4. First, we turn to the questions (i) whether our notion of duality is too weak (Sect. 3.2.2) and (ii) how it relates to Sect. 2.3 topic of interpretation (Sect. 3.2.3).

3.2.2 A Logically Weak but Physically Strong Notion of Duality

In Sect. 2.1, we admitted that our definition of duality is logically weak because there is a duality whenever two models are isomorphic. Thus one might worry that, whenever two given models share some common structure smaller than the model triples, they are dual with respect to the substructures they share. In this Section, we argue that this worry is unfounded, for models that purport to describe physical systems—which is our concern in this paper. Thus the notion of duality is physically stronger than it would at first seem. The point will be to distinguish between a purely formal model versus a physical (although uninterpreted) model. Our schema is intended for the latter: and it is only the model triples, not their specific structure, that is physically significant.

We will illustrate this in an example. Consider, for simplicity, the following model, based on the su(2) algebra:

$$M_0 = \langle \mathcal{S}_0, \mathcal{Q}_0, \mathcal{D}_0 \rangle := \langle \mathcal{S}_J, U(\text{su}(2)), C_2(J) \rangle \; . \tag{8}$$

$\mathcal{S}_J$ is here the Hilbert space of irreducible representations of su(2) with total quantum number J. $U(\text{su}(2))$ is the universal enveloping algebra of su(2), i.e. roughly, all powers of the algebra elements, quotiented by the algebra relations. The dynamics is the Hamiltonian of the model, which we take to be $C_2(J)$, the second Casimir.

Let us compare M_0 with a model M with which it shares a common structure, and which is based on the su(2) $\otimes$ su(2) algebra. The state space is: $\mathcal{S} = \mathcal{S}_J \otimes \mathcal{S}_K$, where $\mathcal{S}_J$ and $\mathcal{S}_K$ are the state spaces of the first and the second su(2), respectively. J is the total quantum number of the first su(2), and K is the total quantum number of the second su(2). We take the dynamics to be given by the sum of the Casimirs of the two su(2)'s, $\mathcal{D} = C_2(J) + C_2(K)$. This model is written as:

$$M = \langle \mathcal{S}, \mathcal{Q}, \mathcal{D} \rangle = \Big\langle \mathcal{S}_J \otimes \mathcal{S}_K, U\left(\text{su}(2) \otimes \text{su}(2)\right), C_2(J) + C_2(K) \Big\rangle, \tag{9}$$

where U again indicates the universal enveloping algebra.

M_0 and M are of course both representations of M_0, i.e. M is a representation of su(2) with as 'extra structure' the second su(2). In fact, M_0 is isomorphic to M for the trivial representation with $K = 0$, i.e. $M_0 \cong M|_{K=0}$.

But M and M_0 are *not* dual for arbitrary values of K [as in Eq. (9)]. To see this, we rewrite M in a way which makes explicit the common structure they share, i.e. the first su(2). So, define:

$$M' := \langle m, \bar{M} \rangle \; , \tag{10}$$

where m contains the first su(2), and the specific structure $\bar{M}$ contains the second su(2). Explicitly, $m = \langle \mathcal{S}_J, U(\text{su}(2)), C_2(J)\rangle$ and $\bar{M} = \langle \mathcal{S}_K, U(\text{su}(2)), C_2(K)\rangle$. Thus, m and $\bar{M}$ are both triples, and they are both isomorphic to M_0, in particular $m \cong M_0$.

We can summarise the above definitions introducing the following short notation:

$$\begin{aligned} M_0 \cong\ m &= \mathbf{J} \\ M &= \mathbf{J} \otimes \mathbf{K} \\ M' &= \langle \mathbf{J}, \mathbf{K}\rangle\ , \end{aligned} \tag{11}$$

where **J** is short for the first factor of the tensor product and **K** for the second.

To reconstruct $M = \mathbf{J} \otimes \mathbf{K}$ from $M' = \langle \mathbf{J}, \mathbf{K}\rangle$, one takes the tensor products of the states in **J** and **K**, takes all the products of the quantities, and adds up the dynamics, to reproduce Eq. (9).

If we were to say that $M = \mathbf{J} \otimes \mathbf{K}$ and its model quadruple counterpart, $M' = \langle \mathbf{J}, \mathbf{K}\rangle$, are 'the same', then since it is true that $M' = \langle \mathbf{J}, \mathbf{K}\rangle$'s model triple and $M_0 \cong \mathbf{J}$ are isomorphic: it would follow that $M = \mathbf{J} \otimes \mathbf{K}$ and M_0 would be dual in the relevant sense.

But notice that M' is *not* the same as M, nor is it isomorphic to it, because they differ in what they regard as physically significant (cf. [15, p. 5]). Only the first model triple, **J**, is physically significant in M', whereas both model triples are physically significant in M. Thus there can be no isomorphism between M and M' as candidate models of physics, for they differ in their physical content.

In other words, the difference is in a tensor product model presented as such (i.e. $M = \mathbf{J} \otimes \mathbf{K}$) versus a model quadruple that has as model triple the first factor **J**, and as its specific structure the second factor **K**.

This argument reinforces the point that it is not necessary, nor desirable, to define *bare theories* as equivalence classes of models. This means that the condition to have a *duality* is only that two model triples be isomorphic: but the model triples need not be isomorphic to the bare theory, only homomorphic to it. And since, as we have just seen, the isomorphism between dual models is essentially unique (i.e. it is not possible to weaken the isomorphism to get dual structures, without changing their physical content), there is no gain in requiring that duality must also involve the theory. If one starts with a bare theory which is weaker than two isomorphic model triples that represent it, it may be possible to strengthen it so as to match the two model triples: but there is no gain in this. So, it is best to keep the notion of *bare theory* physically weak, and the notion of *duality* physically strong.

We end with a contrast of the notions of duality in physics and mathematics. In mathematics, just as in physics, 'duality' does not have a fixed meaning; however, all the examples of duality involve just *two* theories. More precisely, the duality operation generates the two-element group $\mathbb{Z}_2$. This is not so in the physics literature, where duality can involve more than two models, and the duality group can be rich. For example, the S-duality group of electric-magnetic duality is $SL(2, \mathbb{Z})$, and string theories realize so-called U-duality groups, which involve orthogonal and exceptional groups. Our schema allows for dualities among many models, and so it is closer to the notion in physics. This also strengthens the analogy between duality and symmetry, mentioned in Sect. 1.1-(2).

3.2.3 Duality and Interpretation

So far, our discussion of interpretation has concerned a *single* theory or model. Thus recall that Sect. 2.3 introduced interpretation maps I_{Int} and I_{Ext} in a rather informal way, as mapping from a bare i.e. uninterpreted theory or a bare model, to the realm of intension ('Sinn'), or to the realm of extension ('Bed'), respectively. Then Sect. 3.1.2 laid out how I_{Int} and I_{Ext} are to mesh with symmetry maps. This amounted to a commutation condition, i.e. I_{Int} and I_{Ext} forming a commuting diagram with symmetry maps, which for simplicity we only considered as defined on state-spaces: either on a bare state-space, or on a state-space in the realm of intension ('Sinn'), or on a state-space in the realm of extension ('Bed'). (Cf. the diagrams in Figs. 3, 4, 5 and 6.) But again, everything in Sect. 3.1.2 concerned a *single* theory or model.

Since duality is about relations between theories/models, there is, at first sight, little to say about duality and interpretation. That is: interpretation should simply proceed independently on the two sides of the duality—for example, we just require the interpretation-symmetry commuting diagram on both sides of the duality. Indeed: we said already at the start of Sect. 1.1 that in some cases of duality, the two sides were clearly not—nor intended to be—physically or semantically equivalent: e.g. the high and low temperature regimes in Kramers-Wannier duality. And our definition of duality as formal (viz. an isomorphism of model triples) certainly allows this idea of 'distinct but isomorphic sectors of reality'—namely as the codomains of the interpretation maps on the two sides of the duality.

This verdict—'there is little to say'—is true, so far as it goes. And of course, it does not forbid the other sort of case: where the two sides of the duality *are* physically/semantically equivalent, i.e. do describe 'the same sector of reality'. In our schema, this would be modelled by the interpretation maps on the two sides having the same images/values in their codomain—so as to give a *triangular*, rather than *square*, commuting diagram. We shall spell this out as regards the interpretation of (bare) quantities: similar diagrams could of course be drawn for states.

For (bare) quantities being mapped by I_{Int} into the realm of intension 'Sinn', the two sides of a duality describing 'the same sector of reality' amounts to the diagram in Fig. 8.

$$\begin{array}{ccc} \mathcal{Q}_{M_1} & \xrightarrow{d_q} & \mathcal{Q}_{M_2} \\ I_{\text{Int}} \searrow & & \swarrow I_{\text{Int}} \\ & \mathcal{Q}_{\text{Sinn}} & \end{array}$$

Fig. 8 The two sides of the duality describe 'the same sector of reality', in the realm of intension

$$\begin{array}{ccc} \mathcal{Q}_{M_1} & \xrightarrow{d_q} & \mathcal{Q}_{M_2} \\ & {\scriptstyle I_{\text{Ext}}} \searrow \quad \swarrow {\scriptstyle I_{\text{Ext}}} & \\ & \mathcal{Q}_{\text{Bed}} & \end{array}$$

Fig. 9 The two sides of the duality describe 'the same sector of reality', in the realm of extension

Similarly: for (bare) quantities being mapped by I_{Ext} into the realm of extension 'Bed'—relative to some given possible world W with a context rich enough to determine references, of course—the two sides of a duality describing 'the same sector of reality' amounts to Fig. 9.

So far, so straightforward. But the above verdict is a bit quick: there are two further points to make.

(1): *What determines equivalence?*:— First, there is the question what determines whether the two sides of a duality are physically/semantically equivalent, i.e. describe the same 'sector of reality'. De Haro [15] and Dieks et al. [23] have argued that the choice between these two options should depend on whether the models in question are given what was called an 'internal' or an 'external' interpretation. The idea is that for the ranges of the interpretation maps to be distinct, there must be other facts, external to the triples themselves and our use of them, that determine the distinct ranges. Typically, these facts will be other pieces of physics to which the system described by each model triple is coupled—with different pieces of physics on the two sides of the duality. This coupling 'breaks the symmetry' between the two sides, and secures that the two model triples are about distinct, albeit isomorphic, subject matters ('sectors of reality'). In the proposed jargon: the coupling provides an 'external interpretation' of the model triple. On the other hand: sometimes we propose a physical theory as a putative theory of the whole universe, i.e. as a putative cosmology, so that according to the theory there are no physical facts beyond those about the system (viz. universe) it describes. If in such a case, there is a duality—which in our framework, means there are two isomorphic model triples, each putatively describing the whole universe—then there can be no such coupling to other pieces of physics. (Gauge-gravity duality provides, of course, a putative example of such a duality between theories of the universe.) An interpretation of each triple must therefore be what was labelled an 'internal interpretation'; and this prompts the conclusion that the two triples describe the very same 'sector of reality'. That is: the interpretation maps have the same range; and there is a triangular diagram, as in Figs. 8 and 9.

(2): *Interpreting the specific structure*:— Second, there is more to say about the interpretation of a model's specific structure, especially in the latter sort of case, i.e. two sides of a duality describing the same 'sector of reality'.

Recall that a model M is more than the model triple m, by which it realizes a bare theory, and which relates to another model triple in a duality. M also has a specific structure $\bar{M}$: as we stressed at the start of Sect. 3.2.1, this structure is *not* related by the duality to 'the other side'. But the specific structure $\bar{M}$ *does* get interpreted—

it supplies arguments for the interpretation maps I_{Int} and I_{Ext}—just as much as the model triple m gets interpreted.[28] This was emphasised by the other notation for models introduced at the end of Sect. 2.2.3: viz. a model is itself, like a bare theory, a triple. M's states and quantities, being specific to M, are in general 'bigger'/'more structured' than the states and quantities of the bare theory that M models/realizes. We wrote them with a 'bar': thus $M = \langle \bar{\mathcal{S}}, \bar{\mathcal{Q}}, \bar{\mathcal{D}} \rangle$. And recall that then Sect. 2.3 took elements of these 'bigger' sets $\bar{\mathcal{S}}, \bar{\mathcal{Q}}$ as arguments for the interpretation maps I_{Int} and I_{Ext}.

So the first point to make is: our discussion of duality has so far ignored the specific structures $\bar{M}_1$ and $\bar{M}_2$ on the two sides, even though they do get interpreted. This silence is presumably no problem in a case where we agree that the two sides are not physically or semantically equivalent. In such a case, the interpretation of $\bar{M}_1$ and $\bar{M}_2$ just means there are physical facts on each of the two sides, additional to the facts that are isomorphic with (a subset of) facts on the other side.[29] Thus for the case of Kramers-Wannier duality, the obvious examples of such non-matching physical facts would be facts about the value of the temperature: high on one side, and low on the other. In short: these additional facts (in the realms of intension and extension, respectively) are: on one side, in the ranges $I_{\text{Int}}(\bar{M}_1)$ and $I_{\text{Ext}}(\bar{M}_1)$; and on the other side, in the ranges $I_{\text{Int}}(\bar{M}_2)$ and $I_{\text{Ext}}(\bar{M}_2)$.

But what about the other sort of case: where the two sides *do* describe 'the same sector of reality'? Is it really satisfactory to say that there are physical facts that:

(a) are additional to those facts described by the isomorphic model triples, i.e. those 'caught' by the duality/the common bare theory; yet also

(b) fall into two such disparate subsets: one subset expressed by $\bar{M}_1$ and the other subset expressed by $\bar{M}_2$?

In short: this world-picture, combining (i) a set of facts expressed by the two sides in the same way (though this sameness may be not obvious—the duality can be surprising), and (ii) two other sets of facts expressed in very different ways by the two sides, is surprising: and maybe it is odd, or unsatisfactory ...

We saw examples of this in comment (3) of Sect. 2.1, for example in gauge-gravity duality. Here, the set of facts that are the common core, *a la* (i), consists only in a class of asymptotic operators and a conformal class of $(d-1)$-dimensional metrics. And the sets of facts *a la* (ii) include, on the bulk side, gravity (such as: Einstein's equations coupled to matter) in d dimensions expressed by $\bar{M}_1$, and on the boundary side, a conformal field theory (such as: the Yang-Mills equations) in

[28] At least, this is what we would in general expect. Agreed, one might interpret a model without interpreting *all* of the specific structure $\bar{M}$: recall footnote 13 on the need to allow the interpretation maps to be partial, i.e. to deliver no value for certain arguments. For more details, see Sect. 1.1.2.a of De Haro [16].

[29] In Sect. 3.2.2, we emphasised the fact that only the model triples, and not the specific structure, are physically significant. When we now consider external interpretations that do give a physical meaning to the specific structure, we have to say that these interpretations change the physical content of the model (its physical degrees of freedom). This is correct, because external interpretations do not need to preserve the structure of the model as a quadruple.

$d-1$ dimensions expressed by $\bar{M}_2$. See DeHaro [15, Sect. 2.1], [16, Sect. 2.2] for a discussion in the context of our schema.

For our example of bosonization, we will see in Sect. 5.2.4 how external interpretations map to different (sets of) worlds, while the internal interpretation maps to the same (set of) world(s). For the latter interpretation to be possible, we will see that the worlds must contain both bosonic and fermionic facts. So, the internal interpretation does not efface the distinction between bosons and fermions, but distinguishes them and identifies them in the world.

3.2.4 Combining Duality and Symmetries

In this Subsection, we turn to the relations between dualities and symmetries. There are three comments to make. They are not controversial. Indeed, they simply gather some threads from discussions in previous Sections. The first makes the obvious comparison between dualities and symmetries, and notes the conditions for a duality to reduce to being a symmetry. The second is about a duality preserving a symmetry of its model-triples; and so returns us to the contrast between the symmetries of a bare theory, and those of its model-triples. The third returns us to the contrast between duality and gauge, discussed in comment (4) at end of Sect. 2.1.

(1): *Making the comparison precise*:—
Earlier (at the end of Sect. 1.1) we announced that we would endorse a basic analogy between duality and symmetry: 'a duality is like a symmetry, but at the level of theory', so that while a symmetry carries e.g. a state into a 'matching' state, a duality carries a theory into a 'matching' theory.

Indeed, we endorse this analogy—allowing of course for the shift of words from 'theory' to 'model-triple'. This endorsement is clear from:

(a): our discussion of symmetries of theories, taken as triples, and symmetries of their models, and their model-triples (Sect. 3.1.1); and

(b): our definition of duality as an isomorphism of model-triples that (i) makes the values of quantities match, and (ii) is equivariant for the two triples' dynamics (cf. Eq. 7 and Fig. 7 at the end of Sect. 3.2.1).

In particular (as mentioned at the end of Sect. 3.2.1): a duality reduces to a symmetry, in the case where there is just one model, and one model triple, at issue, i.e. $M_1 = M_2$ and $m_1 = m_2$. Spelling this out will use the notion of a dual map (in the pure mathematical sense!), introduced in (i) at the start of Sect. 3.1.1. Recall that this notion is defined by the pairing whereby states $s \in \mathcal{S}$ and quantities $Q \in \mathcal{Q}$ assign each other a value $\langle Q, s\rangle$. Namely: given any map $a : \mathcal{S} \to \mathcal{S}$, we said that its dual map on quantities, $a^* : \mathcal{Q} \to \mathcal{Q}$ is defined by requiring that for any $s \in \mathcal{S}$ and $Q \in \mathcal{Q}$: $\langle a^*(Q), s\rangle := \langle Q, a(s)\rangle$. And similarly, starting with quantities: given any map $a : \mathcal{Q} \to \mathcal{Q}$, we said that its dual map on states, $a^* : \mathcal{S} \to \mathcal{S}$ is defined by requiring for all arguments: $\langle Q, a^*(s)\rangle := \langle a(Q), s\rangle$.

Thus suppose there is just one model triple at issue. Then d_s is an automorphism of $\mathcal{S}_{M_1} \equiv \mathcal{S}_{M_2}$, i.e. of the state-space in the one model triple; and similarly, for d_q on $\mathcal{Q}_{M_1} \equiv \mathcal{Q}_{M_2}$. So duality's condition (i), that the values of quantities match [Eq. (7)], becomes the condition

$$\langle Q_1, s_1 \rangle_1 = \langle d_q(Q_1), d_s(s_1) \rangle_1 \ , \quad \forall Q_1 \in \mathcal{Q}_{M_1}, s_1 \in \mathcal{S}_{M_1} . \tag{12}$$

But d_q induces a dual map d_q^* on states, such that: $\langle d_q(Q_1), d_s(s_1) \rangle_1 = \langle Q_1, d_q^* (d_s(s_1)) \rangle_1$. So we conclude that $(d_q^* \circ d_s) : \mathcal{S}_{M_1} \to \mathcal{S}_{M_1}$ is a symmetry (written, as usual for us, as a map on states rather than quantities). For we have:

$$\langle Q_1, s_1 \rangle_1 = \langle d_q(Q_1), d_s(s_1) \rangle_1 = \langle Q_1, d_q^*(d_s(s_1)) \rangle_1 \ . \tag{13}$$

Finally, the same verdict—that for a single theory, duality reduces to symmetry—applies to dynamics, i.e. to dynamical symmetries. That is: if a duality concerns just one model triple, then Sect. 3.2.1's condition (ii) for duality—that the duality map is equivariant for the two triples' dynamics (i.e. d_s is equivariant for Schrödinger dynamics, and d_q is equivariant for Heisenberg dynamics)—reduces to the condition that the duality is also a dynamical symmetry: for example, that d_s is a dynamical symmetry represented as a map on states.

(2): *On duality preserving a symmetry:—*
It is straightforward to confirm that on Sect. 3.2.1's definition of duality, a duality preserves any symmetry of its model triples. There are two points here. First: there is a commuting square diagram of isomorphisms. Second: there is the issue of the values of a quantity being equal on a given state, and on its transform under a symmetry. The first point will lead in to the second.

First: The duality maps d_s, d_q are not only bijections, but isomorphisms: $d_s : \mathcal{S}_{M_1} \to \mathcal{S}_{M_2}$, and $d_q : \mathcal{Q}_{M_1} \to \mathcal{Q}_{M_2}$. And although we did not have to spell out the exact structures of $\mathcal{S}_{M_i}$, $\mathcal{Q}_{M_i}$ that these isomorphisms are to preserve (but cf. footnote 22), it is obvious from the fact that 'is isomorphic to' is both a symmetric and a transitive relation, that the following diagram, with a understood to be any automorphism of $\mathcal{S}_{M_1}$, commutes (cf. Fig. 10).

And of course, this diagram of isomorphisms is just what we mean by saying a duality d preserves an automorphism of the state-space $\mathcal{S}_{M_1}$ in its domain model triple, and preserves $\mathcal{S}_{M_1}$'s structure. Namely, d carries the automorphism—a map a on $\mathcal{S}_{M_1}$—to a corresponding automorphism of states in the codomain (indeed: range)

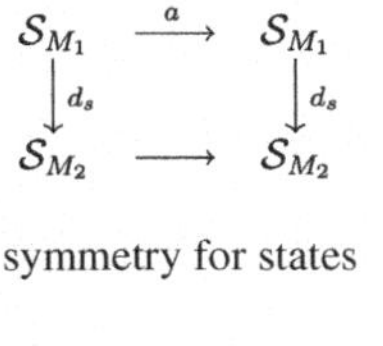

Fig. 10 Commutativity of duality and symmetry for states

$$\begin{array}{ccc} \mathcal{Q}_{M_1} & \xrightarrow{a} & \mathcal{Q}_{M_1} \\ \big\downarrow{\scriptstyle d_q} & & \big\downarrow{\scriptstyle d_q} \\ \mathcal{Q}_{M_2} & \longrightarrow & \mathcal{Q}_{M_2} \end{array}$$

Fig. 11 Commutativity of duality and symmetry for quantities

$$\begin{array}{ccc} \mathcal{S}_{M_1} & \xrightarrow{a} & \mathcal{S}_{M_1} \\ \big\downarrow{\scriptstyle D_{t,t_0}} & & \big\downarrow{\scriptstyle D_{t,t_0}} \\ \mathcal{S}_{M_1} & \xrightarrow{a} & \mathcal{S}_{M_1} \end{array}$$

Fig. 12 Commutativity of symmetry and dynamics

model triple. The diagram defines this corresponding automorphism, i.e. the map forming the fourth side of the square: $d_s \circ a \circ (d_s)^{-1} : \mathcal{S}_{M_2} \to \mathcal{S}_{M_2}$.

There is obviously a corresponding point about quantities, as against states. Since d_q is required to be an isomorphism of quantities, the following diagram, with a now understood to be any automorphism of $\mathcal{Q}_{M_1}$, must commute, cf. Fig. 11.

And again, this diagram is just what we mean by saying a duality d preserves an automorphism of the quantities in its domain model triple, and preserves $\mathcal{Q}_{M_1}$'s structure. Namely, d carries the automorphism—a map a on $\mathcal{Q}_{M_1}$—to a corresponding automorphism of quantities in the codomain (indeed: range) model triple. The diagram defines this corresponding automorphism: $d_q \circ a \circ (d_q)^{-1} : \mathcal{Q}_{M_2} \to \mathcal{Q}_{M_2}$.

Second: But in physics, the notion of symmetry involves more than the notions of automorphism of the state-space, and of the set (usually algebra) of quantities. It involves the pairing whereby states s and quantities Q assign each other a value: $\langle Q, s\rangle$. For these values (for a large and salient set of quantities, though usually not *all* quantities) must be preserved under the symmetry.

But satisfying this is automatic, for a duality as defined at the end of Sect. 3.2.1. That is: For a duality to respect this aspect of symmetry was already built in to our definition of duality: namely in condition (i), that the values are equal between states and quantities that correspond by the duality. Recall Eq. (7), which we here repeat:

$$\langle Q_1, s_1\rangle_1 = \langle d_q(Q_1), d_s(s_1)\rangle_2\,, \quad \forall Q_1 \in \mathcal{Q}_{M_1}, s_1 \in \mathcal{S}_{M_1}. \tag{14}$$

Finally (and just like at the end of (1) above): the same verdict—that a duality preserves any symmetry of its model triples—applies to dynamics, i.e. to dynamical symmetries. Recall from footnote 19 (in Sect. 3.1.1) that a dynamical symmetry is a commutation i.e. equivariance condition. So for the Schrödinger picture of dynamics, the diagram for the 'first' side of a duality, i.e. $m_1 = \langle \mathcal{S}_{M_1}, \mathcal{Q}_{M_1}, \mathcal{D}_{M_1}\rangle$, is, with a the dynamical symmetry, as in Fig. 12.

So we now compose this diagram with Fig. 10, which represents that a duality preserves a symmetry. But since in Fig. 12, the 'first' side, '1', of the duality occurs twice, on both top and bottom rows, we now need to compose Fig. 12 with Fig. 10

$$\begin{array}{ccc}
\mathcal{S}_{M_2} & \longrightarrow & \mathcal{S}_{M_2} \\
\Big\downarrow {\scriptstyle d_s^{-1}} & & \Big\downarrow {\scriptstyle d_s^{-1}} \\
\mathcal{S}_{M_1} & \xrightarrow{\ a\ } & \mathcal{S}_{M_1} \\
\Big\downarrow {\scriptstyle D_{t,t_0}} & & \Big\downarrow {\scriptstyle D_{t,t_0}} \\
\mathcal{S}_{M_1} & \xrightarrow{\ a\ } & \mathcal{S}_{M_1} \\
\Big\downarrow {\scriptstyle d_s} & & \Big\downarrow {\scriptstyle d_s} \\
\mathcal{S}_{M_2} & \longrightarrow & \mathcal{S}_{M_2}
\end{array}$$

Fig. 13 Commutativity of duality, symmetry, and dynamics

twice: both on its bottom row; and also on its top row (with the duality arrow in Fig. 10 reversed). The resulting diagram (Fig. 13) shows that the duality isomorphism on state-spaces d_s carries the dynamical symmetry a on the '1' side of the duality, to a dynamical symmetry on the '2' side: namely, the symmetry $d_s \circ a \circ d_s^{-1}$ (cf. either the top or bottom square). The Schrödinger picture dynamics on $\mathcal{S}_{M_2}$ is (reading down the columns in the Figure): $d_s \circ D_{t,t_0} \circ d_s^{-1}$.

So much by way of showing that a duality always preserves a symmetry of its model triples. In conclusion, we should emphasise again the three summarising comments, (i) to (iii), at the end of Sect. 3.1.1 about the contrasts between the symmetries of a bare theory, and those of its models, and of its model-triples.

(3): *The contrast between duality and gauge*:—
Finally, we should briefly return to our comment (4) at the end of Sect. 2.1. We said there that, although there is some truth in the common remark that two dual theories are like gauge formulations of a single theory, there are two important differences. Sometimes the two duals are agreed to *not* be physically equivalent (as in Kramers-Wannier duality). And anyway, the specific structure in a model is usually not gauge, in the sense of descriptively redundant.

Our discussion since Sect. 2.1 reinforces this comment. For we have seen in more detail the idea of specific structure in a model—-starting with our notation, $\bar{M}$, from Sect. 2.2.3. And by relating duality as isomorphism of model triples to our interpretation maps, we saw that duality allows, but does not entail, physical equivalence (cf. Sect. 3.2.3).

Besides, we have also seen a more specific contrast between duality and gauge, that was not foreshadowed in comment (4) at the end of Sect. 2.1. Namely, we noted in (2) at the end of Sect. 3.1.1 that if a symmetry of a model's specific structure—a symmetry of $\bar{M}$—is well-defined on the model triple, we expect it to be trivial, i.e. the identity map, there. This point implies that we would in general expect gauge (i.e. descriptively redundant) structure to *not* be carried across intact by a duality. And indeed: this has been illustrated in detail in gauge-gravity dualities. De Haro [18] has shown that a certain subgroup of the diffeomorphism group of the gravity model of the theory (roughly, the diffeomorphisms which preserve the asymptotic boundary conditions) is 'invisible' to the gauge model of the theory, in the sense of not representing any difference on that model, and so being trivially represented on the theory (the common core). Similarly for the 'gauge symmetries' of the gauge side of the duality. These are not visible on the gravity side: they are symmetries of the specific structure of the gauge model, and are trivially represented on the theory (the common core); [16, Sect. 2.2.3], [20, Sect. 5.2].

4 The Basic Boson-Fermion Duality

Boson-fermion duality will be our main example of the schema developed in Sects. 2 and 3: for, as we discussed in Sect. 1.2, it lies in the middle of the spectrum between (i) mathematical precision and established physics, (ii) scientific importance.

Boson-fermion duality is a vast field, still an active area of research today, especially in its three- and four-dimensional versions.[30] But even just in two dimensions, there is a large number of examples, which we will discuss in the next Section. In this Section, we will start with the basic case: that is, the equivalence between the free, massless scalar field, and the free, massless Dirac fermion, in two Euclidean dimensions.[31] This case already exhibits all of the interesting and non-trivial features of the more involved dualities, and so we will analyse it in some detail.

Our exposition will be necessarily brief, and will focus on those aspects that best illustrate the schema. Thus we will downplay many other important physical and mathematical aspects of this duality, such as: the existence of classical and quantum soliton solutions in these theories, the integrability of the equations, the notions of Noether and topological charges, the connection with QCD and monopoles. Neglecting these important topics is the price we pay for focussing on illustrating a conceptual schema.

This Section is introductory. We here collect the technical results (especially, about the symmetries, their associated Noether currents, and the algebras the currents generate) that will allow us to illustrate our schema. Section 4.1 introduces the free, massless boson. Section 4.2 introduces the free, massless Dirac fermion. (In Sect. 5, we will present the basic boson-fermion duality, and show how it exemplifies our duality schema from Sects. 2 and 3.)

Our exposition will mainly follow Ginsparg [30] and Frishman and Sonnenschein [28].

4.1 The Free, Massless Boson

In this subsection, we study the free, massless boson. We analyse its symmetries, write down the associated Noether currents, and give relevant details about its quantization,

---

[30] Cf. e.g. [31, 39, 40].

[31] The results in Minkowski signature are readily obtained by a Wick rotation, as we discuss in Sect. 4.1.

in particular we give the algebra that the Noether currents satisfy. This algebra will be the starting point of the comparison, in Sect. 5, with the fermionic model (where we will also justify how the bosonic and fermionic models are 'models', in our sense of Sects. 2.2.2 and 2.2.3).

In two-dimensional quantum field theory, it is very useful to work in complex coordinates. We will use the coordinates $z = x^0 + ix^1$, $\bar{z} = x^0 - ix^1$ parametrising $\mathbb{C} \cong \mathbb{R}^2$, the complex and Euclidean planes, respectively.

We first discuss the classical field and its symmetries, in two points, (i)–(ii), below. A free, massless scalar field Φ satisfies the massless Klein-Gordon equation, which in the complex coordinates chosen in the previous paragraph takes the form: $\partial\bar{\partial}\Phi = 0$ (here, and elsewhere, we use the short-hand $\partial = \partial/\partial z$, $\bar{\partial} = \partial/\partial\bar{z}$). The general solution to this equation allows for much more general functions than it does in higher dimensions, and this is the root of the richness of two-dimensional quantum field theory. The general solution is the sum of a holomorphic and an anti-holomorphic function[32]:

$$\Phi(z, \bar{z}) = \phi(z) + \bar{\phi}(\bar{z}) \,. \tag{15}$$

The holomorphic and anti-holomorphic functions $\phi(z)$ and $\bar{\phi}(\bar{z})$ are often called the left-, respectively right-moving parts of Φ. This is because a holomorphic function depends only on $z = x^0 + ix^1$: and after Wick rotation $x^1 \to -ix$ (with $x^0 = t$), the holomorphic part of Φ induces a function of $t + x$. For any fixed value of $t + x$, this indeed gives motion to the left (the speed is always negative at fixed $t + x$); whereas $\bar{z} \to t - x$, and thus $\bar{f}(\bar{z})$ induces a function of $t - x$, which is right-moving (the speed is always positive at any fixed $t + x$).

The equations of motion (equivalently, the classical action) have two sets of symmetries which will be the starting point of our set-up (once they are generalised to symmetries of the quantum version of the model):

(i) **Conformal transformations.** In two dimensions, the action of a massless scalar field is invariant under a large group of coordinate transformations, namely conformal transformations: these are scale transformations with a variable scale factor, such that angles are preserved. In complex variables $z, \bar{z}$, the conformal transformations are parametrised by arbitrary holomorphic and anti-holomorphic functions:

$$z \to z' = f(z)\,, \quad \bar{z} \to \bar{z}' = \bar{f}(\bar{z})\,. \tag{16}$$

This is the two-dimensional version of the conformal transformations. Unlike the conformal group in higher dimensions, which has an finite number of generators (the generators of the Poincaré group plus additional generators of conformal

[32] Classically, we may indeed require the solutions to be holomorphic and anti-holomorphic functions. Quantum mechanically, there are singularities which are both inevitable and the source of interesting physics, as we will see. Thus we will allow $\phi(z)$ and $\bar{\phi}(\bar{z})$ to have isolated singularities, hence we will allow them to be meromorphic and anti-meromorphic functions (operators, in the quantum version of the model), respectively.

transformations), the above transformations form an group whose corresponding algebra has an infinite number of generators. After taking into account quantum effects, this will be the celebrated Virasoro algebra.

It is easy to see that the above transformations contain, in terms of the Euclidean coordinates $x^\mu = (x^0, x^1)$, in particular: (a) constant translations, $x^\mu \to x^\mu + a^\mu$, (b) SO(2) rotations (in Minkowski signature, these are SO(1, 1) Lorentz transformations), which in complex coordinates induce a U(1) action, (c) dilations (scale transformations) $z' = \lambda\, z$, $\bar{z}' = \bar{\lambda}\, \bar{z}$ $(\lambda \in \mathbb{R})$.

(ii) **Affine current algebra transformations.** These are translations of the field by holomorphic or anti-holomorphic functions,

$$\Phi(z, \bar{z}) \to \Phi(z, \bar{z}) + \varphi(z)\,, \quad \Phi(z, \bar{z}) \to \Phi(z, \bar{z}) + \bar{\varphi}(\bar{z})\,. \tag{17}$$

Again, these transformations generalise the invariance of the action under constant shifts $\Phi \to \Phi + \varphi_0$, and are specific to two dimensions.

The conserved currents associated with these two sets of symmetries are obtained through the Noether procedure. The currents for the affine current algebra transformations (17) are, up to an overall constant:

$$J(z) := \partial\phi(z)\,, \quad \bar{J}(\bar{z}) := \bar{\partial}\bar{\phi}(\bar{z})\,, \tag{18}$$

and they are anti-holomorphically, respectively holomorphically conserved in virtue of their (anti-) holomorphicity. These currents are called 'affine currents' because, in the quantum version of the model, they generate an affine Lie algebra or Kac-Moody algebra.

The conserved currents associated with the conformal transformations (16) are the (holomorphic and anti-holomorphic) components of the stress-energy tensor:

$$T(z) = -\frac{1}{2}\,\partial\phi\,\partial\phi = -\frac{1}{2}\,J^2(z)\,, \quad \bar{T}(\bar{z}) = -\frac{1}{2}\,\bar{\partial}\bar{\phi}\,\bar{\partial}\bar{\phi} = -\frac{1}{2}\,\bar{J}^2(\bar{z})\,. \tag{19}$$

The fact that the components of the stress-energy tensor can be written as squares of the affine algebra currents will be important upon quantisation, since it will link together the Virasoro and the Kac-Moody algebras. In the quantum case, the right-hand side of (19) will contain the normally ordered product, and the relation is then called the *Sugawara construction*.

Our next task is to quantise the model. There are two well-known ways to quantise this model (which we briefly discuss in what follows), but we will settle for a third one, which is the more 'modern approach', called 'radial quantisation'. It is well suited to our perspective because it exploits the conformal symmetry group, and delivers the conformal algebra in the way we want it for Sect. 5.

One can adopt conventional canonical quantization, where one writes down canonical commutation relations for the fields and their canonically conjugate momenta. One can realize these commutation relations by choosing a Fock-space representation of the fields, and writing down the algebra of the creation and annihilation operators, which (in the present case of two dimensions) depend on one-dimensional momenta. One can then write down a Hamiltonian in terms of the creation and annihilation operators, and set up the physical states in the Fock space. See e.g. [28, Sect. 1.6].

The model can also be readily quantised using path integral quantisation. See e.g. Frishman and Sonnenschein [28, Sect. 1.9]. Here, we will adopt the third method, *radial quantisation*, as follows (ibid. Sect. 1.7).

This approach is based on the complex coordinates $z, \bar{z}$. First, we will use the conformal symmetry group to choose more convenient coordinates: mapping $\xi := x^0 + ix^1 \mapsto z := e^{\xi} = e^{x^0+ix^1}$ (for more details, see the Appendix). 'Radial' then refers to the fact that, after this conformal transformation, the equal-time slices $x^0 =$ constant, used for canonical quantisation, become circles of constant radius. The line integrals over space which appear in physical quantities, such as charges, then turn into contour integrals on the complex plane. We will not spell out the details of this procedure, of which there are many good reviews (see e.g. Sect. 2.2 of Ginsparg [30] or Lüst et al. [44, Sect. 4.1]), but the key technique we will mention is the use of radial ordering to define the order of operators which are integrated over a contour in the complex plane.

Remember that our aim for the quantum model of the boson is to obtain its algebra of operators. To this end, we will use the short-distance behaviour of a distinguished set of fields (so-called 'primary fields': cf. next paragraph). The algebra is indeed encoded in the short-distance behaviour of the products of the primary fields among each other and with the stress-energy tensor. This short-distance behaviour of products is called the 'operator product expansion'. To this we now turn.

So let us consider a *primary field* $\Psi(z, \bar{z})$. Primary fields are defined by their transformation properties under conformal transformations (see Eq. (40) in the Appendix).[33] It can be readily shown that the short-distance behaviour of the product of the stress-energy tensor with any primary field is:

$$T(z)\,\Psi(w,\bar{w}) = \frac{h}{(z-w)^2}\,\Psi(w,\bar{w}) + \frac{1}{z-w}\,\partial_w\Psi(w,\bar{w}) + \text{finite terms}, \;\; (z \to w) \tag{20}$$

$$\bar{T}(\bar{z})\,\Psi(w,\bar{w}) = \frac{\bar{h}}{(\bar{z}-\bar{w})^2}\,\Psi(w,\bar{w}) + \frac{1}{\bar{z}-\bar{w}}\,\partial_{\bar{w}}\Psi(w,\bar{w}) + \text{finite terms}, \;\; (\bar{z} \to \bar{w})$$

where $h, \bar{h}$ are the conformal weights of the primary field Ψ, i.e. the powers with which a field scales under meromorphic, respectively anti-meromorphic

[33] They are called 'primary' because all other fields, which are called 'descendants', can be obtained from them, through successive application of derivatives. See De Haro et al. [15, Sect. 3].

transformations (16), as in Eq. (40) of the Appendix.[34] The conformal weights are usually denoted as $(h, \bar{h})$, and they encode how a field transforms under dilatations.[35] For example, taking $\Psi = \partial\phi(w)$, i.e. the (derivative of the) massless left-moving part of ϕ, we have $h = 1$, since this field is a vector, hence the conformal weight is $(1, 0)$. Obviously, this is our primary field of interest in this model. As we will see below, after Eq. (22), the stress-energy tensor is *not* a primary field. On the other hand (cf. Sect. 4.2), a Weyl-Majorana fermion $\Psi = \chi(w)$ has conformal weight $(\frac{1}{2}, 0)$. The anti-holomorphic bosonic field $\bar{\phi}$ has conformal weight $(0, 1)$; and the anti-holomorphic Weyl-Majorana fermion has $(0, \frac{1}{2})$.

The expansion (20) is called an *operator product expansion*. Operator product expansions are important because they tell us almost everything we need to know about a field: in particular, we will derive from them *the algebra of operators* of the model. The form of (20) is a direct consequence of the quantum nature of the operators, together with the assumption of Ψ's being a primary field, i.e. that it satisfies (40). It can be shown that the operator product expansion is equivalent to the canonical commutation relations of the modes of the fields. Eq. (20) thus encodes the short-distance behaviour of Ψ, and is often taken to be an alternative *definition* of a primary field Ψ.

In the classical model, the stress-energy tensor was given, in Eq. (19), by the squares of operators evaluated at the same point. In the quantum version of the model, this gives rise to divergences which need to be (and can be) renormalised. For the two-dimensional quantum field theories which we consider in this paper, the divergences are renormalised, to all orders, by the addition of a *single* counterterm to the action; alternatively, it suffices to define expressions such as (19) by *normal ordering*:

$$T(z) = -\frac{1}{2} : J(z)\, J(z) : \,, \qquad \bar{T}(\bar{z}) = -\frac{1}{2} : \bar{J}(\bar{z})\, \bar{J}(\bar{z}) : \,, \tag{21}$$

where the affine currents are still given, in the free bosonic scalar case, by (18), now as operator equations. The normal ordering is denoted by the colons. For the details of the normal ordering procedure, see e.g. Ginsparg [30, Sect. 2.3].

Similarly to (20), the operator product expansion of the stress-energy tensor with itself can be worked out:

$$T(z)\, T(w) = \frac{c/2}{(z-w)^4} + \frac{2}{(z-w)^2}\, T(w) + \frac{1}{z-w}\, \partial T(w)\,. \tag{22}$$

Here, $c = 1$ for the free scalar (19), and it is called the *central charge*. Comparing the second term with (20), we see that $h = 2$, i.e. the stress-energy tensor T is a

[34]As remarked in footnote 32, in the quantum case we allow for (anti-) *mero*morphic, rather than (anti-) holomorphic, operators.

[35]In more detail: $h + \bar{h}$ is the eigenvalue of the dilatation operator, and $h - \bar{h}$ is the eigenvalue of the (Euclidean) rotation operator. Hence, the conformal weights contain information about the mass and the Euclidean spin of a field.

conformal field of weight (2, 0), as expected from dimensional analysis. However, compared to (20), this expansion has, in addition, the first term. So $T(z)$ is not a primary field in the sense of Eq. (20), nor does it transform as in Eq. (40). The term proportional to the central charge $c = 1$ is obtained making use of the normalized expression (21). For another way to calculate the central charge, see Frishman and Sonnenschein [28, Sect. 1.10].

In the same way, the operator product expansion of two affine currents (18), as well as that between T and J, can be calculated, with the following result:

$$\begin{aligned} J(z)\,J(w) &= \frac{1}{(z-w)^2} + \text{finite terms}, \quad (z \to w) \\ T(z)\,J(w) &= \frac{1}{(z-w)^2}\,J(w) + \frac{1}{z-w}\,\partial J(w)\,. \end{aligned} \tag{23}$$

The second equation between T and J is simply (20) applied to the special case $h = 1$, owing to the fact that $J = \partial\phi$ is indeed a primary field of weight $(1, 0)$.

Like in the canonical formalism, where a classical expansion of the field translates, quantum mechanically, into Fock modes: also in radial quantization it is convenient to introduce modes, which will give rise to creation and annihilation operators of the fields upon quantisation. That this is possible is ensured by (15), which says that the field can be decomposed into meromorphic and anti-meromorphic parts. In particular, we can do a Laurent expansion of the meromorphic and anti-meromorphic parts of the field. Thus, in virtue of (18) and (21), the affine currents and the stress-energy tensor will have their own Laurent expansions:

$$\begin{aligned} J(z) &= \sum_{n\in\mathbb{Z}} \frac{J_n}{z^{n+1}}\,, \qquad \bar{J}(\bar{z}) = \sum_{n\in\mathbb{Z}} \frac{\bar{J}_n}{\bar{z}^{n+1}} \\ T(z) &= \sum_{n\in\mathbb{Z}} \frac{L_n}{z^{n+2}}\,, \qquad \bar{T}(\bar{z}) = \sum_{n\in\mathbb{Z}} \frac{\bar{L}_n}{\bar{z}^{n+2}}\,. \end{aligned} \tag{24}$$

Overall factors of $1/z$ and $1/z^2$ can be extracted from J, respectively T: a fact which will be useful because these currents have $h = 1$ and $h = 2$, respectively, and so J_0 and L_0, as defined by (24), will have special physical significance. The summation range is infinite and therefore this normalization can always be reached, by a simple translation of n.

Because the currents satisfy (18) and (21), it is clear that J_n are linear in the creation and annihilation operators of the meromorphic field ϕ, and the L_n are quadratic in (an infinite sum of) the J_n's. This fact will be built into our considerations in what follows, though we do not work it out explicitly.

Finally, we find the algebras satisfied by J_n and L_n. Again these can be found either by canonical quantization or, in line with the methods we have used so far, they can be obtained directly from the operator product expansions (23) and (22), combined with the radial ordering prescription for the operators which contain circle integrals, mentioned above. Writing δ_{m+n} as short for $\delta_{(m+n)0}$ (defined as usual to be 1 or 0 according as $m+n=0$ or $m+n\neq 0$): the result is:

$$\begin{aligned}
[L_m, L_n] &= (m-n)\,L_{m+n} + \frac{c}{12}\,n(n^2-1)\,\delta_{m+n}\\
[J_m, J_n] &= -m\,\delta_{m+n}\\
[L_m, J_n] &= -n\,J_{m+n}\,.
\end{aligned} \tag{25}$$

and the same algebra is satisfied by the $\bar{J}_n$ and $\bar{L}_n$. Barred and unbarred quantities commute with each other. In the case at hand, $c=1$.

We recognise the first line as the celebrated Virasoro algebra, as expected from the fact that the classical theory has a conformal symmetry group. In the second line, one may recognise the level $k=1$, abelian Kac-Moody algebra. The third line is obtained from the operator product expansion between T and J in (23), and it makes the total algebra into the semi-direct product of the abelian Kac-Moody algebra and the Virasoro algebra. The algebra (25) is called the *enveloping Virasoro algebra* (with $c=1$ and $k=1$). The general enveloping algebra of the affine Lie algebra is given in (41) in the Appendix.[36]

This, i.e. Eq. (25), is the central result from the physics literature which we have been seeking, and will use in Sect. 5. For, together with (21) and the mode expansions (24), the tensor product of the holomorphic enveloping algebra of the affine Lie algebra (25) and its anti-holomorphic copy contains all of the information about the quantum version of the model. This is because the states now live in the vector space on which this algebra acts (a Hilbert space), and the quantities and the dynamics are constructed from the operators satisfying the algebra. We will show this in Sect. 5.

At this point, we notice that the generators $L_\pm$ and L_0 span an SL(2, $\mathbb{R}$) subalgebra. Together with the generators $\bar{L}_\pm$ and $\bar{L}_0$, which satisfy the same algebra, they form the conformal algebra of SL(2, $\mathbb{C}$), which is the symmetry group of the vacuum of this model.

4.2 The Free, Massless Dirac Fermion

In this subsection, we consider our second model: of a free, massless Dirac fermion in two dimensions. Our goal is to rederive the infinite-dimensional algebra (25) in this model. We will follow similar steps as in Sect. 4.1: we will derive the classical symmetries, quantise the model, and obtain operator product expansions. (The

[36]Notice that (25) satisfies the property that the level k can be changed by a rescaling of J. Thus, in the simple case we are dealing with here, in which the affine Lie algebra is based on the commutative Lie group U(1), the level has no real meaning. This is not important for us, since we will not use it: rather, our analysis in Sect. 5 will be based on the fact that we are here dealing with a special case of the general enveloping algebra of the affine Lie algebra.

discussion of how this accommodates to the sense of 'model', in Sects. 2.2.2 and 2.2.3, is postponed to Sect. 5.2.)

The massless Dirac fermion is a two-component, complex spinor which can be decomposed as $\Psi =: (\psi, \tilde{\psi})$, where ψ and $\tilde{\psi}$ are chiral (Weyl) fermions, called left- and right-chiral, respectively. The action is:

$$S_{\text{Dirac}} = \frac{1}{4\pi} \int \mathrm{d}^2x\, \bar{\Psi}\, \partial\!\!\!/ \Psi = \frac{1}{4\pi} \int \mathrm{d}^2z \left(\psi^\dagger\, \bar{\partial}\psi + \tilde{\psi}^\dagger\, \partial\tilde{\psi} \right), \tag{26}$$

where in the middle expression we used the real form of the action, and in the last expression we used complex coordinates. The equations of motion imply that ψ and $\tilde{\psi}$ are, respectively, holomorphic and anti-holomorphic (respectively meromorphic, in the quantum version of the model).

It is customary to further decompose the Dirac fermions into real, i.e. Majorana fermions, as follows: $\Psi = \frac{1}{\sqrt{2}}\, (\Psi_1 + i\, \Psi_2)$. In terms of the chiral (Weyl) fermions, we get $\psi = \frac{1}{\sqrt{2}}\, (\psi_1 + i\psi_2)$, where $\psi_{1,2}$ are Weyl-Majorana fermions, and so the action takes the form of the sum of two copies of a single Weyl-Majorana fermion, which is conventionally called χ (where $\chi = \psi_{1,2}$). The action for a single Weyl-Majorana fermion is:

$$S_{\text{WM}} = \frac{1}{8\pi} \int \mathrm{d}^2z\, (\chi\, \bar{\partial}\chi + \tilde{\chi}\, \partial\tilde{\chi}), \tag{27}$$

and again $\chi, \tilde{\chi}$ are (if extended from the real line to the entire complex plane) meromorphic and anti-meromorphic, respectively.

Like in the case of the free, massless boson studied in Sect. 4.1, this action is invariant under two sets of symmetries:
(i) **Conformal transformations:** $z \to f(z)$, $\bar{z} \to \bar{f}(\bar{z})$, the same transformations on the complex plane that we found in the bosonic model, Sect. 4.1.
(ii) **Left-holomorphic-chiral and right-anti-holomorphic-chiral transformations:** which act on the Weyl-Majorana fermions as follows:

$$\begin{aligned} \psi &\to \psi' = e^{i\,\alpha(z)}\, \psi \\ \tilde{\psi} &\to \tilde{\psi}' = e^{i\,\tilde{\alpha}(\bar{z})}\, \tilde{\psi}\,. \end{aligned} \tag{28}$$

Quantisation now proceeds similarly to what we did for the free, massless boson. One can do canonical quantisation (see [28, Sect. 2.12]) or again use circle integration and radial ordering to calculate operator product expansions, and get from them the commutation relations for the generators.

First, let us notice that the definition of primary fields, Eq. (40) in the Appendix, applies equally well to fermions as it does to bosons. Also, the stress-energy tensor for fermions can be introduced, analogously to Eq. (19) (as we do below).

The Weyl-Majorana fermion is a primary field, as one finds from its operator product expansion with the stress-energy tensor (20). Like before, we have conserved currents (i.e. annihilated by $\bar{\partial}$ for the meromorphic, and by ∂ for the anti-meromorphic current) associated with the sets of symmetries (i) and (ii) above:

$$\begin{aligned} J(z) &= :\psi^\dagger\psi: \,, \qquad \bar{J} =: \tilde{\psi}^\dagger\,\tilde{\psi}: \\ T(z) &= -\frac{1}{2} : J(z)\,J(z) := -\frac{1}{2} : \left(\psi^\dagger\,\partial\psi - \partial\psi^\dagger\,\psi\right): \,, \end{aligned} \tag{29}$$

with a similar expression for $\bar{T}$ in terms of $\tilde{\psi}$. χ has conformal dimension $(\frac{1}{2}, 0)$, and $\tilde{\chi}$ has conformal dimension $(0, \frac{1}{2})$. The central charge (22) is $c = \frac{1}{2}$ (and so, a Dirac fermion, which is the sum of two Majorana-Weyl fermions, has twice this amount, i.e. $c = 1$).

Notice that despite the half-integral values of the conformal dimensions, the currents $J(z)$, $\bar{J}(\bar{z})$ in (29) actually have the same conformal dimensions as the currents (21) in the bosonic model. This is because they are now *quadratic* in the fermions.

As it turns out, the operator product expansion of J with itself is identical to that in (23). Consequently, because T in (29) satisfies the Sugawara construction (cf. after Eq. (19)), the product expansions of T with itself and of T and J are also identical to those in (22) and in (23). Now since also here, J and T are meromorphic operators, we can expand them in coefficients J_n and L_n, as in (24). The resulting algebra is thus the very same as in the bosonic case, (25), i.e. the semi-direct product of the Virasoro algebra with $c = 1$ with the abelian affine algebra at level $k = 1$ (and its anti-meromorphic copy)! Therefore, also the state-space will be the same, since it is constructed as the Hilbert space on which the algebra acts.

The implication of this basic fact—the agreement of the two models on their algebra of currents—will be explored in the next Section. We will see that it naturally leads to the existence of a duality, and thus to the formulation of a theory comprising the two models.

5 Boson-Fermion Duality Illustrates the Schema

In this Section, we show how Sect. 4's bosonic and fermionic models (models in our sense!) illustrate the schema set out in Sects. 2 and 3, leading up to the definition of duality in Sect. 3.2. In Sect. 5.1, we state the basic 'dictionary' of the duality, mapping fields, currents and stress-energy tensors. Then Sect. 5.2 builds on this, to show that the two models are isomorphic, in exactly the sense of our schema: i.e. as regards the whole trio of states, quantities and dynamics. Then in Sect. 5.3, we return to Sect. 2.4's theme: of defining a theory by abstraction from its models. We first note some special features of our case-study, in particular that it has just two isomorphic models; and then we define a theory, a common core, from the two models. In Sect. 5.4, we discuss other ways one might define a theory from these models, i.e. so as to have them be representations of it. Finally, we briefly discuss generalizations of our case-study: to include massive particles, and to include non-abelian degrees of freedom (Sect. 5.5).

5.1 The Duality Dictionary

In Sect. 4, we derived the algebraic structures of the bosonic and the fermionic models. In particular, we saw that both models satisfy the enveloping algebra of the affine algebra (Eqs. (25) and (41) in the Appendix) with $c = 1$ and $k = 1$, for the quantum currents corresponding to the Noether symmetries.

At this point, Frishman and Sonnenschein [28, Sect. 6.1] conclude that the two theories are equivalent. They write:

'Due to the uniqueness of the irreducible unitary $k = 1$ representation of the affine Lie algebras, and the fact that the infinite-dimensional algebraic structure fully determines the theories, we conclude that *in two space-time dimensions the theories of massless free scalar field and Dirac field are equivalent.* The equivalence implies that every operator of one model should have a partner in the other model, in such a way that the operator product expansions of these operators should be identical.' (p. 133: Italics in the original).

While we basically agree with this statement, one of its clauses—that the infinite-dimensional algebraic structure fully determines the theories (in our jargon form Sect. 2.2: the models)—has not been proven, and requires some explanation and justification. Doing that is our plan for this subsection and the next. In this subsection, we give the basic dictionary that Frishman and Sonnenschein refer to in the quoted passage. (Then in Sect. 5.2.1–5.2.3, we will argue that this infinite-dimensional algebraic structure indeed is sufficient to fully specify what we have called a model.)

The duality dictionary is given by the correspondence of the bosonic affine current algebra currents with the corresponding fermionic currents, and between the stress-energy tensors, as follows (cf. [28, Sect. 6.1])[37]:

$$J_{\rm B}(z) = \partial\phi(z) \;\leftrightarrow\; J_{\rm F}(z) = \;:\psi^\dagger(z)\,\psi(z): \tag{30}$$
$$T_{\rm B}(z) = -\frac{1}{2} :\partial\phi(z)\,\partial\phi(z): \;\leftrightarrow\; T_{\rm F}(z) = -\frac{1}{2} : \left(\psi^\dagger(z)\,\partial\psi(z) - \partial\psi^\dagger(z)\,\psi(z)\right) : \;,$$

and similarly for the anti-meromorphic currents. We have already seen that both sides satisfy the same operator product expansion, and therefore they satisfy the same algebra.

Notice that only fermion bilinears appear in (30). This is what we expect, since a single boson ($h = 1$) should correspond to a pair of fermions ($h = \frac{1}{2}$ and $\bar{h} = \frac{1}{2}$). But it is not a priori clear which bosonic field a single fermion should correspond to. One would here expect to take some kind of 'square root' of the boson. But, surprisingly, the dictionary turns out to extend to a single fermion field as follows:

[37] The dictionary thus relates (18) and (21) to (29). We here add the subscripts 'B' and 'F' for 'bosonic' and 'fermionic', respectively.

$$
\begin{aligned}
: e^{i\,\phi(z)} : \quad &\leftrightarrow \quad \psi(z)\,, \qquad : e^{-i\phi(z)} : \quad \leftrightarrow \quad \psi^{\dagger}(z) \\
: e^{-i\bar{\phi}(\bar{z})} : \quad &\leftrightarrow \quad \tilde{\psi}(\bar{z})\,, \qquad : e^{i\,\bar{\phi}(\bar{z})} : \quad \leftrightarrow \quad \tilde{\psi}^{\dagger}(\bar{z})\,,
\end{aligned}
\tag{31}
$$

and again the operator product expansions agree. At first, this is a very surprising result, reminiscent of the construction of a coherent state. However, the operator product expansion shows that the conformal dimension of $: e^{i\alpha\,\phi(z)} :$ is $h = \alpha^2/2$, so the above dictionary indeed reproduces $h = \frac{1}{2}$ and $\bar{h} = \frac{1}{2}$ for the left- and right-chiral Dirac fermions, respectively. This is indeed a purely quantum result, with no straightforward classical analogue.

One of the most surprising features of (31) is that ϕ satisfies canonical commutation relations, while ψ satisfies canonical *anti*-commutation relations. How can this be? To calculate the anticommutator of two ψ's, one uses the formula $e^A e^B = e^{[A,B]}\,e^B e^A$, which holds when $[A, B]$ is a c-number. Using this formula to evaluate the commutator of two exponentials, and using the canonical commutation relations for the boson and its conjugate canonical momentum, one finds that, indeed, the left- and right-chiral fermions anti-commute!

5.2 *Two Isomorphic Model Triples*

In Sect. 5.1, we gave the 'dictionary' between the bosonic and fermionic models. This is a bijection between the basic operators of the theory (the fields). But there is, of course, more to duality than this. We need to show that the two models are *isomorphic,* as triples. Recall our second sense of 'model', in Sect. 2.2.2: as a representation of a theory. So a model is a triple (what we called the 'model triple'), together with some specific structure. And the model triple was not a 'pure copy' of the theory, but a representation of it using the specific structure. The next three subsections deal with the triples of states, quantities, and dynamics. In each of the subsections we show the existence of an isomorphism between the states, quantities, and dynamics of the two models (in the third case, an equivariance relation). This will justify that the boson-fermion equivalence is indeed a duality, in the sense of our schema. The final subsection considers the interpretation of this duality.

5.2.1 States

We begin by showing that the state spaces of the two models are isomorphic. This will form the first item in our model triple. The state spaces of the bosonic and the fermionic models were introduced at the ends of Sects. 4.1 and 4.2, as the Hilbert spaces obtained from the representations of the enveloping algebra of the affine Lie algebra (25) (which, for short, we shall also call the 'enveloping algebra'). We give some more details here.

The enveloping algebra is realized in the bosonic and the fermionic cases in terms of different fields: the bosonic J-currents (18) and the fermionic J-currents (29) are defined in terms of different fields. Consequently, also the stress-energy tensors, given in the two cases by the Sugawara construction, differ, and so do the mode operators J_n and L_n which enter the enveloping algebra of the affine Lie algebra. In other words, we have two representations of the same algebra. Let us call these $\mathcal{A}_{\text{B}}$ for the bosonic representation and $\mathcal{A}_{\text{F}}$ for the bosonic representation.

The state spaces are, as mentioned in Sects. 4.1 and 4.2, the Hilbert spaces obtained from the representations of the enveloping algebra, Eq. (25): or, better, its generalisation, Eq. (41) in the Appendix, to general underlying Lie group and general value of k. The vacuum state is obtained by requiring $J_n^a\,|0\rangle = 0$, for $n \geq 0$, where a labels the generators of the Lie algebra. Here, the generators J_n^a are the ladder operators of the algebra. This condition can also be understood physically as the regularity condition $J(z)\,|0\rangle = 0$ at $z = 0$ for the affine currents.

The non-trivial representations are constructed from the irreducible representations of the algebra, which are uniquely characterised by the highest-weight states (analogous to states of maximal J for SU(2)), obtained by application of a primary field (cf. Eqs. (20) and (40)) to the vacuum. If we denote such a state by $|l, \bar{l}\rangle$, where $l, \bar{l}$ are the representations, the remaining states in the representation (so-called 'descendants') are obtained by appropriately applying generators, and take the generic form: $L_{-m_1} \cdots L_{-m_M}\, \bar{L}_{-\bar{m}_1} \cdots \bar{L}_{-\bar{m}_{\bar{M}}}\, J^{a_1}_{-n_1} \cdots J^{a_N}_{-n_N}\, \bar{J}^{\bar{a}_1}_{-\bar{n}_1} \cdots \bar{J}^{\bar{a}_{\bar{N}}}_{-\bar{n}_{\bar{N}}}\, |l, \bar{l}\rangle$, for some integers $M, \bar{M}, N, \bar{N}$. For instance, on the fermionic model, expanding the fermion into modes[38]:$\psi(z) = \sum_{r \in \mathbb{Z}+\nu} \frac{\psi_r}{z^{r+\frac{1}{2}}}$, the ψ_{-n} operators are creation operators, and the ψ_n are annihilation operators. The fermionic vacuum is defined as $\psi_n\,|0\rangle = 0$ $(n > 0)$, and the states in the Hilbert space are of the type $\psi_{-n_1} \cdots \psi_{-n_k}\,|0\rangle$.

We will use the fact that *the irreducible unitary representations of the enveloping algebra, thus constructed, are unique up to unitary equivalence.* (See e.g. Frishman and Sonnenschein [28, p. 133]).[39]

So, for each of the models, we construct a representation of the algebra on a Hilbert space, $\mathcal{H}_{\text{B}}$ for the bosons and $\mathcal{H}_{\text{F}}$ for the fermions; and these representations are unitarily equivalent. Let us denote the unitary operator in question U.

5.2.2 Quantities

Next, we show that also the quantities of the two models are isomorphic, thus providing the second item of the model triple. What are the relevant physical quantities? We already mentioned in Sect. 3.2.1 that we take the physical quantities to be all the renormalizable, self-adjoint operators constructed from a more basic set of quantities (which in a moment we will identify with the currents) and respecting the appropriate symmetries.

---

[38] Here, $\nu = \frac{1}{2}$ for periodic boundary conditions, and $\nu = 0$ for anti-periodic.

[39] For more details, see Di Francesco et al. [24, Chap. 14] or Kac [37].

Let us look again at the two representations, $\mathcal{A}_{\rm B}$ and $\mathcal{A}_{\rm F}$, of the enveloping algebra of the affine Lie algebra, which we discussed in Sect. 5.2.1. We can say more about these representations: for we already have the 'dictionary', Eqs. (30)–(31), between the fields. At this point we know that this dictionary is a bijection, but we wish to find a condition for it to be an isomorphism. First: notice that this bijection induces a similar bijection between the respective mode operators, i.e. it induces a bijective map between the algebras, $d : \mathcal{A}_{\rm B} \rightarrow \mathcal{A}_{\rm F}$: ($d$ for 'duality')

$$\begin{aligned} d(J_{n,\,\rm B}) &= J_{n,\,\rm F} \\ d(L_{n,\,\rm B}) &= L_{n,\,\rm F}\,. \end{aligned} \tag{32}$$

This is of course just the statement that we have two equivalent representations of the enveloping algebra.

From the previous section, we know that the representation spaces $\mathcal{H}_{\rm B}$ and $\mathcal{H}_{\rm F}$ of these algebras are constructed from the irreducible representations of highest weight, which are unique up to unitary equivalence. Since the construction of the representations of the states is the same for the two models [(i.e. in terms of the algebra generators mapped by Eq. (32)], the algebra generators themselves must be compatible with the bijection which maps the highest weight representations on the two sides. So, for consistency we must require:

$$d(\cdots) = U^\dagger\,(\cdots)\,U\,, \tag{33}$$

where U is the *same* map used to map the states of the highest-weight representations in the previous subsection. If the maps U and d were not related *a la* (33), the structure of the theory would not be preserved by the duality (since an operator acting on a state would not be mapped to the the corresponding operator acting on the corresponding state). In short: once we have fixed the states to be mapped by U, as in Sect. 5.2.1, then the *same* transformation must map the quantities, as in (33).

But Eq. (33) is precisely the condition for the map to be an *isomorphism*, in addition to a bijection. And it now follows from the above that the modes of the currents of the two models are mapped as:

$$\begin{aligned} U^\dagger\,J_{n,\,\rm B}\,U &= J_{n,\,\rm F} \\ U^\dagger\,L_{n,\,\rm B}\,U &= L_{n,\,\rm F}\,. \end{aligned} \tag{34}$$

It also follows that the same relations hold for the currents (30) themselves:

$$\begin{aligned} U^\dagger\,J_{\rm B}(z)\,U &= J_{\rm F}(z) \\ U^\dagger\,T_{\rm B}(z)\,U &= T_{\rm F}(z)\,. \end{aligned} \tag{35}$$

The same map also maps the fermionic operators:

$$U^\dagger : e^{i\,\phi(z)} : U = \psi(z)\,, \qquad U^\dagger : e^{-i\,\phi(z)} : U = \psi^\dagger(z)$$
$$U^\dagger : e^{-i\,\bar\phi(\bar z)} : U = \tilde\psi(\bar z)\,, \qquad U^\dagger : e^{i\,\bar\phi(\bar z)} : U = \tilde\psi^\dagger(\bar z)\,. \tag{36}$$

This is the formalization of (31), which was justified by checking its properties (including the correct statistics, but also all the correct correlation functions). Given (31), the factors of U follow from the fact that $\psi(z)$ creates or annihilates a *fermionic* state in $\mathcal{H}_{\rm F}$, and this state is obtained from the corresponding bosonic state in $\mathcal{H}_{\rm B}$ using the same map U. The operator creating or annihilating this state must therefore transform bilinearly in U. One can also check that the map between the affine currents, the first equation in (34), follows from (36).

We have discussed the maps between the operators of the enveloping algebra of the affine Lie algebra, and the corresponding fields. We now discuss the *physical* quantities, $\mathcal{Q}_{\rm B}$ and $\mathcal{Q}_{\rm F}$, of self-adjoint operators respecting the relevant symmetries. In the fermionic model, all the self-adjoint operators are quadratic in the fermions: and taking into account the chiral symmetry algebra, they must necessarily be powers of the fermionic currents (35) (and their anti-chiral counterparts), appropriately normally-ordered. In fact, arbitrary analytic functionals of the currents are allowed. There can be a mixing between the meromorphic and anti-meromorphic sectors in the analytic functionals, but only such that the chiral symmetry algebra is preserved. Thus, the enveloping algebra indeed contains all of the information about the physical quantities: $\mathcal{Q}_{\rm F}$ consists of arbitrary analytic functionals in the fermionic currents (35) (and their anti-meromorphic counterparts), appropriately normally-ordered. The normal ordering automatically ensures that these correlation functions are well-defined.

One can in principle enlarge the set $\mathcal{Q}_{\rm F}$ to also contain the correlation functions of operators which violate the chiral symmetry algebra, evaluated on the same states, i.e. without changing the Lagrangian of the model. However, one is then changing the symmetries of the model triple $f = \langle \mathcal{H}_{\rm F}, \mathcal{Q}_{\rm F}, \mathcal{D}_{\rm F}\rangle$, and hence one is defining a *new model triple* (and, consequently, if there is a duality for this larger class of models, one is defining a new theory). We will return to this possibility when we discuss sine-Gordon/massive Thirring duality in Sect. 5.5.1.

In the bosonic model, there is a similar structure. Now it is not the mixing of *chiralities* of the fields which the symmetries forbid, but the appearance of the *underived* field $\phi(z)$ in $\mathcal{Q}_{\rm B}$. Namely, the translation symmetry $\Phi \to \Phi + a$ (or, more generally, the affine current symmetry algebra (17)), forbids the appearance of operators which depend on $\phi(z)$ or $\bar\phi(\bar z)$ *directly*, i.e. as opposed to depending on them through their derivatives. When the model is written in terms of the meromorphic and anti-meromorphic parts of Φ, translation symmetry is preserved iff all operators depend on the *derivatives* of the scalar field. This is precisely how the scalar field appears in the currents (35). Thus, once again, the affine current symmetry algebra of the model tells us that all the operators which are physical quantities of $\mathcal{Q}_{\rm B}$, must be analytic functions of the two currents (and their anti-meromorphic counterparts). $\mathcal{Q}_{\rm B}$ consists of all possible analytic functionals of those currents. As for the renormalizability constraint on the physical quantities: as we mentioned before, normal ordering automatically takes care of getting the correct renormalized quantities.

To end this subsection, let us get a possible worry out of the way. Namely, one might be concerned that (36) seems to be introducing complex operators, such as $: e^{i\,\phi(z)} :$, into the set of physical quantities of the bosonic model, which is supposed to be entirely real. But this is not quite right. For (36) are in fact *not* physical operators, on our conception of the term. It is indeed true that $: e^{i\,\phi(z)} :$ is not self-adjoint in the bosonic model, but then neither is $\psi(z)$ self-adjoint in the fermionic model. Neither of these two operators are thus to be taken as physical on either of the two sides, even if we can build physical operators, such as the currents (30), by taking powers of them. Thus, none of the operators in (36) belong to either $\mathcal{Q}_{\rm B}$ or $\mathcal{Q}_{\rm F}$. Rather we should think of such operators as *states* in the Hilbert space, using the state-operator correspondence. Indeed, recall, from Sect. 5.2.1, that the states in the fermionic Hilbert space are of the type: $\psi_{-n_1} \cdots \psi_{-n_k} |0\rangle$. So this explains how $\psi(z)$ can be a physical operator—namely, it creates and annihilates physical states in $\mathcal{H}_{\rm F}$—while it is not self-adjoint and does not belong to the set $\mathcal{Q}_{\rm F}$ of physical quantities.

Perhaps the most surprising aspect of the above is the fact that, in two dimensions, *the Hilbert space of the bosonic model contains fermionic states*, e.g. states with conformal weight $(\frac{1}{2}\,, 0)$. In fact, because the conformal weight of the operator $: e^{i\,\alpha\,\phi(z)} :$ is $h = \alpha^2/2$, the Hilbert space contains a 1-parameter family of states with a continuous range of Euclidean spin values.[40] This feature of the quantised models is indeed surprising—and illustrates our theme of surprise announced in (2) of Sect. 2.1.

5.2.3 Dynamics

Finally, we discuss the equivariance of the dynamics of the two model triples. 'Dynamics' can be understood in different ways in different theories, and even in one formulation of a single theory. Think, for example, of the difference in the dynamics if it is formulated in the Heisenberg or in the Schrödinger pictures of a theory.

First of all, we have formulated our model triples in the Heisenberg picture: operators are generally time-dependent and the states are time-independent. Let us consider the bosonic model first. We are working in Euclidean spacetime, but when analytically continued to Minkowski spacetime the operator $H_{\rm B} := L_{0,\,\rm B} + \bar{L}_{0,\,\rm B}$ (which generates dilatations on the plane) is the generator of time translations, and is to be identified with the Hamiltonian. It is indeed the 00-component of the stress-energy tensor. The same is true in the fermionic model: the 00-component of the stress-energy tensor, which is the generator of time translations, is the operator $H_{\rm F} := L_{0,\,\rm F} + \bar{L}_{0,\,\rm F}$. These two Hamiltonians are mapped to each other by the map U in (34) (and the anti-meromorphic version of that equation). Thus the dynamics is correctly preserved by the duality map: more precisely, it is equivariant with the unitary transformation.

---

[40]Our use of the word 'Euclidean spin' here follows the jargon in the physics literature, for the eigenvalue under Euclidean rotations, as we mentioned in Sect. 4.1. It is questionable whether such jargon is justified by the physical interpretation in 1+1 dimensions. But we will not need to dwell on this point, since our main aim in this Section is formal.

We see that the requirement that the dynamics is correctly mapped between the two theories does not give any additional piece of the duality map, but simply follows from the other two: since all the quantities were already dual. This seems to be a general fact in dualities between quantum field theories, though we do not necessarily expect it to be the case for any duality. Namely, in a quantum field theory model, once we require that all states and physical quantities between the two theories map correctly, i.e. according to the *isomorphism* (33), then all correlation functions of the model automatically map correctly as well. But the correlation functions of a quantum field theory model exhaust all the dynamical information about the model, so that knowing all the correlation functions we can, in principle, reconstruct the dynamics.

Thus, an alternative way to formulate duality is in terms of the correlation functions. This is, in fact, how duality is usually formulated in the string theory literature. However, we find our own conception of duality more illuminating, because it allows us to map the states and the operators individually and directly, without having to unscramble this information from the full set of correlation functions. But either way, the main point is that the dynamics is correctly equivariant with the duality map.

5.2.4 Interpretation

With our boson-fermion duality now in hand, we return to Sect. 3.2.3's theme that a duality should respect the interpretation maps introduced in Sect. 2.3. Our initial point in Sect. 3.2.3 was that since duality relates two model triples, and interpretation maps apply to model triples (strictly speaking: their components, such as the set of quantities), interpretation simply proceeds independently on the two sides of the duality. The range of these interpretation maps could be: either two distinct but isomorphic 'sectors of reality', or the very same sector of reality—in which case there is a triangular rather than square diagram, as in Figs. 8 and 9. We also said that the choice between these cases was a matter of an 'internal' versus an 'external' interpretation.

External interpretations for the models are straightforward to read off from the model triples: they are the bosonic, respectively fermionic, interpretations which the two models come with in the first place (i.e. in this case, their original historical interpretations). There is a natural map I_{Int} which assigns intensions: for example, it maps a bare or abstract bosonic state to the meaning 'boson on a line with such-and-such properties'. And the map I_{Ext} assigns extensions. For example, it maps the abstract expectation value of a bosonic field to a measurable property of a specific boson that is fixed as the reference by the context of use. For example, the property might be the boson's amplitude or probability of being at a specific place at a specific time: we say 'amplitude or probability', since we understand a value or expectation value, written schematically as $\langle Q, s\rangle$, to also represent all the matrix elements $\langle s_1|\hat{Q}|s_2\rangle$ (cf. Sect. 3.2.1-(1)). In the fermionic model, there are similar interpretation maps for the fermionic model triple: where the codomains of the maps are now not the properties of bosons, but the properties of (say) electrons! Thus clearly, on the external interpretation the codomains of the maps do not need to agree. In philosophical jargon: the set of possible worlds where the bosons and the fermions appear need not be the same sets of worlds.

The internal interpretation abstracts from the specific structure of the two models, so that the two model triples receive the same interpretation. This interpretation is therefore best worked out at the level of the theory. Since the duality is exact, each model contains both bosonic and fermionic states, as we discussed in Sect. 5.2.2. This means that the world (or set of worlds) which is the codomain of our interpretation, should contain both bosons and fermions. The interpretation maps, applied to the models, must commute with the duality map d (or, equivalently, U) in Eqs. (32)–(33). The interpretations will map $J(z)$ and $T(z)$ to a current and a stress-energy tensor, respectively; (as intensions or as extensions, as appropriate: in the ensuing discussion, we shall not make this distinction, since everything we say applies equally well to both kinds of maps). The internal interpretation also maps the abstract expectation value of a bosonic field (whether written as $\phi(z)$ or as a fermion bilinear) to the amplitude or probability of some bosonic event. Likewise, the abstract expectation value (matrix element) of a fermionic field between two suitable given states will be mapped to the transition amplitude between two fermionic states.

5.3 *Defining a Theory from the Two Model Triples*

We now return to Sect. 2.4's theme: of defining a theory by abstraction from its model triples. We first discuss how our duality's having just two isomorphic model triples makes this enterprise vulnerable, in two ways (Sect. 5.3.1). Then we undertake to define a theory, a common core, from the two models; and discuss their specific structures (Sect. 5.3.2). Thus the general issues of Sect. 5.3.1—which return us to the Hilbertian themes at the start of Sect. 1—will lead in to the specific details of Sect. 5.3.2.

5.3.1 Two Vulnerabilities

At first sight, there might seem to be no issues about the definition of a theory (in our sense, from Sects. 1.1 and 2.2.1) from a pair of model triples that are isomorphic: or indeed, from any set of two or more isomorphic model triples. Can we not simply define the theory as the structure of which the isomorphic model triples are isomorphic 'concrete' copies? More precisely: here we should clarify the phrase 'as the structure of which', in order to respect Sect. 2.2.2's point that a model (in our sense!) usually realizes a theory by being a representation of it, and representation allows mere homomorphism, rather than isomorphism. So the thought is: can we not simply define the theory as the structure, S say, of which the (two or more) isomorphic model triples are representations—as it happens, isomorphic ones? Talk of S thus defined as a theory will then engender admitting all its representations as models in Sect. 2.2.2's sense.

We agree that this relaxed attitude is tenable. After all, 'theory' is a term of art: so one is at liberty to define it as one sees fit. But in both pure mathematics and theoretical physics, the abstraction ('extraction') of a general pattern or structure from a given set of examples is a matter of judgement, to be made in the light of one's aims and intuitions, including the aim of representing the world as accurately as possible. This means that there are two related worries one might have: that is, two limitations to which the above strategy for defining a theory is vulnerable.

(i): Suppose that—as in our bosonization case-study—the given examples are isomorphic: where 'isomorphic' is itself a term of art, made precise by some mathematical structure we see exactly copied in the examples. Then one worries that there might be non-isomorphic examples which one has not thought of (has not 'been given') that are equally good examples of the general pattern one is trying to write down—that equally fit one's aims and intuitions.

(ii): Suppose the given examples are not isomorphic; (again, in the sense we have in mind, especially initially). Still one worries that they might not be sufficiently varied, so that the pattern one writes down after considering them is too restrictive. That is: the pattern encodes aspect(s) that, given one's aims and intuitions, are really accidental commonalities of the examples. Recall (from the end of Sect. 2.4) the example of colour as an artefact that could beset Frege's definition of direction as an equivalence class of mutually parallel lines.

The general answer to these worries lies in the point, argued in Sect. 3.2.2, that for models which purport to describe the physical world, the distinction between what is in the triple and what is specific structure cannot be blurred.[41] So these worries can be set aside: for a give target system, typically only a single triple will provide a complete description—up to isomorphism, that is.

So much by way of rehearsing the general issues about defining a theory from models: specifically, from model triples. For the purposes of this paper, what matters is how these issues play out in our case-study, bosonization. Section 5.3.2 will give details about this. But to summarise:— The theory we will construct in Sect. 5.3.2 is the simplest one that can be constructed from our bosonic and fermionic model triples, using the isomorphism at hand from Sect. 5.2. But in line with this Subsection's comments, we make no claim that is the only (or even best) way to 'extract a pattern'

[41] As an example, consider the bosonic and fermionic models, but now weakened by the stipulation that the Virasoro algebra belong to the model triple, while the affine Kac-Moody algebra [(and the third line in Eq. (25)] belongs to the specific structure. As we argued in Sect. 3.2.2, this stipulation *changes the physical content* of the models, and so it is not innocuous. The models thus obtained contain different numbers of (uninterpreted) physical degrees of freedom, and so cannot describe the bosons or the fermions of Sect. 4. This is because the boson and the fermion CFTs (even before they are physically interpreted) treat the Kac-Moody degrees of freedom not as 'accidental commonalities', in the sense of Sect. 2.4: but as physical, and related to the Virasoro generators by the Sugawara construction. (For example, if we drop the chiral symmetry on the fermionic model, we lose the reason to restrict to chiral quantities only: cf. Sect. 5.2.2; and likewise for the boson's affine current symmetry algebra.) Thus the boson and fermion models are *not* dual, if based on just the Virasoro algebra. We thank Josh Hunt for bringing up this example.

from the model triples. Indeed, our final two Subsections, Sects. 5.4 and 5.5, *will* consider other such ways: in particular, by taking into account other models.

5.3.2 Defining the Theory

In this subsection, we collect our results from Sect. 5.2 about the two model triples, in order to define our *theory*. We have so far defined two *models:—*

(a) A bosonic model triple, $b := \langle \mathcal{S}_{\text{B}}, \mathcal{Q}_{\text{B}}, \mathcal{D}_{\text{B}} \rangle$, with: $\mathcal{S}_{\text{B}} = \mathcal{H}_{\text{B}}$, the Hilbert space which is constructed from the irreducible highest-weight representations of the enveloping algebra of the affine Lie algebra, represented on the bosons. 'Represented on the bosons' here means that the generators of the affine Lie algebra (25), which proceed from the Laurent expansion (24), are constructed from the *bosonic* fields (18) and (19) (and their anti-holomorphic counterparts). Thus, $\mathcal{Q}_{\text{B}}$ is the set of normally-ordered analytic functions in the bosonic currents $(J(z)_{\text{B}}, \bar{J}(\bar{z})_{\text{B}}, T(z)_{\text{B}}, \bar{T}(\bar{z})_{\text{B}})$; and $\mathcal{D}_{\text{B}} = L_{0,\text{B}} + \bar{L}_{0,\text{B}}$ the dilatation operator of the bosonic model.

(b) A fermionic model triple, $f := \langle \mathcal{S}_{\text{F}}, \mathcal{Q}_{\text{F}}, \mathcal{D}_{\text{F}} \rangle$, with: $\mathcal{S}_{\text{F}}$ the same Hilbert space, represented on the fermions, $\mathcal{H}_{\text{F}}$; $\mathcal{Q}_{\text{F}}$ the set of normally-ordered analytic functions of the currents $(J(z)_{\text{F}}, \bar{J}(\bar{z})_{\text{F}}, T(z)_{\text{F}}, \bar{T}(\bar{z})_{\text{F}})$; and $\mathcal{D}_{\text{F}} = L_{0,\text{F}} + \bar{L}_{0,\text{F}}$ the dilatation operator of the fermionic model.

What structures make up the 'specific structure' $\bar{M}$, of each of the models $B := \langle b, \bar{B} \rangle$ and $F := \langle f, \bar{F} \rangle$, in the sense of our notation in Sect. 2.2.3, and especially Eq. (2)? On the bosonic side, the specific structure $\bar{B}$ clearly contains the field $\phi(z)$ (and functionals of it), together with the symmetry algebra (17) acting on it. This symmetry algebra will, however, manifest itself in the model triple through the affine currents and their algebra. Also the defining relations of the field (equation of motion, etc.) are specific to $\bar{B}$.

On the fermionic side, it is the field $\psi(z)$, with its chirality symmetry (and a different set of defining relations, equation of motion, etc.), which are parts of the specific structure $\bar{F}$. Though $\psi(z)$ defines a state in the Hilbert space, as discussed in Sect. 5.2.2, thinking of this state as created by $\psi(z)$, i.e. a fermion with certain meromorphic and chirality properties, it is part of the specific structure. All the Hilbert space knows about this fermion is that there is a state of conformal weight $(\frac{1}{2}, 0)$.

We are now ready to discuss the *theory* which we can construct from these two models: by discussing the common core of the two models, i.e. the model triples (or roots), to which the theory is isomorphic: $b \cong f \cong T$. It is a theory based on four currents $(J(z), \bar{J}(\bar{z}), T(z), \bar{T}(\bar{z}))$ of conformal dimensions $(1, 0)$, $(0, 1)$, $(2, 0)$, $(0, 2)$, satisfying the enveloping algebra with $c = 1$ and $k = 1$. Its states are the unitary representations of this algebra. The dynamics is given by singling out the Hamiltonian $H = L_0 + \bar{L}_0$.

Sets of symmetries of the theory: The theory has two built-in sets of symmetries: (i) a conformal group and (what we shall call), generated by the stress-energy tensor, (ii) an affine current symmetry algebra, generated by the J-currents.

(i) The conformal symmetry group, Eq. (16), is represented in the same way in the two models—since this is a symmetry group of the background spacetime. This is also related to the fact that the Hamiltonian, and hence the dilations, are related by equivariance between the two theories. Again: $T(z)$, the stress-energy tensor of the theory, is represented in both the models.

(ii) The affine current symmetry algebra is represented very differently in the two models! Namely, as affine current algebra transformations [(Eq. (17) in Sect. 4.1(ii)] on the bosonic model, but as left- and right-chiral symmetry algebra (Eq. (28) in Sect. 4.2(ii)) on the fermionic model. This symmetry algebra restricts the kinds of physical quantities in the theory, as we have explicitly discussed in Sect. 5.2.2, in the same way. Yet the theory does not "see" any of the features from which the affine current algebra arises: which are very different in the two models, viz. the transformations of the fields Eq. (17) versus (28).

We have discussed this duality in some detail because it is a good model to the more general, and technically involved, boson-fermion dualities in two-dimensional conformal field theory: to which we turn in Sect. 5.5.

5.4 Further Abstraction

The discussion in Sects. 5.2.2 and 5.3 illustrates our theme of the limitations of abstraction. We constructed our *theory* from two isomorphic model triples: the model triple of the theory was built from the representations of the enveloping algebra of the affine Lie algebra (25), which give a unique set of *states* (discussed in Sect. 5.2.1), and a basic set of *quantities*: the generators of the algebra themselves. The *dynamics* was a choice of a Hamiltonian among the quantities (Sect. 5.2.3). As we saw in Sect. 5.2.2, the *full* set of quantities $\mathcal{Q}$ of the theory contains more than just the basic set: so that arbitrary analytic functionals of the currents $(J(z), \bar{J}(\bar{z}), T(z), \bar{T}(\bar{z}))$, and their derivatives, are allowed. Compare the discussion of the symmetry algebra (ii) in Sect. 5.3.

But we also know, from Sects. 2.4 and 5.3.1, that there are no general rules, fixed once-and-for-all, for defining theories. So one asks: to what extent is our procedure above unique? It is of course unique if what one wishes to describe is a boson or a fermion, as given systems with known degrees of freedom. But the procedure of abstraction suggests three natural ways in which our theory might be modified. The first way would make for a more *restrictive* theory; the other two entail further abstraction, thus allowing for a more *general* one.

(a) The *conformal symmetry group* (Sect. 5.3.2-(i)) was used to form the enveloping algebra. Hence it is realised by the states and the quantities, in the sense that the states and quantities form representations of this symmetry group. But one can construct a new theory in which the class of operators is *reduced*: namely, by placing restrictions on the conformal transformation properties of the quantities. This leads to a theory with a smaller set of quantities (and subsequently to bosonic and fermionic models with smaller sets of quantities). Alternatively, one can reduce the space of states by placing similar restrictions on them.

(b) The *affine current symmetry algebra* (Sect. 5.3.2-(ii)) limited the set of quantities to those that can be constructed from the currents $(J(z), \bar{J}(\bar{z}), T(z), \bar{T}(\bar{z}))$ (as mentioned at the start of this subsection). But one might decide that this is not a symmetry algebra one wishes to keep, for the physical system of interest; (for example, in the presence of mass terms, this symmetry algbera will be dropped). One then allows a larger class of, or even all, self-adjoint, renormalisable operators constructed from (31), not just the ones that preserve this symmetry algebra. In this way, one clearly gets a richer theory (with a larger set of operators and states), of which Sect. 4's two models are still representations.[42] And as we stressed in Sect. 3.2.2, it is a matter of physical judgment, which kinds of operators one needs to admit in order to describe the physics at hand. In particular: if one wants to add a mass, one is forced to generalise the theory in this way. The boson-fermion duality continues to hold, thanks to the existence of the maps Eqs. (31) and (34). But we get a *more general class of theories*, which will not necessarily be each other's duals: the generality of the class depends on which additional set of operators one takes on board with the quantities. This will be illustrated explicitly in Sect. 5.5.

(c) Though the two models share the *spacetime coordinates* $z, \bar{z}$, these coordinates do not enter the basic considerations that led to building the states, quantities, and dynamics of the theory, in Sects. 5.2.1–5.2.3. Indeed, the basic object of interest, in constructing the triple, is the algebra of the mode operators L_n and J_n (and powers of them): and these are spacetime-independent. Furthermore, these modes contain essentially the same information as the spacetime-dependent currents $(J(z), \bar{J}(\bar{z}), T(z), \bar{T}(\bar{z}))$, i.e. the latter can be reconstructed from the former through the Laurent expansions (24), which take identical form in the fermionic case. So, one might decide that $z, \bar{z}$ are just book-keeping devices with no essential information about the theory. On this view, one can construct a theory just based on the algebra (25) and its representations, without its spatio-temporal realization. The bosonic and fermionic model triples then still form (spatio-temporal) representations of this algebra: but one can envisage the existence of other representations, which are not spatio-temporally realised. This would presumably give rise to non-isomorphic models, in the sense of Sects. 2.4 and 5.3.1. While doing away with spacetime may seem a radical suggestion, it is not so uncommon: think e.g. of spin chains as possible models.

Point (a) is a straightforward modification of our theory, but also of the models. So it should not be seen as illustrating the limitations of abstraction, in the sense of Sects. 2.4 and 5.3.1. Rather, it is a *method to obtain more restrictive theories,* by consistently strengthening the symmetry requirements of the models.

But points (b) and (c) do illustrate our remarks, in Sects. 2.4 and 5.3.1, about the need for models to be ‘sufficiently varied’. By taking, in (b) and (c), some of the

[42]Notice that the ambiguity here is in the best definition of the *theory*, not of the *duality*: cf. footnote 41.

symmetry shared by the models to be accidental, one gets a larger class of models, which is likely to include non-isomorphic ones.

In the next subsection, we will give such examples of dualities between isomorphic models which are more general: either because they have less symmetry, or because they have more fields, with additional symmetries.

5.5 *General Boson-Fermion Dualities*

In this Subsection, we briefly discuss two generalizations of our basic duality. The two generalizations are important for our discussion since they fulfill, even better than our basic duality does, Sect. 1.2's two desiderata for the choice of examples of dualities. Recall that these desiderata were: on the one hand, (i) an example should be mathematically precise and represent established physics; and on the other hand, (ii) an example should be scientifically important (as described more fully in Sect. 2.1).

As we will see, these desiderata will be amply fulfilled by this Subsection's two generalizations. The first generalization especially illustrates scientific importance. The second illustrates, not so much mathematical precision per se: but rather, *mathematical richness and generality*, thus showing that the boson-fermion duality described in this Section is not an isolated 'coincidence' that holds for free, massless models, but part of a very rich class of mathematically interesting (as well as rigorous!) isomorphic models. Thus we will here define a rich class of theories, each of which is an equivalence class of exactly two isomorphic model triples.

We will discuss the two kinds of generalization in turn, in the next two subsections. Section 5.5.1 discusses the duality between the massive Thirring model and the sine-Gordon model. Section 5.5.2 considers non-abelian versions of boson-fermion duality. In both cases, we must be brief and must suppress technical details.

5.5.1 Duality Between the Thirring Model and the Sine-Gordon Model

The basic boson-fermion duality can be extended to include mass terms for the fermions, and interaction terms for both fermions and bosons. This generalisation is important, because it shows that the duality is not a special property that only occurs in the free, massless case, in which the action is conformally invariant. Massive, and interactive, theories are also subject to duality. So this strengthens the scientific importance of duality: it brings duality into the 'real world'. In fact, the Thirring model-sine-Gordon duality is quite important in condensed matter systems. See Giamarchi [29], Altland and Simons [3].

The massless Thirring model generalises the free, massless Dirac fermion by the addition of a quartic interaction term for the fermions, with coupling constant g. This quartic interaction is built from the J-currents, and so preserves the chiral symmetry algebra described in (ii), Sect. 4.2. The model can be solved exactly, and the quantum theory is well-defined only for $g > -\pi$ [9, p. 2094].[43]

The fermionic mass term in the massive Thirring model explicitly breaks the chiral symmetry algebra described in (ii), Sect. 4.2. This is because the mass term which is added to the action of the Dirac fermion, Eq. (26), mixes the left- and right-chiral (Weyl) fermions. It takes the following form:

$$\Delta S_{\text{mass}} = m\left(\tilde{\psi}^\dagger\,\psi + \psi^\dagger\,\tilde{\psi}\right) . \tag{37}$$

Remember that the action (26) did not mix the left-chiral Weyl fermion ψ, $\psi^\dagger$ with the right-chiral Weyl fermion $\tilde{\psi}$, $\tilde{\psi}^\dagger$. The above mass term explicitly mixes the two, and so breaks the chiral symmetry algebra.

The mass term (37) can be translated into a bosonic term using a straightforward generalisation of the dictionary (31). The generalisation is straightforward in that it takes the same form, but now depends on a bosonic coupling β, which is related to the fermionic coupling through (37). The equation of motion of the sine-Gordon model is:

$$\Box\phi + \frac{\mu^2}{\beta} : \sin(\beta\phi) := 0\,, \tag{38}$$

and so the fermionic coupling is related, through (1), to the bosonic coupling β. Notice that, to linear order in β, the above reduces to a standard bosonic mass term, with mass μ. We are here, however, interested in the exact model.

There are divergences that need to be renormalised. The bosonic term corresponding to (37) is:

$$\Delta S_{\text{sine-G}} = \frac{\mu^2}{\beta^2}\ :\cos\beta\phi:\ , \tag{39}$$

where μ is a scale which originates in the normal ordering procedure, and appears in the boson-fermion dictionary as an overall multiplicative constant.[44]

The algebra underlying the model triple of this model is still the enveloping algebra of the affine Lie algebra with $c = 1$, but the level now depends on the coupling: $k = \frac{\beta^2}{4\pi} = \frac{1}{1+g/\pi}$. We see that in the limit of zero fermionic coupling, we reproduce the algebra at level 1.

Our discussion, in Sect. 5.2, of the isomorphic model triples, thus generalises to the massive Thirring and sine-Gordon models, with appropriate modifications. In both cases, the algebra is the enveloping algebra, now with a coupling-dependent level. Therefore, the discussion of the *states* is analogous to the one in Sect. 5.2.1:

[43] Looking at the relation between the couplings (1) (Sect. 1.2), this will correspond to the value $\beta^2 < \infty$ of the bosonic coupling.

[44] The scale μ is already present in the massless theory. But it does not play any important role, since it is just an overall renormalisation constant.

the Hilbert space consists of the highest-weight representations of the enveloping algebra, and their descendants. The *quantities* are constructed from a wider class of operators, compared to our discussion in Sect. 5.2.2. Namely, the quantities now include non-chiral operators (respectively, operators which break affine current algebra transformations: see Sect. 4.1-(ii) for bosons, Sect. 4.2 for fermions) built from the fields: such as (38) for the bosons, and (37) for the fermions. Finally, the dynamics is still given by the Hamiltonian, which is the zero component of the Virasoro generator L_0, in the bosonic or fermionic representation. Thus, we get a theory by abstraction from these two isomorphic triples, as outlined in Sect. 5.3.

As we remarked before, this generalised theory explicitly illustrates our comment (b) in Sect. 5.4, about the contingent nature of the chiral symmetry algebra. By allowing the theory to break the chiral symmetry algebra, we get a wider class of theories (which depend on the coupling and the mass): a class of which the basic free, massless case is just a special case.

5.5.2 Non-abelian Boson-Fermion Dualities

As we mentioned in this Subsection's preamble, the free, massless bosons and fermions can be generalised in another direction [58]: to include non-abelian degrees of freedom. We will not here provide any technical details, but we will simply list some of the important examples of dualities studied in the literature; all of which are conformal field theories, except for (c) and (e):

(a) N free Majorana (real) fermions, in an N-dimensional vector representation of $\mathrm{O}(N)$. They are dual to the bosonic, Wess-Zumino-Witten (WZW) model with an $\mathrm{O}(N)$ symmetry group. The Wess-Zumino-Witten model is a model whose action is built from a bosonic group element (rather than an algebra element), in this case an $N \times N$ matrix of $\mathrm{O}(N)$. The bosonic and fermionic models are both invariant under the affine Lie algebra transformations of $\mathrm{O}_{\mathrm{L}}(N) \times \mathrm{O}_{\mathrm{R}}(N)$ (for left- and right-action, respectively) at level $k = 1$. In both cases, the central charge is $c = N/2$.

(b) N free, massless Dirac (complex) fermions are dual to a bosonic WZW model with group $\mathrm{U}(N)$. The two models satisfy the affine Lie algebra of $\mathrm{SU}_{\mathrm{L}}(N) \times \mathrm{SU}_{\mathrm{R}}(N) \times \mathrm{U}(1)$, with central charge $c = N$ and $k = 1$.

(c) Mass terms can be added to the Dirac fermions in case (b), with modifications in the dictionary and in the bosonic theory which are similar to the ones discussed in the previous subsection.

(d) The Majorana and Dirac fermion models can be endowed with colour and flavour charges. Majorana fermions with N_{F} flavours and N_{C} colours, thus transforming under the group $[\mathrm{O}(N_{\mathrm{F}}) \times \mathrm{O}(N_{\mathrm{C}})]_{\mathrm{L}} \times [\mathrm{O}(N_{\mathrm{F}}) \times \mathrm{O}(N_{\mathrm{C}})]_{\mathrm{R}}$, are dual to the Wess-Zumino-Witten action with two bosonic fields, taking values in $\mathrm{O}(N_{\mathrm{F}})$ and $\mathrm{O}(N_{\mathrm{C}})$. In the same way, $N_{\mathrm{F}} \times N_{\mathrm{C}}$ Dirac fermions can be expressed as the sum of two Wess-Zumino-Witten actions, and a third term for an additional field. The three bosonic fields take values in the group manifolds $\mathrm{SU}(N_{\mathrm{F}})$, $\mathrm{SU}(N_{\mathrm{C}})$, and $\mathrm{U}(1)$. Again, one finds two copies of the affine Lie algebra, now with levels different from one, viz. $k = N_{\text{F}}$ and $k = N_{\text{F}}$, respectively. This duality can also be generalised to other gauge groups.

(e) Mass terms can be added to the theories (d) with flavour and colour charges, with appropriate modifications in the dictionary and in the bosonic theory, as before.

All of the above models end up having model triple structures which are constructed as representations of the enveloping algebra of the affine Lie algebra (41) in the Appendix, for various values of the level, k, and the central charge c, and for different Lie groups. Thus their Hilbert spaces are constructed from the highest-weight representations, which as we already saw in Sect. 5.2.1 are unique up to unitary transformations. Clearly, all of these theories can be given a set of states, as discussed in Sect. 5.2.1, of which the bosonic and fermionic models form two representations. Also, all of these theories are based on the same class of algebras, with generators J_n and L_n as in (25) for the bosonic case, and likewise for the fermionic case in Sect. 4.2 and their anti-holomorphic counterparts (see Eq. (41) in the Appendix). Thus, each of these theories can also be given a set of quantities, in the way discussed in Sect. 5.2.2. For the non-chiral theories (c) and (e), this set of quantities is enlarged by the addition of non-chiral quantities, as we already discussed in detail in the abelian case in Sect. 5.5.1. Finally, the dynamics of these models is as discussed in Sect. 5.2.3.

In conclusion, this large class of examples, based on a general enveloping algebra of an affine Lie algebra, exemplifies our schema for duality in Sects. 2 and 3, just as the basic case did in Sect. 5.2. Namely: a duality is an isomorphism between models. More specifically: it is an isomorphism between model triples; since models also have their own specific structures. And the theory obtained for each of the dualities accords with what we said in Sect. 2.4: equivalence classes of isomorphic model triples give rise to a theory which is itself a triple, in which the models' specific structures have been abstracted away. And finally: our comments about non-isomorphic models (in Sects. 2.4, 5.3.1, and 5.4) are illustrated by the examples (c)–(e). For these models have less symmetry: the theory which then results is more general.

Envoi

In the physics and philosophy of physics literature, a duality in physics is agreed to be a matter of two theories being in some sense 'the same'. In this paper, we have answered the question how this can be made precise, and illustrated our answer with a case-study: bosonization.

We have proposed that a duality is best understood formally, i.e. in terms of uninterpreted theories: hence our term, 'schema'. Namely: there is a bare theory—the common core of the two given theories—which has various models, among which are the two given theories. The duality then consists in the fact that these two models are isomorphic as regards the structure and notions given in the bare theory. (Thus each of the two models also has specific structure of its own, which is unmatched by the other; and the bare theory also has, in general, other non-isomorphic models.)

Often, this isomorphism is a surprising fact, since the two given theories are presented in very different terms.

We spelt out this schema in detail, in Sect. 2 and 3. Among the themes we emphasized are: (i) the distinction between theories and models, (ii) the role of interpretation, (iii) the relations between a duality and the symmetries of the two given theories, and (iv) the presence of non-isomorphic models in physics.

Then in Sects. 4 and 5, we illustrated the schema with bosonization. This is a matter of a quantum field theory of bosons being in some sense 'the same' as a quantum field theory of fermions. Nowadays, many such boson-fermion pairings are known. Our discussion emphasized the simplest, and earliest, case, which is known to hold exactly: a duality between a free, massless bosonic quantum field theory, and a free, massless fermionic theory, both in two spacetime dimensions. But we ended with a brief overview of other examples: involving, in particular, interacting and massive theories. (And there are extensions to higher dimensions: see e.g. [39]; as well as an experimental interest in these systems as realising e.g. one-dimensional spin chains: Giamarchi [29, Chap. 2], Altland and Simons [3, Sects. 4.3 and 9.4.4].)

Our schema, and this illustration of it, of course leaves plenty of work still to be done. As to physics, one should seek other illustrations of the schema: maybe some of these will prompt revision, or at least augmentation, of the schema. As to philosophy, one should ask what light this schema casts on philosophical debates about the interpretation of physical theories, and about such theories' equivalence. But we postpone these topics to another occasion.

Acknowledgements We thank Joseph Kouneiher, not least for his patience! We also thank an anonymous referee for comments, the participants at the Symmetries and Asymmetries in Physics conference in Hannover, and especially Josh Hunt for comments. SDH thanks several audiences: the British Society for the Philosophy of Science 2016 annual conference, the Oxford philosophy of physics group, LSE's Sigma Club, the Munich Center for Mathematical Philosophy, and DICE2016. SDH's work was supported by the Tarner scholarship in Philosophy of Science and History of Ideas, held at Trinity College, Cambridge.

Appendix: Some Elements of Conformal Field Theory

In Sect. 4.1, we used the notion of a primary field. A primary field of conformal weight $(h, \bar{h})$ is defined to transform, under a conformal transformation (16), as follows:

$$\Phi(z, \bar{z}) \to \left(\frac{\partial f}{\partial z}\right)^h \left(\frac{\partial \bar{f}}{\partial \bar{z}}\right)^{\bar{h}} \Phi(f(z), \bar{f}(\bar{z})) \,. \tag{40}$$

This is in analogy with the transformation law for covariant tensors in ordinary QFTs: it takes the transformation property of the field under conformal transformations as defining for the class of primary fields. The physical significance of primary fields is discussed around Eq. (20).

In our analysis in Sects. 4 abd 5, an essential role was played by the enveloping Virasoro algebra (25), with $c = 1$ and $k = 1$. This algebra is a special case of the following general enveloping algebra of an affine Lie algebra:

$$\begin{aligned} [L_n, L_m] &= (n-m)\, L_{n+m} + \frac{c}{12}\, n\,(n^2-1)\, \delta_{n+m} \\ [L_n, J^a_m] &= -m\, J^a_{n+m} \\ [J^a_n, J^b_m] &= i\, f^{ab}_c\, J^c_{n+m} + k\, n\, \delta_{ab}\, \delta_{n+m}\,. \end{aligned} \tag{41}$$

Here, c is the central charge and k is the level, and ${f^{ab}}_c$ are the structure constants of the underlying Lie algebra of the affine Lie algebra. Notice that the above algebra contains, in the first line, the ordinary Virasoro algebra. And the last line is the affine Lie algebra. The middle line gives the commutation relation between generators of the two algebras.[45]

References

1. M. Ammon, J. Erdmenger, *Gauge/Gravity Duality: Foundations and Applications* (Cambridge University Press, Cambridge, 2015)
2. M.F. Atiyah, *Duality in Mathematics and Physics*, lecture delivered at the Institut de Matemàtica de la Universitat de Barcelona: http://www.imub.ub.es (2007)
3. A. Atland, B. Simons, *Condensed Matter Field Theory* (Cambridge University Press, Cambridge, 2010)
4. T.W. Barrett, H. Halvorson, Glymour and quine on theoretical equivalence. J. Philos. Logic **45**(5), 467–483 (2016). http://philsci-archive.pitt.edu/id/eprint/11341
5. T.W. Barrett, H. Halvorson, Morita equivalence. Rev. Symbolic Logic **9**(3), 556–582 (2016a). http://philsci-archive.pitt.edu/id/eprint/11511
6. R. Carnap, *Meaning and Necessity* (University of Chicago Press, Chicago, 1947)
7. E. Castellani, Duality and 'particle' democracy. Stud. Hist. Philos. Mod. Phys. **59**, 100–108 (2017)
8. A. Caulton, The role of symmetry in the interpretation of physical theories. Stud. Hist. Philos. Mod. Phys. **52**, 153–162 (2015)
9. S. Coleman, Quantum sine-Gordon equation as the massive Thirring model. Phys. Rev. D **11**(8), 2088–2097 (1975)
10. D. Corfield, Duality as a category-theoretic concept. Stud. Hist. Philos. Mod. Phys. **59**, 55–61 (2017)
11. L. Corry, Hilbert and physics 1900–1915, in *The Symbolic Universe: Geometry and Physics 1890–1930*, ed. by J. Gray (Oxford University Press, Oxford, 1999), pp. 145–188
12. L. Corry, *David Hilbert and the Axiomatisation of Physics* (Springer-Science, Berlin, 2004)
13. L. Corry, On the origins of Hilbert's sixth problem: physics and the empiricist approach to axiomatization, in *Proceedings of the ICM 2006* (Madrid, Spain, 2006), pp. 1697–1718
14. L. Corry, *Mie's Electromagnetic Theory of Matter and the Background to Hilbert's Unified Foundations of Physics*, this volume (2018)
15. S. De Haro, Spacetime and physical equivalence, in *Space and Time after Quantum Gravity*, ed. by N. Huggett, C. Wüthrich, to appear (2016)

[45] For more on affine Lie algebras, see Di Francesco et al. [24, Chap. 14] or Kac [37].

16. S. De Haro, *Duality and Physical Equivalence*. http://philsci-archive.pitt.edu/id/eprint/12279, the title of this preprint has changed (2016a)
17. S. De Haro, Dualities and emergent gravity: gauge/gravity duality. Stud. Hist. Philos. Mod. Phys. **59**(2017), 109–125 (2017)
18. S. De Haro, Invisibility of diffeomorphisms. Found. Phys. **47**(11), 1464–1497 (2017a)
19. S. De Haro, D. Mayerson, J.N. Butterfield, Conceptual aspects of gauge/gravity duality. Found. Phys. **46**(11), 1381–1425 (2016). https://doi.org/10.1007/s10701-016-0037-4
20. S. De Haro, N. Teh, J.N. Butterfield, Comparing dualities and gauge symmetries. Stud. Hist. Philos. Mod. Phys. **59**, 68–80 (2017)
21. G.F. Dell'Antonio, Y. Frishman, D. Zwanzinger, Thirring model in terms of currents: solution and light-cone expansions. Phys. Rev. D **6**(4), 988–1007 (1972)
22. N. Dewar, Sophistication about symmetries, forthcoming in British Journal for the Philosophy of Science; available at: http://philsci-archive.pitt.edu/id/eprint/12469 (2016)
23. D. Dieks, J. van Dongen, S. de Haro, Emergence in holographic scenarios for gravity. Stud. Hist. Philos. Mod. Phys. **52**(B), 203–216 (2015). https://doi.org/10.1016/j.shpsb.2015.07.007
24. P. Di Francesco, P. Mathieu, D. Sénéchal, *Conformal Field Theory* (Springer, New York, 1997)
25. D. Fraser, Formal and physical equivalence in two cases in contemporary quantum physics. Stud. Hist. Philos. Mod. Phys. **59**, 30–43 (2017). https://doi.org/10.1016/j.shpsb.2015.07.005
26. G. Frege, *The Foundations of Arithmetic*, trans. by J.L. Austin, (Oxford, Blackwell, 1884/1950)
27. G. Frege, Über Sinn und Bedeutung, *Zeitschrift für Philosophie und philosophische Kritik*, pp. 25–50; translated as On Sense and reference, ed. by P.T. Geach and M. Black (1960), Translations from the Philosophical Writings of Gottlob Frege, (Blackwell, Oxford, 1892)
28. Y. Frishman, J. Sonnenschein, *Non-Perturbative Field Theory* (Cambridge University Press, Cambridge, 2010)
29. T. Giamarchi, *Quantum Physics in One Dimension* (Oxford University Press, Oxford, 2003)
30. P. Ginsparg, Applied conformal field theory, in *Fields, Strings, Critical Phenomena: Proceedings*, ed. by E. Brezin, J. Zinn-Justin (North-Holland, Amsterdam, 1990). hep-th/9108028
31. A.O. Gogolin, A.A. Nersesyan, A.M. Tsvelik, *Bosonization and Strongly Correlated Systems* (Cambridge University Press, Cambridge, 2004)
32. I. Grattan-Guinness, A sideways look at Hilbert's twenty-three problems of 1900. Not. Am. Math. Soc. **47**, 752–757 (2000)
33. J. Gray, *The Hilbert Challenge* (Oxford University Press, Oxford, 2000)
34. J. Gray, *Plato's Ghost: The Modernist Transformation of Mathematics* (Princeton University Press, Princeton, 2008)
35. J. Gray, *Henri Poincare, a Scientific Biography* (Oxford University Press, Oxford, 2012)
36. N. Huggett, Target space $\neq$ space. Stud. Hist. Philos. Mod. Phys. **59**, 81–88 (2017). https://doi.org/10.1016/j.shpsb.2015.08.007
37. V. Kac, *Infinite-Dimensional Lie Algebras* (Cambridge University Press, Cambridge, 1990)
38. H. Kennedy, The origins of modern axiomatics: from Pasch to Peano. Am. Math. Mon. **79**, 133–136 (1972)
39. P. Kopietz, *Bosonization of Interacting Fermions in Arbitrary Dimensions* Lecture Notes in Physics Monographs, vol. 48 (Springer, Berlin, 2008)
40. J. Kouneiher, *Where We Stand Today: QFT, Dualities, Integrable Systems, and Supersymmetry*, this volume (2018)
41. D. Lewis, General semantics. Synthese **22**, 18–67 (1970); reprinted in his *Philosophical Papers*, vol. 1 (Oxford University Press, Oxford, 1983)
42. D. Lewis, How to define theoretical terms. J. Philos. **67**, 427–446 (1970a); reprinted in his *Philosophical Papers*, vol. 1 (Oxford University Press, Oxford, 1983)
43. D. Lewis, Index, context and content, in *Philosophy and Grammar*, ed. by S. Kanger, S. Ohman (D. Reidel, Dordrecht, 1980), pp. 79–100
44. D. Lüst, S. Theisen, *Lectures on String Theory*. Lecture Notes in Physics, vol. 346 (Springer, Berlin, 1989)
45. S. Mandelstam, Soliton operators for the quantized sine-Gordon equation. Phys. Rev. D **11**(10), 3026–3030 (1975)

46. K. Matsubara, Realism, underdetermination and string theory dualitites. Synthese **190**, 471–489 (2013)
47. H. Putnam, The analytic and the synthetic, in *Minnesota Studies in the Philosophy of Science*, vol. III, ed. by H. Feigl, G. Maxwell (University of Minnesota Press, Minneapolis, 1962), pp. 358–397
48. D. Rickles, Duality and the emergence of spacetime. Stud. Hist. Philos. Mod. Phys. **44**, 312–320 (2013)
49. D. Rickles, Dual theories: 'same but different' or different but same'? Stud. Hist. Philos. Mod. Phys. **59**, 62–67 (2017). https://doi.org/10.1016/j.shpsb.2015.09.005
50. L. Smolin, *What Are We Missing in Our Search for Quantum Gravity*, this volume (2018)
51. J. Stachel, J. Kouneiher, *Einstein and Hilbert*, this volume (2018)
52. M. Stöltzner, Opportunistic axiomatics Von Neumann on the methodology of mathematical physics, in *John Von Neumann and the Foundations of Quantum Physics* ed. by M. Redei, M. Stöltzner (Springer, Berlin, vol. 8 of the series Vienna Circle Institute Yearbook, 2000/2001), pp. 35–62
53. M. Stöltzner, How Metaphysical is "Deepening the Foundation"? Hahn and Frank on Hilbert's Axiomatic Method, in *History of Philosophy of Science*, ed. by M. Heidelberger, F. Stadler (Springer, Berlin, vol. 9 of the series Vienna Circle Institute Yearbook, 2001), pp. 245–262
54. N.J. Teh, Holography and emergence. Stud. Hist. Philos. Mod. Phys. **44**(3), 300–311 (2013)
55. N. Teh, Gravity and gauge. Br. J. Philos. Sci. **67**(2), 497–530 (2016)
56. J. von Neumann, *Mathematical Foundations of Quantum Mechanics*, English translation published in 1955 (Princeton University Press, Princeton, 1932)
57. J.O. Weatherall, Categories and the foundations of classical field theories. Forthcoming in *Categories for the Working Philosopher*, ed. by E. Landry (Oxford University Press, Oxford, 2015)
58. E. Witten, Nonabelian bosonization in two dimensions. Commun. Math. Phys. **92**, 455 (1984)
59. E. Witten, *What Every Physicist Should Know about String Theory*, this volume (2018)

The Dressing Field Method of Gauge Symmetry Reduction, a Review with Examples

J. Attard, J. François, S. Lazzarini and T. Masson

Abstract Gauge symmetries are a cornerstone of modern physics but they come with technical difficulties when it comes to quantization, to accurately describe particles phenomenology or to extract observables in general. These shortcomings must be met by essentially finding a way to effectively reduce gauge symmetries. We propose a review of a way to do so which we call the dressing field method. We show how the BRST algebra satisfied by gauge fields, encoding their gauge transformations, is modified. We outline noticeable applications of the method, such as the electroweak sector of the Standard Model and the local twistors of Penrose.

PACS numbers: 02.40.Hw · 11.15.-q · 11.25.Hf

1 Introduction

To this day, modern Field Theory framework (either classical or quantum), so successful in describing Nature from elementary particles to cosmology, rests on few keystones, one of which being the notion of gauge symmetry. Elementary fields are subject to local transformations which are required to leave invariant the theory (the Lagrangian). These transformations thus form a so-called local symmetry of the theory, known as gauge symmetry. This notion, originates with Weyl's 1918 unified theory resting on the idea of local *scale*, or *gauge*, invariance [49, 66, 67]. The heuristic appeal of gauge symmetries is that imposing them on a theory of free fields requires, a minima, the introduction of fundamental interactions through minimal

J. Attard · S. Lazzarini · T. Masson (✉)
Centre de Physique Théorique, Aix Marseille Univ, Université de Toulon,
CNRS, CPT, Marseille, France
e-mail: thierry.masson@cpt.univ-mrs.fr

J. François
Institut Élie Cartan de Lorraine, Université de Lorraine, UMR 7502,
Vandœuvre-lès-Nancy F-54506, France

J. Kouneiher (ed.), *Foundations of Mathematics and Physics One Century After Hilbert*,
https://doi.org/10.1007/978-3-319-64813-2_13

coupling. This is the content of the so-called *gauge principle* for Field Theory,[1] captured by Yang's well-known aphorism: "symmetry dictates interaction" [72].[2]

This is one of the major conceptual breakthrough of the century separating us from Hilbert's lectures on the foundations of mathematics and physics. And the story of the interactions between gauge theories and differential geometry is a highlight of the long history of synergy between mathematics and physics.[3]

In spite of their great theoretical appeal, gauge theories come with some shortcomings. Prima facie indeed, gauge symmetries forbid mass terms for (at least) the interaction fields, which was known to be in contradiction with the phenomenology of the nuclear interactions. Also, the quantization of gauge theories via Feynman's path integral has its specific problem because integrating on gauge equivalent fields configurations makes it ill-defined. Finally, it is in general not so straightforward to extract observables from a gauge theory since the physical content must be gauge invariant, e.g. the abelian (Maxwell-Faraday) field strength or Wilson loops. This issue is made acutely clear in General Relativity (GR), where observables must be diffeomorphic invariant. Addressing these shortcomings essentially boils down to finding a way to *reduce effectively gauge symmetries*, in part or completely. One can think of only a few ways to do so, among which we mention the three most prominent.

First, gauge fixing: one selects a representative in the gauge orbit of each gauge field. This is usually the main approach followed to make contact with physical predictions: one only needs to make sure that the results are independent of the choice of gauge. This is also how a sensible quantization procedure is carried on, for example through the Fadeev-Popov procedure. However, a consistent choice of gauge might not necessarily be possible in all circumstances, a fact known as the *Gribov ambiguity* [29, 59].

Second, one can try to implement a spontaneous symmetry breaking mechanism (SSBM). This is famously known to be the standard interpretation of the Brout-Englert-Higgs-Guralnik-Hagen-Kibble (BEHGHK) mechanism [10, 30, 31], which historically solved the masses problem for the weak gauge bosons in the electroweak unification of Glashow-Weinberg-Salam, and, by extension, of the masses of particles in the Standard Model of Particles Physics. We stress that this interpretation presupposes settled the philosophical problem of the ontological status of gauge symmetries: by affirming that a gauge symmetry can be "spontaneously broken", one states that it is a structural feature of reality rather than of our description of it. While this remains quite controversial in philosophy of physics, given the empirical success of the BEHGHK mechanism, a pragmatic mind could consider the debate closed via an inference to the best explanation. We will show here that this conclusion would be hasty.

---

[1] See [45] for a critical discussion of its scope and limits.

[2] Weyl topped this with an even stronger endorsement of the importance of symmetries in physics: "As far as I see, all a priori statements in physics have their origin in symmetry" [68].

[3] See the nice short appendix by S. S. Chern of the book on differential geometry he co-authored [12].

Finally, one can seek to apply the bundle reduction theorem. This is a result of the fiber bundle theory, widely known to be the geometric underpinning of gauge theories, stating the circumstances under which a bundle with a given structure group can be reduced to a subbundle with a smaller structure group. Some authors have recast the BEHGHK mechanism in light of this theorem [60, 63, 65].

In this paper we propose a brief review of another way to perform gauge symmetry reduction which we call the *dressing field method*. It is formalized in the language of the differential geometry of fiber bundles and it has a corresponding BRST differential algebraic formulation. The method boils down to the identification of a suitable field in the geometrical setting of a gauge theory that allows to construct partially of fully gauge invariant variables out of the standard gauge fields. This formalizes and unifies several works and approaches encountered in past and recent literature on gauge theories, whose ancestry can be traced back to Dirac's pioneering ideas [15, 16].

The paper is thus organized. In Sect. 2 we outline the method and state the most interesting results (pointing to the published literature for proofs), one of which being the noticeable fact that the method allows to highlight the existence of gauge fields of a non-standard kind; meaning that these implement the gauge principle but are not of the same geometric nature than the objects usually encountered in Yang-Mills theory for instance.

In Sects. 3 and 4 we illustrate the scheme by showing how it is applied to the electroweak sector of the standard model and to GR. We argue in particular that our treatment provides an alternative interpretation of the BEHGHK mechanism that is more in line with the conclusions of the community of philosophers of physics.

In Sect. 5 we address the substantial example of the conformal Cartan bundle $\mathcal{P}(\mathcal{M}, H)$ with connection ϖ. Standard formulations of the so-called Tractors and Twistors can then be found by applying the dressing field method to this geometry. Furthermore, from this viewpoint they appear to be clear instances of gauge fields of the non-standard kind alluded to above. This fact, as far as we know, has not been recognized.

In our conclusion, Sect. 6, we indicate other possible applications of the method and stress the obvious remaining open questions to be addressed.

2 Reduction of Gauge Symmetries: The Dressing Field Method

As we have stated, the differential geometry of fiber bundles supplemented by the BRST differential algebra are the mathematical underpinning of classical gauge theories. So, this is the language in which we will formalize our approach. Complementary material and detailed proofs can be found in [1, 20, 21, 23].

Let us give the main philosophy of the dressing field method in a few words. From a mathematical point of view, a gauge field theory requires some spaces of fields on which the gauge group acts in a definite way. So, to define a gauge field theory, the spaces of fields themselves are not sufficient: one has to specify the actions of the gauge group on them. This implies that the same mathematical space can be considered as a space of different fields, according to the possible actions of the gauge group.

Generally, the action is related to the way the space of fields is constructed. For instance, in the usual geometrical framework of gauge field theories, the primary structure is a principal fiber bundle $\mathcal{P}$, and the gauge group is its group of vertical automorphisms. Then, the sections of an associated vector bundle to $\mathcal{P}$, constructed using the action ρ of the structure group on a vector space V, support an action of the gauge group which is directly related to ρ.

The physical properties of a gauge field theory are generally encoded into a Lagrangian L written in terms of the gauge fields (and their derivatives): it is required to be invariant when the gauge group acts on all the fields involved in its writing.

The main idea behind the dressing field method is to exhibit a very special field (the dressing field) *out of the gauge fields in the theory*, with a specific gauge action. Then, one performs some change of field variables, very similar to some change of variables in ordinary geometry, by *combining* in a convenient way (through sums and products *when they make sense*) the gauge fields with the dressing field. The resulting "dressed fields" (new fields variables of the theory) are then subject to new actions of the gauge group, *that can be deduced from the combination of fields*. In favorable situations, these dressed fields are invariant under the action of a subgroup of the gauge group: a part of the gauge group does not act anymore on the new fields of the theory, that is, the symmetry has been reduced.

Notice some important facts. Firstly, the dressed fields do not necessarily belong to the original space of fields from which they are constructed. Secondly, in general the dressing (i.e. the combination of the dressing field with a field from the theory) looks like a gauge transformation. But we insist on the fact that it is not a genuine gauge transformation. Finally, the choice of the dressing field relies sometimes on the physical content of the theory, that is on the specific form of the Lagrangian. So, the dressing field method can depend on the mathematical, as well as on the physical content of the theory.

Les us now describe the mathematical principles of the method.

2.1 Composite Fields

Let $\mathcal{P}(\mathcal{M}, H)$ be a principal bundle over a manifold $\mathcal{M}$ equipped with a connection ω with curvature Ω, and let φ be a ρ-equivariant V-valued map on $\mathcal{P}$ to be considered as a section of the associated vector bundle $E = \mathcal{P} \times_H V$.

The group of vertical automorphisms of $\mathcal{P}$,

$$\mathrm{Aut}_v(\mathcal{P}) := \{\Phi : \mathcal{P} \to \mathcal{P} \mid \forall h \in H, \forall p \in \mathcal{P}, \Phi(ph) = \Phi(p)h \text{ and } \pi \circ \Phi = \Phi\}$$

is isomorphic to the gauge group $\mathcal{H} := \{\gamma : \mathcal{P} \to H \mid R_h^*\gamma(p) = h^{-1}\gamma(p)h\}$, the isomorphism being $\Phi(p) = p\gamma(p)$. The composition law of $\text{Aut}_v(\mathcal{P})$, $\Phi_1 \circ \Phi_2$, corresponds to the product $\gamma_1\gamma_2$.

In this geometrical settings, the gauge group $\mathcal{H} \simeq \text{Aut}_v(\mathcal{P})$ acts on fields via pullbacks. It acts on itself as $\eta^\gamma := \Phi^*\eta = \gamma^{-1}\eta\gamma$, and on connections ω, curvatures Ω and (ρ, V)-tensorial forms φ as,

$$\begin{aligned} \omega^\gamma &:= \Phi^*\omega = \gamma^{-1}\omega\gamma + \gamma^{-1}d\gamma, & \varphi^\gamma &:= \Phi^*\varphi = \rho(\gamma^{-1})\varphi, \\ \Omega^\gamma &:= \Phi^*\Omega = \gamma^{-1}\Omega\gamma, & (D\varphi)^\gamma &:= \Phi^*D\varphi = D^\gamma\varphi^\gamma = \rho(\gamma^{-1})D\varphi. \end{aligned} \tag{1}$$

These are *active* gauge transformations, formally identical but to be conceptually distinguished from *passive* gauge transformations relating two local descriptions of the same global objects in local trivializations of the fiber bundle described as follows. Given two local sections σ_1, σ_1 of $\mathcal{P}$, related as $\sigma_2 = \sigma_1 h$, either over the same open set $\mathcal{U}$ of $\mathcal{M}$ or over the overlap of two open sets $\mathcal{U}_1 \cap \mathcal{U}_2$, one finds

$$\begin{aligned} \sigma_2^*\omega &= h^{-1}\sigma_1^*\omega\, h + h^{-1}dh, & \sigma_2^*\varphi &= \rho(h^{-1})\sigma_1^*\varphi, \\ \sigma_2^*\Omega &= h^{-1}\sigma_1^*\Omega\, h, & \sigma_2^*D\varphi &= \rho(h^{-1})\sigma_1^*D\varphi. \end{aligned} \tag{2}$$

This distinction between active and passive gauge transformations is reminiscent of the distinction between diffeomorphisms and coordinates transformations in GR.

The main idea of the dressing field approach to gauge symmetry reduction is stated in the following

Proposition 1 [20] *Let K and G be subgroups of H such that $K \subseteq G \subset H$. Note $\mathcal{K} \subset \mathcal{H}$ the gauge subgroup associated with K. Suppose there exists a map*

$$u : \mathcal{P} \to G \quad \textit{satisfying the } K\textit{-equivariance property} \quad R_k^*u = k^{-1}u. \tag{3}$$

Then this map u, that will be called a dressing field, *allows to construct through $f : \mathcal{P} \to \mathcal{P}$ defined by $f(p) = pu(p)$, the following* composite fields

$$\omega^u := f^*\omega = u^{-1}\omega u + u^{-1}du, \qquad \varphi^u := f^*\varphi = \rho(u^{-1})\varphi. \tag{4}$$

which are $\mathcal{K}$-invariant and satisfy

$$\begin{aligned} \Omega^u &:= f^*\Omega = u^{-1}\Omega u = d\omega^u + \tfrac{1}{2}[\omega^u, \omega^u], \\ D^u\varphi^u &:= f^*D\varphi = \rho(u^{-1})D\varphi = d\varphi^u + \rho_*(\omega^u)\varphi^u. \end{aligned}$$

These composite fields are K-horizontal and thus project on the quotient $\mathcal{P}/K$.

The $\mathcal{K}$-invariance of the composite fields (4) is most readily proven. Indeed from the definition (3) one has $f(pk) = f(p)$ so that f factorizes through a map $\mathcal{P} \to \mathcal{P}/K$ and given $\Phi(p) = p\gamma(p)$ with $\gamma \in \mathcal{K} \subset \mathcal{H}$, one has $\Phi^* f^* = (f \circ \Phi)^* = f^*$.

The dressing field can be *equally defined* by its $\mathcal{K}$-gauge transformation: $u^\gamma = \gamma^{-1}u$, for any $\gamma \in \mathcal{K} \subset \mathcal{H}$. Indeed, given Φ associated to $\gamma \in \mathcal{K}$ and (3): $(u^\gamma)(p) := \Phi^* u(p) = u(\Phi(p)) = u(p\gamma(p)) = \gamma(p)^{-1}u(p) = (\gamma^{-1}u)(p)$.

Several comments are in order. First, (4) looks *algebraically* like (1): this makes easy to check algebraically that the composite fields are $\mathcal{K}$-invariant. Indeed, let $\chi \in \{\omega, \Omega, \varphi, \dots\}$ denote a generic field when performing an operation that applies equally well to any specific one. For two maps α, α' with values in H, if one defines χ^α *algebraically* as in (1), then one has the left action $(\chi^\alpha)^{\alpha'} = \chi^{\alpha\alpha'}$. This is for instance the usual way to compose the actions of two elements of the gauge group. But this relation is independent of the specific action of the gauge group on α and α', which could belong to different representation spaces of $\mathcal{H}$. Then $(\chi^u)^\gamma = (\chi^\gamma)^{u^\gamma} = (\chi^\gamma)^{\gamma^{-1}u} = \chi^u$, where the last (and essential) equality is the one emphasized above.

Second, if $K = H$, then the composite fields (4) are $\mathcal{H}$-invariant, the gauge symmetry is fully reduced, and they live on $\mathcal{P}/H \simeq \mathcal{M}$. This shows that the existence of a global dressing field is a strong constraint on the topology of the bundle $\mathcal{P}$: a K-dressing field means that the bundle is trivial along the K-subgroup, $\mathcal{P} \simeq \mathcal{P}/K \times K$, while a H-dressing field means its triviality, $\mathcal{P} \simeq \mathcal{M} \times H$ [20, Prop. 2].

Third, in the event that $G \supset H$, then one has to assume that the H-bundle $\mathcal{P}$ is a subbundle of a G-bundle, and *mutatis mutandis* the proposition still holds. Such a situation occurs for instance when $\mathcal{P}$ is a reduction of a frame bundle (of unspecified order), see Sect. 4 for an example.

Notice that despite the formal similarity with (1) [or (2)], the composite fields (4) are not gauge transformed fields. Indeed, the defining equivariance property (3) of the dressing field implies $u \notin \mathcal{H}$, and $f \notin \mathrm{Aut}_v(\mathcal{P})$. As a consequence, in general the composite fields do not belong to the gauge orbits of the original fields: $\chi^u \notin \mathcal{O}(\chi)$. Another consequence is that the dressing field method must not be confused with a simple gauge fixing.

2.2 Residual Gauge Symmetry

Suppose there is a normal subgroup K and a subgroup J of H such that any $h \in H$ can be uniquely written as $h = jk$ for $j \in J$ and $k \in K$. Then $H = JK$ and $J \simeq H/K$, whose Lie algebra is denoted by $\mathfrak{j}$. Such a situation occurs for instance with $H = J \times K$. Several examples are based on this structure, see for instance Sects. 3 and 5.

The quotient bundle $\mathcal{P}/K$ is then a J-principal bundle $\mathcal{P}' = \mathcal{P}'(\mathcal{M}, J)$, with gauge group $\mathcal{J} \simeq \mathrm{Aut}_v(\mathcal{P}')$. The residual gauge symmetry of the composite fields depends, on the one hand, on that of the gauge fields, and, on the other hand, on that of the dressing field. A classification of the numerous possible situations is impractical, but below we provide the general treatment of two most interesting cases.

2.2.1 The Composite Fields as Genuine Gauge Fields

With the previous decomposition of H, our first case is summarized in this next result.

Proposition 2 *Let u be a K-dressing field on $\mathcal{P}$. Suppose its J-equivariance is given by*

$$R_j^* u = Ad_{j^{-1}} u, \quad \text{for any } j \in J. \tag{5}$$

Then, the dressed connection ω^u is a J-principal connection on $\mathcal{P}'$. That is, for $X \in \mathfrak{j}$ and $j \in J$, ω^u satisfies: $\omega^u(X^v) = X$ and $R_j^*\omega^u = Ad_{j^{-1}}\omega^u$. Its curvature is given by Ω^u. Also, φ^u is a (ρ, V)-tensorial map on $\mathcal{P}'$ and can be seen as a section of the associated bundle $E' = \mathcal{P}' \times_J V$. The covariant derivative on such sections is given by $D^u = d + \rho(\omega^u)$.

From this we immediately deduce the following

Corollary 3 *The transformation of the composite fields under the residual $\mathcal{J}$-gauge symmetry is found in the usual way to be*

$$\begin{aligned}
(\omega^u)^{\gamma'} &:= \Phi'^*\omega^u = \gamma'^{-1}\omega^u\gamma' + \gamma'^{-1}d\gamma', & (\varphi^u)^{\gamma'} &:= \Phi'^*\varphi^u = \rho(\gamma'^{-1})\varphi^u, \\
(\Omega^u)^{\gamma'} &:= \Phi'^*\Omega^u = \gamma'^{-1}\Omega^u\gamma', & (D^u\varphi^u)^{\gamma'} &:= \Phi'^* D^u\varphi^u = \rho(\gamma'^{-1})D^u\varphi^u,
\end{aligned} \tag{6}$$

with $\Phi' \in \text{Aut}(\mathcal{P}') \simeq \mathcal{J} \ni \gamma'$.

A quick way to convince oneself of this is to observe that for $\Phi' \in \text{Aut}_v(\mathcal{P}')$ one has, using (5), $\left(u^{\gamma'}\right)(p) := (\Phi'^*u)(p) = u(\Phi'(p)) = u(p\gamma'(p)) = \gamma'(p)^{-1} u(p)\gamma'(p) = (\gamma'^{-1}u\gamma')(p)$. So, using again the generic variable χ one finds that $(\chi^u)^{\gamma'} = (\chi^{\gamma'})^{u^{\gamma'}} = (\chi^{\gamma'})^{\gamma'^{-1}u\gamma'} = \chi^{u\gamma'}$, which proves (6). In field theory, the relation $u^{\gamma'} = \gamma'^{-1}u\gamma'$ can be preferred to (5) as a condition on the dressing field u.

The above results show that when (5) holds, the composite fields (4) are $\mathcal{K}$-invariant but genuine $\mathcal{J}$-gauge fields with residual gauge transformations given by (6). It may then be possible to perform a further dressing operation provided a suitable dressing field exists and satisfies the compatibility condition of being invariant under the $\mathcal{K}$-gauge subgroup just erased. The extension of this scheme to any number of dressing fields can be found in [21]. Let us now turn to our next interesting case.

2.2.2 The Composite Fields as Twisted-Gauge Fields

To define these gauge fields with a new behavior under the action of the gauge group, we need to introduce some definitions. Let $G' \supset G$ be a Lie group for which the representation (ρ, V) of G is also a representation of G'. Let us assume the existence of a C^∞-map $C : \mathcal{P} \times J \to G'$, $(p, j) \mapsto C_p(j)$, satisfying

$$C_p(jj') = C_p(j)C_{pj}(j'). \tag{7}$$

From this we have that $C_p(e) = e$, with e the identity in both J and G', and $C_p(j)^{-1} = C_{pj}(j^{-1})$. Its differential is

$$dC_{|(p,j)} = dC(j)_{|p} + dC_{p|j} : T_p\mathcal{P} \oplus T_jJ \to T_{C_p(j)}G',$$

where $\ker dC(j) = T_jJ$ and $\ker dC_p = T_p\mathcal{P}$, and where $dC(j)$ (resp. dC_p) uses the differential on $\mathcal{P}$ (resp. J). Notice that $C_p(j)^{-1}dC_{|(p,j)} : T_p\mathcal{P} \oplus T_jJ \to T_eG' = \mathfrak{g}'$. We then state the following result.

Proposition 4 *Let u be a K-dressing field on $\mathcal{P}$. Suppose its J-equivariance is given by*

$$(R_j^*u)(p) = j^{-1}u(p)C_p(j), \quad \text{with } j \in J \text{ and } C \text{ a map as above}. \tag{8}$$

Then ω^u satisfies

1. $\omega_p^u(X_p^v) = c_p(X) := \frac{d}{dt}C_p(e^{tX})|_{t=0}$, *for $X \in \mathfrak{j}$ and $X_p^v \in V_p\mathcal{P}'$.*
2. $R_j^*\omega^u = C(j)^{-1}\omega^u C(j) + C(j)^{-1}dC(j)$.

The dressed curvature Ω^u is J-horizontal and satisfies $R_j^*\Omega^u = C(j)^{-1}\Omega^u C(j)$. Also, φ^u is a $\rho(C)$-equivariant map, $R_j^*\varphi^u = \rho\left(C(j)\right)^{-1}\varphi^u$. The first order differential operator $D^u := d + \rho_*(\omega^u)$ is a natural covariant derivative on such φ^u so that $D^u\varphi^u$ is a $(\rho(C), V)$-tensorial form: $R_j^*D^u\varphi^u = \rho\left(C(j)\right)^{-1}D^u\varphi^u$ and $(D^u\varphi^u)_p(X_p^v) = 0$.

This proposition shows that ω^u behaves "almost as a connection": we call it a C-twisted connection 1-form. There is a natural geometric structure to interpret the dressed field φ^u. Omitting the representation ρ of G' on V to simplify notations, we can define the following equivalence relation on $\mathcal{P} \times V$:

$$(p, v) \sim (pj, C_p(j)^{-1}v) \text{ for any } p \in \mathcal{P}, v \in V, \text{ and } j \in J.$$

Using the properties of the map C, it is easy to show that this is indeed an equivalence relation. In particular, one has $(pjj', C_p(jj')^{-1}v) \sim (pjj', C_{pj}(j')^{-1}C_p(j)^{-1}v) \sim (pj, C_p(j)^{-1}v) \sim (p, v)$. Then one can define the quotient vector bundle over $\mathcal{M}$

$$E = \mathcal{P} \times_{C(J)} V := (\mathcal{P} \times V)/\sim \tag{9}$$

that we call a $C(J)$-twisted associated vector bundle to $\mathcal{P}$. Notice that when $J = \{e\}$, one has $E = \mathcal{P} \times V$. Adapting standard arguments in fiber bundle theory, one can show that sections of E are $C(J)$-equivariant maps

$$\varphi : \mathcal{P} \to V \text{ such that } \varphi(pj) = C_p(j)^{-1}\varphi(p) \text{ for any } p \in \mathcal{P},\ \mathrm{j} \in J.$$

The dressing field φ^u is then a section of E satisfying $\varphi^u(pk) = \varphi^u(p)$ for any $p \in \mathcal{P}$ and $k \in K$ by construction.

We can now deduce the transformations of the composite fields under the residual gauge group $\mathcal{J}$. Consider $\Phi \in \mathrm{Aut}_v(\mathcal{P}') \simeq \mathcal{J} \ni \gamma$, where $\gamma : \mathcal{P} \to J$ satisfies $\gamma(pk) = \gamma(p)$ and $\gamma(pj) = j^{-1}\gamma(p)j$ for any $p \in \mathcal{P}, k \in K$ and $j \in J$, and define the map $C(\gamma) : \mathcal{P} \to G'$, $p \mapsto C_p(\gamma(p))$, given by the compositions

$$\begin{array}{ccccccc} \mathcal{P} & \xrightarrow{\Delta} & \mathcal{P} \times \mathcal{P} & \xrightarrow{\mathrm{id}\times\gamma} & \mathcal{P} \times J & \xrightarrow{C} & G' \\ p & \longmapsto & (p, p) & \longmapsto & (p, \gamma(p)) & \longmapsto & C_p(\gamma(p)) \end{array}$$

Its differential $dC(\gamma)_{|p} : T_p\mathcal{P} \to T_{C_p(\gamma(p))}G'$ is given by $dC(\gamma) = dC \circ (\mathrm{id} \oplus d\gamma) \circ d\Delta$, and we have $C_p(\gamma(p))^{-1} dC(\gamma)_{|p} : T_p\mathcal{P} \to T_eG' = \mathfrak{g}'$. The residual gauge transformation of the dressing field is then $\big(u^\gamma\big)(p) := (\Phi^*u)(p) = u(p\gamma(p)) = \gamma(p)^{-1}u(p)C_p\big(\gamma(p)\big) = \big(\gamma^{-1}uC(\gamma)\big)(p)$, that is

$$u^\gamma = \gamma^{-1}uC(\gamma). \tag{10}$$

This relation can be taken as an alternative to (8) as a condition on the dressing field u. We have then the following proposition.

Proposition 5 *Given $\Phi \in \mathrm{Aut}_v(\mathcal{P}') \simeq \mathcal{J} \ni \gamma$, the residual gauge transformations of the composite fields are*

$$\begin{aligned} (\omega^u)^\gamma &:= \Phi^*\omega^u = C(\gamma)^{-1}\omega^u C(\gamma) + C(\gamma)^{-1}dC(\gamma), \\ (\varphi^u)^\gamma &:= \Phi^*\varphi^u = \rho\left(C(\gamma)^{-1}\right)\varphi^u, \\ (\Omega^u)^\gamma &:= \Phi^*\Omega^u = C(\gamma)^{-1}\Omega^u C(\gamma), \\ (D^u\varphi^u)^\gamma &:= \Phi^* D^u\varphi^u = \rho\left(C(\gamma)^{-1}\right) D^u\varphi^u. \end{aligned} \tag{11}$$

This shows that the composite fields (4) behave as *gauge fields of a new kind*, on which the implementation of the *gauge principle* is factorized through the map C. Given (10) and the usual $\mathcal{J}$-gauge transformations for the standard gauge fields χ, the above results can be obtained by a direct algebraic calculation: $(\chi^u)^\gamma = (\chi^\gamma)^{u^\gamma} = (\chi^\gamma)^{\gamma^{-1}uC(\gamma)} = \chi^{uC(\gamma)}$.

Under a further gauge transformation $\Psi \in \mathrm{Aut}_v(\mathcal{P}') \simeq \eta \in \mathcal{J}$, there are two ways to compute the composition $\Psi^*(\Phi^*u)$ of the two actions: first we use the composition inside the gauge group, $(\Phi \circ \Psi)(p) = p\gamma(p)\eta(p)$, so that $\left(\Psi^*(\Phi^*u)\right)(p) = \left((\Phi \circ \Psi)^*u\right)(p) = u\left(p\gamma(p)\eta(p)\right) = \eta(p)^{-1}\gamma(p)^{-1}u(p)C_p\big(\gamma(p)\eta(p)\big)$; secondly, we compute the actions successively,

$$\begin{aligned} \left(\Psi^*(\Phi^*u)\right)(p) &= \left(\gamma^{-1}uC(\gamma)\right)(\Psi(p)) = \gamma\left(p\eta(p)\right)^{-1} u\left(p\eta(p)\right) C_{p\eta(p)}\left(\gamma(p\eta(p)\right) \\ &= \eta(p)^{-1}\gamma(p)^{-1}\eta(p) \cdot \eta(p)^{-1}u(p)C_p\left(\eta(p)\right) \cdot C_{p\eta(p)}\left(\eta(p)^{-1}\gamma(p)\eta(p)\right) \\ &= \eta(p)^{-1}\gamma(p)^{-1}u(p)C_p\left(\gamma(p)\eta(p)\right). \end{aligned}$$

In both cases, $\Psi^*(\Phi^* u) = \eta^{-1}\gamma^{-1}\, u\, C\,(\gamma\eta)$, which secures the fact that the actions (11) of the residual gauge symmetry on the composite fields are well behaved as representations of the residual gauge group, *even if C is not a group morphism.*

Ordinary connections correspond to $C_p(j) = j$ for any $p \in \mathcal{P}'$ and $j \in J$, in which case, it is a group morphism.

The case of 1-α-cocycles. For a $p \in \mathcal{P}'$, suppose given $C_p : J \to G'$ satisfying $C_p(jj') = C_p(j)\ \alpha_j[C_p(j')]$ for $\alpha : J \to \text{Aut}(G')$ a continuous group morphism. Such an object appears in the representation theory of crossed products of C^*-algebras and is known as a *1-α-cocycle* (see [50, 69]).[4] Then, defining $C_{pj}(j') := \alpha_j[C_p(j')]$, one has an example of (7), and the above result applies to the 1-α-cocycle C. As a particular case, consider the following

Proposition 6 *Suppose J is abelian and let A_p, $B : J \to GL_n$ be group morphisms where $R_j^* A_p(j') = B(j)^{-1} A_p(j') B(j)$. Then $C_p := A_p B : J \to GL_n$ is a 1-α-cocyle with $\alpha : J \to \text{Aut}(GL_n)$ defined by $\alpha_j[g] = B(j)^{-1} g B(j)$ for any $g \in GL_n$.*

Using the commutativity of J through $B(j)B(j') = B(jj') = B(j'j) = B(j')B(j)$, the proposition is proven as $C_p(jj') = A_p(jj')B(jj') = A_p(j)A_p(j')B(j)B(j') = A_p(j)B(j)\ B(j)^{-1}[A_p(j')B(j')]B(j) = C_p(j)\ B(j)^{-1}[C_p(j')]B(j)$. Notice also that we have $C_p(jj') = C_p(j'j) = C_p(j')\ B(j')^{-1}[C_p(j)]B(j')$. Such 1-$\alpha$-cocycles will appear in the case of the conformal Cartan geometry and the associated Tractors and Twistors in Sect. 5.

2.3 *Application to the BRST Framework*

2.3.1 The BRST Differential Algebra

The BRST differential algebra captures the infinitesimal version of (1). Abstractly (see for instance [17]) it is a bigraded differential algebra generated by $\{\omega, \Omega, v, \zeta\}$ where v is the so-called ghost and the generators are respectively of degrees $(1, 0)$, $(2, 0)$, $(0, 1)$ and $(1, 1)$. It is endowed with two nilpotent antiderivations d and s, homogeneous of degrees $(1, 0)$ and $(0, 1)$ respectively, with vanishing anticommutator: $d^2 = 0 = s^2$, $sd + ds = 0$. The algebra is equipped with a bigraded commutator $[\alpha, \beta] := \alpha\beta - (-)^{\deg[\alpha]\deg[\beta]}\beta\alpha$. The action of d is defined on the generators by: $d\omega = \Omega - \frac{1}{2}[\omega, \omega]$ (Cartan structure equation), $d\Omega = [\Omega, \omega]$ (Bianchi identity), $dv = \zeta$ and $d\zeta = 0$. The action of the BRST operator on the generators gives the usual defining relations of the BRST algebra,

$$s\omega = -dv - [\omega, v], \quad s\Omega = [\Omega, v], \quad \text{and} \quad sv = -\tfrac{1}{2}[v, v]. \tag{12}$$

[4] In the general theory the group G' is replaced by a C^*-algebra A.

When the abstract BRST algebra is realized in a differential geometrical framework, the bigrading is according to the de Rham form degree and ghost degree: d is the de Rham differential on $\mathcal{P}$ (or $\mathcal{M}$ if one works in local trivializations) and s is the de Rham differential on $\mathcal{H}$. The ghost is the Maurer-Cartan form on $\mathcal{H}$ so that $v \in \bigwedge^1(\mathcal{H}, \text{Lie}\mathcal{H})$, and given $\xi \in T\mathcal{H}$, $v(\xi) : \mathcal{P} \to \mathfrak{h} \in \text{Lie}\mathcal{H}$ [7]. So in practice the ghost can be seen as a map $v : \mathcal{P} \to \mathfrak{h} \in \text{Lie}\mathcal{H}$, a placeholder that takes over the role of the infinitesimal gauge parameter. Thus the first two relations of (12) (and (13) below) reproduce the infinitesimal gauge transformations of the gauge fields (1), while the third equation in (12) is the Maurer-Cartan structure equation for the gauge group $\mathcal{H}$. The BRST transformations of the section φ (of degrees $(0, 0)$) and its covariant derivative are

$$s\varphi = -\rho_*(v)\varphi, \quad \text{and} \quad sD\varphi = -\rho_*(v)D\varphi. \tag{13}$$

where ρ_* is the representation of the Lie algebra induced by the representation ρ of the group.

The BRST framework provides an algebraic characterization of relevant quantities in gauge theories, such as admissible Lagrangian forms, observables and anomalies, all of which are required to belong to the s-cohomology group $H^{*,*}(s)$ of s-closed but not s-exact quantities.

2.3.2 Modified BRST Differential Algebra

Since the BRST algebra encodes the infinitesimal gauge transformations of the gauge fields, it is expected that the dressing field method modifies the former. To see how, let us first consider the following

Proposition 7 *Consider the BRST algebra* (12)–(13) *on the initial gauge variables and the ghost $v \in Lie\mathcal{H}$. Introducing the* dressed ghost

$$v^u = u^{-1}vu + u^{-1}su, \tag{14}$$

the composite fields (4) *satisfy the* modified BRST algebra*:*

$$s\omega^u = -D^u v^u = -dv^u - [\omega^u, v^u], \qquad s\varphi^u = -\rho_*(v^u)\varphi^u,$$
$$s\Omega^u = [\Omega^u, v^u], \qquad sv^u = -\tfrac{1}{2}[v^u, v^u].$$

This result does not rest on the assumption that u is a dressing field.

The result is easily found by expressing the initial gauge variable $\chi = \{\omega, \Omega, \varphi\}$ in terms of the dressed fields χ^u and the dressing field u, and re-injecting in the initial BRST algebra (12)–(13). At no point of the derivation does su need to be explicitly known. It then holds regardless if u is a dressing field or not.

If the ghost v encodes the infinitesimal initial $\mathcal{H}$-gauge symmetry, the dressed ghost v^u encodes the infinitesimal residual gauge symmetry. Its concrete expression depends on the BRST transformation of u.

Under the hypothesis $K \subset H$, the ghost decomposes as $v = v_\mathfrak{k} + v_{\mathfrak{h}/\mathfrak{k}}$, and the BRST operator splits accordingly: $s = s_\mathfrak{k} + s_{\mathfrak{h}/\mathfrak{k}}$. If u is a dressing field its BRST transformation is the infinitesimal version of its defining transformation property: $s_\mathfrak{k} u = -v_\mathfrak{k} u$. So the dressed ghost is

$$\begin{aligned} v^u &= u^{-1} v u + u^{-1} s u = u^{-1}(v_\mathfrak{k} + v_{\mathfrak{h}/\mathfrak{k}})u + u^{-1}(-v_\mathfrak{k} u + s_{\mathfrak{h}/\mathfrak{k}} u) \\ &= u^{-1} v_{\mathfrak{h}/\mathfrak{k}} u + u^{-1} s_{\mathfrak{h}/\mathfrak{k}} u. \end{aligned}$$

The Lie$\mathcal{K}$ part of the ghost, $v_\mathfrak{k}$, has disappeared. This means that $s_\mathfrak{k} \chi^u = 0$, which expresses the $\mathcal{K}$-invariance of the composite fields (4).

Residual BRST symmetry If $K \subset H$ is a normal subgroup, then $H/K = J$ is a group with Lie algebra $\mathfrak{h}/\mathfrak{k} = \mathfrak{j}$. We here provide the BRST treatment of the two cases detailed in Sect. 2.2.

Suppose the dressing field satisfies the condition (5), whose BRST version is $s_\mathfrak{j} u = [u, v_\mathfrak{j}]$. The dressed ghost is then

$$v^u = u^{-1} v_\mathfrak{j} u + u^{-1} s_\mathfrak{j} u = u^{-1} v_\mathfrak{j} u + u^{-1}(u v_\mathfrak{j} - v_\mathfrak{j} u) = v_\mathfrak{j}. \tag{15}$$

This in turn implies that the new BRST algebra is

$$\begin{aligned} s\omega^u &= -D^u v_\mathfrak{j} = -dv_\mathfrak{j} - [\omega^u, v_\mathfrak{j}], & s\varphi^u &= -\rho_*(v_\mathfrak{j})\varphi^u, \\ s\Omega^u &= [\Omega^u, v_\mathfrak{j}], & s v_\mathfrak{j} &= -\tfrac{1}{2}[v_\mathfrak{j}, v_\mathfrak{j}]. \end{aligned} \tag{16}$$

This is the BRST version of (6), and reflects the fact that the composite fields (4) are genuine $\mathcal{J}$-gauge fields, in particular that ω^u is a J-connection.

Suppose now that the dressing field satisfies the condition (8), whose BRST version is $s_\mathfrak{j} u = -v_\mathfrak{j} u + u c_p(v_\mathfrak{j})$. The dressed ghost is then

$$v^u = u^{-1} v_\mathfrak{j} u + u^{-1} s_\mathfrak{j} u = u^{-1} v_\mathfrak{j} u + u^{-1}\left(-v_\mathfrak{j} u + u c_p(v_\mathfrak{j})\right) = c_p(v_\mathfrak{j}). \tag{17}$$

This in turn implies that the new BRST algebra is

$$\begin{aligned} s\omega^u &= -dc_p(v_\mathfrak{j}) - [\omega^u, c_p(v_\mathfrak{j})], & s\varphi^u &= -\rho_*(c_p(v_\mathfrak{j}))\varphi^u, \\ s\Omega^u &= [\Omega^u, c_p(v_\mathfrak{j})], & sc_p(v_\mathfrak{j}) &= -\tfrac{1}{2}[c_p(v_\mathfrak{j}), c_p(v_\mathfrak{j})]. \end{aligned} \tag{18}$$

This is the BRST version of (11), and reflects the fact that the composite fields (4) instantiate the gauge principle in a satisfactory way.

To conclude we mention that the dressing operation is compatible with Stora's method of altering a BRST algebra so that it describes the action of infinitesimal diffeomorphisms of the base manifold on the gauge fields, in addition to their gauge transformations, as described in [36, 61] for instance: details can be found in [24].

2.4 Local Construction and Physics

Until now, we have been focused on the global aspects of the dressing approach on the bundle $\mathcal{P}$ to emphasize the geometric nature of the composite fields obtained. Most notably we showed that the composite field can behave as "generalized" gauge fields. But to do physics we need the local representatives on an open subset $\mathcal{U} \subset \mathcal{M}$ of global dressing and composite fields. These are obtained in the usual way from a local section $\sigma : \mathcal{U} \to \mathcal{P}$ of the bundle. The important properties they thus retain are their gauge invariance and their residual gauge transformations.

If it happens that a dressing field is defined locally on $\mathcal{U}$ first, and not directly on $\mathcal{P}$, then the local composite fields χ^u are defined in terms of the local dressing field u and local gauge fields χ by (4). The gauge invariance and residual gauge transformations of these local composite fields are derived from the gauge transformations of the local dressing field under the various subgroups of the local gauge group $\mathcal{H}_{\text{loc}}$ according to $(\chi^u)^\gamma = (\chi^\gamma)^{u^\gamma}$. The BRST treatment for the local objects mirrors exactly the one given for global objects.

This being said, note $A = \sigma^*\omega$ and $F = \sigma^*\Omega$ for definiteness but keep u and φ to denote the local dressing field and sections of the associated vector bundle E. Suppose that the base manifold is equipped with a (r, s)-Lorentzian metric allowing for a Hodge star operator, and that V is equipped with an inner product $\langle\, ,\, \rangle$. We state the final proposition dealing with gauge theory.

Proposition 8 *Given the geometry defined by a bundle $\mathcal{P}(\mathcal{M}, H)$ endowed with ω and the associated bundle E, suppose we have a gauge theory given by the prototypical $\mathcal{H}_{loc}$-invariant Yang-Mills Lagrangian*

$$L(A, \varphi) = \tfrac{1}{2}\,\text{Tr}(F \wedge *F) + \langle D\varphi,\ *D\varphi\rangle - U(\|\varphi\|)\,\text{vol},$$

where vol *is the metric volume form on $\mathcal{M}$, $\|\varphi\| := |\langle\varphi, \varphi\rangle|^{1/2}$ and U is a potential term.*[5] *If there is a local dressing field $u : \mathcal{U} \to G \subset H$ with $\mathcal{K}_{loc}$-gauge transformation $u^\gamma = \gamma^{-1}u$, then the above Lagrangian is actually a "$\mathcal{H}_{loc}/\mathcal{K}_{loc}$-gauge theory" defined in terms of $\mathcal{K}_{loc}$-invariant variables since we have*

$$L(A, \varphi) = L(A^u, \varphi^u) = \tfrac{1}{2}\,\text{Tr}(F^u \wedge *F^u) + \langle D^u\varphi^u,\ *D^u\varphi^u\rangle - U(\|\varphi^u\|)\,\text{vol}\,.$$

The relation $L(A, \varphi) = L(A^u, \varphi^u)$ is satisfied since, as already noticed, relations (4) look algebraically like gauge transformations (1) under which L is supposed to be invariant in a formal way.

---

[5]For instance, such a term is the one for a spontaneous symmetry breaking mechanism.

The terminology "$\mathcal{H}_{\text{loc}}/\mathcal{K}_{\text{loc}}$-gauge theory" means that the Lagrangian is written in terms of fields which are invariant under the action of $\gamma : \mathcal{U} \to K$. Since the quotient H/K needs not be a group, the remaining symmetries of the fields might not be described in terms of a group action.

Notice that since u is a dressing field, $u \notin \mathcal{H}_{\text{loc}}$, so the dressed Lagrangian $L(A^u, \varphi^u)$ ought not to be confused with a gauge-fixed Lagrangian $L(A^\gamma, \varphi^\gamma)$ for some chosen $\gamma \in \mathcal{H}_{\text{loc}}$, even if it may happen that $\gamma = u$ as fields if one forgets about the corresponding representations of the gauge group, a fact that might go unnoticed. As we have stressed in Sect. 2, the dressing field approach is distinct from both gauge-fixing and spontaneous symmetry breaking as a means to reduce gauge symmetries.

Let us highlight the fact that a dressing field can often be constructed by requiring the gauge invariance of a prescribed "gauge-like condition". Such a condition is given when a local gauge field χ (often the gauge potential) transformed by a field u with value in the symmetry group H, or one of its subgroups, is required to satisfy a functional constraint: $\Sigma(\chi^u) = 0$. Explicitly solved, this makes u a function of χ, $u(\chi)$, thus sometimes called *field dependent gauge transformation*. However this terminology is valid if and only if $u(\chi)$ transforms under the action of $\gamma \in \mathcal{H}_{\text{loc}}$ as $u(\chi)^\gamma := u(\chi^\gamma) = \gamma^{-1}u(\chi)\gamma$, in which case $u(\chi) \in \mathcal{H}_{\text{loc}}$. But if the functional constraint still holds under the action of $\mathcal{H}_{\text{loc}}$, or of a subgroup thereof, it follows that $(\chi^\gamma)^{u^\gamma} = \chi^u$ (or equivalently that $s\chi^u = 0$). This in turn suggests that $u^\gamma = \gamma^{-1}u$ (or $su = -vu$) so that $u \notin \mathcal{H}_{\text{loc}}$ but is indeed a dressing field.

This, and the above proposition, generalize the pioneering idea of Dirac [15, 16] aiming at quantizing QED by rewriting the classical theory in terms of gauge-invariant variables. The idea was rediscovered several times and sometimes termed *Dirac variables* [37, 54]. They reappeared in various contexts in gauge theory, such as QED [38], quarks theory in QCD [40], the proton spin decomposition controversy [22, 42, 43]. The dressing field approach thus gives a unifying and clarifying framework for these works, and others concerning the BRST treatment of anomalies in QFT [28, 44], Polyakov's "partial gauge fixing" for $2D$-quantum gravity [41, 55] or the construction of the Wezz-Zumino functionnal [3].

In the following we provide examples of significant applications of the dressing field approach in various contexts: the electroweak sector of the Standard Model, the tetrad *vs* metric formulation of GR, and tractors and twistors obtained from conformal Cartan geometry.

3 The Electroweak Sector of the Standard Model

The aim of the electroweak model is to give a gauge theoretic account of the fact that there is one long range interaction mediated by a massless boson, electromagnetism, together with a short range interaction mediated by massive bosons, the weak interaction. Here we discard the spinors (matter fields) of the theory and consider only the theory describing the gauge potentials and the scalar field. The spinors could be treated along the lines of the following exposition. More details can be found in [21, 46].

3.1 Reduction of the $\mathcal{SU}(2)$-symmetry via Dressing

The principal bundle of the model is $\mathcal{P}(\mathcal{M}, U(1) \times SU(2))$ and it is endowed with a connection whose local representative is $A = a + b$. Its curvature is $F = f_a + g_b$. The defining representation of the structure group is $(\mathbb{C}^2, \ell)$, with ℓ the left matrix multiplication. The associated vector bundle is $E = \mathcal{P} \times_\ell \mathbb{C}^2$ and we denote by $\varphi : \mathcal{U} \subset \mathcal{M} \to \mathbb{C}^2$ a (local) section. The covariant derivative is $D\varphi = d\varphi + (g'a + gb)\varphi$, with g', g the coupling constants of $U(1)$ and $SU(2)$ respectively. The action of the gauge group $\mathcal{H} = \mathcal{U}(1) \times \mathcal{SU}(2)$ (we drop the subscript "loc" from now on) is,

$$
\begin{aligned}
a^\alpha &= a + \tfrac{1}{g'}\alpha^{-1}d\alpha, & b^\alpha &= b, & \varphi^\alpha &= \alpha^{-1}\varphi,\\
a^\beta &= a, & b^\beta &= \beta^{-1}b\beta + \tfrac{1}{g}\beta^{-1}d\beta, & \varphi^\beta &= \beta^{-1}\varphi,
\end{aligned}
$$

where $\alpha \in \mathcal{U}(1)$ and $\beta \in \mathcal{SU}(2)$. The structure of direct product group is clear. The $\mathcal{H}$-invariant Lagrangian form of the theory is,

$$
\begin{aligned}
L(a, b, \varphi) &= \tfrac{1}{2}\operatorname{Tr}(F \wedge *F) + \langle D\varphi,\ *D\varphi\rangle - U(\|\varphi\|)\,\mathrm{vol},\\
&= \tfrac{1}{2}\operatorname{Tr}(f_a \wedge *f_a) + \tfrac{1}{2}\operatorname{Tr}(g_b \wedge *g_b)\\
&\qquad + \langle D\varphi,\ *D\varphi\rangle - \big(\mu^2\langle\varphi, \varphi\rangle + \lambda\langle\varphi, \varphi\rangle^2\big)\,\mathrm{vol},
\end{aligned}
\tag{19}
$$

where $\mu, \lambda \in \mathbb{R}$. This gauge theory describes the interaction of a doublet scalar field φ with two gauge potentials a and b. As it stands, nor a nor b can be massive, and indeed L contains no mass term for them. It is not a problem for a since we expect to have at least one massless field to carry the electromagnetic interaction. But the weak interaction is short range, so its associated field must be massive. Hence, it is necessary to reduce the $SU(2)$ gauge symmetry in the theory in order to allow a mass term for the weak field. Of course we, know that this can be achieved via SSBM. Actually, the latter is used in conjunction with a gauge fixing, the so-called *unitary gauge*, see e.g. [6]. Some authors have given a more geometrical account of the mechanism based on the bundle reduction theorem, see [60, 63, 65].

We now show that the $\mathcal{SU}(2)$ symmetry can be erased via the dressing field method. Given the gauge transformations as above, we define a dressing field out of the doublet scalar field φ by using a polar decomposition $\varphi = u\eta$ in $\mathbb{C}^2$ with

$$
u \in SU(2) \quad \text{and} \quad \eta := \begin{pmatrix} 0 \\ \|\varphi\| \end{pmatrix} \in \mathbb{R}^+ \subset \mathbb{C}^2, \quad \text{so that} \quad u^\beta = \beta^{-1}u, \tag{20}
$$

as can be checked explicitly. Then u is a $\mathcal{SU}(2)$-dressing field that can be used to apply Prop. 1 and to construct the $\mathcal{SU}(2)$-invariant composite fields

$$\begin{aligned}
&\widehat{A} = u^{-1}Au + \tfrac{1}{g}u^{-1}du = a + (u^{-1}bu + \tfrac{1}{g}u^{-1}du) =: a + B,\\
&\widehat{F} = u^{-1}Fu = f_a + u^{-1}g_b u =: f_a + G, \qquad \text{with } G = dB + gB^2,\\
&\widehat{\varphi} = u^{-1}\varphi = \eta, \qquad \text{and} \qquad \widehat{D}\widehat{\varphi} = u^{-1}D\varphi = \widehat{D}\eta = d\eta + (g'a + gB)\eta.
\end{aligned} \tag{21}$$

By virtue of Prop. 8, we conclude that the theory defined by the electroweak Lagrangian (19) is actually a $\mathcal{U}(1)$-gauge theory described in terms of the above composite fields,

$$\begin{aligned}
L(a, B, \eta) &= \tfrac{1}{2}\operatorname{Tr}(\widehat{F} \wedge *\widehat{F}) + \langle \widehat{D}\eta, *\widehat{D}\eta\rangle - U(\eta)\,\text{vol},\\
&= \tfrac{1}{2}\operatorname{Tr}(f_a \wedge *f_a) + \tfrac{1}{2}\operatorname{Tr}(G \wedge *G) + \langle \widehat{D}\eta, *\widehat{D}\eta\rangle - \left(\mu^2\eta^2 + \lambda\eta^4\right)\text{vol}\,.
\end{aligned} \tag{22}$$

Notice that by its very definition $\eta^\beta = \eta^\alpha = \eta$, so it is already a fully gauge invariant scalar field which then qualifies as an observable.

3.2 Residual $\mathcal{U}(1)$-symmetry

Is a mass term allowed for the $\mathcal{SU}(2)$-invariant field B? To answer the question, one needs to check its $\mathcal{U}(1)$-residual gauge transformation B^α, which depends on the $\mathcal{U}(1)$-gauge transformation of the dressing field u. One can check that

$$u^\alpha = u\widetilde{\alpha}, \qquad \text{where} \qquad \widetilde{\alpha} = \begin{pmatrix} \alpha & 0 \\ 0 & \alpha^{-1} \end{pmatrix}.$$

We therefore have

$$\begin{aligned}
B^\alpha &= (b^\alpha)^{u^\alpha} = \widetilde{\alpha}^{-1}u^{-1}bu\widetilde{\alpha} + \tfrac{1}{g}\widetilde{\alpha}^{-1}(u^{-1}du)\widetilde{\alpha} + \tfrac{1}{g}\widetilde{\alpha}^{-1}d\widetilde{\alpha} = \widetilde{\alpha}^{-1}B\widetilde{\alpha} + \tfrac{1}{g}\widetilde{\alpha}^{-1}d\widetilde{\alpha},\\
G^\alpha &= (g_b^\alpha)^{u^\alpha} = \widetilde{\alpha}^{-1}u^{-1}g_{bu}\widetilde{\alpha} = \widetilde{\alpha}^{-1}G\widetilde{\alpha}.
\end{aligned}$$

In view of this, it would seem that B still cannot have mass terms. But given the decomposition $B = B_a\sigma^a$ where σ^a are the Hermitian Pauli matrices and $B_a \in i\mathbb{R}$, so that $\bar{B}_a = -B_a$, we have explicitly

$$B = B_a\sigma^a = \begin{pmatrix} B_3 & B_1 - iB_2 \\ B_1 + iB_2 & -B_3 \end{pmatrix} =: \begin{pmatrix} B_3 & W^- \\ W^+ & -B_3 \end{pmatrix},$$

and

$$B^\alpha = \begin{pmatrix} B_3 + \frac{1}{g}\alpha^{-1}d\alpha & \alpha^{-2}W^- \\ \alpha^2 W^+ & -B_3 - \frac{1}{g}\alpha^{-1}d\alpha \end{pmatrix}.$$

The fields $W^\pm$ transform tensorially under $\mathcal{U}(1)$, and so they can be massive: they are the ($U(1)$-charged) particles detected in the SPS collider in January 1983. The field B_3 transforms as a $U(1)$-connection, making it another massless field together with the genuine $U(1)$-connection a. Considering (a, B_3) as a doublet, one can perform a natural change of variables

$$\begin{pmatrix} A \\ Z^0 \end{pmatrix} := \begin{pmatrix} \cos\theta_W & \sin\theta_W \\ -\sin\theta_W & \cos\theta_W \end{pmatrix}\begin{pmatrix} a \\ B_3 \end{pmatrix} = \begin{pmatrix} \cos\theta_W a + \sin\theta_W B_3 \\ \cos\theta_W B_3 - \sin\theta_W a \end{pmatrix},$$

where the so-called Weinberg (or weak mixing) angle is defined by $\cos\theta_W = {}^{g}/_{\sqrt{g^2+g'^2}}$ and $\sin\theta_W = {}^{g'}/_{\sqrt{g^2+g'^2}}$. By construction, it is easy to show that the 1-form Z^0 is then fully gauge invariant and can therefore be both massive and observable: it is the neutral weak field whose boson has been detected in the SPS collider in May 1983. Now, still by construction, we have $A^\beta = A$ and $A^\alpha = A + \frac{1}{e}\alpha^{-1}d\alpha$ with coupling constant $e := {}^{gg'}/_{\sqrt{g^2+g'^2}} = g'\cos\theta_W = g\sin\theta_W$. So A is a $U(1)$-connection: it is the massless carrier of the electromagnetic interaction and e is the elementary electric charge.

The electroweak theory (22) is then expressed in terms of the gauge invariant fields η, Z^0 and of the $U(1)$-gauge fields $W^\pm$, A:

$$\begin{aligned}
L(A, W^\pm, Z^0, \eta) &= \tfrac{1}{2}\operatorname{Tr}(\widehat{F}\wedge *\widehat{F}) + \langle D\eta, *D\eta\rangle - U(\eta)\,\mathrm{vol} \\
&= dZ^0\wedge *dZ^0 + dA\wedge *dA + dW^-\wedge *dW^+ \\
&\quad + 2g\Big\{\sin\theta_W\big(dA\wedge *(W^-W^+) + \cos\theta_W\big(dZ^0\wedge *(W^-W^+) \\
&\qquad + dW^-\wedge *(W^+A) + dW^-\wedge *(W^+Z^0) \\
&\qquad + dW^+\wedge *(AW^-)\big) + dW^+\wedge *(Z^0W^-)\big)\Big\} \\
&\quad + 4g^2\Big\{\sin^2\theta_W\ AW^-\wedge *(W^+A) \\
&\qquad + \cos^2\theta_W\ Z^0W^-\wedge *(W^+Z^0) \\
&\qquad + \sin\theta_W\cos\theta_W\ AW^-\wedge *(W^+Z^0) \\
&\qquad + \sin\theta_W\cos\theta_W\ Z^0W^-\wedge *(W^+A) \\
&\qquad + \frac{1}{4}\ W^-W^+\wedge *(W^-W^+)\Big\} \\
&\quad + d\eta\wedge *d\eta - g^2\eta^2\ W^+\wedge *W^- - (g^2+g'^2)\eta^2\ Z^0\wedge *Z^0 \\
&\quad - \big(\mu^2\eta^2 + \lambda\eta^4\big)\,\mathrm{vol}\,.
\end{aligned} \tag{23}$$

We can read off all possible interactions between the four electroweak fields. Notice that there is no coupling between the fields A and Z^0, showing the electric neutrality of the Z^0.

The next natural step is to expand the $\mathbb{R}^+$-valued scalar field η around its *unique* configuration η_0 minimizing the potential $U(\eta)$, the so-called Vacuum Expectation Value (VEV), as $\eta = \eta_0 + H$ where H is the gauge invariant Higgs field. True mass terms for Z^0, $W^\pm$ and H depending on η_0 then appear from the couplings of the electroweak fields with η and from the latter's self interaction. The absence of coupling between η and A indicates the masslessness of the latter (the two photons decay channel of the Higgs boson involves intermediary leptons, not treated here).

The theory has two qualitatively distinct phases. In the phase where $\mu^2 > 0$, the VEV vanishes and so do all masses, while in the phase where $\mu^2 < 0$, the VEV is non-vanishing, $\eta_0 = \sqrt{-\mu^2/2\lambda}$. The masses of the fields Z^0, $W^\pm$ and H are then $m_{Z_0} = \eta_0\sqrt{(g^2 + g'^2)}$, $m_{W^\pm} = \eta_0 g$, with ratio $\frac{m_{W^\pm}}{m_{Z^0}} = \cos\theta_W$, and $m_H = \eta_0\sqrt{2\lambda}$. In this latter case, (23) becomes the electroweak Lagrangian form of the Standard Model in the so-called unitary gauge. But keep in mind that, as a result of the dressing field method, neither the gauge fixing nor the SSBM is involved to obtain it.

3.3 Discussion

It is worth stressing some differences with the usual viewpoint. The SSBM is usually constructed as follows. At high energy (i.e. in the phase $\mu^2 > 0$) the symmetric VEV $\varphi_0 = \binom{0}{0}$ of $\varphi \in \mathbb{C}^2$ respect the full $\mathcal{SU}(2) \times \mathcal{U}(1)$ gauge symmetry group, so that no gauge potential in the theory can be massive. At low energy (i.e. in the phase $\mu^2 < 0$) the field φ must fall somewhere in the space of configurations that minimize the potential $U(\varphi)$. A space which is a circle in $\mathbb{C}^2$ defined by $M_0 = \{\varphi \in \mathbb{C}^2 \mid \bar{\varphi}_1\varphi_1 + \bar{\varphi}_2\varphi_2 = -\mu^2/\lambda\}$, whose individual points are not invariant under $\mathcal{SU}(2)$. Then, once an arbitrary minimum $\varphi_0 \in M_0$ is *randomly* selected, the gauge group is broken down to $\mathcal{U}(1)$ and mass terms for $SU(2)$-gauge potentials are generated. See e.g. [73]. This usual interpretation takes place in the *history* of the Universe, and this "phase transition" is a contingent phenomenon, since it selects by chance one specific value in M_0. The Standard Model of Particles Physics (SMPP) then relies on two strong foundations: one is structural (in the mathematical way), it is the Lagrangian of the theory; the other one is contingent, it is the historical aspect of the SSBM.

The dressing field approach allows to clearly distinguish the *erasure of* $\mathcal{SU}(2)$ and the *generation of mass terms* as two distinct operations, the former being a prerequisite of the latter but not its direct *cause*, as the textbook interpretation would have it. Notice also that the relevant $\mathcal{SU}(2)$-invariant variables, corresponding to the physical fields (fermions fields are treated in the same manner, see [46]), are identified at the mathematical level of the theory in both phases (i.e. independently of the sign of μ^2). The transition between these phases, characterized by different electroweak vacua, remains a dynamical process parametrized by the sign of μ^2.[6] But we stress that in our scheme there is no arbitrariness in the choice of VEV for $\eta \in \mathbb{R}^+$ since it is now *unique*: $\eta_0 = \sqrt{-\mu^2/2\lambda}$ when $\mu^2 < 0$. In particular, all the bosons Z^0, A, W^+, W^- (and fermions fields) can be identified at the level of the theory, without requiring any historical contingent process. In that respect, the contingent aspect of the SMPP is dispelled to the benefit of its unique structural foundation.

The arbitrariness of the polar decomposition $\varphi = u\eta$ is discussed in [46]: imposing that the final $U(1)$ charges are clearly identified, the field content of the Lagrangian in the new variables is the same up to global transformations involving some rigid transformations of the fields. This implies that the content of the theory in terms of $\mathcal{SU}(2)$-invariant fields takes place at an ontological level, since it does not require any historical arguments.

According to [65], the very meaning of the terminology "*spontaneous* symmetry breaking" lies in the fact that M_0 is not reduced to a point. Granting this reasonable observation, the dressing field approach would then lead to deny the soundness of this terminology to characterize the electroweak model. First because the symmetry reduction is not related to the choice of a VEV in M_0, then because the latter is reduced to a point. A better characterization would emphasize the link between mass generation and electroweak vacuum phase transition: "mass generation through electroweak vacuum phase transition".

The fact that the dressing field approach to the electroweak model allows to dispense with the idea of spontaneous breaking of a gauge symmetry is perfectly in line with the so-called Elitzur theorem stating that in lattice gauge theory a gauge symmetry cannot be spontaneously broken. An equivalent theorem for gauge field theory has not been proven yet, but no reason has been given as to why it would fail either.

Furthermore, as mentioned in the introduction, the status of gauge symmetries is a disputed question in philosophy of physics. A well argued position considers gauge symmetries as "surplus structures", as philosopher of physics Michael Redhead calls it, that is a redundancy in our mathematical description of reality. They would then have an epistemological status. The idea of a spontaneous breakdown of a gauge symmetry on the other hand, insofar as it implies observable qualitative physical effects (particles acquire masses in a historical process), supports an ontological view of gauge symmetries, making them a structural feature of reality rather than of our description of it. And indeed, the part of the philosophy of physics community interested in this problem has struggled to reconcile the empirical success of the electroweak model with their analysis of gauge symmetries (see e.g. [8, 9]). Often a workaround is proposed in arguing that a gauge fixing removes the local dependence of the symmetry and that only a global one remains to be broken spontaneously. The latter, by the Goldstone theorem, generates Goldstone bosons.

---

[6] In fact, it could even be reduced to a technical step useful to perform the usual field quantization procedure, which relies heavily on the identification of propagators and mass terms in the Lagrangian.

These efforts of interpretation are enlightened once it is recognized that the notion of spontaneous breaking of gauge symmetry is not pivotal to the empirical success of the electroweak model. Higgs had a glimpse of this fact [32], and Kibble saw it clearly [34] (see the paragraph just before the conclusion of his paper). Both had insights by working on toy models, just before the electroweak model was proposed by Weinberg and Salam in 1967. The invariant version of the model was first given in [27] in 1981 (compare Sect. 6 with our exposition above), but was rediscovered independently by others [13, 19, 33, 39, 46]. The dressing field approach provides a general unifying framework for these works, and achieves the conceptual clarity philosophers of physics have been striving for [25, 62, 64].

4 From Tetrad to Metric Formulation of General Relativity

Einstein teaches us that gravitation is the dynamics of space-time, the base manifold itself. It deals with spatio-temporal degrees of freedom, not "inner" ones like in Yang-Mills-type gauge theories. In the most general case there exists a notion of *torsion*, a concept absent in Yang-Mills theories. There are more possible invariants one can use in a Lagrangian due to index contractions impossible in Yang-Mills theories: the actual Lagrangian form for GR is not of Yang-Mills type.

All these issues from the existence in gravitational theories of the *soldering form*, also known as (co-)tetrad field, which realizes an isomorphism between the tangent space at each point of space-time and the Minkowski space [63]. The soldering form can be seen as the formal implementation of Einstein's "happiest thought", the *Equivalence Principle*, which is the key specific physical feature distinguishing the gravitational interaction from the three others (Yang-Mills) gauge interactions.

So, while Yang-Mills fields are described by Ehresmann connections (principal connections) on a principal bundle, the gravitational field is described by both an Ehresmann connection, the Lorentz/spin connection, *and* a soldering form. In 1977, McDowell and Mansouri treated the concatenation of the connection and of the soldering form as a single gauge potential [47]. The mathematical foundation of this move is Cartan geometry [70, 71]: the third additional axiom defining a Cartan connection, and distinguishing it from a principal connection, defines an absolute parallelism on $\mathcal{P}$. This in turn induces, in simple cases, a soldering form [58]. In other words, the geometry of the bundle $\mathcal{P}$ is much more tightly related to the geometry of the base manifold. One can then convincingly argue that Cartan geometry is a very natural framework for classical gravitational theories.

In the following we recast the tetrad formulation of GR in terms of the adequate Cartan geometry, and show that switching to the metric formulation can be seen as an application of the dressing field method.

4.1 Reduction of the Lorentz Gauge Symmetry

The relevant Cartan geometry is based on the Klein model (G, H) given by $G = SO(1,3) \ltimes \mathbb{R}^{1,3}$, the Poincaré group, and $H = SO(1,3)$, the Lorentz group, so that the associated homogeneous space is $G/H = \mathbb{R}^{1,3}$, the Minkowski space. The infinitesimal Klein pair is $(\mathfrak{g}, \mathfrak{h})$ with $\mathfrak{g} = \mathfrak{so}(1,3) \oplus \mathbb{R}^{1,3}$ and $\mathfrak{h} = \mathfrak{so}(1,3)$. The principal bundle of this Cartan geometry is $\mathcal{P}(\mathcal{M}, SO(1,3))$. The local Cartan connection and its curvature are respectively the 1-form $\varpi \in \bigwedge^1(\mathcal{U}, \mathfrak{g})$ and the 2-form $\Omega \in \bigwedge^2(\mathcal{U}, \mathfrak{g})$, which can be written in matrix form

$$\varpi = \begin{pmatrix} A & \theta \\ 0 & 0 \end{pmatrix}, \qquad \Omega = \begin{pmatrix} R & \Theta \\ 0 & 0 \end{pmatrix} = \begin{pmatrix} dA + A \wedge A & d\theta + A \wedge \theta \\ 0 & 0 \end{pmatrix},$$

where $A \in \bigwedge^1(\mathcal{U}, \mathfrak{so})$ is the spin connection with Riemann curvature 2-form R and torsion $\Theta = D\theta$, and $\theta \in \bigwedge^1(\mathcal{U}, \mathbb{R}^{1,3})$ is the soldering form. In other words, this Cartan geometry is just the usual Lorentz geometry (with torsion). We can thus consider the Cartan connection ϖ as the gravitational gauge potential. The local gauge group is $\mathcal{SO} := \mathcal{SO}(1,3)$ and its action by an element $\gamma : \mathcal{U} \to SO$, assuming the matrix form $\gamma = \left(\begin{smallmatrix} S & 0 \\ 0 & 1 \end{smallmatrix}\right)$, is

$$\varpi^\gamma = \gamma^{-1}\varpi\gamma + \gamma^{-1}d\gamma = \begin{pmatrix} S^{-1}AS + S^{-1}dS & S^{-1}\theta \\ 0 & 0 \end{pmatrix},$$

$$\Omega^\gamma = \gamma^{-1}\Omega\gamma = \begin{pmatrix} S^{-1}RS & S^{-1}\Theta \\ 0 & 0 \end{pmatrix}.$$

Given these geometrical data, the associated Lagrangian form of GR is given by,

$$L_{\text{Pal}}(A, \theta) = \frac{-1}{32\pi \mathsf{G}} \operatorname{Tr}\left(R \wedge *(\theta \wedge \theta^t)\right) = \frac{-1}{32\pi \mathsf{G}} \operatorname{Tr}\left(R \wedge *(\theta \wedge \theta^T \eta)\right), \tag{24}$$

with η the metric of $\mathbb{R}^{1,3}$ and G the gravitational constant. Given $\mathcal{S} = \int L$, variation w.r.t. θ gives Einstein's equation in vacuum and variation w.r.t. A gives an equation for the torsion which in the vacuum vanishes (even in the presence of matter, the torsion does not propagate).

Looking for a dressing field liable to neutralize the $\mathcal{SO}$-gauge symmetry, given the gauge transformation of the Cartan connection, the tetrad field $e = e^a$ in the soldering form $\theta^a = e^a{}_\mu dx^\mu$ is a natural candidate: $\theta^S = S^{-1}\theta$ implies $e^S = S^{-1}e$, so that we can define

$$u = \begin{pmatrix} e & 0 \\ 0 & 1 \end{pmatrix} \quad \text{and we get } u^\gamma = \gamma^{-1}u.$$

Then u is a $\mathcal{SO}$-dressing field, and notice that its target group $G = GL$ is bigger than the structure group which happens to be also its equivariance group, $H = K = SO$.[7] We can apply Prop. 1 and construct the $\mathcal{SO}$-invariant composite fields,

$$\widehat{\varpi} = u^{-1}\varpi u + u^{-1}du = \begin{pmatrix} e^{-1}Ae + e^{-1}de & e^{-1}\theta \\ 0 & 0 \end{pmatrix} =: \begin{pmatrix} \Gamma & dx \\ 0 & 0 \end{pmatrix},$$
$$\widehat{\Omega} = u^{-1}\Omega u = \begin{pmatrix} e^{-1}Re & e^{-1}\Theta \\ 0 & 0 \end{pmatrix} =: \begin{pmatrix} \mathsf{R} & T \\ 0 & 0 \end{pmatrix},$$

where $\Gamma = \Gamma^{\mu}{}_{\nu} = \Gamma^{\mu}{}_{\nu,\rho}dx^{\rho}$ is the linear connection 1-form on $\mathcal{U} \subset \mathcal{M}$, and R and T are the Riemann curvature and torsion 2-forms written in the coordinate system $\{x^{\mu}\}$ on $\mathcal{U}$. We can get their explicit expressions as functions of the components of the dressed Cartan connection $\widehat{\varpi}$ on account of,

$$\widehat{\Omega} = d\widehat{\varpi} + \widehat{\varpi} \wedge \widehat{\varpi} = \begin{pmatrix} d\Gamma & d^2x \\ 0 & 0 \end{pmatrix} + \begin{pmatrix} \Gamma \wedge \Gamma & \Gamma \wedge dx \\ 0 & 0 \end{pmatrix} = \begin{pmatrix} d\Gamma + \Gamma \wedge \Gamma & \Gamma \wedge dx \\ 0 & 0 \end{pmatrix}.$$

We clearly see that if Γ is symmetric on its lower indices, the torsion vanishes.

A Cartan connection always induces a metric on the base manifold $\mathcal{U} \subset \mathcal{M}$ by $g(X, Y) = \eta\big(\theta(X), \theta(Y)\big)$, with $X, Y \in T_x\mathcal{U}$. In component this reads $g_{\mu\nu} = e_{\mu}{}^{a}\eta_{ab}e^{b}{}_{\nu}$, or in index free notation $g = e^T\eta e$. Notice that by definition g is $\mathcal{SO}$-gauge-invariant. It is easy to show that in this formalism, the metricity condition is necessarily satisfied: $\widehat{D}g := \nabla g = dg - \Gamma^T g - g\Gamma = -e^T\big(A^T\eta + \eta A\big)e = 0$. Therefore if $T = 0$, Γ is the Levi-Civita connection associated to g.

Now, by application of Prop. 8 we see that the classic calculation that allows to switch from the $\mathcal{SO}$-gauge formulation to the metric formulation can be seen as an example of the dressing field method,

$$\begin{aligned} L_{\text{Pal}}(A, \theta) &= \frac{-1}{32\pi \mathsf{G}}\,\text{Tr}\,\big(R \wedge *(\theta \wedge \theta^t)\big) = \frac{-1}{32\pi \mathsf{G}}\,\text{Tr}\,\big(\mathsf{R}g\big) \wedge *(dx \wedge dx) \\ &= \frac{1}{16\pi \mathsf{G}}\sqrt{|g|}d^m x\ \mathsf{R}_{\text{icc}} =: L_{\text{EH}}(\Gamma, g). \end{aligned}$$

The last equation defines the Einstein-Hilbert Lagrangian form, depending on the $\mathcal{SO}$-invariant composite fields Γ and g.

4.2 Residual Symmetry

The $\mathcal{SO}$-invariant fields g, $\widehat{\varpi} = (\Gamma, dx)$ and $\widehat{\Omega} = (\mathsf{R}, T)$ belong to the *natural geometry* of the base manifold $\mathcal{M}$, i.e. the geometry defined only in terms of its frame bundle and its associated vector bundles. The only residual transformations these

[7] While in the previous example we had $G = K = SU(2) \subset H = U(1) \times SU(2)$.

fields can display are coordinate transformations. On the overlap of two patches of coordinates $\{x^\mu\}$ and $\{y^\mu\}$ in a trivializing open set $\mathcal{U} \subset \mathcal{M}$, the initial gauge fields ϖ and Ω, as differential forms, are well defined and invariant. But obviously $\theta = edx = e'dy$ implies that the tetrad undergoes the transformation $e' = eG$, with $G = G^\mu{}_\nu = \frac{\partial x^\mu}{\partial y^\nu}$. The dressing fields then transform as $u' = uG$, with $G = \left(\begin{smallmatrix} G & 0 \\ 0 & 1 \end{smallmatrix}\right)$, so the composite fields have coordinate transformations,

$$
\begin{aligned}
\widehat{\varpi}' &= u'^{-1}\varpi u' + u'^{-1}du' = G^{-1}\widehat{\varpi}G + G^{-1}dG \\
&= \begin{pmatrix} G^{-1}\Gamma G + G^{-1}dG & G^{-1}dx \\ 0 & 0 \end{pmatrix} =: \begin{pmatrix} \Gamma' & dy \\ 0 & 0 \end{pmatrix}, \\
\widehat{\Omega}' &= u'^{-1}\Omega u' = G^{-1}\widehat{\Omega}G = \begin{pmatrix} G^{-1}\mathsf{R}G & G^{-1}T \\ 0 & 0 \end{pmatrix} =: \begin{pmatrix} \mathsf{R}' & T' \\ 0 & 0 \end{pmatrix}, \\
g' &= e'^T \eta e' = G^T g G.
\end{aligned}
$$

This gives the well known transformations of the linear connection, of the metric, of the Riemann and torsion tensors under general changes of coordinates. Of course the Lagrangian form, L_{EH}, is invariant.

4.3 Discussion

The tetrad as a dressing field does not belong to the gauge group $\mathcal{SO}$ of the theory. So, strictly speaking, the invariant composite field $\widehat{\varpi}$ is not a gauge transformation of the Cartan connection ϖ. In particular this means that, contrary to what is sometimes said, Γ is not a gauge transformation of the Lorentz connection A. Indeed Γ is an $\mathcal{SO}$-invariant $\mathfrak{gl}$-valued 1-form on $\mathcal{M}$, clearly it does not belong to the initial space of connections of the theory. Even if one considers that the gauge symmetry of GR are the coordinate changes, thinking of it as a gauge theory on the frame bundle $L\mathcal{M}$ with gauge group $\mathcal{GL}$, the tetrad $e^a{}_\mu$ still doesn't belong to $\mathcal{GL}$. So one cannot view Γ and A as gauge related. To obtain A from Γ one needs the bundle reduction theorem, which allows to reduce $L\mathcal{M}$ to the subbundle $\mathcal{P}(\mathcal{M}, SO(1,3))$. To recover Γ from A, one needs to think in terms of the dressing field method.

5 Conformal Cartan Geometry, Tractors and Twistors

In this Section we show how tractors and twistors, which are conformal calculi for torsionless manifolds [5, 53], can be derived from the conformal Cartan geometry via the dressing field method. We thus start by a brief description of this geometry and then we deal with tractors and twistors.

5.1 Conformal Cartan Geometry in a Nutshell

A conformal Cartan geometry $(\mathcal{P}, \varpi)$ can be defined over n-manifolds $\mathcal{M}$ for any $n \geq 3$ and signature (r, s) according to the group $SO(r+1, s+1)$. We will admit that the base manifold is such that a corresponding spinorial version $(\bar{\mathcal{P}}, \bar{\varpi})$ exists, based on the group $\text{Spin}(r+1, s+1)$, so that we have the two-fold covering $\bar{\mathcal{P}} \xrightarrow{2:1} \mathcal{P}$. Since we seek to reproduce twistors in signature $(1, 3)$, as spinors corresponding to tractors, we are here interested in conformal Cartan geometry over 4-manifolds, and thus take advantage of the accidental isomorphism $\text{Spin}(2, 4) \simeq SU(2, 2)$.

We then treat in parallel the conformal Cartan geometry $(\mathcal{P}(\mathcal{M}, H), \varpi)$ modeled on the Klein model (G, H) and its naturally associated vector bundle E, as well as the spinorial version $(\bar{\mathcal{P}}(\mathcal{M}, \bar{H}), \bar{\varpi})$ modeled on the Klein model $(\bar{G}, \bar{H})$ and its naturally associated vector bundle E. For simplicity we designate them as the real and complex cases respectively. By dressing, the real case will yield tractors and the complex case will yield twistors.

In the real case, we have

$$G = PSO(2,4) = \left\{M \in GL_6(\mathbb{R}) \mid M^T \Sigma M = \Sigma, \det M = 1\right\} / \{\pm \text{id}\}$$

with $\Sigma = \begin{pmatrix} 0 & 0 & -1 \\ 0 & \eta & 0 \\ -1 & 0 & 0 \end{pmatrix}$ the group metric, η the Minkowski flat metric of signature $(1, 3)$, and H is a parabolic subgroup comprising Lorentz, Weyl and conformal boost symmetries: it has the following matrix presentation [11, 58], with $W := \mathbb{R}^*_+$ (Weyl dilation group),

$$H = K_0\, K_1 = \left\{ \begin{pmatrix} z & 0 & 0 \\ 0 & S & 0 \\ 0 & 0 & z^{-1} \end{pmatrix} \begin{pmatrix} 1 & r & \frac{1}{2} r r^t \\ 0 & \mathbb{1}_4 & r^t \\ 0 & 0 & 1 \end{pmatrix} \middle| \begin{array}{l} z \in W, \\ S \in SO(1,3), \\ r \in \mathbb{R}^{4*} \end{array} \right\},$$

where K_0 (resp. K_1) corresponds to the matrices on the left (resp. right) in the product. Clearly $K_0 \simeq CO(1, 3)$ via $(S, z) \mapsto zS$, and K_1 is the abelian group of conformal boosts. Here ^{T} is the usual matrix transposition, $r^t = (r\eta^{-1})^T$ stands for the η-transposition, and $\mathbb{R}^{4*}$ is the dual of $\mathbb{R}^4$.

The corresponding Lie algebras $(\mathfrak{g}, \mathfrak{h})$ are graded: $[\mathfrak{g}_i, \mathfrak{g}_j] \subseteq \mathfrak{g}_{i+j}$, $i, j = 0, \pm 1$, with the abelian Lie subalgebras $[\mathfrak{g}_{-1}, \mathfrak{g}_{-1}] = 0 = [\mathfrak{g}_1, \mathfrak{g}_1]$. They decompose respectively as, $\mathfrak{g} = \mathfrak{g}_{-1} \oplus \mathfrak{g}_0 \oplus \mathfrak{g}_1 \simeq \mathbb{R}^4 \oplus \mathfrak{co}(1, 3) \oplus \mathbb{R}^{4*}$, with $\mathfrak{co}(1, 3) = \mathfrak{so}(1, 3) \oplus \mathbb{R}$, and $\mathfrak{h} = \mathfrak{g}_0 \oplus \mathfrak{g}_1 \simeq \mathfrak{co}(1, 3) \oplus \mathbb{R}^{4*}$. In matrix notation we have,

$$\mathfrak{g} = \left\{ \begin{pmatrix} \varepsilon & \iota & 0 \\ \tau & s & \iota^t \\ 0 & \tau^t & -\varepsilon \end{pmatrix} \middle| \begin{array}{l} (s - \varepsilon \mathbb{1}_4) \in \mathfrak{co}(1,3), \\ \tau \in \mathbb{R}^4, \\ \iota \in \mathbb{R}^{4*} \end{array} \right\} \supset \mathfrak{h} = \left\{ \begin{pmatrix} \varepsilon & \iota & 0 \\ 0 & s & \iota^t \\ 0 & 0 & -\varepsilon \end{pmatrix} \right\}.$$

The graded structure of the Lie algebras is automatically handled by the matrix commutator.

In order to introduce the complex case, let us first consider the canonical isomorphism of vector spaces between Minkowski space $\mathbb{R}^{1,3}$ and Hermitian 2×2 matrices $\text{Herm}(2, \mathbb{C}) = \{M \in M_2(\mathbb{C}) \mid M^* = M\}$, where * means trans-conjugation: $\mathbb{R}^4 \to \text{Herm}(2, \mathbb{C})$, $x \mapsto \bar{x} = x^a \sigma_a$ ($\sigma_0 = \mathbb{1}_2$ and $\sigma_{i=\{1,2,3\}}$ are the Pauli matrices). There is a corresponding double covering group morphism $SL(2, \mathbb{C}) \xrightarrow{2:1} SO(1, 3)$, $\bar{S} \mapsto S$ (so that $S^{-1}x \mapsto \bar{S}^{-1}\bar{x}\bar{S}^{-1*}$ and $x^t S \mapsto \bar{S}^*\bar{x}^t\bar{S}$), and its associated Lie algebra isomorphism $\mathfrak{so}(1, 3) \simeq \mathfrak{sl}(2, \mathbb{C})$ is denoted by $s \mapsto \bar{s}$. In the following, the bar notation will relate the "real" and "complex" cases in a natural way by using the same letters, so generalizing the above maps.

For the complex case, we have then $\bar{G} = SU(2, 2) \simeq \text{Spin}(2, 4)$, which is the group preserving the metric $\bar{\Sigma} = \left(\begin{smallmatrix} 0 & \mathbb{1}_2 \\ \mathbb{1}_2 & 0 \end{smallmatrix}\right)$, and $\bar{H}$ is given in matrix notation by

$$\bar{H} = \bar{K}_0\bar{K}_1 := \left\{ \begin{pmatrix} z^{1/2}\bar{S}^{-1*} & 0 \\ 0 & z^{-1/2}\bar{S} \end{pmatrix} \begin{pmatrix} \mathbb{1}_2 & -i\bar{r} \\ 0 & \mathbb{1}_2 \end{pmatrix} \,\middle|\, \begin{matrix} z \in W,\ \bar{S} \in SL(2, \mathbb{C}), \\ \bar{r} \in \text{Herm}(2, \mathbb{C}) \end{matrix} \right\}. \tag{25}$$

There is a double covering $\bar{H} \xrightarrow{2:1} H$ which reduces to a double covering $\bar{K}_0 \xrightarrow{2:1} K_0$ and a natural isomorphism $\bar{K}_1 \simeq K_1$. Using the bar notation, the Lie algebra isomorphism $\mathfrak{so}(2, 4) = \mathfrak{g} \to \mathfrak{su}(2, 2) = \bar{\mathfrak{g}}$ is explicitly given by

$$\begin{aligned} \bar{\mathfrak{g}} = \bar{\mathfrak{g}}_{-1} + \bar{\mathfrak{g}}_0 + \bar{\mathfrak{g}}_1 &= \left\{ \begin{pmatrix} -(\bar{s}^* - \frac{\varepsilon}{2}\mathbb{1}_2) & -i\bar{\iota} \\ i\bar{\tau} & \bar{s} - \frac{\varepsilon}{2}\mathbb{1}_2 \end{pmatrix} \,\middle|\, \begin{matrix} \varepsilon \in \mathbb{R},\ \bar{s} \in \mathfrak{sl}(2, \mathbb{C}) \\ \bar{\tau}, \bar{\iota} \in \text{Herm}(2, \mathbb{C}) \end{matrix} \right\} \\ &\supset \bar{\mathfrak{h}} = \bar{\mathfrak{g}}_0 + \bar{\mathfrak{g}}_1. \end{aligned} \tag{26}$$

Once given two Cartan bundles such that $\bar{\mathcal{P}}(\mathcal{M}, \bar{H}) \xrightarrow{2:1} \mathcal{P}(\mathcal{M}, H)$, we endow $\mathcal{P}(\mathcal{M}, H)$ with a conformal Cartan connection whose local representative on $\mathcal{U} \subset \mathcal{M}$ is $\varpi \in \bigwedge^1(\mathcal{U}, \mathfrak{g})$, with curvature $\Omega \in \bigwedge^2(\mathcal{U}, \mathfrak{g})$. In matrix presentation, one has

$$\varpi = \begin{pmatrix} a & P & 0 \\ \theta & A & P^t \\ 0 & \theta^t & -a \end{pmatrix} \quad \text{and} \quad \Omega = d\varpi + \varpi^2 = \begin{pmatrix} f & C & 0 \\ \Theta & W & C^t \\ 0 & \Theta^t & -f \end{pmatrix}.$$

In the same way, $\bar{\mathcal{P}}(\mathcal{M}, \bar{H})$ is endowed with a spinorial Cartan connexion

$$\bar{\varpi} = \begin{pmatrix} -(\bar{A}^* - \frac{a}{2}\mathbb{1}_2) & -i\bar{P} \\ i\bar{\theta} & \bar{A} - \frac{a}{2}\mathbb{1}_2 \end{pmatrix} \quad \text{and} \quad \bar{\Omega} = \begin{pmatrix} -(\bar{W}^* - \frac{f}{2}\mathbb{1}_2) & -i\bar{C} \\ i\bar{\Theta} & \bar{W} - \frac{f}{2}\mathbb{1}_2 \end{pmatrix}.$$

The soldering part of ϖ is $\theta = e \cdot dx$, i.e. $\theta^a := e^a{}_\mu dx^\mu$.[8] Denote by g the metric of signature $(1, 3)$ on $\mathcal{M}$ induced (as already seen) from η via ϖ according to

[8] Notice that from now on we shall make use of "·" to denote Greek indices contractions, while Latin indices contraction is naturally understood from matrix multiplication.

$g(X, Y) := \eta(\theta(X), \theta(Y)) = \theta(X)^T \eta\theta(Y)$, or in a way more familiar to physicists $g := e^T \eta e$, so that $g_{\mu\nu} = e_\mu{}^a \eta_{ab} e^b{}_\nu$. The action of $\mathcal{H}$ on ϖ induces, through θ, a conformal class of metrics $c := [g]$ on $\mathcal{M}$. But $(\mathcal{P}, \varpi)$ is not equivalent to $(\mathcal{M}, c)$. Nevertheless, there is a distinguished choice, the so-called *normal* conformal Cartan connection ϖ_N, which is unique in satisfying the conditions $\Theta = 0$ and $W^a{}_{bad} = 0$ (which in turn, through the Bianchi identity, implies $f = 0$), so that $(\mathcal{P}, \varpi_\text{N})$ is indeed equivalent to a conformal manifold $(\mathcal{M}, c)$.

Still, it would be hasty to identify A in ϖ or ϖ_N with the Lorentz connection one is familiar with in physics, and by a way of consequence to take $R := dA + A^2$ and P as the Riemann and Schouten tensors. Indeed, contrary to expectations, A is invariant under Weyl rescaling and neither R nor P have the well-known Weyl transformations. It turns out that one recovers the spin connection and the aforementioned associated tensors only after a dressing operation, as shown in [1].

Using the natural representation of H on $\mathbb{R}^6$, we can introduce the associated vector bundle $E = \mathcal{P} \times_H \mathbb{R}^6$. A section of E is a H-equivariant map on $\mathcal{P}$ whose local expression is $\varphi : \mathcal{U} \subset \mathcal{M} \to \mathbb{R}^6$, given explicitly as column vectors

$$\varphi = \begin{pmatrix} \rho \\ \ell \\ \sigma \end{pmatrix}, \quad \text{with } \ell = \ell^a \in \mathbb{R}^4, \text{ and } \rho, \sigma \in \mathbb{R}.$$

The covariant derivative induced by the Cartan connection is $D\varphi = d\varphi + \varpi\varphi$, with $D^2\varphi = \Omega\varphi$. The group metric Σ defines an invariant bilinear form on sections of E: for any $\varphi, \varphi' \in \Gamma(E)$, one has $\langle \varphi, \varphi' \rangle = \varphi^T \Sigma \varphi' = -\sigma\rho' + \ell^T \eta \ell' - \rho\sigma'$. The covariant derivative D preserves this bilinear form since ϖ is $\mathfrak{g}$-valued: $D\Sigma = d\Sigma + \varpi^T \Sigma + \Sigma\varpi = 0$.

We now follow the same line of constructions in the complex case, using the natural representation $\mathbb{C}^4$ of $\bar{H}$ to define the associated vector bundle $\mathsf{E} = \bar{\mathcal{P}} \times_{\bar{H}} \mathbb{C}^4$. A section of E is a $\bar{H}$-equivariant map on $\bar{\mathcal{P}}$ whose local expression is $\psi : \mathcal{U} \subset \mathcal{M} \to \mathbb{C}^4$ given as

$$\psi = \begin{pmatrix} \pi \\ \omega \end{pmatrix}, \quad \text{with } \pi, \omega \in \mathbb{C}^2 \text{dual Weyl spinors.}$$

The covariant derivative is now $\bar{D}\psi = d\psi + \bar{\varpi}\psi$, with $\bar{D}^2\psi = \bar{\Omega}\psi$. The group metric $\bar{\Sigma}$ defines an invariant bilinear form on sections of E: for any $\psi, \psi' \in \Gamma(\mathsf{E})$, one has $\langle \psi, \psi' \rangle = \psi^* \bar{\Sigma} \psi' = \pi^*\omega' + \omega^*\pi'$. Again, the covariant derivative $\bar{D}$ preserves this bilinear form.

The gauge groups $\mathcal{H} = \mathcal{K}_0\mathcal{K}_1$ and $\bar{\mathcal{H}} = \bar{\mathcal{K}}_0\bar{\mathcal{K}}_1$ act on the gauge variables, with $\gamma \in \mathcal{H}$ and $\bar{\gamma} \in \bar{\mathcal{H}}$, as

$$\varpi^\gamma = \gamma^{-1}\varpi\gamma + \gamma^{-1}d\gamma, \quad \varphi^\gamma = \gamma^{-1}\varphi, \quad \bar{\varpi}^{\bar{\gamma}} = \bar{\gamma}^{-1}\bar{\varpi}\bar{\gamma} + \bar{\gamma}^{-1}d\bar{\gamma}, \quad \psi^{\bar{\gamma}} = \bar{\gamma}^{-1}\psi.$$

This induces the actions $\Omega^\gamma = \gamma^{-1}\Omega\gamma$, $\bar{\Omega}^{\bar{\gamma}} = \bar{\gamma}^{-1}\bar{\Omega}\bar{\gamma}$, $(D\varphi)^\gamma = \gamma^{-1}D\varphi$, and $(\bar{D}\psi)^{\bar{\gamma}} = \bar{\gamma}^{-1}\bar{D}\psi$. Given $\gamma_0 \in \mathcal{K}_0$, the soldering part of the gauge transformed

Cartan connection ϖ^{γ_0} is $\theta^{\gamma_0} = zS^{-1}\theta$, so that the metric induced by ϖ^{γ_0} is $g' = z^2 g$. On the other hand, $\theta^{\gamma_1} = \theta$ for $\gamma_1 \in \mathcal{K}_1$. So, as mentioned above, the action of the gauge group induces a conformal class of metric c on $\mathcal{M}$.

5.2 *Tractors and Twistors: Constructive Procedure via Dressing*

It has been noticed that tractor and twistor vector bundles are associated to the conformal Cartan bundle, and that tractor and twistor connections are related to the conformal Cartan connection [5, 26]. However as it stands, the gauge transformations above obtained show that φ is not a tractor and that ψ is not a twistor. It turns out that to recover tractors and twistors one needs to erase the conformal boost symmetry $\mathcal{K}_1 \simeq \bar{\mathcal{K}}_1$. We outline the procedure below and give the important results. Details can be found in [1, 2].

Given the decompositions $H = K_0 K_1$ and $\bar{H} = \bar{K}_0 \bar{K}_1$, the most natural choice of dressing field to erase the conformal boost gauge symmetry is $u_1 : \mathcal{U} \to K_1$ in the real case and its corresponding element $\bar{u}_1 : \mathcal{U} \to \bar{K}_1 \simeq K_1$ in the complex case, given by

$$u_1 = \begin{pmatrix} 1 & q & \frac{1}{2}qq^t \\ 0 & \mathbb{1}_4 & q^t \\ 0 & 0 & 1 \end{pmatrix}, \qquad \bar{u}_1 = \begin{pmatrix} \mathbb{1}_2 & -i\bar{q} \\ 0 & \mathbb{1}_2 \end{pmatrix}.$$

It turns out that u_1 can be defined via the "gauge-like" constraint $\Sigma(\varpi^{u_1}) := \text{Tr}(A^{u_1} - a^{u_1}) = -na^{u_1} = 0$. Indeed, this gives the equation $a - q\theta = 0$, which once solved for q gives $q_a = a_\mu e^\mu{}_a$, or in index free notation $q = a \cdot e^{-1}$.[9] Now, from ϖ^{γ_1} one finds that $q^{\gamma_1} = a^{\gamma_1} \cdot (e^{\gamma_1})^{-1} = (a - re) \cdot e^{-1} = q - r$. One then checks easily that the constraint $\Sigma(\varpi^{u_1}) = 0$ is $\mathcal{K}_1$-invariant and that u_1 is a dressing field for K_1: from $q^{\gamma_1} = q - r$ one shows that $u_1^{\gamma_1} = \gamma_1^{-1} u_1$. In the same way, one has $\bar{u}_1^{\bar{\gamma}_1} = \bar{\gamma}_1^{-1} \bar{u}_1$.

With these $\mathcal{K}_1$-dressing fields, we can apply (the local version of) Prop. 1 and form the $\mathcal{K}_1 \simeq \bar{\mathcal{K}}_1$-invariant composite fields in the real and complex cases:

$$\varpi_1 := \varpi^{u_1} = u_1^{-1}\varpi u_1 + u_1^{-1}du_1 = \begin{pmatrix} 0 & P_1 & 0 \\ \theta & A_1 & P_1^t \\ 0 & \theta^t & 0 \end{pmatrix}, \qquad \bar{\varpi}_1 = \bar{\varpi}^{\bar{u}_1} = \begin{pmatrix} -\bar{A}_1^* & -i\bar{P}_1 \\ i\bar{\theta} & \bar{A}_1 \end{pmatrix}$$

$$\Omega_1 := \Omega^{u_1} = u_1^{-1}\Omega u_1 = d\varpi_1 + \varpi_1^2, \qquad \bar{\Omega}_1 = \bar{\Omega}^{\bar{u}_1} = \bar{u}_1^{-1}\bar{\Omega}\bar{u}_1,$$

[9]Beware of the fact that in this index free notation a is the set of components of the 1-form a. This should be clear from the context.

$$\varphi_1 := u_1^{-1}\varphi,$$
$$D_1\varphi_1 = d\varphi_1 + \varpi_1\varphi_1 = \begin{pmatrix} d\rho_1 + P_1\ell_1 \\ d\ell_1 + A_1\ell_1 + \theta\rho_1 + P_1^t\sigma \\ d\sigma + \theta^t\ell_1 \end{pmatrix} = \begin{pmatrix} \nabla\rho_1 + P_1\ell_1 \\ \nabla\ell_1 + \theta\rho_1 + P_1^t\sigma \\ \nabla\sigma + \theta^t\ell_1 \end{pmatrix},$$

$$\psi_1 := \bar{u}_1^{-1}\psi,$$
$$\bar{D}_1\psi_1 = d\psi_1 + \bar{\varpi}_1\psi_1 = \begin{pmatrix} d\pi_1 - \bar{A}_1^*\pi_1 - i\bar{P}_1\omega_1 \\ d\omega_1 + \bar{A}_1\omega_1 + i\bar{\theta}\pi_1 \end{pmatrix} = \begin{pmatrix} \bar{\nabla}\pi_1 - i\bar{P}_1\omega_1 \\ \bar{\nabla}\omega_1 + i\bar{\theta}\pi_1 \end{pmatrix},$$

with obvious notations. As expected, $D_1^2\varphi_1 = \Omega_1\varphi_1$ and $\bar{D}_1^2\psi_1 = \bar{\Omega}_1\psi_1$. Notice also that $f_1 = P_1 \wedge \theta$ is the antisymmetric part of the tensor P_1.

We claim that φ_1 is a tractor and that the covariant derivative D_1 is a "generalized" tractor connection [5]. In the same way, we assert that ψ_1 is a twistor and that the covariant derivative $\bar{D}_1$ is a generalized twistor connection [53]. Both assertions are supported by the analysis of the residual gauge symmetries.

Residual gauge symmetries. Being by construction $\mathcal{K}_1 \simeq \bar{\mathcal{K}}_1$-invariant, the composite fields collectively denoted by χ_1 are expected to display $\mathcal{K}_0$-residual and $\bar{\mathcal{K}}_0$-residual gauge symmetries. The group K_0 breaks down as a direct product of the Lorentz and Weyl groups, $K_0 = SO(1,3)W$, and in the same way, $\bar{K}_0 = SL(2,\mathbb{C})W$, with respective matrix presentations

$$K_0 = \left\{ \mathsf{SZ} := \begin{pmatrix} 1 & 0 & 0 \\ 0 & S & 0 \\ 0 & 0 & 1 \end{pmatrix} \begin{pmatrix} z & 0 & 0 \\ 0 & \mathbb{1}_4 & 0 \\ 0 & 0 & z^{-1} \end{pmatrix} \,\middle|\, z \in W,\ S \in SO(1,3) \right\} \tag{27}$$
$$\bar{K}_0 = \left\{ \bar{\mathsf{S}}\bar{\mathsf{Z}} := \begin{pmatrix} \bar{S}^{-1*} & 0 \\ 0 & \bar{S} \end{pmatrix} \begin{pmatrix} z^{1/2} & 0 \\ 0 & z^{-1/2} \end{pmatrix} \,\middle|\, z \in W,\ \bar{S} \in SL(2,\mathbb{C}) \right\} \tag{28}$$

We focus on Lorentz symmetry first, then only bring our attention to Weyl symmetry. In the following, we will use the above matrix presentations S and $\bar{\mathsf{S}}$ for elements of the Lorentz gauge group $\mathcal{SO}$ and the $SL(2,\mathbb{C})$-gauge group $\mathcal{SL}$. The residual gauge transformations of the composite fields under $\mathcal{SO}$ is inherited from that of the dressing field u_1. Using ϖ^{γ_0} to compute $q^{\mathsf{S}} = a^{\mathsf{S}} \cdot (e^{\mathsf{S}})^{-1} = qS$, one easily finds that $u_1^{\mathsf{S}} = \mathsf{S}^{-1}u_1\mathsf{S}$, and correspondingly, $\bar{u}_1^{\bar{\mathsf{S}}} = \bar{\mathsf{S}}^{-1}\bar{u}_1\bar{\mathsf{S}}$. This is a local instance of Prop. 2, which then allows to conclude that the composite fields χ_1 are *genuine* gauge fields (see Sect. 2.2.1), w.r.t. Lorentz gauge symmetry. Hence, from Cor. 3 it follows that the residual $\mathcal{SO}$-gauge and $\mathcal{SL}$-gauge transformations are:

$$\varpi_1^{\mathsf{S}} = \mathsf{S}^{-1}\varpi_1\mathsf{S} + \mathsf{S}^{-1}d\mathsf{S} = \begin{pmatrix} 0 & P_1 S & 0 \\ S^{-1}\theta & S^{-1}A_1 S + S^{-1}dS & S^{-1}P^t \\ 0 & \theta^t S & 0 \end{pmatrix}, \tag{29}$$

$$\bar{\varpi}_1^{\bar{\mathsf{S}}} = \bar{\mathsf{S}}^{-1}\bar{\varpi}_1\bar{\mathsf{S}} + \bar{\mathsf{S}}^{-1}d\bar{\mathsf{S}} = \begin{pmatrix} -\left(\bar{S}^*\bar{A}_1\bar{S}^{-1*} + d\bar{S}^*\bar{S}^{-1*}\right) & -i\,\bar{S}^*\bar{P}_1\bar{S} \\ i\,\bar{S}^{-1}\bar{\theta}\bar{S}^{-1*} & \bar{S}^{-1}\bar{A}_1\bar{S} + \bar{S}^{-1}d\bar{S} \end{pmatrix}, \tag{30}$$

and

$$\Omega_1^{\mathsf{S}} = \mathsf{S}^{-1}\Omega_1\mathsf{S}, \qquad \varphi_1^{\mathsf{S}} = \mathsf{S}^{-1}\varphi_1, \qquad (D_1\varphi_1)^{\mathsf{S}} = \mathsf{S}^{-1}D_1\varphi_1, \tag{31}$$

$$\bar{\Omega}_1^{\bar{\mathsf{S}}} = \bar{\mathsf{S}}^{-1}\bar{\Omega}_1\bar{\mathsf{S}}, \qquad \psi_1^{\bar{\mathsf{S}}} = \bar{\mathsf{S}}^{-1}\psi_1, \qquad (\bar{D}_1\psi_1)^{\bar{\mathsf{S}}} = \bar{\mathsf{S}}^{-1}\bar{D}_1\psi_1. \tag{32}$$

See [1, 2] for details. Notice that φ_1 and ψ_1 transform as sections of the $SO(1,3)$-associated bundle $E_1 = E^{u_1} = \mathcal{P}\times_{SO}\mathbb{R}^6$ and the $SL(2,\mathbb{C})$-associated bundle $\mathsf{E}_1 = \mathsf{E}^{\bar{u}_1} = \bar{\mathcal{P}}\times_{SL}\mathbb{C}^4$ respectively.

We exactly repeat the same procedure to analyze the Weyl gauge symmetry, using again the matrix notations defined in (27) and (28) for Z in the Weyl group $\mathcal{W}\subset\mathcal{K}_0$ and $\bar{\mathsf{Z}}$ in its complex counterpart $\bar{\mathcal{W}}\subset\bar{\mathcal{K}}_0$. We first compute the action of $\mathcal{W}$ on the dressing field: using ϖ^{γ_0} to compute $q^{\mathsf{Z}} = a^{\mathsf{Z}}\cdot(e^{\mathsf{Z}})^{-1}$, one easily finds that $u_1^{\mathsf{Z}} = \mathsf{Z}^{-1}u_1C(z)$, where $C: W\to K_1W\subset H$ is defined by

$$C(z) := k_1(z)\mathsf{Z} = \begin{pmatrix} 1 & \Upsilon & \tfrac{1}{2}\Upsilon^2 \\ 0 & \mathbb{1}_4 & \Upsilon^t \\ 0 & 0 & 1\end{pmatrix}\begin{pmatrix} z & 0 & 0 \\ 0 & \mathbb{1}_4 & 0 \\ 0 & 0 & z^{-1}\end{pmatrix} = \begin{pmatrix} z & \Upsilon & \tfrac{z^{-1}}{2}\Upsilon^2 \\ 0 & \mathbb{1}_4 & z^{-1}\Upsilon^t \\ 0 & 0 & z^{-1}\end{pmatrix} \tag{33}$$

where explicitly $\Upsilon = \Upsilon_a = \Upsilon_\mu e^\mu{}_a$, with $\Upsilon_\mu := z^{-1}\partial_\mu z$, and $\Upsilon^2 = \Upsilon_a\eta^{ab}\Upsilon_b$. The corresponding complex case is $\bar{u}_1^{\bar{\mathsf{Z}}} = \bar{\mathsf{Z}}^{-1}\bar{u}_1\bar{C}(z)$, where $\bar{C}: W\to\bar{K}_1W\subset\bar{H}$ is defined by, with $\bar{\Upsilon} = \Upsilon_a\sigma^a$,

$$\bar{C}(z) := \bar{k}_1(z)\bar{\mathsf{Z}} = \begin{pmatrix} \mathbb{1}_2 & -i\bar{\Upsilon} \\ 0 & \mathbb{1}_2\end{pmatrix}\begin{pmatrix} z^{1/2}\mathbb{1}_2 & 0 \\ 0 & z^{-1/2}\mathbb{1}_2\end{pmatrix} = \begin{pmatrix} z^{1/2}\mathbb{1}_2 & -i\,z^{-1/2}\bar{\Upsilon} \\ 0 & z^{-1/2}\mathbb{1}_2\end{pmatrix}. \tag{34}$$

The map C is not a group morphism, $C(z)C(z')\neq C(zz')$, but is a local instance of a 1-α-cocycle satisfying Prop. 6: $C(zz') = C(z'z) = C(z')\,\mathsf{Z}'^{-1}C(z)\mathsf{Z}'$. Under a further $\mathcal{W}$-gauge transformation and due to $e^{\mathsf{Z}} = ze$, one has $k_1(z)^{\mathsf{Z}'} = \mathsf{Z}'^{-1}k_1(z)\mathsf{Z}'$, which implies $C(z)^{\mathsf{Z}'} = \mathsf{Z}'^{-1}C(z)\mathsf{Z}'$. So, if u_1 undergoes a further $\mathcal{W}$-gauge transformation Z', we get $\left(u_1^{\mathsf{Z}}\right)^{\mathsf{Z}'} = \left(\mathsf{Z}^{\mathsf{Z}'}\right)^{-1}u_1^{\mathsf{Z}'}C(z)^{\mathsf{Z}'} = \mathsf{Z}^{-1}\,\mathsf{Z}'^{-1}u_1C(z')\,\mathsf{Z}'^{-1}C(z)\mathsf{Z}' = (\mathsf{Z}\mathsf{Z}')^{-1}u_1C(zz')$. *Mutadis mutandis*, all this is true for $\bar{C}$ in (34) and for $\bar{u}_1$ as well. We have then a well-behaved action of the gauge groups $\mathcal{W}$ and $\bar{\mathcal{W}}$ in the real and complex cases.

From this we conclude that the composite fields χ_1 are instances of generalized gauge fields described in Sect. 2.2.2. By Prop. 5, the residual $\mathcal{W}$-gauge and $\bar{\mathcal{W}}$-gauge transformations are $\varpi_1^{\mathsf{Z}} = C(z)^{-1}\varpi_1C(z) + C(z)^{-1}dC(z)$ and $\bar{\varpi}_1^{\bar{\mathsf{Z}}} = \bar{C}(z)^{-1}\bar{\varpi}_1\bar{C}(z) + \bar{C}(z)^{-1}d\bar{C}(z)$, explicitly given by

$$\varpi_1^Z = \begin{pmatrix} 0 & z^{-1}\big(P_1 + \nabla\Upsilon - \Upsilon\theta\Upsilon + \tfrac{1}{2}\Upsilon^2\theta^t\big) & 0 \\ z\theta & A_1 + \theta\Upsilon - \Upsilon^t\theta^t & * \\ 0 & z\theta^t & 0 \end{pmatrix}, \tag{35}$$

$$\bar{\varpi}_1^{\bar{Z}} = \begin{pmatrix} -\bar{A}_1^* - (\bar{\Upsilon}\bar{\theta})_0 & -i\, z^{-1}\left[\bar{P}_1 + \big(d\bar{\Upsilon} - \bar{\Upsilon}\bar{A}_1 - \bar{A}_1^*\bar{\Upsilon}\big) - \bar{\Upsilon}\bar{\theta}\bar{\Upsilon}\right] \\ i\, z\bar{\theta} & \bar{A}_1 + (\bar{\theta}\bar{\Upsilon})_0 \end{pmatrix}, \tag{36}$$

where $(\bar{\theta}\bar{\Upsilon})_0$ is the $\mathfrak{sl}(2,\mathbb{C})$ part of $\bar{\theta}\bar{\Upsilon} = (\bar{\theta}\bar{\Upsilon})_0 + \frac{\Upsilon\theta}{2}\mathbb{1}_2$. And (see [1, 2] for details):

$$\Omega_1^Z = C(z)^{-1}\Omega_1 C(z) \qquad \bar{\Omega}_1^{\bar{Z}} = \bar{C}(z)^{-1}\bar{\Omega}_1\bar{C}(z) \tag{37}$$

$$\varphi_1^Z = C(z)^{-1}\varphi_1 = \begin{pmatrix} z^{-1}\left(\rho_1 - \Upsilon\ell_1 + \frac{\sigma}{2}\Upsilon^2\right) \\ \ell_1 - \Upsilon^t\sigma \\ z\sigma \end{pmatrix}, \qquad (D_1\varphi_1)^Z = C(z)^{-1}D_1\varphi_1, \tag{38}$$

$$\psi_1^{\bar{Z}} = \bar{C}(z)^{-1}\psi_1 = \begin{pmatrix} z^{-1/2}\left(\pi_1 + i\bar{\Upsilon}\omega_1\right) \\ z^{1/2}\omega_1 \end{pmatrix}, \qquad (\bar{D}_1\psi_1)^{\bar{Z}} = \bar{C}(z)^{-1}\bar{D}_1\psi_1. \tag{39}$$

From (35), we see that A_1 exhibits the known Weyl transformation for the Lorentz connection, and P_1 transforms as the Schouten tensor (in an orthonormal basis). But, actually, the former genuinely reduces to the latter only when one restricts to the dressing of the *normal* Cartan connection $\varpi_{N,1}$, so that A_1 is a function of θ and $\mathsf{P}_1 = P_1(A_1)$ is the genuine symmetric Schouten tensor. So f_1 vanishes and we have,

$$\Omega_{N,1} = d\varpi_{N,1} + \varpi_{N,1}^2 = \begin{pmatrix} 0 & \mathsf{C}_1 & 0 \\ 0 & \mathsf{W}_1 & \mathsf{C}_1^t \\ 0 & 0 & 0 \end{pmatrix}, \tag{40}$$

$$\Omega_{N,1}^Z = C(z)^{-1}\Omega_{N,1}C(z) = \begin{pmatrix} 0 & z^{-1}(\mathsf{C}_1 - \Upsilon\mathsf{W}_1) & 0 \\ 0 & \mathsf{W}_1 & * \\ 0 & 0 & * \end{pmatrix}, \tag{41}$$

$$\bar{\Omega}_{N,1} = d\bar{\varpi}_{N,1} + \bar{\varpi}_{N,1}^2 = \begin{pmatrix} -\bar{\mathsf{W}}_1^* & -i\bar{\mathsf{C}}_1 \\ 0 & \bar{\mathsf{W}}_1 \end{pmatrix}, \tag{42}$$

$$\bar{\Omega}_{N,1}^{\bar{Z}} = \bar{C}(z)^{-1}\bar{\Omega}_{N,1}\bar{C}(z) = \begin{pmatrix} -\bar{\mathsf{W}}_1^* & -i\, z^{-1}\big(\bar{\mathsf{C}}_1 - \bar{\Upsilon}\bar{\mathsf{W}}_1 - \bar{\mathsf{W}}_1^*\bar{\Upsilon}\big) \\ 0 & \bar{\mathsf{W}}_1 \end{pmatrix}. \tag{43}$$

We see that $\mathsf{C}_1 = \nabla\mathsf{P}_1$ is the Cotton tensor, and indeed transforms as such, while W_1 is the invariant Weyl tensor.

From φ_1^Z in (38), we see that the dressed section φ_1 is a section of the $C(W)$-twisted vector bundle $E_1 = E^{u_1} = \mathcal{P} \times_{C(W)} \mathbb{R}^6$ (see (9)). The latter is nothing but the defining Weyl of a tractor field as derived in [5]. Then E_1 is the so-called *standard tractor bundle*. Since $C(z) \in K_1W \subset H$, we have $(C(z)^{-1})^T\Sigma C(z)^{-1} = \Sigma$. So the bilinear form on E defined by the group metric Σ is also defined on E_1: $\langle\varphi_1, \varphi_1'\rangle = \varphi_1^T\Sigma\varphi_1'$.

This is otherwise known as the *tractor metric*. Furthermore, $(D_1\varphi_1)^{\mathsf{Z}}$ in (38) shows that the operator $D_1 := d + \varpi_1$ is a generalization of the *tractor connection* [5, 14]. The term "connection", while not inaccurate, could hide the fact that ϖ_1 is no more a geometric connection w.r.t. Weyl symmetry. So we shall prefer to call D_1 a generalized tractor *covariant derivative*. The standard tractor covariant derivative is recovered by restriction to the dressing of the normal Cartan connection, $D_{\text{N},1} = d + \varpi_{\text{N},1}$, and $\Omega_{\text{N},1}$ in (40) is known as the *tractor curvature*.

In the same way, $\psi_1^{\bar{\mathsf{Z}}}$ in (39) shows that the dressed section ψ_1 is a section of the $\bar{C}(W)$-twisted vector bundle $\mathsf{E}_1 = \mathsf{E}^{\bar{u}_1} = \bar{\mathcal{P}} \times_{\bar{C}(W)} \mathbb{C}^4$. The latter is also, modulo the z factors, the defining Weyl transformation of a local twistor as given by Penrose [53]. So E_1 is identified with the *local twistor bundle*. It is endowed with a bilinear form defined by the group metric $\bar{\Sigma}$ of $SU(2,2)$: $\langle \psi_1, \psi_1' \rangle = \psi_1^T \bar{\Sigma} \psi_1'$. It is well-defined since, in view of $\bar{C}(z) \in \bar{K}_1 W \subset \bar{H}$, we have $(C(z)^{-1})^* \bar{\Sigma} C(z)^{-1} = \bar{\Sigma}$. In the twistor literature, the quantity $\frac{1}{2}\langle \psi_1, \psi_1 \rangle$ is known as the *helicity* of the twistor field ψ_1 [51, 52]. Also, $(\bar{D}_1\psi_1)^{\bar{\mathsf{Z}}}$ in (39) shows that the operator $\bar{D}_1 := d + \bar{\varpi}_1$ is a generalization of the *twistor connection*. For the reason stated above, we shall prefer to call $\bar{D}_1$ a generalized twistor covariant derivative. The usual twistor covariant derivative is recovered by restriction to the normal case, $\bar{D}_{\text{N},1} = d + \bar{\varpi}_{\text{N},1}$, and $\bar{\Omega}_{\text{N},1}$ in (42) is known as the *twistor curvature*.

Remark that the actions of the Lorentz/$SL(2,\mathbb{C})$ and Weyl gauge groups on the composite fields χ_1 commute. In the real case for instance, we have $\mathsf{S}^{\mathcal{W}} = \mathsf{S}$ so that $(\chi_1^{\mathcal{SO}})^{\mathcal{W}} = (\chi_1^{\mathsf{S}})^{\mathcal{W}} = (\chi_1^{\mathcal{W}})^{\mathsf{S}^{\mathcal{W}}} = (\chi_1^{C(z)})^{\mathsf{S}} = \chi_1^{C(z)\mathsf{S}}$. But we also have $C(z)^{\mathcal{SO}} = \mathsf{S}^{-1}C(z)\mathsf{S}$, so we get $(\chi_1^{\mathcal{W}})^{\mathcal{SO}} = (\chi_1^{C(z)})^{\mathcal{SO}} = (\chi_1^{\mathcal{SO}})^{C(z)^{\mathcal{SO}}} = (\chi_1^{\mathsf{S}})^{\mathsf{S}^{-1}C(z)\mathsf{S}} = \chi_1^{C(z)\mathsf{S}}$. Our notations for the tractor and twistor bundles can then be refined to reflect this point: $E_1 = \mathcal{P} \times_{C(W)\cdot SO} \mathbb{R}^6$ and $\mathsf{E}_1 = \mathcal{P} \times_{\bar{C}(W)\cdot SL} \mathbb{C}^4$.

Following the ending considerations of Sect. 2.2.1, the fact that the composite fields ϖ_1, φ_1 are genuine Lorentz-gauge fields satisfying (29) and (31) suggests that a further dressing operation aiming at erasing Lorentz symmetry is possible. In [1] we showed that in the case of tractors, the vielbein $e = e^a{}_\mu$ could be used to this purpose since it has the transformation $e^S = S^{-1}e$, characteristic of a $\mathcal{SO}$-dressing field. This is the same process as in the example of GR, treated in Sect. 4. The difference is that in GR one erases Lorentz symmetry and ends-up with "nothing", that is no gauge symmetry but only coordinate transformations characteristic of geometric objects living on $\mathcal{M}$, while in the tractor case one ends-up with Weyl rescalings as residual gauge symmetry in addition to coordinate transformations. Computing the residual Weyl symmetry after this second dressing displays a slightly different C-map to be used to perform the transformation of the composite fields, see [1]. As a matter of fact, in the literature two kinds of transformation law for tractors can be found, which in our framework corresponds to either erasing only the K_1-symmetry [56, 57], or to erasing both K_1 and Lorentz-symmetries [5, 14].

Since there is no finite dimensional spin representation of GL, in the twistor case the vielbein cannot be used as a second dressing field. By the way, looking at the $SL(2,\mathbb{C})$ gauge transformation of the vielbein, one sees that it is unsuited as a

$\mathcal{SL}$-dressing field. So, as far as twistors are concerned, the process of symmetry reduction ends here.

BRST treatment The gauge group of the initial Cartan geometries are $\mathcal{H}$ and $\bar{\mathcal{H}}$. The associated ghost $v \in \text{Lie}\mathcal{H}$ and $\bar{v} \in \text{Lie}\bar{\mathcal{H}}$ split along the grading of $\mathfrak{h}$ and $\bar{\mathfrak{h}}$,

$$v = v_0 + v_\iota = v_\varepsilon + v_{\text{s}} + v_\iota = \begin{pmatrix} \varepsilon & \iota & 0 \\ 0 & s & \iota^t \\ 0 & 0 & -\varepsilon \end{pmatrix},$$

$$\bar{v} = \bar{v}_0 + \bar{v}_\iota = \bar{v}_\varepsilon + \bar{v}_{\text{s}} + \bar{v}_\iota = \begin{pmatrix} -(\bar{s}^* - {}^{\varepsilon}\!/\!{}_2) & -i\bar{\iota} \\ 0 & \bar{s} - {}^{\varepsilon}\!/\!{}_2 \end{pmatrix}.$$

The BRST operator splits accordingly as $s = s_0 + s_1 = s_{\text{W}} + s_{\text{L}} + s_1$. The algebra satisfied by the gauge fields $\chi = \{\varpi, \Omega, \varphi, \bar{\varpi}, \bar{\Omega}, \psi\}$, noted BRST, is

$$\begin{aligned} s\varpi &= -Dv = -dv - [\varpi, v], & s\Omega &= [\Omega, v], & sv &= -v^2, \\ s\bar{\varpi} &= -D\bar{v} = -d\bar{v} - [\bar{\varpi}, \bar{v}], & s\bar{\Omega} &= [\bar{\Omega}, \bar{v}], & s\bar{v} &= -\bar{v}^2, \\ s\varphi &= -v\varphi, & s\psi &= -\bar{v}\psi, \end{aligned}$$

From Sect. 2.3, the composite fields $\chi_1 = \{\varpi_1, \Omega_1, \varphi_1, \psi_1\}$ satisfy a modified BRST algebra, formally similar but with composite ghost $v_1 := u_1^{-1} v u_1 + u_1^{-1} s u_1$. From the finite gauge transformations of u_1, and the linearizations $\gamma_1 \simeq \mathbb{1} + v_\iota$ and $\mathsf{S} \simeq \mathbb{1} + v_{\text{s}}$, the BRST actions of $\mathcal{K}_1$ and $\mathcal{SO}$ are found to be: $s_1 u_1 = -v_\iota u_1$ and $s_{\text{L}} u_1 = [u_1, v_{\text{s}}]$. This shows that the Lorentz sector is an instance of the general result (15). Using the linearizations $\mathsf{Z} \simeq \mathbb{1} + v_\varepsilon$ and $k_1(z) \simeq \mathbb{1} + \kappa_1(\varepsilon)$, so that $C(z) = k_1(z)\mathsf{Z} \simeq \mathbb{1} + c(\varepsilon) = \mathbb{1} + \kappa_1(\varepsilon) + v_\varepsilon$, the BRST action of $\mathcal{W}$ is $s_{\text{W}} u_1 = -v_\varepsilon u_1 + u_1 c(\varepsilon)$. This shows that the Weyl sector is an instance of the general result (17). After a straightforward computation and a similar analysis for the complex case, we get the composite ghosts

$$v_1 = c(\varepsilon) + v_{\text{s}} = \begin{pmatrix} \varepsilon & \partial\varepsilon & 0 \\ 0 & s & \partial\varepsilon^t \\ 0 & 0 & -\varepsilon \end{pmatrix}, \quad \bar{v}_1 = \bar{c}(\varepsilon) + \bar{v}_{\text{s}} = \begin{pmatrix} -\left(\bar{s}^* - \frac{\varepsilon}{2}\mathbb{1}_2\right) & -i\bar{\partial}\varepsilon \\ 0 & \bar{s} - \frac{\varepsilon}{2}\mathbb{1}_2 \end{pmatrix},$$

where $\partial\varepsilon := \partial_a\varepsilon = \partial_\mu\varepsilon\, e^\mu{}_a$. The ghost of conformal boosts, ι, has disappeared from these new ghosts, replaced by the first derivative of the Weyl ghost. This means that $s_1\chi_1 = 0$, which reflects the $\mathcal{K}_1$-gauge invariance of the composite fields χ_1. The composite ghost v_1 only depends on v_{s} and ε: it encodes the residual $\mathcal{K}_0$-gauge symmetry. The algebra satisfied by the composite fields χ_1, denoted by $\mathsf{BRST}_{\text{W,L}}$, is then simply

$$s\varpi_1 = -D_1 v_1 = -dv_1 - [\varpi_1, v_1], \qquad s\Omega_1 = [\Omega_1, v_1], \qquad sv_1 = -v_1^2,$$
$$s\bar{\varpi}_1 = -D_1 \bar{v}_1 = -d\bar{v}_1 - [\bar{\varpi}_1, \bar{v}_1], \qquad s\bar{\Omega}_1 = [\bar{\Omega}_1, \bar{v}_1], \qquad s\bar{v}_1 = -\bar{v}_1^2,$$
$$s\varphi_1 = -v_1 \varphi_1, \qquad s\psi_1 = -\bar{v}\psi_1,$$

and reproduces the infinitesimal version of (29)–(32) (Lorentz/$SL(2, \mathbb{C})$ sector) and (35)–(39) (Weyl sector). Explicit results are obtained via simple matrix calculations, we refer to [1, 2] for all the details.

Since $v_1 = c(\varepsilon) + v_s$, $\mathsf{BRST}_{\text{W,L}}$ splits naturally as Lorentz and Weyl subalgebras, $s = s_{\text{W}} + s_{\text{L}}$. The Lorentz sector (s_{L}, v_s) shows the composite fields χ_1 to be genuine Lorentz gauge fields. While the Weyl sector $(s_{\text{W}}, c(\varepsilon))$ shows χ_1 to be generalized Weyl gauge fields.

5.3 *Discussion*

Today, tractors and twistors are terms whose meaning extends beyond their original context of definition, conformal (and projective) geometry, and are quite broad concepts in the theory of parabolic geometries [11]. In their original meaning, most often tractor and local twistor bundles are constructed in a "bottom-up" way, starting with a conformal manifold $(\mathcal{M}, c)$ and building a gauge structure on top of it.

First, one poses a defining differential equation on $(\mathcal{M}, c)$. In the case of tractors, this is the *almost Einstein equation* (AE)

$$\nabla_\mu \nabla_\nu \sigma - \mathsf{P}_{\mu\nu}\sigma - \tfrac{g_{\mu\nu}}{n}(\Delta\sigma - \mathsf{P}\sigma) = 0,$$

with σ a 1-conformal density ($\widehat{\sigma} = z^{-1}\sigma$), ∇ the Levi-Civita connection associated to a choice of metric $g_{\mu\nu} \in c$, $\Delta := g^{\mu\nu}\nabla_\mu\nabla_\nu$, and $\mathsf{P} := g^{\mu\nu}\mathsf{P}_{\mu\nu}$. For twistors, one defines the *twistor equation*

$$\nabla^{(A}{}_{A'}\omega^{B)} = 0, \quad \text{or equivalently} \quad \nabla_{AA'}\omega^B - \tfrac{1}{2}\delta_A^B\, \nabla_{CA'}\omega^C = 0,$$

where $\omega^B : \mathcal{M} \to \mathbb{C}^2$ is a Weyl spinor. Then one prolongs these equations, recast them as first order systems. These are interpreted as first order differential operators acting on multiplets: $\nabla^{\mathcal{T}} V = 0$ and $\nabla^{\mathsf{T}} Z = 0$ respectively, where $V = (\sigma, \ell_\mu, \rho)$ and $Z = (\omega^A, \pi_{A'})$. The transformations of the components of V and Z under Weyl rescaling of the metric is given either by definition, when the components are functions of the metric (V), or by choice (Z). This takes some algebra to prove. With still more algebra, one shows that these transformation laws also apply to $\nabla^{\mathcal{T}} V$ and $\nabla^{\mathsf{T}} Z$. But then V and Z are interpreted as parallel sections of some vectors bundles over $\mathcal{M}$, the *standard tractor bundle* $\mathcal{T}$ and *local twistor bundle* T respectively, which are endowed with their linear connections, the *tractor connection* $\nabla^{\mathcal{T}}$ and *twistor connection* ∇^{T} (hence the notation). Their commutators $[\nabla^{\mathcal{T}}, \nabla^{\mathcal{T}}]V = \kappa V$ and $[\nabla^{\mathsf{T}}, \nabla^{\mathsf{T}}]Z = \mathsf{K}Z$ are said to define respectively the tractor and twistor curvatures.

Thus, starting from differential equations on $(\mathcal{M}, c)$, one ends-up with a gauge structure on top of it in the form of the tractor and twistor bundles and their connections. The latter provide natural conformally covariant calculi for torsion-free conformal manifolds. We refer the reader to [5, 14] for detailed calculations of this bottom-up procedure in the tractor case, and to the classic [53, Sect. 6.9] for the twistor case. See also [18, Sect. 6.1], which extends the twistor construction to para-conformal manifolds. It has been noticed that the tractor and twistor bundles can be seen as associated bundles to the principal Cartan bundle $\mathcal{P}(\mathcal{M}, H)$, and a link between the normal conformal Cartan connection and the twistor 1-form was drawn by Friedrich [26]. Nevertheless, the construction via prolongation has been deemed more explicit in [18], and more intuitive and direct in [5], than the viewpoint in terms of associated vector bundles.

However, our procedure present several advantages. Starting from a "bigger" gauge structure over $\mathcal{M}$ controlled by the conformal Cartan bundle $\mathcal{P}$ and a double cover complex version $\bar{\mathcal{P}}$, we obtain the vectors bundles endowed with covariant derivatives (E_1, D_1) and $(\mathsf{E}_1, \bar{D}_1)$ in a very straightforward and systematic way by symmetry reduction. So, our constructive procedure via the dressing method is "top-down" and involves much less calculations than the usual "bottom-up" approach outlined above, and is arguably more direct and intuitive.

Furthermore, these bundles reduce to the usual tractor and twistor bundles and their respective covariant derivatives when restricted to the *normal* Cartan geometry, and one gets $(E_1, D_{\text{N},1}) = (\mathcal{T}, \nabla^{\mathcal{T}})$ and $(\mathsf{E}_1, \bar{D}_{\text{N},1}) = (\mathsf{T}, \nabla^{\mathsf{T}})$. So, here we effortlessly generalize the tractor and twistor derivatives, providing essentially tractor and twistor calculi for conformal manifolds with torsion. It follows that if $\varpi_{\text{N},1}$ and $\bar{\varpi}_{\text{N},1}$ are the genuine tractor and twistor 1-forms, then ϖ_1 and $\bar{\varpi}_1$ may be labeled as *generalized tractor and twistor* 1*-forms*.

Our approach allows to clearly highlight the fact that, while tractors, twistors, and the associated (generalized) 1-forms and curvatures are genuine Lorentz/$SL(2, \mathbb{C})$ gauge fields, they are gauge fields of generalized kind w.r.t. Weyl rescaling gauge symmetry, transforming using a 1-α-cocycle on the Weyl group. A fact that, as far as we know, has never been noticed.

Let us finally notice that in this framework, one can easily write a Yang-Mills-type Weyl-invariant Lagrangian and compute the corresponding field equations. It turns out that this Lagrangian reproduces Weyl gravity if one restricts to a normal Cartan connection, as was shown in [4]. This by the way explains the equivalence between the Bach equation and the Yang-Mills equation for the normal conformal Cartan connection [35] or the twistor 1-form [48].

6 Conclusion

The dressing field method of gauge symmetry reduction is a fourth way, beside gauge fixing, SSBM, and the bundle reduction theorem, to handle challenges one faces in gauge theories. As a matter of fact, as mentioned at the end of Sect. 2.4, it is relevant in many places in gauge fields theories, from QCD to anomalies in QFT. In this review paper we have outlined the main general results of the method concerning the construction of partially gauge invariant composite fields out of the usual gauge variables, and discussed two important cases where their residual gauge transformations can be treated on a general ground. Interestingly, we saw a case in which the composite fields are gauge fields of an unusual geometric nature, so that we label them "generalized" gauge fields.

We have shown that the method applies to the BEHGHK mechanism pivotal to the electroweak model. In doing so, we highlighted the fact that the notion of spontaneously broken gauge symmetry, which have long raised doubts among both philosophers of science and lattice gauge theorists (in view of the Elitzur theorem), is dispensable and anyway unnecessary for the empirical success of the Standard Model. This result is thus satisfying from a philosophical standpoint, and does not question the heuristic power of the gauge principle.

We have argued that the usual switching between the tetrad and metric formulations of GR is a simple application of the dressing field method. In doing so, we have stressed that, contrary to what is sometimes said, the linear connection Γ and the Lorentz connection A are not mutual gauge transformations, even if one considers GR as a gauge theory on the frame bundle $L\mathcal{M}$. Actually, to recover A from Γ one needs the bundle reduction theorem, and to get Γ from A one needs the dressing field method. So that, in this instance, these are reciprocal operations.

The method applied to the conformal Cartan geometry and its spinorial version allows to obtain generalizations of the tractor and twistor calculi for conformal manifolds, extending them to manifolds with torsion, in a very straightforward "top-down" way. It happens to be computationally much more economical than the usual "bottom-up" approach by prolongation of the Almost Einstein and twistor equations, and arguably more direct and intuitive. Also, we have seen that tractors and twistors, while being genuine Lorentz gauge fields, are generalized gauge fields as far as Weyl rescaling symmetry is concerned.

One suspects that still more instances of the dressing field method could be found in the literature on gauge theories. Furthermore, its simplicity may put within reach results otherwise difficult to obtain by other approaches; the example of tractor calculi for various parabolic geometries and their application to physics comes to mind. It is our hope that this approach could contribute to clarify and enrich some aspects of gauge field theories in physics.

In its present form, the method relies on the defining (structural) relations for gauge transformations: as already mentioned, while the field contents are different, definitions (4) look *algebraically* like gauge transformations (1).[10] This is a key ingredient of the method. One can raise the question about some possible other routes one could elaborate to define dressed fields on which a part of the gauge symmetry is erased, but not using gauge transformation-like relations.

---

[10]Let us mention here how it is has been difficult, in several occasions, to convince some colleagues that these relations are not mathematically on the same footing.

Finally, to make the dressing field method a full-fledged approach to gauge QFT, the question of its compatibility with quantization must be addressed. In particular, do the operations of quantization and of reduction by dressing commute? So far, the question has not been fully addressed. One can find in [46] some hints that the problem is not easy and straightforward, mainly because we may first face the problem of the definition of a mathematically sound, let alone unique, quantization scheme. A rich topic in itself, that again exemplifies the fruitful cross-fertilization between physics and mathematics.

References

1. J. Attard, J. François. Tractors and Twistors from Conformal Cartan Geometry: A Gauge Theoretic Approach I. Tractors. Adv. Theor. Math. Phys. 2016. arXiv:1609.07307
2. J. Attard, J. François. Tractors and Twistors from Conformal Cartan Geometry: A Gauge Theoretic Approach II. Twistors. Class. Quantum Grav. **34**(8), 085004, 2017. arxiv:1611.03891
3. J. Attard, S. Lazzarini. A note on Weyl invariance in gravity and the Wess-Zumino functional. Nucl. Phys. B, 2016. ISSN 0550-3213. https://doi.org/10.1016/j.nuclphysb.2016.07.016
4. J. Attard, J. François, S. Lazzarini. Weyl gravity and Cartan geometry. Phys. Rev. D, **93**, 085032, 2016. https://doi.org/10.1103/PhysRevD.93.085032
5. T. Bailey, M. Eastwood, A.R. Gover. Thomas's structure bundle for conformal, projective and related structures. Rocky Mt. J. Math. **24**(4), 1191–1217 (1994). https://doi.org/10.1216/rmjm/1181072333
6. C.M. Becchi, G. Ridolfi, *An Introduction to Relativistic Processes and the Standard Model of Electroweak Interactions* (Springer, Berlin, 2006)
7. L. Bonora, P. Cotta-Ramusino, Some remark on brs transformations, anomalies and the cohomology of the Lie algebra of the group of gauge transformations. Comm. Math. Phys. **87**, 589–603 (1983)
8. K. Brading, E. Castellani, *Symmetries in Physics: Philosophical Reflections* (Cambridge University Press, Cambridge, 2003)
9. K. Brading, E. Castellani. *Symmetry and Symmetry Breaking*, 2013. http://plato.stanford.edu/archives/spr2013/entries/symmetry-breaking/
10. R. Brout, F. Englert, Broken symmetry and the mass of the gauge vector mesons. Phys. Rev. Lett. **13**, 321–323 (1964)
11. A. Cap, J. Slovak. Parabolic geometries I: background and general theory, in *Mathematical Surveys and Monographs*, vol. 1 (American Mathematical Society, 2009)
12. S.S. Chern, W.H. Chen, K.S. Lam. *Lectures on differential geometry*. Series on University Mathematics (Book 1) (World Scientific Publishing Company, 1999)
13. M.N. Chernodub, L. Faddeev, A.J. Niemi. Non-Abelian supercurrents and electroweak theory. JHEP, **12**, 014 (2008). https://doi.org/10.1088/1126-6708/2008/12/014
14. S. Curry, A.R. Gover. *An Introduction to Conformal Geometry and Tractor Calculus, with a View to Applications in General Relativity*, 2014. arXiv:1412.7559
15. P.A.M. Dirac, Gauge-invariant formulation of quantum electrodynamics. Can. J. Phys. **33**, 650–660 (1955)
16. P.A.M. Dirac, *The Principles of Quantum Mechanics*, 4th edn. (Oxford University Press, Oxford, 1958)
17. M. Dubois-Violette, The Weil-BRS algebra of a Lie algebra and the anomalous terms in gauge theory. J. Geom. Phys. **3**, 525–565 (1987)
18. M. Eastwood, T. Bailey, Complex paraconformal manifolds - their differential geometry and twistor theory. Forum Mathematicum **3**(1), 61–103 (1991)

19. L. Faddeev, in *An Alternative Interpretation of the Weinberg-Salam Model* (Springer, Dordrecht, 2009). pp. 3–8. ISBN 978-90-481-2287-5. https://doi.org/10.1007/978-90-481-2287-5_1
20. C. Fournel, J. François, S. Lazzarini, T. Masson, Gauge invariant composite fields out of connections, with examples. Int. J. Geom. Methods Mod. Phys. **11**(1), 1450016 (2014)
21. J. François, *Reduction of Gauge Symmetries: A New Geometrical Approach* (Aix-Marseille Université, Thesis, 2014)
22. J. François, S. Lazzarini, T. Masson. Nucleon spin decomposition and differential geometry. Phys. Rev. D, **91**, 045014 (2015). https://doi.org/10.1103/PhysRevD.91.045014
23. J. François, S. Lazzarini, T. Masson. Residual Weyl symmetry out of conformal geometry and its BRST structure. JHEP **09**, 195 (2015). https://doi.org/10.1007/JHEP09(2015)195
24. J. François, S. Lazzarini, T. Masson. Becchi-Rouet-Stora-Tyutin structure for the mixed Weyl-diffeomorphism residual symmetry. J. Math. Phys. **57**(3), 033504 (2016). https://doi.org/10.1063/1.4943595
25. S. Friederich. Gauge symmetry breaking in gauge theories—in search of clarification. Eur. J. Philos. Sci. **3**(2), 157–182 (2013). ISSN 1879-4920. https://doi.org/10.1007/s13194-012-0061-y
26. H. Friedrich. Twistor connection and normal conformal Cartan connection. Gen. Relativ. Gravit. **8**(5), 303–312 (1977). ISSN 1572-9532. https://doi.org/10.1007/BF00771141
27. J. Frohlich, G. Morchio, F. Strocchi. Higgs phenomenon without symmetry breaking order parameter. Nucl. Phys. B **190**(3), 553–582 (1981). ISSN 0550-3213. https://doi.org/10.1016/0550-3213(81)90448-X
28. D. Garajeu, R. Grimm, S. Lazzarini, W-gauge structures and their anomalies: an algebraic approach. J. Math. Phys. **36**, 7043–7072 (1995)
29. V.N. Gribov, Quantization of non-abelian gauge theories. Nucl. Phys. B **139**, 1–19 (1978)
30. G.S. Guralnik, C.R. Hagen, T.W.B. Kibble, Global conservation laws and massless particles. Phys. Rev. Lett. **13**, 585–587 (1964)
31. P.W. Higgs, Broken symmetry and the mass of gauge bosons. Phys. Rev. Lett. **13**, 508–509 (1964)
32. P.W. Higgs. Spontaneous symmetry breakdown without massless bosons. Phys. Rev. **145**, 1156–1163 (1966). https://doi.org/10.1103/PhysRev.145.1156
33. A. Ilderton, M. Lavelle, D. McMullan, Symmetry breaking, conformal geometry and gauge invariance. J. Phys. A: Math. Theor. **43**(31), 312002 (2010)
34. T.W.B. Kibble. Symmetry breaking in non-abelian gauge theories. Phys. Rev. **155**, 1554–1561 (1967). https://doi.org/10.1103/PhysRev.155.1554
35. M. Korzyński, J. Lewandowski, The normal conformal Cartan connection and the Bach tensor. Class. Quantum Grav. **20**(16), 3745 (2003)
36. F. Langouche, T. Schücker, R. Stora, Gravitational anomalies of the Adler-Bardeen type. Phys. Lett. B **145**, 342–346 (1984)
37. L.D. Lantsman, Dirac fundamental quantization of gauge theories is the natural way of reference frames in modern physics. Fizika B **18**, 99–140 (2009)
38. M. Lavelle, D. McMullan. Nonlocal symmetry for QED. Phys. Rev. Lett. **71**, 3758–3761 (1993). https://doi.org/10.1103/PhysRevLett.71.3758
39. M. Lavelle, D. McMullan. Observables and gauge fixing in spontaneously broken gauge theories. Phys. Lett. B **347**(1), 89–94 (1995). ISSN 0370-2693. https://doi.org/10.1016/0370-2693(95)00046-N
40. M. Lavelle, D. McMullan, Constituent quarks from QCD. Phys. Rep. **279**, 1–65 (1997)
41. S. Lazzarini, C. Tidei. Polyakov soldering and second-order frames: the role of the Cartan connection. Lett. Math. Phys. **85**(1), 27–37 (2008). ISSN 1573-0530. https://doi.org/10.1007/s11005-008-0253-8
42. E. Leader, C. Lorcé, The angular momentum controversy: what is it all about and does it matter? Phys. Rep. **514**, 163–248 (2014)
43. C. Lorcé, Geometrical approach to the proton spin decomposition. Phys. Rev. D **87**, 034031 (2013)

44. J. Mañes, R. Stora, B. Zumino, Algebraic study of chiral anomalies. Comm. Math. Phys. **102**, 157–174 (1985)
45. C.A. Martin, Gauge principles, gauge arguments and the logic of nature. Proc. Philos. Sci. Assoc. **3**, 221–234 (2002)
46. T. Masson, J.-C. Wallet. A remark on the spontaneous symmetry breaking mechanism in the standard model, 2011. arXiv:1001.1176
47. S.W. McDowell, F. Mansouri, Unified geometric theory of gravity and supergravity. Phys. Rev. Lett. **38**, 739–742 (1977)
48. S.A. Merkulov, The twistor connection and gauge invariance principle. Comm. Math. Phys. **93**(3), 325–331 (1984)
49. L. O'Raifeartaigh, *The Dawning of Gauge Theory* (Princeton University Press, Princeton Series in Physics, Princeton, 1997)
50. G.K. Pedersen, *C-star Algebras and Their Automorphisms Groups* (London Mathematical Society Monographs, Academic Press Inc., Cambridge, 1979)
51. R. Penrose, The central programme of twistor theory. Chaos Solitons Fractals **10**, 581–611 (1999)
52. R. Penrose, M. MacCallum. Twistor theory: an approach to the quantisation of fields and space-time. Phys. Rep. **6**(4), 241–316 (1973). ISSN 0370-1573. https://doi.org/10.1016/0370-1573(73)90008-2
53. R. Penrose, W. Rindler, *Spinors and Space-Time*, vol. 2 (Cambridge University Press, Cambridge, 1986)
54. V. Pervushin. Dirac variables in gauge theories. Lecture notes in DAAD Summerschool on Dense Matter in Particle - and Astrophysics, JINR, Dubna, Russia, August 20–31, 2001. arXiv:hep-th/0109218v2
55. A.M. Polyakov. Gauge transformations and diffeomorphisms. Int. J. Mod. Phys. **A5**, 833 (1990). https://doi.org/10.1142/S0217751X90000386
56. A.R. Gover, A. Shaukat, A. Waldron. Tractors, mass and Weyl invariance. Nucl. Phys. B **812**(3), 424–455 (2009)
57. A.R. Gover, A. Shaukat, A. Waldron. Weyl invariance and the origins of mass. Phys. Lett. B, **675**(1), 93–97 (2009)
58. R.W. Sharpe. Differential geometry: Cartan's generalization of Klein's Erlangen Program, in *Graduate Text in Mathematics*, vol. 166 (Springer, Berlin, 1996)
59. I.M. Singer, Some remark on the Gribov ambiguity. Comm. Math. Phys. **60**, 7–12 (1978)
60. S. Sternberg, *Group Theory and Physics* (Cambridge University Press, Cambridge, 1994)
61. R. Stora. The Wess Zumino consistency condition: a paradigm in renormalized perturbation theory. Fortsch. Phys. **54**, 175–182 (2006). https://doi.org/10.1002/prop.200510266
62. W. Struyve. Gauge invariant accounts of the Higgs mechanism. Stud. Hist. Philos. Sci. B: Stud. Hist. Philos. Modern Phys. **42**(4), 226–236 (2011). ISSN 1355-2198. https://doi.org/10.1016/j.shpsb.2011.06.003
63. A. Trautman, *Fiber Bundles, Gauge Field and Gravitation, in General Relativity and Gravitation*, vol. 1 (Plenum Press, New-York, 1979)
64. S. van Dam. Spontaneous symmetry breaking in the Higgs mechanism. *PhiSci-Archive*, 2011
65. C.V. Westenholz, On spontaneous symmetry breakdown and the Higgs mechanism. Acta Phys. Acad. Sci. Hung. **48**, 213–224 (1980)
66. H. Weyl. Gravitation and electricity. Sitzungsber. Preuss. Akad. Wiss. Berlin (Math. Phys.) **1918**, 465 (1918)
67. H. Weyl. A new extension of relativity theory. Annalen Phys. **59**, 101–133 (1919); Annalen Phys. **364**, 101 (1919)
68. H. Weyl, *Symmetry* (Princeton University Press, Princeton, 1952)
69. D.P. Williams. Crossed Products of C-star Algebras, in *Mathematical Surveys and Monographs*, vol. 134 (American Mathematical Society, 2007)
70. D.K. Wise, Symmetric space, Cartan connections and gravity in three and four dimensions. SIGMA **5**, 080–098 (2009)

71. D.K. Wise, MacDowell-Mansouri gravity and Cartan geometry. Class. Quantum Grav. **27**, 155010 (2010)
72. C. Yang, *Selected Papers (1945–1980), with Commentary* (World Scientific Publishing Company, Singapore, 2005)
73. J. Zinn-Justin, *Quantum Field Theory and Critical Phenomena*, 4th edn. (Oxford Science Publications, Oxford, 2011)

Syntactic Phylogenetic Trees

Kevin Shu, Sharjeel Aziz, Vy-Luan Huynh, David Warrick and Matilde Marcolli

Abstract In light of recent controversies surrounding the use of computational methods for the reconstruction of phylogenetic trees of language families (especially the Indo-European family), a possible approach based on syntactic information, complementing other linguistic methods, appeared as a promising possibility, largely developed in recent years in Longobardi's Parametric Comparison Method. In this paper we identify several serious problems that arise in the use of syntactic data from the SSWL database for the purpose of computational phylogenetic reconstruction. We show that the most naive approach fails to produce reliable linguistic phylogenetic trees. We identify some of the sources of the observed problems and we discuss how they may be, at least partly, corrected by using additional information, such as prior subdivision into language families and subfamilies, and a better use of the information about ancient languages. We also describe how the use of phylogenetic algebraic geometry can help in estimating to what extent the probability distribution at the leaves of the phylogenetic tree obtained from the SSWL data can be considered reliable, by testing it on phylogenetic trees established by other forms of linguistic analysis. In simple examples, we find that, after restricting to smaller language subfamilies and considering only those SSWL parameters that are fully mapped for the whole subfamily, the SSWL data match extremely well reliable phylogenetic trees, according to the evaluation of phylogenetic invariants. This is a promising sign for the use of SSWL data for linguistic phylogenetics. We also argue how dependencies and nontrivial geometry/topology in the space of syntactic parameters would have to

K. Shu · S. Aziz · V.-L. Huynh · D. Warrick · M. Marcolli (✉)
Division of Physics, Mathematics, and Astronomy, California Institute of Technology, 1200 E. California Blvd, Pasadena, CA 91125, USA
e-mail: matilde@caltech.edu

K. Shu
e-mail: kshu@caltech.edu

S. Aziz
e-mail: saziz@caltech.edu

V.-L. Huynh
e-mail: vhuynh@caltech.edu

D. Warrick
e-mail: warrick.david58@gmail.com

J. Kouneiher (ed.), *Foundations of Mathematics and Physics One Century After Hilbert*,
https://doi.org/10.1007/978-3-319-64813-2_14

be taken into consideration in phylogenetic reconstructions based on syntactic data. A more detailed analysis of syntactic phylogenetic trees and their algebro-geometric invariants will appear elsewhere [33].

1 Introduction

This paper is based on a talk given by the last author at the workshop "Phylogenetic Models: Linguistics, Computation, and Biology" organized by Robert Berwick at the CSAIL department of MIT in May 2016.

The reconstruction of phylogenetic trees of language families is a crucial problem in the field of Historical Linguistics. The construction of an accurate family tree for the Indo-European languages accompanied and originally motivated the development of Historical Linguistics, and has been a focus of attention for linguists for the span of two centuries. In recent years, Historical Linguistics has seen a new influx of mathematical and computational methods, originally developed in the context of mathematical biology to deal with species phylogenetic trees, see for instance [5, 10, 12, 23, 28, 36]. A considerable amount of controversy arose recently in relation to the accuracy and effectiveness of these methods and the related problem of phylogenetic inference. In particular, claims regarding the phylogenetic tree of the Indo-European languages made in [6] were variously criticized by historical linguists, see the detailed discussion in [27].

Most of the literature dealing with computational phylogenetic trees in the context of Linguistics focused on the use of lexical data, in the form of Swadesh lists of words, and the encoding as binary data of the counting of cognate words, see for instance the articles in [12]. Other reconstructions used phonetic data and sound change, as in [5], or a combination of several types of linguistic data (referred to as "characters"), including phonetic, lexical, and morphological properties, as in [3, 36]. A different approach to linguistic phylogenetic reconstruction, based on syntactic parameters, was developed recently in [13, 17–21]. This method is known as Parametric Comparison Method (PCM). A coding theory perspective on the PCM was given in [22, 32].

The notion of syntactic parameters arises in Generative Linguistics, within the Principles and Parameters model developed by Chomsky in [7, 8]. A more expository account of syntactic parameters is given in [2]. Syntactic parameters are conceived as binary variables that express syntactic features of natural languages. The notion of syntactic parameters has undergone changes, reflecting changes in the modeling of generative grammar: for a recent overview of the parametric modeling of morphosyntactic features see [30]. A main open problems in the parametric approach for comparative generative grammar is understanding the space of syntactic parameters, identifying dependence relations between parameters and possibly identifying a fundamental set of such variables that would represent a good system of coordinates for the space of languages. Recently, the use of mathematical methods for the study of the space of syntactic parameters of world languages was proposed in [26, 29, 34].

At present, the only available extensive database of binary parameters describing syntactic features is the SSWL database [37], which collects data of 115 parameters over 253 world languages. It is debatable whether the binary variables collected in SSWL represent fundamental syntactic parameters: surface orders, for instance, are often confounded with the deep underlying parameter values. Moreover, SSWL does not record any dependence relations between parameters. Different data of syntactic parameters have been used in [20, 21], with dependence relations taken into account, and more data are being collected by these authors and will hopefully be available soon. For the purpose of this paper, we will use the terminology "syntactic parameters" loosely for any collection of binary variables describing syntactic features of natural languages. We work with the SSWL data, simply because it is presently the most extensive database available of syntactic structures.

In Sect. 2 of this paper we show that just using the Hamming distance between vectors of binary variables extracted from the SSWL data and the Neighborhood-Joining Method for phylogenetic inference gives very poor results as far as linguistic phylogenetic trees are concerned. We identify several different sources of problems, some inherent to the SSWL data, some to the inference methodology, and some more generally related to the use of syntactic parameters for phylogenetic linguistics.

In the Sect. 4 we review the method of Phylogenetic Algebraic Geometry of [24] and the main results of [1, 35] on phylogenetic ideals and phylogenetic invariants that we need for applications to the analysis of syntactic phylogenetic trees. In Sect. 5 we show how one can use techniques from Phylogenetic Algebraic Geometry to test the reliability of syntactic parameter data for phylogenetic linguistics, by using known phylogenetic trees that are considered reliable, and to test the reliability of candidate phylogenetic trees assuming a certain degree of reliability of the syntactic data.

In Sect. 6 we argue that dependencies between the syntactic variables recorded in the SSWL database should be taken into consideration in order to improve the reliability of these data for phylogenetic reconstruction. In particular, the presence of geometry/topology in this set of data and the presence of different degrees of recoverability of some of the SSWL syntactic variables in Kanerva network tests indicate that an appropriated weighted use of the data that accounts for these phenomena may improve the results.

2 PHYLIP Analysis of SSWL

We discuss here the problems that occurs in a naive analysis of the SSWL database using the phylogenetic tree algorithm PHYLIP. We identify the main types of errors that occur and the possible sources of the problems. We will discuss in Sect. 4 how one can eliminate some of the problems and obtain more accurate phylogenetic trees from SSWL data, using different methods.

2.1 Data and Code

We acquired the syntactic language data from the SSWL database with two different methods, one consisting of downloading the data as a *.csv* file directly, with the results separated in the format *"language|property|value"*, and one achieved by scraping the data into a *.json* file, formatted as a list of lists of binary variables, in the format "*'language' : {'parameters' : 'values'}*". This was done with a python script `data_obtainer.py` which went through all of SSWL and dumped the data as desired.

The SSWL data, stored in a more convenient `.json` file format produced by the first author, are available as the file `full_langs.json` which can be downloaded at the URL address http://www.its.caltech.edu/~matilde/PhylogeneticSSWL2.

We created, for each language in the database, a vector of binary variables representing the syntactic traits of that language as recorded in the SSWL database, with value 1 indicated that the language possessed the respective trait, and value 0 indicating that the language does not possess the trait.

One of the main sources of problems regarding the use of SSWL data arises already at this stage: not all languages in the database have all the same parameters mapped. The lack of information about a certain number of parameters for certain languages alters the counting of the Hamming distances, as it requires a choice of normalization of the string length, with additional entries added representing lack of information. This clearly generates problems, as this inconsistency generates mistakes in the counting of Hamming distances and in the tree reconstruction. In Sect. 2.2 we will illustrate specific examples where this problem occurs.

The Hamming distance algorithm `HF.py` takes two equal-length binary sequences, throwing an error if this length requirement is violated, and returns the sum of all bitwise XORs between them, or the total number of differences. In this way, we construct with `distance_matrix_checker.py` the Hamming distance matrix $M_{ab} = d_H(\ell_a, \ell_b)$, whose entries are the Hamming distances between the vectors of binary syntactic parameters of languages ℓ_a and ℓ_b.

For example, Germanic languages on average have normalized Hamming distance in the range 0.3–0.4. Old Saxon and Old English have a Hamming distance of 0.17 from German, while Swiss German has distance 0.09. Modern English has below average differences at 0.27. While these distances may appear reasonable, one can detect easily another major source of problems in the use of SSWL data for phylogenetic reconstruction. Many languages belonging to very different families have small Hamming distance: for example, the Indo-European Hindi (60% mapped in SSWL) and the Sino-Tibetan Mandarin (87% mapped in SSWL) receive a normalized distance of 0.12. This is certainly in large part due to the different level of accuracy with which the two languages are mapped in the same database. However, one can also observe syntactic similarities between languages belonging to different families, which are not due to poor recording of the respective data, but are a genuine consequence of the syntactic properties being described.

This 253 × 253 matrix of Hamming distances was then given as input to the PHYLIP package[1] for phylogenetic tree reconstruction, which is widely used in Mathematical Biology. Given the Hamming distance matrix $M_{ab} = d_H(\ell_a, \ell_b)$, the PHYLIP software provides several options for tree construction from distance matrix data: additive tree model, ultrametric model, neighbor joining method, and average linkage clustering (UPGMA). The resulting tree produced by PHYLIP, containing all 253 languages in the SSWL database, is contained in `outfile`, where the tree in the text file is drawn with dashes and exclamation points. The information of the output tree and distances is also given in the output file `outtree` in Newick format, with parentheses and commas. The accompanying file `key.txt` contains the key that indicates the full language name that corresponds to each two-letter string in outfile. The output files can be opened in any text editor.

The python code and the output files, prepared by the second, third and fourth authors of this paper, are available at http://www.its.caltech.edu/~matilde/Phylogenetic SSWL.

2.2 *Main Problems in the Resulting Tree*

A quick inspection of the output file obtained by running PHYLIP on the SSWL data immediately reveals that there are many problems with the resulting phylogenetic tree. We will give explicit examples here that illustrate some of the main type of problems one encounters. There are many more such examples one can easily find by inspecting the output tree available in the repository at the URL indicated above.

2.3 *Sources of Problems*

An important problem in computational phylogenetic reconstruction is how to validate statistically the model. There are well known problem inherent in using the Hamming distance as a source for phylogenetic trees. Estimating tree branch lengths is a hard problem. Distance matrices can be non-additive due to error, and it is typically difficult to distinguish distances that deviate from additivity due to change from deviations due to error. This problem is significant even in the context of Biology, where the use of DNA data is more reliable than the use of vectors of binary variables coming from linguistic properties [31]. For a discussion of some of these issues in Biology see [9]. For a comparison of phylogenetic methods (not including syntactic parameters) in Linguistics, see [3].

As we discuss with individual specific examples in the subsections that follow, there are several different source of problems that combine to create different kinds of errors in the resulting phylogenetic tree. The main problems are the following:

[1] http://evolution.genetics.washington.edu/phylip/software.html.

(1) inherent problems in the computational method based on Hamming distances, as discussed above;
(2) problems with non-uniform coverage of syntactic data across different languages and language families in the SSWL database;
(3) the nature of the syntactic variables recorded in the SSWL database (for instance with respect to surface versus deep structure) and the presence of relations between these variables;
(4) the existence of languages belonging to unrelated linguistic families that can be similar at the level of syntactic structures.

Clearly, some of these problems are of linguistic nature, like the last one listed, while others are of computational nature, like the first one, while others depend on the nature and accuracy of the SSWL data. It is difficult to disentangle the effects of each individual problem on the output tree, but the examples listed below illustrate cases where one can identify one of the problems listed here as the most likely origin of the mistakes one sees in the resulting phylogenetic tree.

2.3.1 Misplacement of Languages Within the Correct Subfamily Tree

This type of problem occurs when a group of languages are correctly identified as belonging to the same subfamily of a given historical-linguistic family, but the internal structure of the subfamily tree appears inconsistent with the structure generally agreed upon based on other linguistic data.

In the naive PHYLIP analysis of the SSWL database we see an example of this kind by considering the subtree of the Latin languages within the Indo-European family. The shape of this subtree, as it appears in in the output file, is of the form illustrated in Fig. 1. We see here that, although these languages are correctly grouped together as belonging to the same subfamily, the relative position within the subtree does not agree with what historical linguistic methods have established. Indeed, one can easily see, for instance, that the position of Portuguese in the subtree is incorrectly placed closer to Italian and Sicilian, than to Spanish and Catalan. This example is interesting because the error does not appear to be due to the poor mapping of parameters for these languages: Italian and Sicilian are 100% mapped in SSWL and Spanish, Catalan, and Portuguese are 84% mapped. So these are among some of the best recorded languages in the database, and still their respective position in the phylogenetic tree does not agree with reliable reconstructions from Historical Linguistics. It is interesting to compare the reconstruction obtained in this way, with the one obtained, on a different set of syntactic data, by Longobardi's Parametric Comparison in [20], which has Italian and French as a pair of two nearby branches, and Spanish and Portuguese as another pair of nearby branches. This example appears to outline an issue arising from the way syntactic variables are classified in the SSWL (as opposed to the different list of syntactic parameters used in [20]). We discuss in Sect. 6 below some of the problems of dependencies between the SSWL syntactic variables that may be at the sources of this kind of problem.

Fig. 1 PHYLIP output subtree of Latin languages: jq = Spanish, jb = Catalan, gz = Sicilian, bq = Italian, je = Portuguese, fn = French

```
              +jq
           +-250
           !  +jb
        +-58
        !  !     +gz
        !  !  +-181
     +-143  +-238  +bq
     !  !      !
     !  !      +-je
     !  !
     !  +---fn
     !
```

2.3.2 Placement of Languages in the Wrong Subfamily Tree

Another type of mistake one finds in the naive phylogenetic tree reconstruction from SSWL syntactic data is illustrated by the Germanic languages in Fig. 2. In this case, we find that most of the languages in this subtree are correctly grouped together as Germanic, but a language that clearly belongs to a different subfamily is also placed in the same group. It is very puzzling why Ancient Neapolitan ends up incorporated in the tree of Germanic languages rather then near Italian and the other dialects of Italian in the subtree of Latin languages of Fig. 1. Linguistically, one could perhaps argue that Ancient Neapolitan did in fact have several Germanic influences due to the Ostrogoths, but it is more reasonable to expect such influences to appear at the lexical rather than syntactic

Fig. 2 PHYLIP output subtree of Germanic languages: dm = Norwegian, cw = Faroese, hj = Italian Ancient Neapolitan, fd = Icelandic, jc = Afrikaans, ey = West Flemish, ia = Dutch, hc = German, gi = Swedish, cg = English

```
                 +-dm
              +-90
           +-191  +-cw
           !  !
        +-133  +-hj
        !  !
        !  +--fd
        !
     +-164     +-jc
     !  !  +-236
     !  !  !  +-ey
     !  !  !
     !  +-128        +ia
     !     !      +-208
   ·74     !  +-184  +hc
     !     !  !  !
     !     +-235  +-gi
     !        !
     !        +-cg
```

level. Moreover, the specific placement within the Germanic tree near Faroese, Norwegian and Icelandic, does not necessarily reflect this hypothesis. In terms of the accuracy with which these languages are recorded in the SSWL database, Ancient Neapolitan is 83% mapped, while its nearest neighbor on this PHYLIP output tree have Norwegian, which is also mapped with a similar accuracy of 84%, and Faroese and Icelandic with a lower accuracy of 69%. It is possible that this example already reflects a problem with the different accuracy of mapping of different languages in the SSWL database, or it may be a problem with the algorithmic reconstruction method itself. There are several similar instances in the output tree, which point to a problem that is systematic, hence likely generated by the method of phylogenetic reconstruction adopted in this naive analysis.

2.3.3 Proximity of Languages from Unrelated Families

Another type of problem that occurs frequently in the output tree of this naive analysis is the case of completely unrelated languages (from completely different language families) that are placed in adjacent positions in the tree. We see an example in Fig. 3, where the Mayan K'iche' language and Georgian (Kartvelian family) are placed next to each other in the tree. Both K'iche' and Georgian are 69% mapped in the SSWL database. Although this is not as accurate a mapping as some of the languages we discussed in the previous examples, it is nonetheless the same level of precision available, for instance, for some of the Germanic languages in the previous example, which were at least placed correctly in the Germanic subtree. Thus, the type of problem we see in this example is not entirely due to poor mapping of the languages involved. It must be also an effect of other factors like the computational reconstruction method used, as in the previous class of examples. However, there can also be some purely linguistic factors involved. Namely, there are known cases of languages belonging to unrelated historical linguistic families that may appear close at the syntactic level. This type of phenomenon may be responsible for at least part of the cases where one finds unrelated languages placed in close proximity in the output tree. This is an indication that one should not rely on syntactic data alone, without accompanying them with other linguistic data, that can provide, for example, a prior subdivision of languages into language families. Using the same method of phylogenetic tree reconstruction on data already grouped into linguistic families, with individual family trees separately constructed, improves the accuracy of the resulting trees. Other combinations of syntactic and lexical/morphological data can be used to improve accuracy.

Fig. 3 Misplaced proximity: io = K'iche', dj = Georgian

```
   +---io
+-222
!  +---dj
!
.
```

2.3.4 The Position of Ancient Languages in the Tree

Finally, there is an additional problem one encounters in the naive phylogenetic reconstruction based on the SSWL data, namely the position of the ancient languages in the tree. Clearly, the algorithm assumes that all the data correspond to leaves of the tree and that the inner nodes are hidden variables, while the fact that we do have knowledge of some of the ancient languages and that several are recorded in the SSWL database means that some of the inner nodes should in fact carry some of the data. This problem can be resolved if the inner languages would be placed as a single leaf attached to the corresponding inner node. By inspecting the resulting output tree we see that sometimes this is the case, and the inner node to which the corresponding ancient language is attached reasonably with respect to the modern languages that derived from it. One such example is the position of Old English with respect to the tree of the Germanic languages in Fig. 4. However, in other cases, ancient languages are correctly placed in proximity of each other, but in the wrong position, in the tree, with respect to the resulting modern languages. This is the case with Ancient Greek and Latin (see Fig. 5). In this case, the algorithm correctly captures the close syntactic proximity between Ancient Greek and

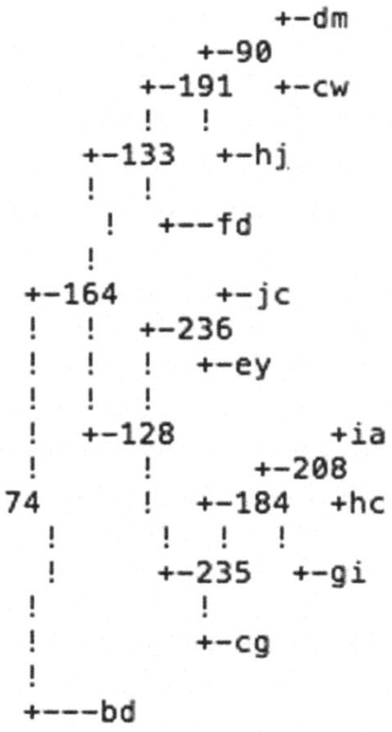

Fig. 4 The position of Old English with respect to the Germanic languages:bd= Old English

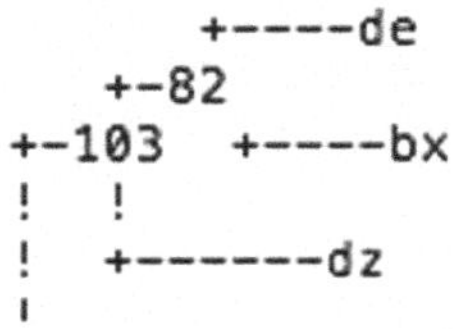

Fig. 5 Proximity of Ancient Greek and Latin: de=Latin, bx=Ancient Greek, dz=Medieval Greek

Latin, but it does not place these two languages correctly with respect to either the tree of Latin languages nor the modern part of the Hellenic branch. This problem can be improved by first subdividing the data into language families and smaller subfamilies and then perform the phylogenetic tree reconstruction on the subfamilies separately, so that the corresponding ancient language is placed correctly, and then related the resulting trees by proximity of the ancient languages. However, this method clearly applies only where enough other linguistic information is available, in addition to the syntactic data. It should be noted, moreover, that, while Ancient Greek is correctly placed in proximity to Latin, Homeric Greek is entirely misplaced in the PHYLIP tree reconstruction and does not appear in proximity of the Ancient Greek of the classical period, even though both Homeric and Ancient Greek are mapped with the best possible accuracy (100% mapped) in the SSWL database.

2.4 The Indo-European Tree

Although the many problems illustrated above render a phylogenetic reconstruction based solely on SSWL data unreliable, it is still worth commenting on what one obtains with this method regarding some of the controversial early branchings of the Indo-European tree. Again, the same type of systematic problems illustrated above occur repeatedly when one analyzes these regions of the output tree.

For example, Tocharian A and B are treated by the PHYLIP reconstruction as modern languages leaves of the tree and placed in immediate proximity of Hittite and in close proximity of some of the modern Indo-Iranic languages, like Pashto and Punjabi, and a further step away from some Turkic languages like Tuvan. The proximity of Tocharian and Hittite suggests here a Tocharian-Anatolian branching. The placement of the Indo-Iranic languages in proximity of this Tocharian-Anatolian branching is likely arising from the fact that the Indo-Iranic branch of the Indo-European family is very poorly mapped in the SSWL database, with the ancient languages entirely missing and very few of the modern languages recorded, hence the reconstructed tree necessarily skips over all these missing data. The complete absence of Sanskrit from the current version of the SSWL database (the entry in the database is just an empty place holder) in particular causes the phylogenetic reconstruction to miss entirely the proximity of the Indo-Iranic and the Hellenic branches. Near the subtree shown in Fig. 6 one finds several instances of misplaced languages of the type discussed in Sect. 2.3.3.

The situation with the Armenian branch is very problematic in the PHYLIP analysis of the SSWL data. There are three entries recorded in the database: Western Armenian is 68% mapped, while Eastern Armenian appears as two different entries in the database, one 84% mapped and the other only 52% mapped. Classical Armenian only appears as an empty place holder with no data in the current version of the database. These three data points are not placed in proximity of one another in the PHYLIP reconstruction. Western Armenian ends up completely misplaced (it appears in proximity of Korean and Japanese). This misplacement may be corrected if one first subdivides data by language families and then runs the phylogenetic reconstruction only on the

Fig. 6 Tocharian–Anatolian branching: gb=Tocharian A, hn=Tocharian B, bv=Hittite, fk=Pashto, iy=Panjabi, cx=Tuvan (Turkic)

```
             +hn
          +-195
       +-157  +gb
         !  !
    +-140  +-bv
     !  !
 +-232  +--fk
  !  !
67  +--iy
 !
 +-cx
```

Fig. 7 Eastern Armenian: cy = Eastern Armenian (84%), ee = Pima (misplaced Uto-Aztecan), ai = Ossetic Digor, dh= Ossetic Iron

```
             +dh
          +-85
       +-108  +ai
        !  !
    +-76  +ee
      !  !
     !  +-cy
     !
```

Indo-European data. The better mapped entry for Eastern Armenian is placed in proximity of the subtree of Fig. 6 containing the Tocharian–Anatolian branch and some Indo-Iranian languages (plus some other misplaced languages from other families). The nearest neighbors that appear in this region of the tree are Digor Ossetic and Iron Ossetic: again this is likely an effect of the poor mapping of the Indo-Iranic branch of the Indo-European family, as in the case of Fig. 6. Another error due to misplacement from an entirely different family occurs, with the Uto-Aztecan Pima placed in this same subtree, see Fig. 7. This subtree is placed adjacent to a subtree containing a group of Balto-Slavic languages (and some misplaced languages) with both of these branches then connecting to the subtree of Fig. 6. The poorly mapped Eastern Armenian entry (52%) is placed as single leaf attached to an otherwise deep inner node of the tree. Another language that is often difficult to position in the Indo-European tree, Albanian (68% mapped), is misplaced in the PHYLIP reconstruction, and placed next to Gulf Arabic (69% mapped).

These examples confirm the fact that a naive phylogenetic analysis of the SSWL database cannot deliver any reliable information on the question of the early branchings of the Indo–European tree.

3 Phylogenetic Networks

We verified that the same types of problems illustrated in the previous subsections occur when the SSWL data are analyzed using phylogenetic networks instead of the PHYLIP phylogenetic trees.

We compiled the SSWL data [37], using only the Indo-European languages, which have more complete parameter information as a sample set. As in the tree analysis discussed before, we input the syntactic parameters as a sequence of binary strings into the phylogenetic networks programs.

The `Splitstree 4` program[2] generated a split tree, which is intuitively a confidence interval on trees. The farther from 'tree-like' the generated tree, the less any given tree is able to describe the evolution of the languages. The output of this program indicated that the phylogenetics of languages analyzed on the basis of SSWL syntactic parameters diverges strongly from being tree-like. As discussed before, this may be regarded as further indication of systematic problems that create high uncertainties in the candidate trees. These are again an illustration of the effect of a combination of the factors (1)–(4) listed in Sect. 2.3.

We also fed the same data to the `Network 5` program.[3] This generated a phylogenetic network, using the median-joining algorithm which represents all of the shortest-path length (maximum parsimony) trees which are possible given the data.

We discuss below some of the aspects of the network generated by `Splittree 4` in comparison to some of the outputs described above obtained with the PHYLIP phylogenetic trees. Figure 8 illustrates a large region of the phylogenetic network produced by `Splittree 4` using the entire set of SSWL data. It is evident that some of the same problems we have discussed before occur in this case as well, in particular the misplacement of the ancient languages with respect to their modern descendent (see the position of Latin and Ancient Greek, for example). However, with respect to the PHYLIP results discussed above, we see less instances of languages that get completely misplaced within the wrong family. For example, as one can see from Figs. 9 and 10, Ancient Neapolitan now appears correctly placed in the Latin languages (and near Spanish) rather than misplaced among the Germanic languages as in Fig. 2. However, one can see that other problems that occurred in the PHYLIP reconstructions for this group of languages are still present in the `Splittree 4` network. For example, as in Fig. 1, Portuguese appears closer to Italian than to Spanish in the network of Fig. 9, contrary to the general understanding of the phylogenetic tree of the Latin languages. (We will discuss the case of the subtree of the Latin languages more in detail in Sect. 5.) Misplacements of languages within these smaller subfamilies are still occurring, however: one can see that, for example, in the positioning of the Romance language Occitan in the region of the phylogenetic network in proximity of Germanic languages like Old Norse and Icelandic in Fig. 10.

[2] http://ab.inf.uni-tuebingen.de/data/software/splitstree4/download/manual.pdf.

[3] http://www.fluxus-engineering.com/Network5000_user_guide.pdf.

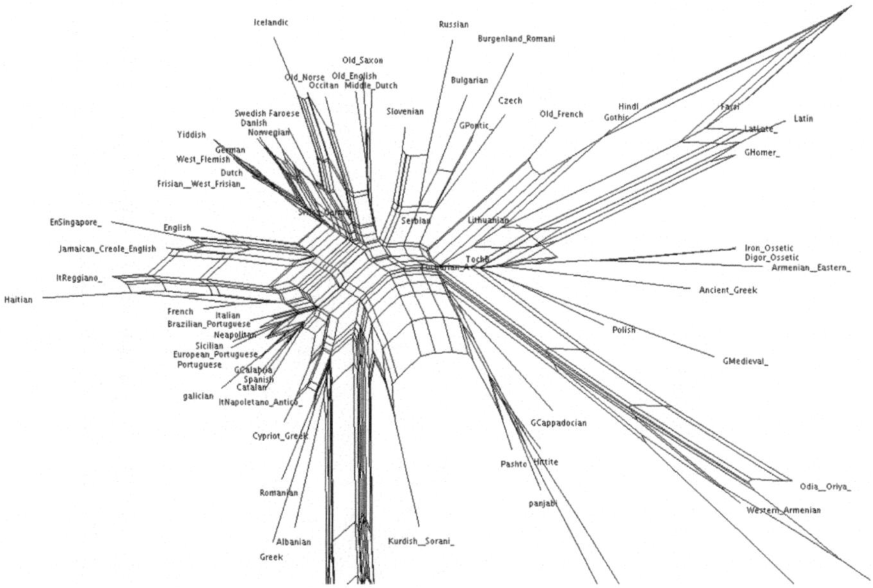

Fig. 8 Phylogenetic network produced by `Splittree 4` on the entire SSWL database

Fig. 9 Latin languages region of the phylogenetic network

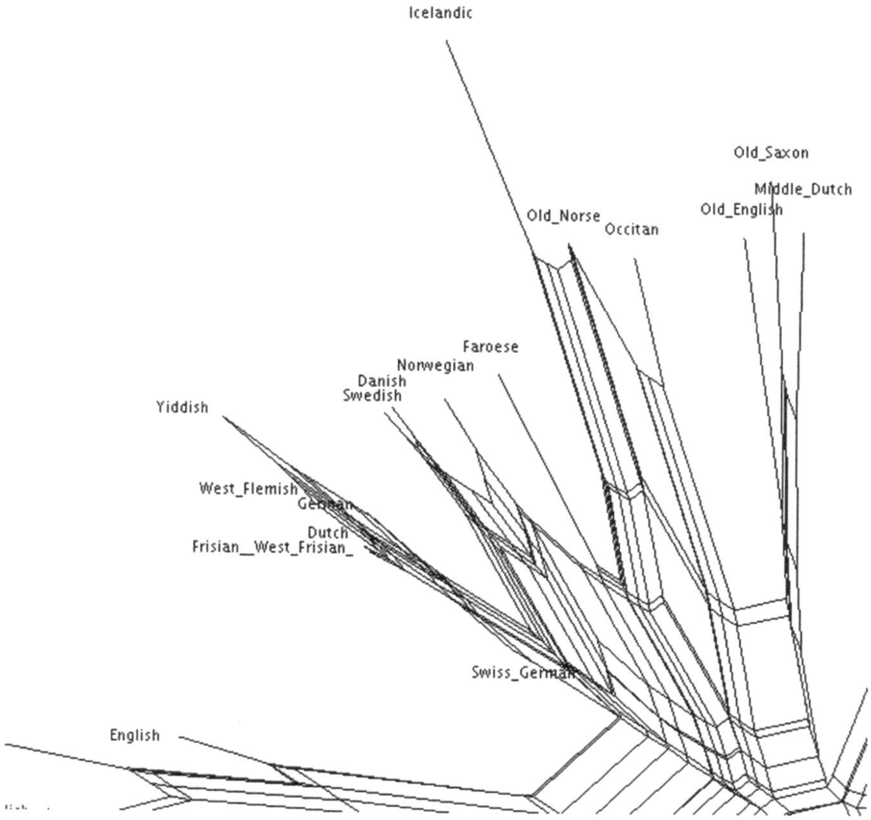

Fig. 10 Germanic languages region of the phylogenetic network

The results of the `Splittree 4` phylogenetic networks analysis of the Indo-European languages are available as the file `Indo_Euro.nex`, which can be downloaded at the URL http://www.its.caltech.edu/~matilde/PhylogeneticSSWL2.

4 Phylogenetic Algebraic Geometry

Given the unsatisfactory results one obtains in analyzing the SSWL database with software aimed at phylogenetic reconstructions, one can turn the problem on its head and try to obtain specific quantitative estimates of the level of reliability or unreliability of specific subsets of the SSWL data for the purpose of phylogenetic, by relying on existing reconstructions of linguistic phylogenetic trees, obtained by other linguistic methods and other sources of data, which are considered reliable reconstructions. The problem is then to test the distribution at the leaves of the tree obtained from the SSWL data with specific polynomial invariants associated to a given reliable tree. Such invariants would be vanishing on a probability distribution at the leaves obtained from an evolutionary process modeled by a Markov model on the tree, hence we can use the estimate of how far the values are from zero as a numerical estimate of a degree of unreliability of the data for phylogenetic reconstruction. Again, this does not identify explicitly the source of the problem, among the various possible causes outlined above, but it still gives a numerical estimate that can be useful in trying to improve the results. We propose here to use methods from phylogenetic algebraic geometry developed by Pachter, Sturmfels, et al. [11, 24, 25, 35] to achieve this goal. We first give a quick review of the main setting of phylogenetic algebraic geometry and then we illustrate in some specific examples how we intend to use these techniques for the purpose described here.

4.1 Phylogenetic Models: General Assumptions

The basic setup for linguistic phylogenetic models consists of a *dynamical process* of language change (which in our case means change of syntactic parameters), considered as a Markov process on a *binary tree* (a finite tree with all internal vertices of valence 3).

It can be argued whether trees really give the best account of language change based on syntactic data, rather than more general non-simply-connected graphs (generally referred to as "networks"). We will return to discuss some empirical reasons in favor of phylogenetic networks instead of trees in Sect. 6. The mathematics of phylogenetic networks is discussed at length in [14, 15]. About the use of phylogenetic networks in Linguistics, see [23].

Another general assumption of phylogenetic models, which requires careful examination in the case of applications to Linguistics, is the usual assumption that the variables (binary variables in the case of syntactic parameters) behave like *independent* identically distributed variables, whose dynamics evolves according to *the same* Markov process. This assumption is especially problematic when dealing with syntactic parameters because of the presence of relations between parameters that are not entirely understood, so that it is currently extremely hard to ensure one is using a set of independent binary variables. Moreover, while acceptable in first approximation, even the assumption that the underlying Markov model driving the change should be the same for all syntactic parameters appears problematic. The fact that different syntactic parameters have very different frequencies of occurrence among world languages certainly suggests otherwise. We will return to this point in Sect. 6 and suggest a possible approach, based on the results of [26], to correct, at least in part, for this problem.

The leaves of the tree correspond to the modern languages with observed values of the parameters giving a joint probability distribution

$$\mathbb{P}(X_{\ell_1} = i_1, \ldots, X_{\ell_n} = i_n) = p_{i_1,\ldots,i_n}, \tag{4.1}$$

with $i_k \in \{0, 1\}$, and with n the number of leaves. Here the quantity $p_{i_1,\dots,i_n}$ represents the frequency with which syntactic parameters of the languages $\ell_1, \dots, \ell_n$ at the leaves of the tree have values $(i_1, \dots, i_n) \in \{0, 1\}^n$, respectively.

In the usual setting of Markov models for phylogenetic reconstructions, one further assumes that all the *inner nodes* are hidden variables and that only the distribution at the leaves of the tree is known. Here again we encounter a problem with respect to applications to Linguistics. In certain language families, like the Indo-European family, several ancient languages have known parameters. In the SSWL database, for instance, Ancient Greek is one of the very few languages that are 100% mapped with respect to their list of 115 parameters. Thus, one needs to consider some of the inner vertices as known rather than hidden. One way to do that is to consider a single leaf coming out of some of the inner vertices that will correspond to the known values of the parameters at that vertex. As we discussed in Sect. 2.2, one encounters problems with the placement of the ancient languages in the PHYLIP reconstruction of the syntactic phylogenetic trees, which should be corrected for. Better results are obtained when one first separates out the data into language families and subfamilies and builds trees for smaller subfamilies first, including the known data about the ancient languages, and then combines these trees into a larger tree. This procedure avoids the type of problem mentioned in Sect. 2.2, by which the greater syntactic similarity between some of the ancient Indo-European languages like Latin and Ancient Greek is detected correctly, but in turn prevents their respective placement in the correct position with respect to the modern languages that originated from them.

For a given set of n leaves, there are

$$\tau_n = \frac{(2n-4)!}{(n-2)!2^{n-2}}$$

different possible binary tree topologies. Clearly, it is not a computationally efficient strategy to analyze all of them. However, one would like to have some computable invariants that one can associate to a given candidate tree T, which estimates how accurate T is as a phylogenetic tree, among all the τ_n possible choices, given knowledge of the joint probability distribution (4.1) at the leaves. The Phylogenetic Algebraic Geometry approach (see [24, 25] and the survey [4]) aims at constructing such phylogenetic invariants using Algebraic Geometry and Commutative Algebra. We review the main ideas in the next subsection.

4.2 Phylogenetic Varieties and Ideals

We consider here the Jukes–Cantor model describing a Markov process on a binary rooted tree T with n leaves. The stochastic behavior of the model is determined by the datum of a probability distribution $(\pi, 1-\pi)$ at the root vertex (the frequency of expression of the 0 and 1 values of the syntactic parameters at the root) and the datum of a bistochastic matrix

$$M^e = \begin{pmatrix} 1 - p_e & p_e \\ p_e & 1 - p_e \end{pmatrix}$$

along each edge of the tree. These data (π, M^e) are often referred to in the literature as parameters of the model. In order to avoid confusion with our use of the term parameter for the syntactic binary variables, we will refer to the (π, M^e) as "stochastic parameters". For a tree T with n leaves, and variables with k states, the number of stochastic parameters is

$$N = (2n - 3)k(k - 1) + k - 1.$$

In our case, with binary variables, we have $k = 2$ and the number of stochastic parameters of the model is simply $N = 4n - 5$.

Phylogenetic invariants are polynomial functions ϕ that vanish on all the expected distributions $p_{i_n,\ldots,i_n}$ at the tails of the tree T, for all values of the stochastic parameters (π, M^e).

The simplest example of such an invariant is the linear polynomial

$$\phi(z_{i_n,\ldots,i_n}) = -1 + \sum_{i_n,\ldots,i_n} z_{i_n,\ldots,i_n},$$

since the joint distribution at the leaves is normalized by $\sum_{i_n,\ldots,i_n} p_{i_n,\ldots,i_n} = 1$. This invariant is uninteresting, in the sense that it is independent of the tree T, hence it does not provide any information about distinguishing between candidate phylogenetic trees. In general one seeks other, more interesting, phylogenetic invariants ϕ_T, and the minimum number of such invariants required for phylogenetic inference. An answer to this question is provided by Algebraic Geometry, as shown in [1, 24, 25, 35].

Consider the polynomial ring $\mathbb{C}[z_{i_1,\ldots,i_n}]$, where n is the number of leaves of the tree and and $i_k \in \{0, 1\}$ for all $k = 1, \ldots, n$. The phylogenetic invariants are defined by the vanishing $\phi_T(p_{i_1,\ldots,i_n}) = 0$. This condition determines an ideal $\mathcal{I}_T$ in the polynomial ring. For a Markov model as above, with $N = 4n - 5$ stochastic parameters (π, M^e), one obtains a polynomial map

$$\Phi : \mathbb{C}^{4n-5} \to \mathbb{C}^{2^n}$$

that assigns $\Phi(\pi, M^e) = p_{i_1,\ldots,i_n}$. This is, more explicitly, of the form

$$p_{i_1,\ldots,i_n} = \Phi(\pi, M^e) = \sum_{w_v \in \{0,1\}} \pi_{w_{v_r}} \prod_e M^e_{w_{s(e)}, w_{t(e)}},$$

with a sum over "histories" (paths in the tree) consistent with the data at the leaves. This determines an algebraic variety, the *phylogenetic variety*, given by the Zariski closure

$$V_T = \overline{\Phi(\mathbb{C}^{4n-5})} \subset \mathbb{C}^{2^n}.$$

Dually we have a map

$$\Psi : \mathbb{C}[z_{i_1,\ldots,i_n}] \to \mathbb{C}[x_1, \ldots, x_{4n-5}]$$

with $\mathrm{Ker}\Psi = \mathcal{I}_T$, where $\mathcal{I}_T$ is the *phylogenetic ideal*.

One can use phylogenetic invariants to select between candidate phylogenetic trees in the following way. Suppose one obtains, through some phylogenetic algorithm, a candidate phylogenetic tree T. One also has available the joint probability distribution (4.1) of the binary variables at the leaves. By evaluating phylogenetic invariants $\phi_T \in \mathcal{I}_T$ at the observed distribution $p_{i_n,\ldots,i_n}$, one can check whether the candidate tree T satisfies

$$\left|\phi_T(p_{i_n,\ldots,i_n})\right| < \epsilon \tag{4.2}$$

for all phylogenetic invariants $\phi_T \in \mathcal{I}_T$, and for a fixed error size ϵ. The candidate tree T is an acceptable phylogenetic tree if and only if the estimate (4.2) is satisfied. Geometrically, the test (4.2) can be rephrased as the property that the point $p_{i_1,\ldots,i_n} \in \mathbb{C}^{2^n}$ is ϵ-close to the phylogenetic variety V_T if and only if T is an acceptable phylogenetic tree. Computationally, this method requires obtaining a set of explicit generators for the phylogenetic ideal $\mathcal{I}_T$.

In the case of the Jukes–Cantor model with $k = 2$, it was proved in [35] that the phylogenetic ideal $\mathcal{I}_T$ is generated by polynomials of degree two. A completely explicit set of generators for the Jukes–Cantor model with $k = 2$ was obtained in [1], where it is proved that phylogenetic ideal $\mathcal{I}_T$ generated by the 3×3-minors of all *edge flattenings* of the tensor $P = (p_{i_1,\ldots,i_n})$. The edge flattenings are defined by the following procedure. Start with a tree T with Markov model (π, M^e) and with $P \in \mathbb{C}^{2^n}$ the joint probability distribution $P = (p_{i_1,\ldots,i_n})$ at the n leaves. The choice of an edge e in a tree T with n leaves determines two connected components of $T \setminus \{e\}$, hence two sets of leaves $\{\ell_1, \ldots, \ell_r\}$ and $\{\ell_{r+1}, \ldots, \ell_n\}$. Thus, the 2^n binary variables at the n leaves are partitioned into a set of 2^r variables and a set of 2^{n-r} variables, and the joint distribution $P = (p_{i_1,\ldots,i_n})$ determines a $2^r \times 2^{n-r}$-matrix $Flat_{e,T}(P)$ specified by setting

$$Flat_{e,T}(P)(u, v) = P(u_1, \ldots, u_r, v_1, \ldots, v_{n-r}).$$

It can be shown that the rank of this matrix is $\mathrm{rank}(Flat_{e,T}(P)) \leq 2$ (for binary variables, $k = 2$), hence all 3×3 minors of the matrix must vanish. It is shown in [1] that, for $k = 2$ any number n of leaves, the phylogenetic ideal $\mathcal{I}_T$ is generated by the 3×3 minors of the matrices $Flat_{e,T}(P)$ of all edge flattenings. It is easy to see that, even for small trees, there is a very large number of these 3×3 minors, hence the number of generators of the phylogenetic ideal grows rapidly with the size of the tree.

Note that, while for the purpose of validating a candidate phylogenetic tree T it would be necessary to check that all these generators of the phylogenetic ideal vanish [or nearly vanish as in (4.1)], in order to invalidate a candidate tree it sufficed to find at least one of these 3×3 minors for one of the flattenings that evaluates on the observed joint distribution $P = (p_{i_1,\ldots,i_n})$ to a value larger than the allowed error size ϵ.

5 Phylogenetic Invariants and Syntactic Trees

In this section we show how phylogenetic invariants can be used to improve the phylogenetic tree reconstructions based on SSWL syntactic data.

5.1 Phylogenetic Invariants of Small Syntactic Trees

We focus here on sufficiently small subtrees of the syntactic phylogenetic tree of languages compiled from the SSWL data, for which the computation of phylogenetic invariants becomes feasible. Using phylogenetic invariants, we compare the small trees obtained in this way with phylogenetic trees obtained by other linguistic methods and considered reliable, so as to estimate the validity of the joint distribution at the leaves obtained from SSWL data.

We present here an example, based on the subtree of the Latin languages within the Indo–European family. A more detailed analysis of other subtrees of the Indo–European family will be presented elsewhere.

We have seen in Sect. 2.3.1 that the naive PHYLIP analysis of the entire SSWL database misplaces Portuguese in the subtree of the Indo-European languages that collects the Latin languages. We have also seen in Sect. 2.3.4 that the same analysis misplaces Latin, separating it from the tree of the modern languages that originated from it.

We now perform a more accurate analysis, still using only the SSWL data, but where we use the *a priori* knowledge of the grouping of certain languages into a subfamily. Thus, we select only the languages *Latin, Italian, French, Spanish, Portuguese*.

The phylogenetic tree that is generally agreed, through other linguistic reconstructions, to best represent the relative position of these languages would be a tree topology as illustrated in Fig. 11. Note that this is also the tree reconstruction for this group of languages obtained in [20] using a set of syntactic parameters different from those recorded in the SSWL database.

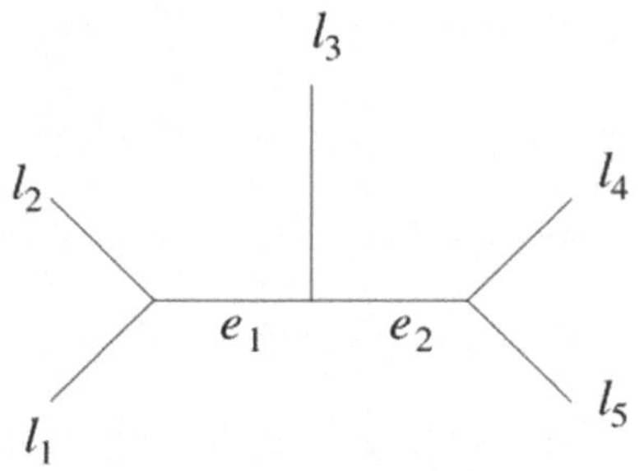

Fig. 11 Tree topology for the phylogenetic tree of the Latin languages, with ℓ_1 = French, ℓ_2 = Italian, ℓ_3 = Latin, ℓ_4 = Spanish, ℓ_5 = Portuguese

The tree of Fig. 11 has two possible splits: $\{\ell_1, \ell_2\} \cup \{\ell_3, \ell_4, \ell_5\}$ and $\{\ell_1, \ell_2, \ell_3\} \cup \{\ell_4, \ell_5\}$. The corresponding flattenings are given by the matrices

$$\mathrm{Flat}_{e1}(P) = \begin{pmatrix} p_{00000} & p_{00001} & p_{00010} & p_{00011} & p_{00100} & p_{00101} & p_{00110} & p_{00111} \\ p_{01000} & p_{01001} & p_{01010} & p_{01011} & p_{01100} & p_{01101} & p_{01110} & p_{01111} \\ p_{10000} & p_{10001} & p_{10010} & p_{10011} & p_{10100} & p_{10101} & p_{10110} & p_{10111} \\ p_{11000} & p_{11001} & p_{11010} & p_{11011} & p_{11100} & p_{11101} & p_{11110} & p_{11111} \end{pmatrix}$$

$$\mathrm{Flat}_{e2}(P) = \begin{pmatrix} p_{00000} & p_{00001} & p_{00010} & p_{00011} \\ p_{00100} & p_{00101} & p_{00110} & p_{00111} \\ p_{01000} & p_{01001} & p_{01010} & p_{01011} \\ p_{01100} & p_{01101} & p_{01110} & p_{01111} \\ p_{10000} & p_{10001} & p_{10010} & p_{10011} \\ p_{10100} & p_{10101} & p_{10110} & p_{10111} \\ p_{11000} & p_{11001} & p_{11010} & p_{11011} \\ p_{11100} & p_{11101} & p_{11110} & p_{11111} \end{pmatrix}$$

where the p_{i_1,i_2,i_3,i_4,i_5} are the frequencies of the observed binary variables at the ends, under the assumption that these behave like independent equally distributed random variables, evolving according to the same Markov model on the tree.

Using the data of SSWL parameters for these five languages reported in the Appendix, we obtain matrices $\mathrm{Flat}_{e_1}(P)$ and $\mathrm{Flat}_{e_2}(P)$ of the form

$$\mathrm{Flat}_{e_1}(P) = \begin{pmatrix} \frac{31}{106} & \frac{1}{106} & \frac{1}{106} & 0 & \frac{23}{106} & \frac{3}{106} & 0 & \frac{1}{53} \\ \frac{1}{106} & 0 & 0 & \frac{1}{106} & 0 & \frac{1}{106} & 0 & \frac{3}{106} \\ \frac{5}{106} & 0 & \frac{1}{53} & 0 & 0 & 0 & 0 & 0 \\ \frac{1}{53} & 0 & \frac{1}{106} & \frac{4}{53} & 0 & 0 & 0 & \frac{21}{106} \end{pmatrix}$$

$$\mathrm{Flat}_{e_2}(P) = \begin{pmatrix} \frac{31}{106} & \frac{1}{106} & \frac{1}{106} & 0 & 0 \\ \frac{23}{106} & \frac{3}{106} & 0 & \frac{1}{53} & 0 \\ \frac{1}{106} & 0 & 0 & \frac{1}{106} & 0 \\ 0 & \frac{1}{106} & 0 & \frac{3}{106} & 0 \\ \frac{5}{106} & 0 & \frac{1}{53} & 0 & 0 \\ 0 & 0 & 0 & 0 & 0 \\ \frac{1}{53} & 0 & \frac{1}{106} & \frac{4}{53} & 0 \\ 0 & 0 & 0 & \frac{21}{106} & 0 \end{pmatrix}.$$

Evaluating all the 3×3 minors of these matrices with `Maple` and selecting the maximum absolute value of the resulting phylogenetic invariants gives

$$\max \left|\phi_T(p_{i_1,\ldots,i_5})\right| = \frac{2415}{1191016} = 0.0020277. \tag{5.1}$$

The fact that for the tree of Fig. 11 the distribution at the leaves determined by the SSWL parameters is extremely close to being a zero of all the phylogenetic invariants implies that the SSWL parameters are in fact in very good agreement with the recognized correct topology of the phylogenetic tree, but only when the set of languages is previously restricted to a smaller subfamily and *only the SSWL parameters that are fully mapped for that subfamily are taken into account*.

This result seems to indicate that the main source of the problems we encounter when doing a naive phylogenetic analysis using the entire SSWL database are not necessarily due to an intrinsic problem with the SSWL data [that is, it is not primarily due to problem number (3) in the list in Sect. 2.3]. It seems rather that the problems encountered above stem from a combination of problems (1), (2), and (4). The use of the phylogenetic invariants method bypasses problem (1), while the prior restriction to a smaller subfamily bypasses problems (2) and (4). A more detailed analysis of this approach with phylogenetic invariants and computations of likelihood via Euclidean distances, applied to other language subfamilies using SSWL data will be carried out more extensively elsewhere [33].

6 Dependencies and Geometry

As we already mentioned above, the problem of the construction of reliable syntactic phylogenetic trees is closely related to the problem of relations and dependencies between syntactic parameters. Are there universal relations that hold across all languages? Are there relations that depend on language families? Can these relations be expressed geometrically, as is the case with relations between continuous coordinates that give rise to topological or differentiable manifolds? Are there different geometries associated to different language families? How detectable are relations between syntactic parameters computationally? Recently, a mathematical approach to these questions was proposed in [22, 26, 29, 34].

In [29], it was shown, again using SSWL data, that syntactic parameters of different language families have different persistent homology. The persistent generators of H_0 appear to correspond to a subdivisions of a given language family into major subfamilies, such as, for example, the Indo-Iranic and the European subfamilies of the Indo-European family, or the Mande, Atlantic-Congo, and Kordofanian subfamilies of the Niger-Congo family. A persistent generator of the H_1 was found in the case of the Indo-European family. It appears to be related to the position of the Hellenic branch in the Indo-European family. It is presently unclear whether this reflects the effect of a genuine historical-linguistic phenomenon, such as an influence of Ancient Greek, at the syntactic level, upon some other European languages (such as some of the Slavic languages), whether it detects the presence of homeoplasy in syntactic parameters, or whether it is due to the nature and format of the syntactic data collected in the SSWL database. However, the presence of non-trivial persistent generators of the H_1 in the persistent homology of the data set is a strong indicator that networks (non-simply-connected graphs) and not trees may provide a better topology for syntactic phylogenetic linguistics.

In [26], it was shown that, to some extent, the presence of dependencies between the syntactic parameters listed in the SSWL database can be detected using Kanerva networks. The latter were introduced in [16] as sparse distributed memories aimed at modeling associative memory in neuroscience. It is well known that, in fact, Kanerva networks are very useful for reconstructing corrupted data and detecting the degree of recoverability of certain parts of the data as a function of the remaining ones. In particular, this makes them suitable for detecting the presence of relations between data. It was shown in [26] that different syntactic parameters in the SSWL database exhibit different degrees of recoverability in a Kanerva network. An overall effect can be identified, which depends on the frequency with which a certain syntactic parameter is expressed across world languages. This effect can be reproduced using random data with the same frequencies. However, there is an additional effect that can be detected normalizing with respect to the frequency and that should be a genuine expression of the level of dependence of a particular syntactic parameter upon the remaining ones. The resulting normalized score computed in [26] is therefore a numerical estimate of the degree of dependence/independence of a given binary syntactic variable. The presence of these computationally detectable dependence relations affects some of the fundamental assumptions of the Markov models of phylogenetic trees, in particular the assumption that all the binary variables are independent, identically distributed variables. A possible way to compensate for this problem in the model it to consider a weighted version of the joint probability distribution $P = p_{i_1,\dots,i_n}$ at the leaves of the phylogenetic tree, where the frequency of expression of the parameters is computed in such a way that each parameter is weighted according to the corresponding normalized degree of recoverability in a Kanerva network, in such a way that the independent variables are weighted more than the dependent ones. This restores the fact that the independent variables assumption of the Markov model can be at least approximately satisfied.

Acknowledgements The first author is supported by a Summer Undergraduate Research Fellowship at Caltech. Part of this work was performed as part of the activities of the last author's Mathematical and Computational Linguistics lab and CS101/Ma191 class at Caltech. The last author is partially supported by NSF grants DMS-1201512 and PHY-1205440 and DMS-1707882.

Appendix: The SSWL Parameters of the Latin languages

The phylogenetic invariants for the tree of Latin languages of Fig. 11 are evaluated at the probability distribution p_{i_1,i_2,i_3,i_4,i_5} at the leaves, based on the SSWL parameters for this group of languages. There are 106 parameters in the SSWL database that are completely mapped for all of these five languages. We have excluded from the list all those SSWL parameters that are only mapped for some but not all of the languages in this group. With the notation ℓ_1 = French, ℓ_2 = Italian, ℓ_3 = Latin, ℓ_4 = Spanish, and

$\ell_5 =$ Portuguese, the syntactic parameters are given by the following list. The column on the left lists the SSWL parameters P as labeled in the database, [37].

P	$[\ell_1,\ell_2,\ell_3,\ell_4,\ell_5]$
01	[1,1,1,1,1]
02	[0,1,1,1,1]
03	[1,1,1,1,1]
04	[0,0,1,0,0]
05	[1,1,1,1,1]
06	[0,0,1,0,0]
07	[0,0,1,0,0]
08	[0,0,1,0,0]
09	[0,0,1,0,0]
10	[0,0,1,0,0]
11	[1,1,1,1,1]
12	[0,0,1,0,0]
13	[1,1,1,1,1]
14	[1,1,1,1,1]
15	[1,1,1,1,1]
16	[0,0,1,0,0]
17	[1,1,1,1,1]
18	[0,0,1,0,0]
19	[0,0,1,0,0]
20	[1,1,1,1,1]
21	[1,1,1,1,1]
22	[0,0,1,1,1]
A01	[1,1,1,1,1]
A02	[1,1,1,1,1]
A03	[1,1,1,1,1]
A04	[0,1,1,0,1]
Aux Sel 01	[1,1,0,0,0]
C01	[1,1,1,1,1]
C02	[0,0,0,0,0]
C03	[1,1,1,1,1]
C04	[0,0,0,0,0]
EE	[1,1,0,1,0]

N2 01	[1,1,1,1,1]
N2 02	[0,0,1,0,0]
N2 03	[1,1,0,1,1]
N2 04	[0,0,1,0,0]
N2 05	[1,1,0,1,1]
N2 06	[1,1,1,1,1]
N2 07	[0,0,0,0,0]
N2 08	[0,0,0,0,0]
N2 09	[0,0,0,0,0]
N2 10	[0,0,0,0,0]
N2 11	[0,0,0,0,0]
Neg 01	[1,1,1,1,1]
Neg 02	[1,0,0,0,0]
Neg 03	[0,0,0,0,0]
Neg 04	[0,0,0,0,0]
Neg 05	[0,0,0,0,0]
Neg 06	[0,0,0,0,0]
Neg 07	[0,0,0,0,0]
Neg 08	[0,0,0,0,0]
Neg 09	[0,0,0,0,0]
Neg 10	[0,0,0,0,0]
Neg 11	[0,0,0,0,0]
Neg 12	[0,0,0,0,0]
Neg 13	[0,0,0,0,0]
Neg 14	[0,0,0,0,0]

Order N3 01	[1,1,1,1,1]
Order N3 02	[1,1,1,1,1]
Order N3 03	[0,0,0,0,0]
Order N3 04	[0,0,1,0,0]
Order N3 05	[0,0,1,0,0]
Order N3 07	[1,1,1,1,1]
Order N3 08	[0,0,1,0,0]
Order N3 11	[0,0,1,0,0]
Q01	[1,0,0,0,0]
Q02	[0,0,0,0,0]
Q03	[0,0,1,0,0]
Q04	[1,1,0,1,1]
Q05	[1,0,0,1,0]
Q06	[1,0,0,1,0]
Q07	[0,0,0,0,0]
Q08	[1,1,0,1,1]
Q09	[0,0,1,0,1]
Q10	[0,0,0,0,1]
Q11	[0,0,0,0,0]
Q12	[0,0,0,0,0]
Q13	[0,0,0,0,0]
Q14	[0,0,1,0,1]
Q15	[0,0,0,0,0]
Q16	[1,1,0,0,0]
Q17	[0,0,1,1,1]
Q18	[0,0,1,0,1]
Q21	[1,0,0,0,0]
Q22	[0,0,0,0,0]

V2 01	[0,0,0,0,0]
V2 02	[0,0,0,1,0]
w01a	[0,1,1,1,1]
w01b	[1,0,0,0,0]
w01c	[0,1,0,0,0]
w02a	[0,0,1,0,0]
w02b	[1,1,0,1,1]
w02c	[0,0,0,0,0]
w03a	[0,0,1,0,0]
w03b	[1,1,0,1,1]
w03c	[0,0,0,0,0]
w03d	[0,0,0,0,0]
w04a	[0,0,1,0,0]
w04b	[1,1,0,1,1]
w04c	[0,0,1,0,0]
w05a	[0,1,1,1,1]
w05b	[1,0,0,0,0]
w05c	[0,1,0,1,1]
w06a	[0,0,1,0,0]
w06b	[1,1,0,1,1]
w06c	[0,0,1,0,0]

One can see by inspecting the different groups of parameters in this list that several parameters within the "same group" tend to behave in the same way (e.g. all the *Neg* parameters) or in more highly correlated way than across groups of parameters. This observation is consistent with the more general observation of dependencies observed through the Kanerva networks method in [26]. Thus, in order to better fit this set of binary variables with the hypothesis of independent equally distributed variables in Markov processes, it may be better to select a subset of the SSWL parameters that cuts across the various groups of more closely correlated variables. We will discuss this aspect more in details elsewhere.

The probability p_{i_1,i_2,i_3,i_4,i_5} is then computed by counting the frequencies of occurrence of binary vectors $[i_1, i_2, i_3, i_4, i_5] \in \{0, 1\}^5$ among the 106 vectors of SSWL parameters above. The only nonzero frequencies are

$$p_{0,0,0,0,0} = \frac{31}{106}, \quad p_{0,0,0,0,1} = \frac{1}{106}, \quad p_{0,0,0,1,0} = \frac{1}{106}, \quad p_{0,0,1,0,0} = \frac{23}{106},$$

$$p_{0,0,1,0,1} = \frac{3}{106}, \quad p_{0,0,1,1,1} = \frac{2}{106}, \quad p_{0,1,0,0,0} = \frac{1}{106}, \quad p_{0,1,0,1,1} = \frac{1}{106},$$

$$p_{0,1,1,0,1} = \frac{1}{106}, \quad p_{0,1,1,1,1} = \frac{3}{106}, \quad p_{1,0,0,0,0} = \frac{5}{106}, \quad p_{1,0,0,1,0} = \frac{2}{106},$$

$$p_{1,1,0,1,0} = \frac{1}{106}, \quad p_{1,1,0,0,0} = \frac{2}{106}, \quad p_{1,1,0,1,1} = \frac{8}{106}, \quad p_{1,1,1,1,1} = \frac{21}{106}.$$

Note how these frequencies confirm some well known facts about the Latin languages. Syntactic parameters (as recorded in SSWL) are very likely to have remained the same across all five languages in the family, with a higher probability of a feature not allowed in Latin remaining not allowed in the other languages (31/106) than of a feature allowed in Latin remaining allowed in the other languages (21/106). It is also very likely that a feature is the same in all the modern ones but different from Latin, with a much higher incidence of cases of a feature allowed in Latin becoming disallowed in all the other languages (23/106) than the other way around (8/106). Among the remaining possibilities, we see incidences where French has an allowed feature that is missing in the other languages (5/106) of disallowed (3/106) and cases where Latin and Portuguese have the same feature allowed, which is disallowed in the other languages (3/106): all other nonzero entries have only two or less occurrences. The resulting matrices for the edge flattenings of the tree of Fig. 11 are then as computed in Sect. 5.

References

1. E. Allman, J. Rhodes, Phylogenetic ideals and varieties for general Markov models. Adv. Appl. Math. **40**, 127–148 (2008)
2. M. Baker, *The Atoms of Language* (Basic Books, USA, 2001)
3. F. Barbançon, S.N. Evans, L. Nakhleh, D. Ringe, T. Warnow, An experimental study comparing linguistic phylogenetic reconstruction methods. Diachronica **30**(2), 143–170 (2013)
4. C. Bocci, Topics in phylogenetic algebraic geometry. Expo. Math. **25**, 235–259 (2007)
5. A. Bouchard-Côté, D. Hall, T.L. Griffiths, D. Klein, Automated reconstruction of ancient languages using probabilistic models of sound change. Proc. Natl. Acad. Sci. (PNAS) **110**(11), 4224–4229 (2013)
6. R. Bouckaert, P. Lemey, M. Dunn, S.J. Greenhill, A.V. Alekseyenko, A.J. Drummond, R.D. Gray, M.A. Suchard, Q.D. Atkinson, Mapping the origins and expansion of the Indo-European language family. Science **337**, 957–960 (2012)
7. N. Chomsky, *Lectures on Government and Binding* (Foris Publications, Dordrecht, 1982)
8. N. Chomsky, H. Lasnik, The theory of Principles and Parameters. in "*Syntax: An International Handbook of Contemporary Research*", (de Gruyter, 1993), pp. 506–569
9. M. DeGiorgio, J.H. Degnan, Robustness to divergence time underestimation when inferring species trees from estimated gene trees. Syst. Biol. **63**(1), 66–82 (2014)
10. A. Delmestri, N. Cristianini, Linguistic phylogenetic inference by PAM-like matrices. J. Quant. Linguist. **19**, 95–120 (2012)
11. N. Eriksson, K. Ranestad, B. Sturmfels, S. Sullivant, Phylogenetic algebraic geometry, in "*Projective Varieties with Unexpected Properties*", pp.237–255, Walter de Gruyter, 2005
12. P. Forster, C. Renfrew, *Phylogenetic Methods and the Prehistory of Language* (McDonald Institute Monographs, 2006)
13. C. Galves (ed.), *Parameter Theory and Linguistic Change* (Oxford University Press, Oxford, 2012)
14. D. Gusfield, *ReCombinatorics* (MIT Press, Cambridge, 2014)
15. D.H. Huson, R. Rupp, C. Scornavacca, *Phylogenetic Networks: Concepts* (Cambridge University Press, Algorithms and Applications, 2010)

16. P. Kanerva, *Sparse Distributed Memory* (MIT Press, Cambridge, 1988)
17. G. Longobardi, Methods in parametric linguistics and cognitive history. Linguist. Var. Yearb. **3**, 101–138 (2003)
18. G. Longobardi, L. Bortolussi, M.A. Irimia, N. Radkevich, A. Ceolin, C. Guadagno, D. Michelioudakis, A. Sgarro, Mathematical modeling of grammatical diversity supports the historical reality of formal syntax. in "*Proceedings of the Leiden Workshop on Capturing Phylogenetic Algorithms for Linguistics*" (2016)
19. G. Longobardi, S. Ghirotto, C. Guardiano, F. Tassi, A. Benazzo, A. Ceolin, G. Barbujani, Across language families: genome diversity mirrors linguistic variation within Europe. Am. J. Phys. Anthropol. **157**(4), 630–640 (2015)
20. G. Longobardi, C. Guardiano, Evidence for syntax as a signal of historical relatedness. Lingua **119**, 1679–1706 (2009)
21. G. Longobardi, C. Guardiano, G. Silvestri, A. Boattini, A. Ceolin, Towards a syntactic phylogeny of modern Indo-European languages. J. Hist. Linguist. **3**(1), 122–152 (2013)
22. M. Marcolli, Syntactic parameters and a coding theory perspective on entropy and complexity of language families. Entropy **18**, 110 [17 pages] (2016)
23. L. Nakhleh, D. Ringe, T. Warnow, Perfect phylogenetic networks: a new methodology for reconstructing the evolutionary history of natural languages. Language **81**(2), 382–420 (2005)
24. L. Pacher, B. Sturmfels, The mathematics of phylogenomics. SIAM Rev. **49**(1), 3–31 (2007)
25. L. Pacher, B. Sturmfels, Tropical geometry of statistical models. Proc. Natl. Acad. Sci. (PNAS) **101**46, 16132–16137 (2004)
26. J.J. Park, R. Boettcher, A. Zhao, A. Mun, K. Yuh, V. Kumar, M. Marcolli, Prevalence and recoverability of syntactic parameters in sparse distributed memories. in *Geometric Science of Information*. Third International Conference GSI 2017. Lecture Notes in Computer Science, vol. 10589 (Springer, 2017), pp. 265–272
27. A. Perelysvaig, M.W. Lewis, *The Indo-European Controversy: Facts and Fallacies in Historical Linguistics* (Cambridge University Press, Cambridge, 2015)
28. F. Petroni, M. Serva, Language distance and tree reconstruction. J. Stat. Mech. **2008**, P08012 [16 pages] (2008)
29. A. Port, I. Gheorghita, D. Guth, J.M. Clark, C. Liang, S. Dasu, M. Marcolli, Persistent Topology of Syntax. Math. Comput. Sci. **12**(1), 33–50 (2018)
30. L. Rizzi, *On the Format and Locus of Parameters: The Role of Morphosyntactic Features*, preprint, 2016
31. N. Saitou, M. Nei, The neighbor-joining method: a new method for reconstructing phylogenetic trees. Mol. Biol. Evol. **4**(4), 406–425 (1987)
32. K.Shu, M.Marcolli, Syntactic Structures and Code Parameters, Math. Comput. Sci. **11**(1), 79–90 (2017)
33. K. Shu, A. Ortegaray, R. C. Berwick, M. Marcolli, Phylogenetics of Indo-European Language families via an Algebro-Geometric Analysis of their Syntactic Structures. arXiv:1712.01719
34. K. Siva, J. Tao, M. Marcolli, Syntactic Parameters and Spin Glass Models of Language Change. Linguist. Anal. **41**(3–4), 559–608 (2017)
35. B. Sturmfels, S. Sullivant, Toric ideals of phylogenetic invariants. J. Comput. Bio. **12**(2), 204–228 (2005)
36. T. Warnow, S.N. Evans, D. Ringe, L. Nakhleh, Stochastic Models of Language Evolution and an Application to the Indo-European Family of Languages. Available at http://www.stat.berkeley.edu/users/evans/659.pdf
37. SSWL Database of Syntactic Parameters: http://sswl.railsplayground.net/

The manufacturer's authorised representative in the EU is Springer Nature Customer Service Centre GmbH, Europaplatz 3, 69115 Heidelberg, Germany. If you have any concerns regarding our products, please contact ProductSafety@springernature.com

Printed and bound by CPI Group (UK) Ltd, Croydon, CR0 4YY
30/09/2026
02225434-0016